ıs and Scope

, series *Green Chemistry and Sustainable Technology* aims to present cutting-e research and important advances in green chemistry, green chemical ,ineering and sustainable industrial technology. The scope of coverage includes ıt is not limited to):

- Environmentally benign chemical synthesis and processes (green catalysis, green solvents and reagents, atom-economy synthetic methods etc.)
- Green chemicals and energy produced from renewable resources (biomass, carbon dioxide etc.)
- Novel materials and technologies for energy production and storage (biofuels and bioenergies, hydrogen, fuel cells, solar cells, lithium-ion batteries etc.)
- Green chemical engineering processes (process integration, materials diversity, energy saving, waste minimization, efficient separation processes etc.)
- Green technologies for environmental sustainability (carbon dioxide capture, waste and harmful chemicals treatment, pollution prevention, environmental redemption etc.)

The series *Green Chemistry and Sustainable Technology* is intended to provide an accessible reference resource for postgraduate students, academic researchers and industrial professionals who are interested in green chemistry and technologies for sustainable development.

More information about this series at http://www.springer.com/series/11661

Green Chemistry and Sustainable Te

Feng-Shou Xiao • Xiangju Meng
Editors

Zeolites in Sustainable Chemistry

Synthesis, Characterization and Catalytic Applications

Springer

Editors
Feng-Shou Xiao
Department of Chemistry
Zhejiang University
Hangzhou, China

Xiangju Meng
Department of Chemistry
Zhejiang University
Hangzhou, China

ISSN 2196-6982 ISSN 2196-6990 (electronic)
Green Chemistry and Sustainable Technology
ISBN 978-3-662-47394-8 ISBN 978-3-662-47395-5 (eBook)
DOI 10.1007/978-3-662-47395-5

Library of Congress Control Number: 2015951340

Springer Heidelberg New York Dordrecht London

Printed on acid-free paper

Springer-Verlag GmbH Berlin Heidelberg is part of Springer Science+Business Media (www.springer.com)

Preface

Crystalline microporous zeolites have been considered as mineral curiosities for a long time since their discovery in 1756. Currently, there are myriads of applications of zeolites and related porous crystals in the fields of industrial, environmental, and social relevance. Particularly, after revolutionizing the refinement of crude oil and the petrochemical industry as a whole by vastly enhancing the efficiencies of the existing process, great efforts have been devoted to the hydrothermal synthesis of zeolites and their properties, giving many breakthrough achievements. Summaries of these exciting results have already led to the publication of some great and successful books.

In recent years, with the development of green chemistry and shortage of energy around the world, there has been a major leap for the synthesis, characterization, and practical applications of zeolite, in terms of both its fundamental and industrial aspects. For instance, hierarchically porous zeolites with excellent mass transfer have been templated; solvent-free route for synthesis of zeolites has been achieved; interlayer expansion methodology has been established and created many new zeolite structures; the great strides made in modern techniques such as electron micrography, solid NMR spectroscopy, and X-ray diffraction have significantly advanced our understanding of the syntheses and structures of zeolites; sustainable and important processes such as methanol to light olefins (MTO) and selective catalytic reduction of NOx with ammonia (NH_3-SCR) catalyzed by zeolite catalysts have been commercialized already. Therefore, it is time to collect the works recently done by the outstanding scientists active in this field to establish an essential handbook.

This book mainly contains three parts, devoting to novel strategies for synthesizing zeolites, new developments in characterizations of zeolites, and emerging applications of zeolites for sustainable chemistry, respectively. In the first part, my colleague Dr. Xiangju Meng and I briefly summarize the synthesis of zeolites via sustainable routes (Chap. 1). Prof. Zhijian Tian from the Dalian Institute of Chemical Physics introduces in detail the ionothermal synthesis of zeolites (Chap. 2). Prof. Toshiyuki Yokoi and Prof. Takashi Tatsumi from the Tokyo Institute of Technology provide a detailed review of the interlayer expansion of the layered zeolites (Chap. 3). Prof. Ryong Ryoo and his colleagues describe the synthesis of mesostructured zeolites (Chap. 4). In the second part, Prof. Xiaodong Zou and her colleague from Stockholm University present the different electron crystallographic techniques and their applications on structure determination of zeolites (Chap. 5). Prof. Hermann Gies and his colleague from Ruhr University Bochum elucidate the solution and refinement of zeolite structures (Chap. 6). Prof. Feng Deng and his colleague from the Wuhan Institute of Physics and Mathematics introduce the solid state NMR method for structural characterization of zeolites (Chap. 7). In the third part, Dr. Bilge Yilmaz and Dr. Ulrich Muller and their colleagues from BASF review the refinery applications (Chap. 8) and catalytic reactions (Chap. 14) of zeolites in industry. Prof. Weiguo Song from the Institute of Chemistry and Prof. Zhongmin Liu and Prof. Yingxu Wei from the Dalian Institute of Chemical Physics demonstrate the conversion process of methanol to light olefins over zeolites (Chap. 9). Prof. Emiel Hensen from TU/e discusses the application of zeolites as catalysts in the conversion of biomass into fuels and chemicals (Chap. 10). My colleague Dr. Liang Wang and I provide a concise review of the new developments of titanosilicate zeolites and their applications in various oxidations (Chap. 11). Prof. Hong He and his colleague from the Research Centre for Eco-Environmental Science explore the emerging applications of zeolites in environmental catalysis (Chap. 12). Prof. Zhengbo Wang from Zhejiang University and Prof. Yushan Yan from the University of Delaware summarize the recent progress in preparation and applications of zeolite thin films and membranes (Chap. 13). In the last Chapter (Chap. 15), Dr. Xiangju Meng and I also give a brief summary of the opportunities and challenges in the research and development of zeolites.

This book provides a comprehensive and an in-depth coverage of this rapidly evolving field from both academic and industrial points of view. We believe it can be used as an essential reference for the researchers who are working in the field of zeolites and related areas. It can also be used as a textbook as well as one of the key references for graduate and undergraduate students in chemistry, chemical engineering, and materials science.

Finally, we, the editors, would like to express our heartfelt gratitude to the authors for their contributions to this book.

Hangzhou, China Feng-Shou Xiao

Contents

Contributors

Kanghee Cho Center for Nanomaterials and Chemical Reactions, Institute for Basic Science (IBS), Daejeon, Republic of Korea

Melissa Clough BASF Refining Catalysts, Houston, TX, USA

Feng Deng State Key Laboratory of Magnetic Resonance and Atomic and Molecular Physics, Wuhan Center for Magnetic Resonance, Wuhan Institute of Physics and Mathematics, The Chinese Academy of Sciences, Wuhan, China

Hermann Gies Department of Geology, Mineralogy and Geophysics, Ruhr University Bochum, Bochum, Germany

Hong He Research Center for Eco-Environmental Sciences, Chinese Academy of Sciences, Beijing, People's Republic of China

Emiel J.M. Hensen Inorganic Materials Chemistry Group, Eindhoven University of Technology, Eindhoven, The Netherlands

Vasileios Komvokis BASF Refining Catalysts, Cheadle, United Kingdom

Shenhui Li State Key Laboratory of Magnetic Resonance and Atomic and Molecular Physics, Wuhan Center for Magnetic Resonance, Wuhan Institute of Physics and Mathematics, The Chinese Academy of Sciences, Wuhan, China

Lynne Xin Lin Tan BASF Refining Catalysts, Suntec-1, Singapore

Fudong Liu Research Center for Eco-Environmental Sciences, Chinese Academy of Sciences, Beijing, People's Republic of China

Hao Liu Dalian National Laboratory for Clean Energy, Dalian Institute of Chemical Physics, Chinese Academy of Sciences, Dalian, China

Zhongmin Liu Dalian Institute of Chemical Physics, Chinese Academy of Sciences, Dalian, People's Republic of China

Bernd Marler Department of Geology, Mineralogy and Geophysics, Ruhr University Bochum, Bochum, Germany

Xiangju Meng Department of Chemistry, Zhejiang University, Hangzhou, China

Filipe Marques Mota, Center for Nanomaterials and Chemical Reactions, Institute for Basic Science (IBS), Daejeon, Republic of Korea

Ulrich Müller BASF, Process Research and Chemical Engineering, Ludwigshafen, Germany

Shuyang Shaun Pan BASF Refining Catalysts, Iselin, NJ, USA

Evgeny A. Pidko Inorganic Materials Chemistry Group, Eindhoven University of Technology, Eindhoven, The Netherlands

Institute of Complex Molecular Systems, Eindhoven University of Technology, Eindhoven, The Netherlands

Ryong Ryoo Center for Nanomaterials and Chemical Reactions, Institute for Basic Science (IBS), Daejeon, Republic of Korea,

Department of Chemistry, Korea Advanced Institute of Science and Technology (KAIST), Daejeon, Republic of Korea

Xiaoyan Shi Research Center for Eco-Environmental Sciences, Chinese Academy of Sciences, Beijing, People's Republic of China

Weiguo Song Institute of Chemistry, Chinese academy of Sciences, Beijing, People's Republic of China

Takashi Tatsumi Chemical Resources Laboratory, Tokyo Institute of Technology, Midori-ku, Yokohama, Japan

Zhi-Jian Tian Dalian National Laboratory for Clean Energy, Dalian Institute of Chemical Physics, Chinese Academy of Sciences, Dalian, China

State Key Laboratory of Catalysis, Dalian Institute of Chemical Physics, Chinese Academy of Sciences, Dalian, China

William N.P. van der Graaff Inorganic Materials Chemistry Group, Eindhoven University of Technology, Eindhoven, The Netherlands

Liang Wang Department of Chemistry, Zhejiang University, Hangzhou, China

Zhengbao Wang College of Chemical and Biological Engineering, and MOE Engineering Research Center of Membrane and Water Treatment Technology, Zhejiang University, Hangzhou, People's Republic of China

Yingxu Wei Dalian Institute of Chemical Physics, Chinese Academy of Sciences, Dalian, People's Republic of China

Tom Willhammar Berzelii Center EXSELENT on Porous Materials, and Inorganic and Structural Chemistry, Department of Materials and Environmental Chemistry, Stockholm University, Stockholm, Sweden

Feng-Shou Xiao Department of Chemistry, Institute of Catalysis, Zhejiang University, Hangzhou, China

Lijuan Xie Research Center for Eco-Environmental Sciences, Chinese Academy of Sciences, Beijing, People's Republic of China

Yushan Yan Department of Chemical and Biomolecular Engineering, University of Delaware, Newark, DE, USA

Bilge Yilmaz BASF Refining Catalysts, 25 Middlesex-Essex Turnpike, Iselin, NJ, USA

Toshiyuki Yokoi Chemical Resources Laboratory, Tokyo Institute of Technology, Midori-ku, Yokohama, Japan

Xiaodong Zou Berzelii Center EXSELENT on Porous Materials, and Inorganic and Structural Chemistry, Department of Materials and Environmental Chemistry, Stockholm University, Stockholm, Sweden

Part I
Novel Strategies for Synthesizing Zeolites

Chapter 1
Sustainable Routes for Zeolite Synthesis

Xiangju Meng, Liang Wang, and Feng-Shou Xiao

Abstract The modern synthesis of zeolites mainly involves the use of organic templates, the addition of solvent, the preparation of starting gels, and the heating of the gels. Each step could be made greener in the future. This chapter presents a brief overview on the recently reported green routes for synthesizing zeolites, mainly focusing on the reduction or elimination of organic templates as well as the complete elimination of solvent. To overcome the disadvantages of using organic templates, nontoxic templates and template recycling steps have been employed in the zeolite syntheses. In addition, organotemplate-free synthesis has become a popular and universal methodology for synthesizing zeolites. Particularly, seed-directed synthesis in the absence of organic templates is a general route for synthesizing a series of zeolites. From an economic and environmental standpoint, solvent-free synthesis is a great move toward "green" synthesis of zeolite due to the following: high yields, high efficiency, low waste, low pollution, low pressure, hierarchical porosity, and simple and convenient procedure. Combining the advantages of solvent-free and organotemplate-free synthesis would particularly open the pathway to a highly sustainable zeolite synthesis protocol in industry.

Keywords Zeolites • Sustainable template • Template recycling • Organotemplate-free synthesis • Solvent-free synthesis

1.1 Introduction

Hydrothermal synthesis of zeolites from silicate or aluminosilicate gels in alkaline media has occupied an important position in zeolite synthesis science, where the temperature is ranged from 60 to 240 °C and the pressure is about 0.1–2 MPa [1, 2]. R. M. Barrer and R. M. Milton, the founders of zeolite synthesis science, started their studies in zeolite synthesis in the 1940s, successfully synthesizing a series of

X. Meng (✉) • L. Wang • F.-S. Xiao
Department of Chemistry, Zhejiang University, Hangzhou 310028, China
e-mail: mengxj@zju.edu.cn

F.-S. Xiao, X. Meng (eds.), *Zeolites in Sustainable Chemistry*,
Green Chemistry and Sustainable Technology,
DOI 10.1007/978-3-662-47395-5_1

artificial zeolites such as P, Q, A, and X [1, 3–6]. Later, a milestone for zeolite synthesis is the introduction of organic quaternary ammonium cations in the hydrothermal synthesis, which opens a door to synthesize novel zeolites [1, 7, 8]. Up to now, more than 200 types of zeolites have been hydrothermally synthesized in the presence of organic templates.

Although hydrothermal synthesis of zeolites has been widely used for decades, it does not meet the critical terms of sustainable chemistry that refers to reduce or eliminate negative environmental impacts, involving the reduction of wastes and improvement of efficiency, due to the use of organic templates and a large amount of water [9].

Currently, organic templates play very important roles in the zeolite synthesis due to the templating of the assembly pathway, filling the pore space, and balancing the charges [1, 9]. However, most organic templates are toxic, which potentially threaten human health. In addition, removal of these templates normally requires high-temperature combustion that produces hazardous greenhouse gases such as NOx and CO_2. On the other hand, water is always regarded as the "greenest" solvent, but a large amount of the water used in industries still results in a series of shortcoming such as waste of polluted water, high autogenous pressure, and consequently safe issues [9].

To solve these problems caused by conventional hydrothermal synthesis, sustainable routes for zeolite synthesis have been developed recently. In this chapter, several novel sustainable routes will be systemically illustrated.

1.2 Synthesis of Zeolites Using Sustainable Templates

Organic quaternary ammonium cations were first introduced into the zeolite synthesis by Barrer and Denny in 1961, and they have successfully synthesized several pure siliceous and high-silica zeolites [1, 7–9]. Different from the inorganic cations, organics play an additional role for templating or structure directing in the zeolite synthesis. Thus, these organics are called templates or structure-directing agents (SDAs). Conventional organic templates mainly include amines, amides, pyrrolidines, quaternary ammonium cations, and metal chelate complex [1, 2, 9].

1.2.1 Synthesis of Zeolites Using Low-Toxicity Templates

EMT zeolite is of great importance in fluid catalytic cracking (FCC) industry, due to its excellent catalytic performance compared with commercial catalyst Y zeolite [10, 11]. However, EMT zeolite is normally prepared in the presence of costly and toxic template of 18-crown-6, which greatly limited its wide applications in industry [12, 13]. Recently, Liu et al. reported successful synthesis of EMT-rich faujasites using polyquaternium-6 as a template, a component of shampoo, which is nontoxic and inexpensive since its extensive use in daily human life [14].

Wang et al. reported another successful example for the preparation of zeolite using nontoxic template [15]. They prepared a family of microporous aluminophosphate zeolite with AFI structure (AlPO-5) using tetramethylguanidine (TMG) as template. Guanidine and its derivatives with relatively low toxicity and low cost are biologically and industrially important chemicals, which could be found in the products of animal metabolism and classified as sustainable templates [16]. Notably, guanidines, containing three nitrogen atoms, might offer stronger coordination ability to aluminum species than conventional amines (e.g. triethylamine) with only one nitrogen atom [16]. As a consequence, the crystallization rate of AlPO-5 in the presence of TMG is much higher than that using triethylamine as templates, and the crystallinity reaches nearly 100 % only after 5 h. Moreover, this kind of sustainable template is not limited to prepare AlPO-5; heteroatom-substituted AlPO-5 crystals such as SAPO-5, MnAPO-5, together with CoAPO-5, and other microporous aluminophosphate (e.g. AlPO-21 with AWO structure) can also be synthesized using TMG as a template [15].

1.2.2 Synthesis of Zeolites Using Low-Cost Templates

Zones et al. have developed a new approach for the synthesis of zeolites, in which a minor amount of SDA is used to specify the nucleation product, and then a larger amount of a nonspecific amine is used to provide both pore-filling and basicity capacities in the synthesis [17]. The concept used in this method was to have the SDA provide the initial nucleation selectivity and then hope that a cheaper, less selective molecule could provide the pore-filling aspect as the crystal continuously grows. For example, various small amines including even ammonia and methylamine were shown to function in conjunction with the imidazole SDA to produce SSZ-32. A number of zeolites including SSZ-13 (CHA), SSZ-33 (CON), SSZ-35 (STF), SSZ-42 (IFR), and SSZ-47 can be prepared in the same manner [18]. There are a number of cost-saving benefits described for this synthesis route including reduced structure-directing agent cost, waste stream cleanup costs, and time in reactor and reagent flexibility.

Similar to this concept, UOP scientists have developed the charge density mismatch (CDM) approach to prepare zeolites via addition of alkali and alkaline Earth cations at low levels, which cooperate with organic templates [19–21]. Such cooperation allows the use of commercial available organic templates for a new material discovery. For example, they prepared hexagonal 12-ring zeolites UZM-4 (BPH) and UZM-22 (MEI) using choline-Li-Sr template system based on the charge density mismatch approach. Notably, the CDM approach to zeolite synthesis was initially proposed as a cheaper alternative to the trend of using ever more complicated quaternary ammonium species.

Ren et al. have designed a copper complex of Cu–tetraethylenepentamine (Cu–TEPA) as candidate for synthesizing CHA-type aluminosilicate zeolite (SSZ-13) [22, 23], which is generally directed by the expensive template of *N,N,N-trimethyl-1*-1-adamantammonium hydroxide, due to (1) good match between the stable

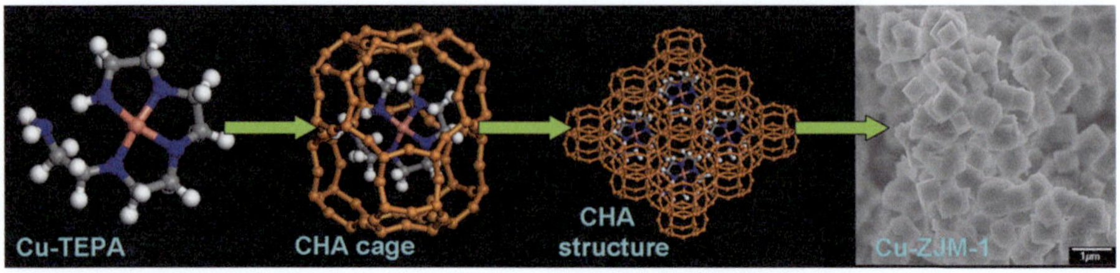

Fig. 1.1 Mechanism on Cu–TEPA-templated Cu-SSZ-13 zeolites (Reprinted with permission from Ref. [22]. Copyright 2011 Royal Society of Chemistry)

molecular configuration of Cu–TEPA with CHA cages, (2) strong interaction between the template molecule and negatively charged silica species, and (3) high stability in strongly alkaline media. They reported rational one-pot synthesis of Cu-SSZ-13 zeolites with molar ratio of SiO_2/Al_2O_3 at 8–15, designated as Cu-ZJM-1, from using Cu–TEPA as template (Fig. 1.1). Compared with the traditional Cu^{2+} ion-exchange method, Cu-ZJM-1 shows much higher copper content and better dispersion of copper cations. More importantly, Cu-ZJM-1 exhibits excellent catalytic properties in SCR of NO_x by NH_3 [22].

1.2.3 Synthesis of Zeolites Using Recyclable Templates

Davis et al. have performed pioneer works in the field of extracting organic templates from micropores of zeolites [24–29]. Firstly, they reported that TEA^+ cations could be easily extracted from CIT-6 zeolite (BEA-type structure) with acetic acid-containing solution [24], because of the weak interaction between the TEA^+ cations and CIT-6 framework. The ease of liberation of charge-balancing tetraethylammonium (TEA) cations from the various metallosilicates was shown to be Zn>B>Al [28]. This method can also be utilized in pure-silica MFI zeolite. They also pointed out that the amount of organic templates removed by extraction was strongly dependent on the size of the organic templates and the strength of interaction between the templates and the zeolites [28].

Later, they reported a complete recycle of an organic template in the synthesis of ZSM-5 [29]. They chose a cyclic ketal as organic template that would remain intact at zeolite synthesis conditions (high pH) and be cleavable at conditions that would not destroy the assembled zeolite (Fig. 1.2). The ^{13}C CP/MAS NMR spectrum showed that the as-synthesized zeolite material contains intact 8,8-dimethyl-1,4-dioxa-8-azaspiro [4, 5] decane (**1**). When the ZSM-5 was treated with 1 M HCl solution at 80 °C for 20 h, the ^{13}C CP/MAS NMR spectrum obtained was consistent with the presence of the ketone fragment, suggested that **1** could be cleaved into the desired pieces inside the zeolite pore space. After ion-exchange treatment by a mixture of 0.01 M NaOH and 1 M NaCl at 100 °C for 72 h, 1,1-dimethyl-4-oxo-

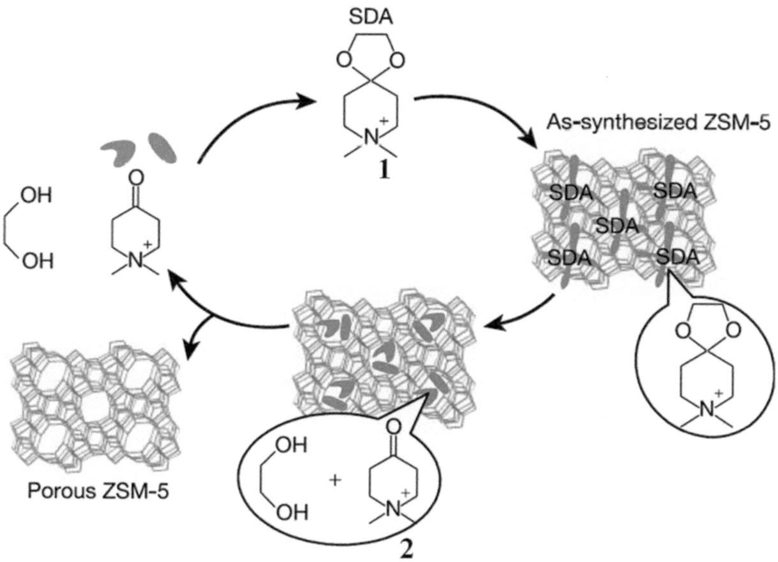

Fig. 1.2 Schematic representations of synthetic methodology for ZSM-5 using **1** as template. Step 1: assemble the SDA with silica precursor, H_2O, alkali metal ions, and so on, for zeolite synthesis. Step 2: cleave the organic molecules inside the zeolite pores. Step 3: remove the fragments. Step 4: recombine the fragments into the original SDA molecule (Reprinted with permission from Ref. [29]. Copyright 2003 Nature Publishing Group)

piperidinium (**2**) could be completely removed as shown in ^{13}C CP/MAS NMR spectrum. Conceptually, this strategy is to assemble an organic template from at least two components using covalent bonds and/or non-covalent interactions that are able to survive the conditions for assembly of the zeolite and yet be reversed inside the microporous void space. The fragments formed from the organic template in the zeolite can then be removed from the inorganic framework and be recombined for use again. Other zeolites such as ZSM-11 and ZSM-12 can also be synthesized using the same manner, suggesting that it can be used as a generalized methodology in the field of zeolite preparation [29].

1.3 Synthesis of Zeolites Without Using Organic Templates

Recently, organotemplate-free synthesis of zeolites has been the hot topic in zeolite area, since it completely avoids the use of organic templates and consequently disadvantages [9, 30]. Several groups have devoted to synthesize a series of zeolites in the absence of organic templates by adjusting molar ratios of the starting gels, addition of zeolite seed solution, and addition of zeolite crystal seeds.

1.3.1 MFI Zeolite

The discovery of ZSM-5 was regarded as a milestone in the history of hydrothermal synthesis of zeolites [1, 2, 31]. The ZSM-5 is the most widely studied zeolite due to its special features (e.g., morphology, zigzag channels, Si/Al ratio) and its importance in petrochemical and fine chemical industry [1, 2]. Notably, ZSM-5 is the first example for organotemplate-free synthesis of high-silica zeolites. In the initial stage of synthesis of ZSM-5, it was widely accepted that ZSM-5 could only be made using a suitable organic template (usually TPA^+) [1, 9, 31]. Grose and Flanigen prepared well-crystallized ZSM-5 zeolite from the Na_2O–SiO_2–Al_2O_3–H_2O in the absence of organics and seeds for the first time [32–34]. Later, Shiralkar and Clearfield reported that the factors of adjusted Si/Al and Na/Al ratios are keys for the organotemplate-free synthesis of ZSM-5 zeolite [35].

1.3.2 BEA Zeolite

Beta zeolite was successfully synthesized using tetraethylammonium cation as the templates in 1967 [36]. In the past 40 years, there is a belief that beta zeolite can only be synthesized in the presence of suitable organic templates [9, 35, 37]. However, in 2008, Xie et al. reported an organotemplate-free and fast route for synthesizing beta zeolite by the addition of calcined beta crystals as seeds in the starting aluminosilicate gel in the absence of any organic templates for the first time [37]. Nitrogen sorption isotherms of as-synthesized sample exhibited a steep increase in the curve at a relative pressure $10^{-6} < P/P_0 < 0.01$, characteristic of Langmuir adsorption due to the filling of micropores, which confirmed that as-synthesized sample had opened micropores already, and therefore the combustion of the sample could be avoided. Later, Kamimura et al. systemically studied various parameters on the seed-directed synthesis of beta zeolite in the absence of organic templates, such as the molar ratios of SiO_2/Al_2O_3, H_2O/SiO_2, and Na_2O/SiO_2 in the starting gels, amount and Si/Al ratios of seeds, and crystallization time [38]. They found that beta zeolite can be successfully synthesized with a wide range of chemical compositions of the initial Na^+–aluminosilicate gel ($SiO_2/Al_2O_3 = 40$–100, $Na_2O/SiO_2 = 0.24$–0.325, and $H_2O/SiO_2 = 20$–25) by adding calcined beta seeds with the Si/Al ratios in the range of 7.0–12.0. Very importantly, such seed-directed beta seed crystals can be used as renewable seed crystals to establish a completely organotemplate-free process for the production of beta zeolite, which is a vital development from the viewpoint of green chemistry. Thus, this kind of seed-directed beta was termed as "green beta zeolite" by the authors.

In a recent report, Zhang et al. reported a rational synthesis of beta-SDS at 120 °C (beta-SDS_{120}) with good crystallinity and improved zeolite quality in the presence of a very small amount of beta seeds (as low as 1.4 %) by decreasing zeolite crystallization rate [39]. X-ray diffraction patterns show that calcination at 550 °C for 4 h

Table 1.1 Textural parameters of as-synthesized beta-SDS and calcined beta-TEA zeolites

Sample	BET surface area (m^2/g)	Micropore area (m^2/g)	Micropore volume (cm^3/g)	HK pore size (nm)
Beta-SDS_{140}	450	386	0.18	0.70
Beta-SDS_{120-1}	655	545	0.25	0.70
Cal-beta-TEA	577	447	0.21	0.66

Reprinted with permission from Ref. [39]. Copyright 2013 Elsevier

results in the loss of crystallinity at 8.0 and 15.8 for beta-SDS_{120} and beta-SDS_{140}, respectively, suggesting that beta-SDS_{120} has higher thermal stability than beta-SDS140. N_2 sorption isotherms show that beta-SDS_{120} has much higher surface area (655 m^2/g) and micropore volume (0.25 cm^3/g) than beta-SDS_{140} (450 m^2/g, 0.18 cm^3/g) (Table 1.1). These phenomena are reasonably assigned to that beta-SDS120 samples have much less framework defects such as terminal Si–OH groups than beta-SDS_{140}. The beta-SDS_{120} samples with good crystallinity, high thermal stability, and large surface area and pore volume offer a good opportunity for their industrial applications as efficient and low-cost catalytic and adsorptive materials.

The mechanism on seed-directed synthesis of beta zeolite has been independently discussed by Xiao and Okubo's groups at nearly the same time [40, 41]. By using a series of modern techniques (XRD, TEM, SEM, XPS, Raman, MAS NMR), Xie et al. have extensively investigated seed-directed synthesis of beta-SDS under various conditions, suggesting that seed-directed beta zeolites are grown from solid beta seeds, and final beta-SDS crystals are mainly alike core–shell structure [40]. The core part of beta seeds has relatively high Si/Al ratios, and the shell part grew from aluminosilicate gels has relatively low Si/Al ratios (Fig. 1.3).

De Baerdemaeker et al. have systemically investigated the catalytic performance of beta-SDS in various reactions, and they found that beta-SDS has different properties than the usual commercial beta zeolites [42]. Part of the differences can be explained by the higher aluminum content and different crystal size. The high aluminum content leads to a large number of acid sites of considerable strength resulting in an active ethylation catalyst even at 150 °C. The large crystal size of beta-SDS makes them sensitive to deactivation through pore blocking. In alkylation reactions with propene and 1-dodecene, this resulted in low activities. An appropriate dealumination treatment can improve the accessibility and delay the deactivation. The high aluminum content also leads to a high framework polarity which is a cause for fast deactivation in acylation reactions. This can be prevented by dealumination where an activity optimum is obtained between framework polarity and acid site concentration. The high amount of strong acid sites also leads to a high yield of cracked products in the *n*-decane hydroconversion at very low temperatures (Fig. 1.4). Clearly, more Pt should be added to improve the balance between the acid sites and the (de)hydrogenation sites. A reduction in the amount of acid sites by dealumination at constant Pt loadings resulted in higher isomerization yields. Yilmaz et al. also pointed out that beta-SDS possesses a high density of active sites with exceptional stability and distinctively ordered nature, useful in, e.g., ethylation

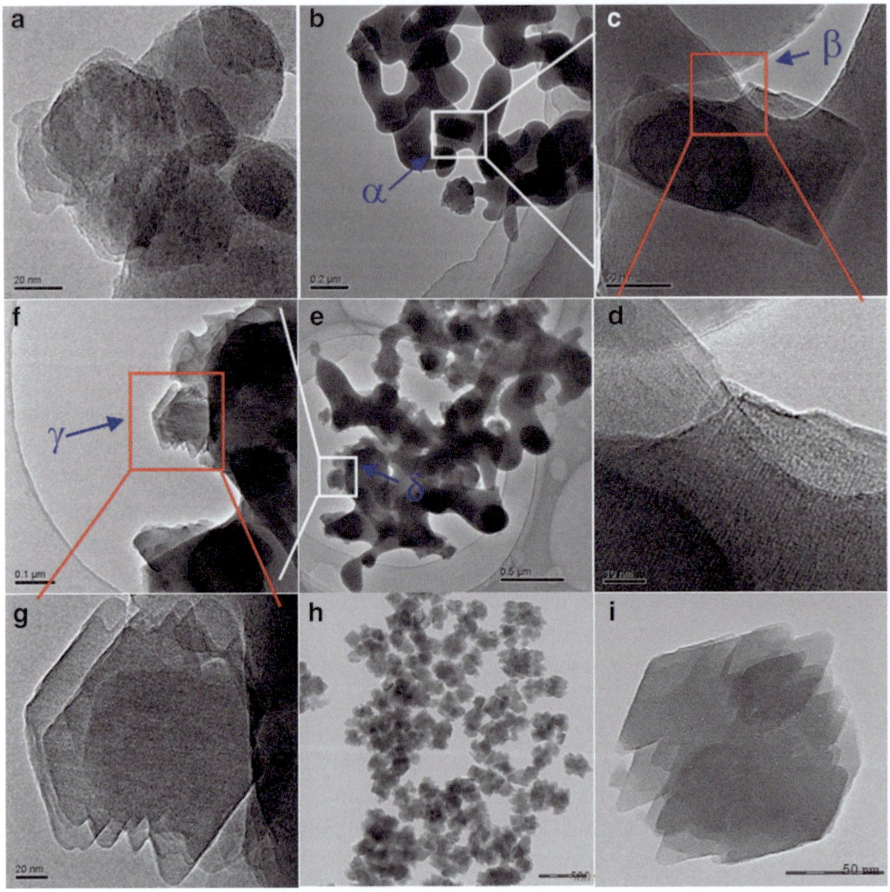

Fig. 1.3 TEM images of beta-SDS samples crystallized for (**a**) 1, (**b–d**) 4, (**e–g**) 8, and (**h** and **i**) 18.5 h at a temperature of 140 °C by addition of 10.3 % beta seeds (Si/Al = 10.2) in the starting aluminosilicate gels. Areas of *a*, *b*, *d*, and *g* in (**b**), (**c**), (**e**), and (**f**) are enlarged as (**c**), (**d**), (**f**), and (**g**), respectively (Reprinted with permission from Ref. [40], Copyright 2011 Royal Society of Chemistry)

of benzene; after dealumination and/or other post-synthesis treatments, catalysts with varying Si/Al ratios, suitable, e.g., for acylation of anisole, are obtained [43]. The ability to manipulate the framework aluminum content in a very broad range, while maintaining structural integrity, proves that beta-SDS zeolites constitute a powerful toolbox for designing new acid catalysts.

Notably, heterogeneous atoms can also be incorporated into the framework of BEA via SDS route [44]. Zhang et al. have demonstrated that an organotemplate-free and seed-directed route has been successfully applied for synthesizing Fe-beta zeolite with good crystallinity, high surface area, uniform crystals, and tetrahedral Al^{3+} and Fe^{3+} species. Catalytic tests for the direct decomposition of nitrous oxide indicate that the Fe-beta exhibits excellent catalytic performance.

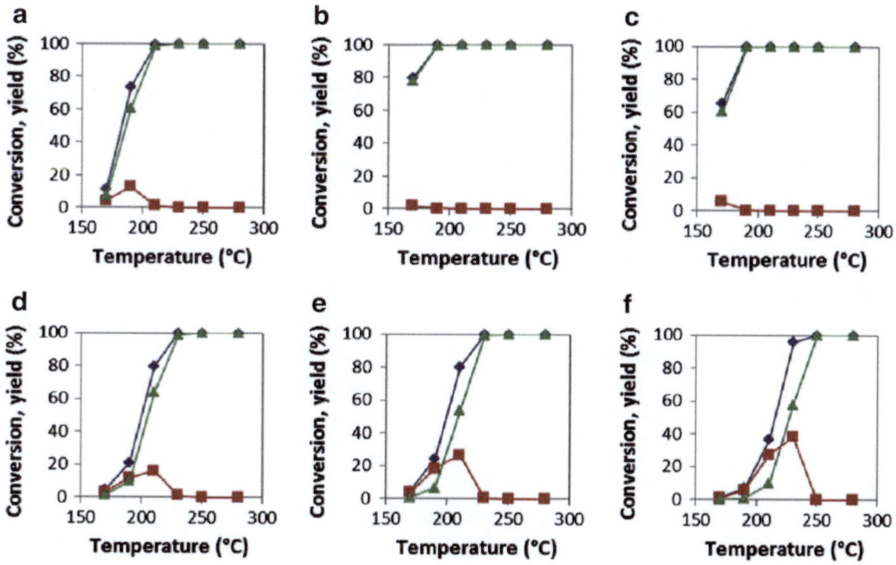

Fig. 1.4 Catalytic results from the n-decane hydroconversion:n-decane conversion (♦), yield of isomerization products (■) and yield of cracking products (▲) for Beta-1 (**a**), OF-Beta (**b**), OF-Beta-ST (**c**), OF-Beta-ST-0.1 (**d**), and OF-Beta-ST-0.5 (**e**) and OF-Beta-ST-6.0 (**f**) (Reprinted with permission from ref 42. Copyright| 2013 Elsevier)

1.3.3 EMT Zeolite

Zeolite EMT is a hexagonal polymorph of faujasite-type zeolites, with one of the lowest framework densities for microporous zeolites. Similar to the FAU zeolite, the EMT framework topology has a three-dimensional large (12-membered ring) pore system. The cubic FAU polymorph features only one type of supercage (with a volume of 1.15 nm^3), but a different stacking of faujasite sheets creates two cages in the EMT zeolite: a hypocage (0.61 nm^3) and a hypercage (1.24 nm^3) [45]. EMT zeolite showed excellent catalytic performance as FCC catalyst, but its high cost precludes its practical applications, compared with Y zeolite [10, 11]. An expensive and toxic template of 18-crown-6 is the most used template for EMT zeolite. Recently, Ng et al. reported organotemplate-free synthesis of ultrasmall hexagonal EMT zeolite nanocrystals (6–15 nm in the sizes) at very low temperature from Na-rich precursor suspensions [46]. Notably, the ratios between different compounds, nucleation temperature and times, and type of heating should be carefully controlled to avoid phase transformations (e.g., to FAU and SOD) and to stabilize the EMT zeolite crystals at a small particle size. The author proposed that under appropriate conditions the EMT was the first kinetic, metastable product in this synthesis field, followed by conversion into the more stable cubic FAU and more dense SOD structures [46].

1.3.4 MTW Zeolite

ZSM-12 is the type zeolite with the framework of MTW with one-dimensional, non-interpenetrating 12-ring pores (with the size of 5.6×6.0 Å along b-axis), which was first reported by Rosinski and Rubin in 1974 [47]. Since then, ZSM-12 has attracted much attention because of its excellent catalytic properties in the cracking of hydrocarbons or in other petroleum refining processes. The conventional synthesis of ZSM-12 has been achieved by using tetraalkylammonium cations such as methyltriethylammonium ($MTEA^+$), tetraethylammonium hydroxide (TEA^+) as organic SDAs [47–51]. Kamimura et al. have reported the synthesis of highly crystalline, pure MTW-type zeolite which has been studied by the addition of calcined ZSM-12 seeds [52, 53]. They have systemically investigated the various parameters on the seed-directed synthesis of MTW zeolite in the absence of organic templates, such as the molar ratios of SiO_2/Al_2O_3, H_2O/SiO_2, and Na_2O/SiO_2 in the starting gels, amount of seeds, and crystallization time. They found that MTW zeolite can be successfully synthesized in a wide range of the initial OSDA-free sodium aluminosilicate gel compositions: $SiO_2/Al_2O_3=60–120$, $Na_2O/SiO_2=0.1–0.2$, and $H_2O/SiO_2=8.25–13.3$ (Table 1.2). Notably, SDS-MTW samples are rodlike crystals with well-defined morphology, which is quite different from the round-shaped, irregularly aggregated morphology of the seeds. Additionally, the crystal size of SDS-MTW is in the range of 0.2–1.5 μm in length and 50–200 nm in diameter, which is larger than the size of the seeds. The solid yield of SDS-MTW was ca. 47 %, which is obviously higher than that in the case of the organotemplate-free synthesis of beta. More importantly, the green production of MTW-type zeolite referred as “Green MTW” is achieved for the first time, by using the product of OSDA-free synthesis as seeds [53].

Interestingly, Kamimura et al. found that pure MTW-type zeolites can also be prepared in the presence of beta zeolite seeds instead of ZSM-12 seeds [54]. To understand the crystallization behavior and the role of beta seeds in the present organotemplate-free Na-aluminosilicate gel systems, the crystallization processes were carefully studied by XRD. Before 55 h, small diffraction peaks of beta seeds were clearly observed and then became smaller possibly because of the partial dissolution of beta seeds, and the diffraction peaks corresponding to the MTW phase simultaneously appeared, suggesting the formation of MTW zeolite. The intensity of the MTW phase gradually increased, indicating the growth of MTW zeolite crystals. Finally, complete crystallization of MTW-type zeolite was obtained after 96 h of heating. Such phenomenon can be explained by that ZSM-12 and beta zeolites possess very similar topology in which their a–c projection viewed along and perpendicular to the 12R straight channels. This fact indicates that beta seeds would possibly provide a specific growth surface for the crystallization of the MTW phase through their structural similarity. Also, as evidenced by the crystallization behavior of MTW, beta seeds were partially dissolved in the course of the hydrothermal treatment. Hence, the fragments from partially dissolved seeds with BEA structure might have a role to induce the crystal growth of MTW phase, although it is still difficult to evaluate and observe the amount of dissolved seeds and fragments under highly alkaline condition. Moreover, the crystallization of MTW is induced by not only the structural similarity between seeds and target zeolite but also the chemical composition of the non-seeded, organotemplate-free gel.

Table 1.2 Chemical compositions of the initial sodium aluminosilicate gel, synthesis conditions, and characteristic properties of the products in the seed-assisted, OSDA-free synthesis of MTW-type zeolite

Sample	SiO_2/Al_2O_3[a]	Na_2O/SiO_2[a]	H_2O/SiO_2[a]	Amount of seeds (wt.%)[b]	Time (h)[c]	Phase[d]	Crystallinity (%)[e]	Si/Al ratio[f]
MTW-No. 1	20	0.100	10	10	72	MTW+Arm	10	–
MTW-No. 2	40	0.150	10	10	72	MTW+Arm	30	–
MTW-No. 3	60	0.100	10	10	72	MTW	100	–
MTW-No. 4	60	0.150	10	10	55	MTW	100	14.5
MTW-No. 5	60	0.200	13.3	10	36	MTW	100	11.7
MTW-No. 6	80	0.125	8.25	10	72	MTW+Cri	100	–
MTW-No. 7	80	0.175	11.5	10	36	MTW	100	16.2
MTW-No. 8	80	0.175	11.5	1	70	MTW	100	–
MTW-No. 9	100	0.100	10	10	72	MTW+Cri	100	23.4
MTW-No. 10	100	0.150	10	10	65	MTW+Cri	100	25.7
MTW-No. 11	100	0.200	13.2	10	65	MTW	100	14.5
MTW-No. 12	120	0.125	8.3	10	96	MTW+Cri	100	33.0
MTW-No. 13	120	0.175	11.7	10	96	MTW+Cri	100	26.7

[a]Chemical composition of the initial reactant gel
[b]Weight ratios of the seeds relative to the silica source
[c]Time for the hydrothermal treatment at 165 °C
[d]Phase of the solid product. The phase shown in the parenthesis indicates the relatively small amount of impurity. *Arm* amorphous, *Cri* cristobalite
[e]Crystallinity of the MTW phase
[f]Si/Al ratio of the product determined by ICP-AES

1.3.5 TON Zeolite

TON-type zeolites including ZSM-22, Theta-1, Nu-10, KZ-2, and ISI-1 have a one-dimensional 10-membered ring pore system with medium-sized pores of ca. 0.47×0.55 nm [55, 56]. The channels run along the longest dimension of the crystals (crystallographic *c* direction). The unique structure of TON zeolites offers superior catalytic performance in petrochemical processes such as isomerization, hydroisomerization dewaxing, and propylene oligomerization. Generally, TON zeolite can be hydrothermally synthesized from aluminosilicate gels using a series of oxygen- or nitrogen-containing linear organics as SDAs such as amines, long-chain polyamines, and quaternary ammonium compounds [55–60]. Recently, Wang et al. have reported a successful seed-directed and organotemplate-free synthesis of TON zeolites (denoted as ZJM-4) [61]. XRD pattern of ZJM-4 sample synthesized in the presence of ZSM-22 seeds without using organic templates under rotation conditions showed a series of characteristic peaks associated with TON structure. Furthermore, SEM image shows that ZJM-4 has uniform rodlike crystals with length at 2–4 μm and width at 100–200 nm, in good agreement with the typical morphology of TON-type zeolites reported previously. Ar sorption isotherms of as-synthesized ZJM-4 exhibited a steep increasing in the curve at a relative pressure $10^{-6} < P/P_0 < 0.01$, which is characteristic of Langmuir adsorption due to the filling of micropores, confirming that as-synthesized sample had opened micropores (Fig. 1.5).

It is worth mentioning that the seed-directed synthesis of ZJM-4 has very high silica utilization, compared with the seed-directed synthesis of beta zeolite. For example, the silica utilization for seed-directed synthesis of ZJM-4 is 88 %, much

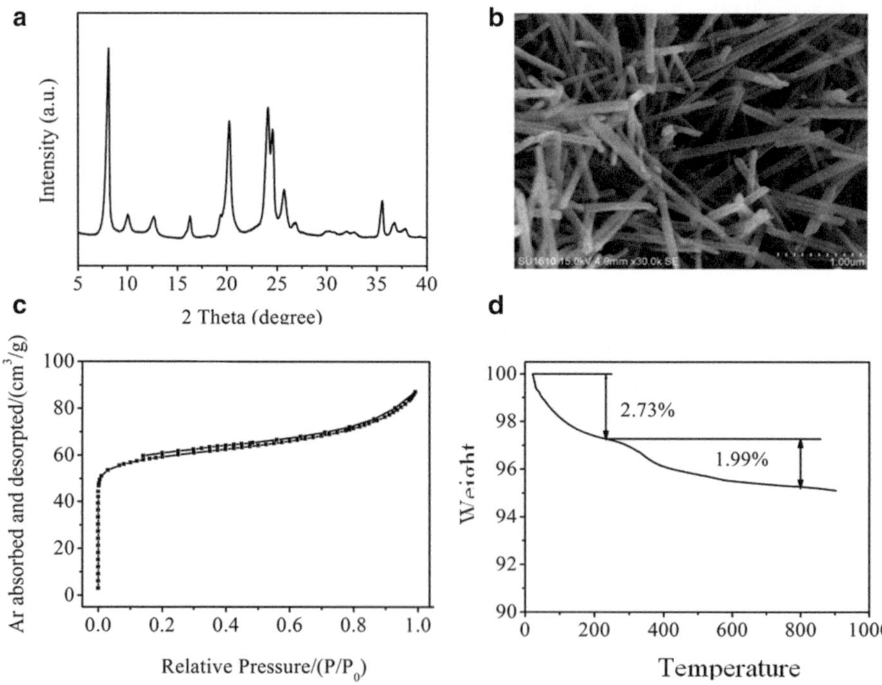

Fig. 1.5 (**a**) XRD and (**b**) SEM image of the as-synthesized ZJM-4 sample, (**c**) Ar sorption isotherms of the H-form of the ZJM-4 sample, and (**d**) TG curve of the as-synthesized ZJM-4 sample (Reprinted with permission from Ref. [61], Copyright 2014 Elsevier)

higher than that of beta-SDS (*ca* 30 %). The high silica utilization in the synthesis of ZJM-4 might be resulted from similar Si/Al ratios of the product with the starting gels.

1.3.6 MTT Zeolite

MTT zeolite family including ZSM-23, KZ-1, EU-13, ISI-6, and SSZ-32 possess a teardrop-shaped channel system with dimensions 0.52×0.45 nm [62–67]. The organic templates for MTT zeolites mainly includes pyrrolidine, diquaternary ammonium cations, isopropylamine, dimethylamine, and N,N-dimethylformamide (DMF) [62–69]. Recently, Wu et al. reported an organotemplate-free, seed-directed, and rapid synthesis of Al-rich MTT zeolite (Si/Al ratio at 20, denoted as ZJM-6) in the presence of ZSM-23 seeds [70]. Similar to ZJM-4 (TON-SDS), ZJM-6 exhibits a series of characteristic XRD peaks associated with MTT structure and rodlike crystal morphology (length at 1–2 μm and diameter at about 100 nm), in good agreement with those of ZSM-23 zeolites reported previously.

It is worth mentioning that the crystallization time of ZJM-6 is very short (5 h at 170 °C), compared with conventional ZSM-23 zeolite synthesized in the presence of organic templates. Generally, conventional hydrothermal synthesis of ZSM-23 zeolite in the presence of pyrrolidine template under rotation still takes 43 h at 180 °C (82 h at 160 °C) in the presence of 10 % seeds to achieve full crystallization [71]. Thus, it is believable that the rapid crystallization of ZJM-6 with MTT structure is reasonably attributed to the unique crystallization process. Generally, hydrothermal synthesis of zeolites includes induction and crystallization periods. However, there is nearly no inductive period in the crystallization of ZJM-6 (Fig. 1.6). In contrast, the inductive period for synthesizing ZSM-23 zeolite in the presence of DMF template is quite long, requiring at least 12–34 h. The addition of ZSM-23 seeds in the synthesis system containing DMF template significantly shortens the induction period, but it still takes 6–15 h. The presence of DMF template in the synthesis system induces the interaction with silica species, forming the zeolite nuclei. The formation of zeolite nuclei in the synthesis will delay the crystallization because the formation of zeolite nuclei takes some time (induction period). Very interestingly, it is observed that, under the same temperature, various samples (ZJM-6, ZSM-23, ZSM-23-S) have very similar crystallization time, but their inductive periods are quite different. These results suggest that rate-determined step for crystallization process of MTT zeolite is the induction period, which significantly reduces the crystallization time in the organotemplate-free and seed-directed synthesis, compared with conventional ZSM-23 synthesis.

ICP analysis shows that Si/Al ratio of ZJM-6 is about 20, which is much lower than conventional ZSM-23 zeolite (ca. 32–62), indicating that ZJM-6 is more Al sites than conventional ZSM-23. This feature would benefit the catalytic performance in acid-catalyzed reactions. The catalytic performance of MTT zeolites has been investigated in isomerization of *m*-xylene. ZJM-6 and ZSM-23 samples show

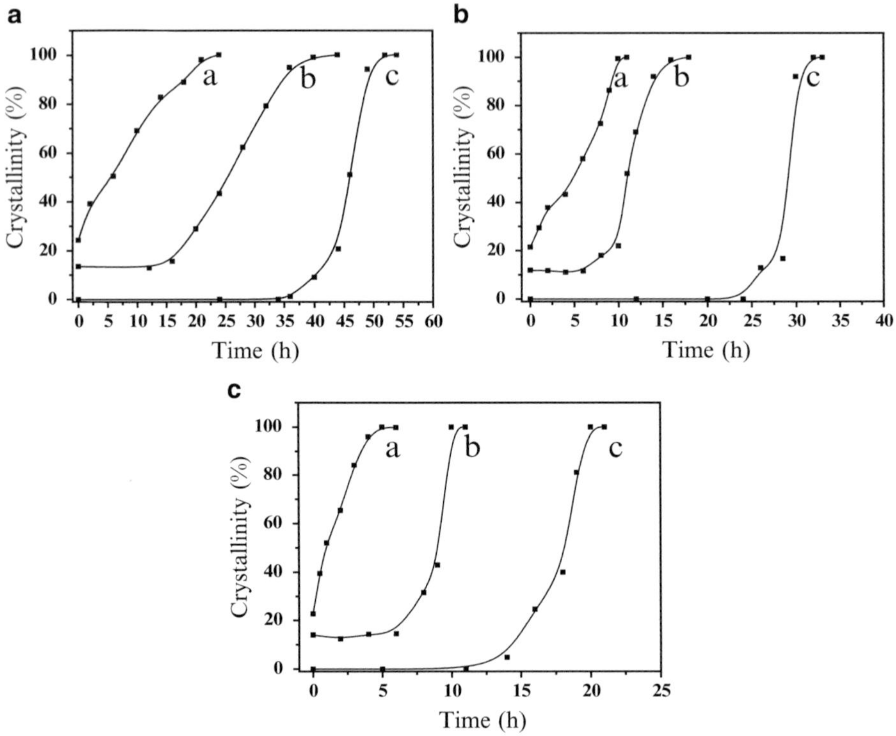

Fig. 1.6 The dependences of crystallinity on the crystallization time of (*a*) ZJM-6, (*b*) ZSM-23 synthesized in the presence of both DMF template and ZSM-23 seeds (ZSM-23-S), and (*c*) ZSM-23 synthesized in the presence of DMF template at (**A**) 150 °C, (**B**) 160 °C, and (**C**) 170 °C, respectively (Reprinted with permission from Ref. [70], Copyright 2014 Elsevier)

very high selectivity for *p*-xylene (ca. 86 %), but ZJM-6 exhibits higher conversions (10.4 %) than those (4.1–9.4 %) of ZSM-23 zeolites, which should be assigned to the contribution of more Al species in the framework of ZJM-6 (Fig. 1.7).

1.3.7 RTH Zeolite

RTH-type zeolite (e.g., borosilicate RUB-13, aluminosilicate SSZ-50) consisted of RTH cages with 8MR openings and two-dimensional channels with aperture size of 0.41×0.38 and 0.56×0.25 nm, parallel to the *a* and *c* axis, respectively [72, 73]. RTH-type zeolites have shown excellent catalytic properties in methanol-to-olefins (MTO) reaction. However, the synthesis of RUB-13 and SSZ-50 always requires using organic templates such as 1,2,2,6,6-pentamethylpiperidine (PMP), ethylenediamine (EDA), or N-ethyl-N-methyl-5,7,7-trimethylazoniumbicyclo[4.1.1]octane cation [72, 73]. Yokoi et al. have reported a successful synthesis of RTH-type zeolites (denoted as TTZ-1) without using organic templates by addition of calcined

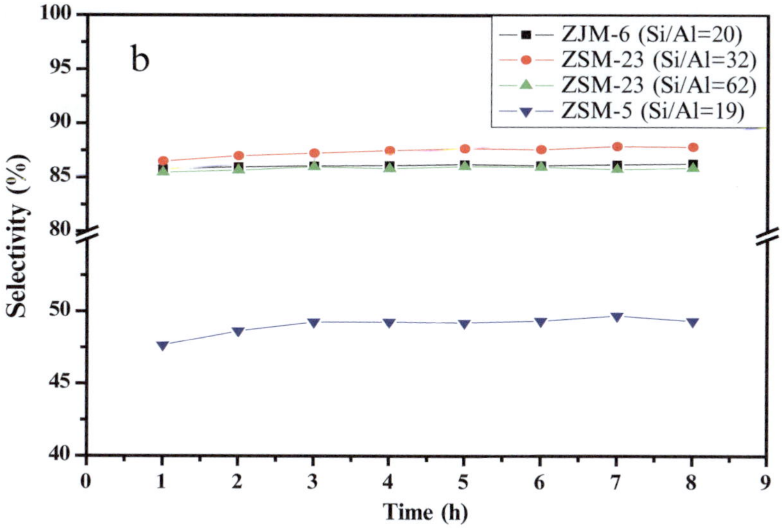

Fig. 1.7 Catalytic performance in isomerization of *m*-xylene to *p*-xylene over ZSM-5 (Si/Al = 19), ZJM-6 (Si/Al = 20), and ZSM-23 (Si/Al = 62 and 32) catalysts as a function of time (Reprinted with permission from Ref. [70], Copyright 2014 Elsevier)

B-RUB-13 as seeds shortly after the discovery of SDS-beta [74]. Direct introduction of Al and Ga heteroatoms into the RTH framework during crystallization of B-TTZ-1 in the absence of organic templates had also been successfully performed. NMR spectra confirmed the tetrahedrally coordinated heteroatoms in the framework. The catalytic properties in MTO over these SDS-RTH zeolites have also been tested. The selectivity for propene was obviously higher than that of SAPO-34 and ZSM-5 zeolites, and the catalytic life of RTH-type zeolites was much longer, which should be assigned to their unique structure.

1.3.8 FER Zeolite

Ferrierite (FER) zeolite, with an anisotropic framework composed of two-dimensional straight channels including a 10MR channel (0.42 × 0.54 nm) along [001] direction and a 8MR channel (0.35 × 0.48 nm) along [010] direction, has been carefully studied, due to its excellent catalytic performance [75–78]. Notably, FER zeolite with low ratios of Si/Al could be synthesized in the absence of organic templates [79], but high-silica FER zeolite (ZSM-35) is always prepared in the presence of organic templates. Zhang et al. have demonstrated successful synthesis of high-silica FER zeolite (Si/Al at 14.5) from the introduction of RUB-37 zeolite (CDO structure) in the absence of organic templates (designated as ZJM-2) [80]. To understand the crystallization behavior and the role of RUB-37 zeolite seeds, the crystallization processes were carefully studied by XRD patterns. Before 12 h, small diffraction peaks of RUB-37 zeolite seeds were clearly observed at 9.6° and then became smaller until they disappear after 24 h possibly because of the dissolution of RUB-37 seeds. At the same time, the diffraction peaks corresponding to the FER phase simultaneously appeared at 9.4°, suggesting the formation of FER zeolite. The intensity of the FER phase gradually increased, indicating the growth of FER zeolite crystal. Finally, highly crystallized FER-type zeolite was obtained after 72 h of heating. It is well known that the building units of FER and CDO are the same, and their difference is only a shift of layers in the horizontal direction. Therefore, it is reasonable to use the building units of RUB-37 zeolite to induce the crystallization of FER-type zeolite, which has been confirmed by UV-Raman spectroscopy of the samples.

1.3.9 LEV Zeolite

Levyne (LEV) zeolite is a typical small-pore zeolite, with relative smaller pore size (3.6×4.8 Å) and low framework density (15.2 T/1000 $Å^3$), characterized by $4^9 6^5 8^3$ heptadecahedral cavity [81]. The natural levyne zeolite with typical composition at $Ca_9(Al_{18}Si_{36}O_{108})\cdot 50H_2O$ was first discovered in 1825 [82]. Synthetic LEV zeolite named ZK-20 was synthesized from an aluminosilicate gel using 1-methyl-1-azonia-4-azabicyclo[2.2.2]octane cation as SDA [83]. Subsequently, other aluminosilicate LEV zeolites were successfully prepared by using a series of organic compounds as SDAs, including N-methylquinuclidinium cation, diethyldimethylammonium, N,N′-bis-dimethylpentanediyldiammonium, N,N-dimethylpiperidine chloride, and choline hydroxide [84–88]. Additionally, phosphate-based LEV-type zeolites were also obtained in the presence of tropone hydroxide, quinuclidine, and 2-methyl-cyclohexylamine; boron-containing LEV-type zeolites were reported by using organic compounds of 3-azabicyclo[3.2.2]nonane and quinuclidine as templates [89–92]. Recently, Zhang et al. reported organotemplate-free and seed-directed synthesis of LEV zeolite (SDS-LEV) in the presence of RUB-50 seeds with the aid of a small amount of alcohol [93]. In this synthesis, the alcohol plays an important role in the synthesis of highly pure SDS-LEV zeolite.

To understand the role of alcohol in the synthesis, various alcohols (e.g., methanol, ethanol, *n*-propanol, and *n*-butanol) were added into the starting aluminosilicate gels. Notably, without using any alcohol in the starting aluminosilicate gel, the product contained impurity phase of MOR zeolite in addition to the LEV product. In contrast, after addition of a small amount of alcohol in the synthesis, the samples showed pure phase of Na-LEV-SDS zeolite (Fig. 1.8). These results indicate that the alcohols strongly prevent the formation of MOR zeolite in this seed-directed synthesis. FTIR spectra and C/N/H elemental analysis confirmed that alcohol molecules did not exist in the micropores of SDS-LEV zeolites. The addition of alcohols could delay the nucleation of MOR zeolite, reducing crystallization rate of MOR zeolite. As a consequence, highly pure SDS-LEV zeolite could be obtained.

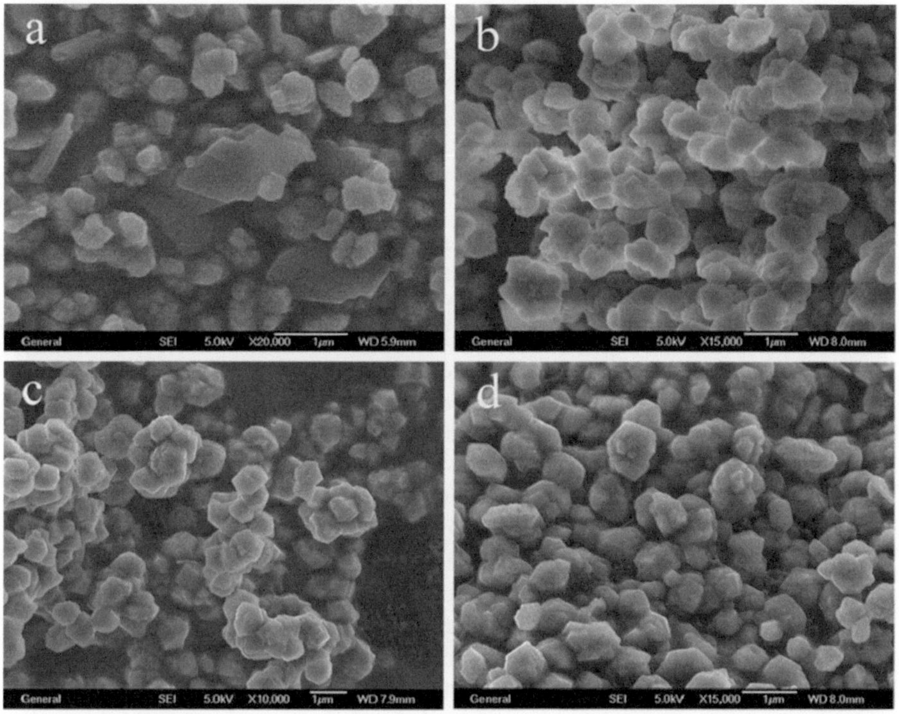

Fig. 1.8 SEM images of Na-LEV-SDS zeolites synthesized (**a**) in the absence of alcohol and in the presence of (**b**) methanol, (**c**) *n*-propanol, and (**d**) *n*-butanol, respectively (Reprinted with permission from Ref. [93], Copyright 2012 Elsevier)

1.3.10 SZR Zeolite

SUZ-4 zeolite (SZR) is an aluminosilicate zeolite with the three-dimensional topology consisting of 5-, 6-, 8-, and 10MRs, which was first reported using TEAOH and quinuclidine as SDAs under rotation conditions [94]. Later, it was successfully synthesized in the presence of N,N,N,N,N,N-hexaethylpentanediammonium bromide (Et_6-diquat-5) [95]. Zhang et al. reported an organotemplate-free route for hydrothermally synthesizing zeolite SUZ-4 under static conditions by adding the calcined SUZ-4 seeds in the starting aluminosilicate gels [96]. To further understand the crystallization of SUZ-4 in the absence of organic templates, different crystallization stages of the crystallization process have been carefully studied by SEM and TEM techniques. The observations suggest that the addition of the seed crystals into the organotemplate-free crystallization mixture causes the deposition of amorphous particles formed by depletion of the heterogeneous hydrogels, and then the crystallization starts by the fast agglomeration of the small-sized particles from the seed-amorphous interface in alkaline medium at a high temperature.

1.3.11 ZSM-34 Zeolite

Aluminosilicate zeolite of ZSM-34 is an intergrowth of offretite (OFF) and erionite (ERI) zeolites containing zeolitic building units of cancrinite (CAN) cages [97–100]. ZSM-34 zeolite was first discovered by Rubin et al. using organic template of choline [$(CH_3)_3NCH_2CH_2OH$] [97], and later ZSM-34 samples were successfully synthesized in the presence of different diamines ($NH_2C_nH_{2n}NH_2$, $n=4, 6, 8, 10$) [101]. Wu et al. for the first time reported the organotemplate-free synthesis of ZSM-34 zeolite assisted by L zeolite seed solution [102]. Various parameters on organotemplate-free synthesis of ZSM-34 zeolite have been systemically investigated including SiO_2/Al_2O_3, H_2O/SiO_2, and Na_2O/SiO_2 in the starting gels, amount of zeolite L seed solution, and crystallization temperature [103]. The amount of zeolite L seed solution and molar ratios of SiO_2/Na_2O in the starting aluminosilicate gels were regarded as the key factors for preparation of ZSM-34 zeolite. Furthermore, heteroatom-substituted ZSM-34 (B, Ga, and Fe) can also be prepared via the same route [103]. UV–vis and NMR spectroscopy confirmed that these heteroatoms had been located in the framework of ZSM-34 zeolite.

Recently, Yang et al. have reported seed-directed synthesis of ZSM-34 zeolite with a very short crystallization time (2–6 h) and smaller crystal sizes (ca. 0.5–3 μm) [104]. SEM images of the ZSM-34 samples prepared via different routes including ZSM-34-C (conventional ZSM-34), HZSM-34-L (assisted by L seed solution), and HZSM-34-S (SDS) exhibit a pure phase of crystals, but quite different crystal sizes and morphologies. ZSM-34-C shows bulky round particles with size at 10–20 μm, and ZSM-34-L is rodlike with the size at 20–40 μm. ZSM-34-S samples are short rodlike with the size at 0.5–3 μm (Fig. 1.9). Notably, in the seed-directed synthesis route, the temperature can be as high as 180 °C, and the crystallization only requires 2 h. On the contrary, in the organotemplate-free synthesis of ZSM-34 zeolite assisted by L zeolite seed solution, the crystallization temperature should be lower than 130 °C; otherwise, PHI zeolite or orthoclase would appear in the products as impure phase. As a consequence, the crystallization time always takes as long as 7 days [103]. This phenomenon should be attributed to the features of addition of seeds: suppress the formation of impurity and acceleration of crystallization.

Catalytic tests of the MTO reaction show that HZSM-34-S has a very high selectivity for propylene (55.2 %), which is even higher than that (47.0 %) of SAPO-34 under the same conditions [104]. Moreover, the hydrothermal treatment of HZSM-34-S significantly improves the catalyst life for MTO due to decreasing acidic concentration and increasing anti-deactivation. The combination of a "green" synthesis

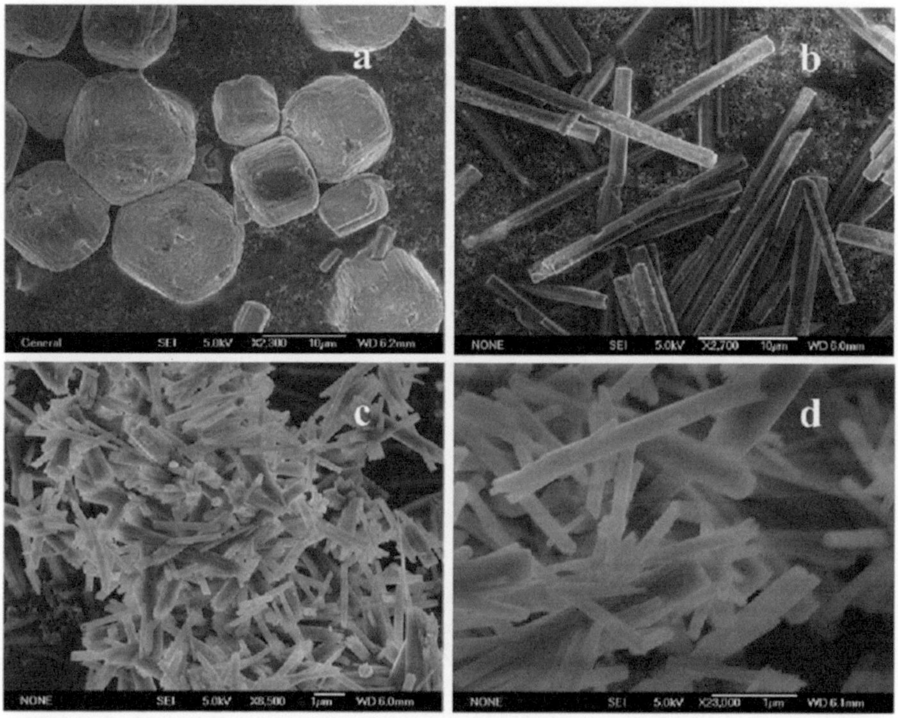

Fig. 1.9 SEM images of (**a**) HZSM-34-C, (**b**) HZSM-34-L, (**c**) HZSM-34-S, and (**d**) HZSM-34-HT samples (Reprinted with permission from Ref. [104], Copyright 2012 Royal Society of Chemistry)

and the good catalytic performance of ZSM-34-S would be potentially important for the highly effective conversion of methanol, which can be easily obtained from coal, natural gas, or biomass at a large scale.

1.3.12 ECR-1 Zeolite

Large-pore aluminosilicate zeolite of ECR-1 is an intimate twin of the mordenite-like sheets between layers of mazzite-like cages, which is first discovered using the organic template of bis-(2-hydroxyethyl)dimethylammonium chloride [105, 106]. Later, other organic templates such as adamantine-containing diquaternary alkylammonium iodides and tetramethylammonium (TMA^+) can also be used in synthesis of ECR-1 [107, 108]. Song et al. synthesized aluminosilicate zeolite of ECR-1 under hydrothermal conditions at 100–160 °C for 1–14 days by carefully adjusting the molar ratios of Na_2O/SiO_2 in the absence of organic template for the first time [109]. The molar ratio of Na_2O/SiO_2 in the synthesis significantly influences the final products of zeolites. Later, this transformation of Y zeolite to ECR-1 has been reported [110]. Based on the XRD patterns and SEM images, it can be found that the products at 7 days are Y zeolites with low crystallinity, and the crystallinity increased together with the appearance of ECR-1 after 9 days. Then 11 days later, most products are ECR-1 zeolite together with a small amount of Y zeolite. After crystallization for 13 days, pure phase of ECR-1 with high crystallinity can be obtained. Recently, Ren et al. reported a fast route to prepare ECR-1 at 120 °C for 4 days compared with the above report at 100 °C for 14 days, which is attributed to the different mechanism. The crystallization at 120 °C is a spontaneous nucleation process, while crystallization at 100 °C is a crystal transformation process (Y to ECR-1).

1.3.13 ZSM-5/ZSM-11

Organotemplate-free synthesis is not limited to single zeolites; intergrowth can also be prepared in the absence of organic templates. Recently, Zhang et al. reported organotemplate-free synthesis of ZSM-5/ZSM-11 zeolite intergrowth with different SiO_2/Al_2O_3 ratios, ZSM-5 percentages, and various morphologies by adjusting compositions of the starting gels [111]. This organotemplate-free system is favorable to the aluminum-rich zeolite. With the increase of initial SiO_2/Al_2O_3 ratios, the ZSM-5 percentage in the ZSM-5/ZSM-11 co-crystalline zeolite increases as well as the crystal size, and the morphology of ZSM-5/ZSM-11 co-crystalline zeolite prepared from the colloidal silica–$NaAlO_2$ solution system changes gradually from nanorod aggregation, micro-spindle to single hexagon, and then to twinned hexagon crystals (Fig. 1.10). Moreover, Na^+ and OH^- in the initial materials can promote the nucleation of the ZSM-5/ZSM-11 co-crystalline zeolite significantly and are beneficial to the formation of crystals with relatively low length/width ratios, while K^+ species postpone the crystallization process seriously.

1.3.14 MEL Zeolite

Zeolite ZSM-11 (MEL) is one of the end-member of the pentasil zeolite family with the same building unit as MFI (e.g., ZSM-5). Framework structures of these two zeolites are closely related to each other, and their framework structures can be described using a stacking manner of pentasil sheets. Unlike MFI, which has zigzag channels along the a-axis and straight channels along the b-axis, MEL has straight channels along both the a- and b-axes. Generally, ZSM-11 was synthesized in the presence of tetrabutylammonium hydroxide (TBAOH) as template.

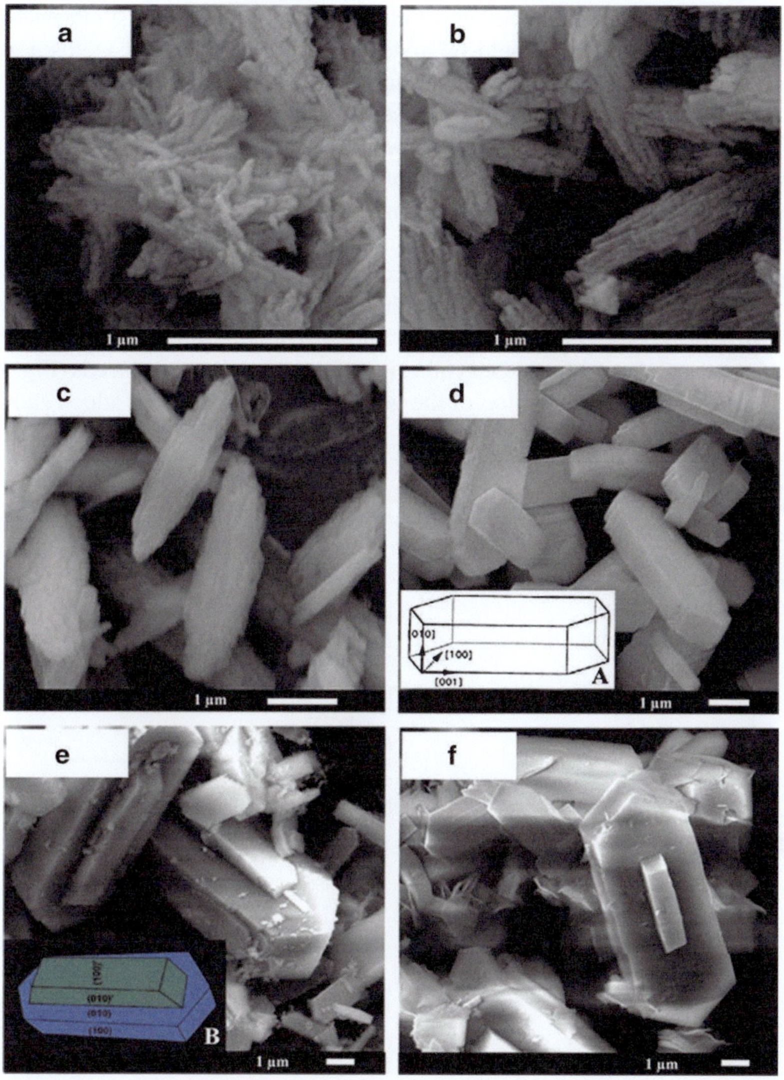

Fig. 1.10 SEM images of sample organotemplate-free synthesis with batch composition of $3.3Na_2O{:}30/rAl_2O_3{:}30SiO_2{:}1350H_2O$, where r represents the initial SiO_2/Al_2O_3 ratio, and the morphology schematic drawings of MFI-type (*A*) single hexagon crystals and (*B*) twinned hexagon crystals. (**a**) $r = 23$, (**b**) $r = 25$, (**c**) $r = 30$, (**d**) $r = 40$, (**e**) $r = 50$, (**f**) $r = 60$ (Reprinted with permission from Ref. [111], Copyright 2012 Elsevier)

Itabashi et al. showed a brief result of seed-directed and organotemplate-free synthesis of ZSM-11 in the presence of ZSM-11 seeds [112]. Typical SEM image of SDS-ZSM-11 crystals shows aggregation of 50–150 nm crystals with tetragonal morphology. The solid yield of SDS-ZSM-11 was about 18 %, a little lower than that of SDS-beta.

Okubo et al. proposed a hypothesis for the organotemplate-free synthesis of zeolite [112]; a target zeolite should be added as seeds to a gel that yields a zeolite containing the common composite building units when the gel is heated without seeds. The requirements for a successful seed-assisted, organotemplate-free synthesis of zeolites were summarized as follows: (1) the spontaneous nucleation should not occur prior to the completion of the crystal growth of the target zeolite; (2) the precursor, the common composite building units, must access the top surface of the seed zeolite; (3) the zeolite seeds should not completely dissolve prior to the onset of crystal growth during the hydrothermal treatment, and the SiO_2/Al_2O_3 ratios of the seeds should be optimized; (4) when the framework structure of the target and the seed zeolites is the same, the seeds should have at least one common composite building unit with the zeolite to be synthesized from the gel without seeds (Fig. 1.11); and (5) the chemical composition of the gel to which the seeds are added should be

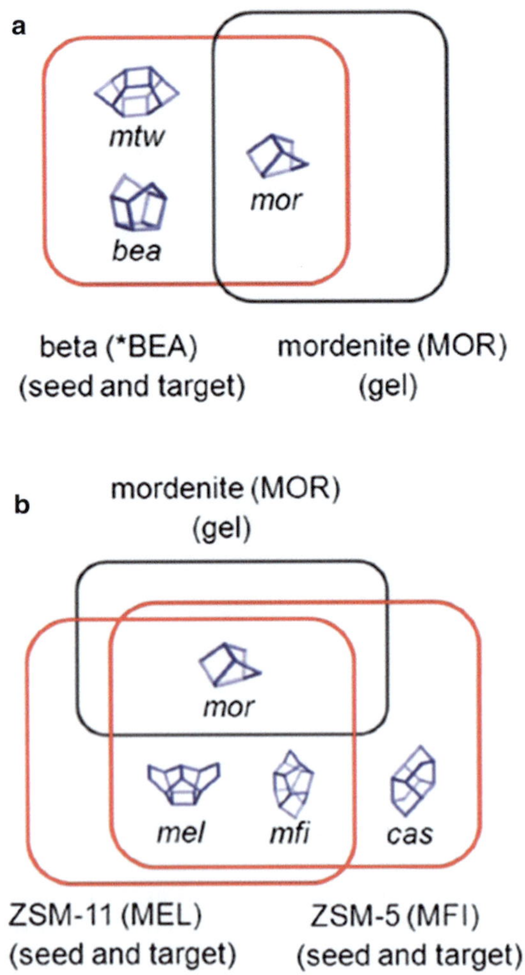

Fig. 1.11 Correlation of common composite building unit between (**a**) MOR and *BEA and (**b**) MOR, MFI, and MEL (Reprinted with permission from Ref. [112], Copyright 2012 American Chemical Society)

optimized. Based on the above hypothesis, not only MEL zeolite but also PAU (ECR-18) zeolite was also successfully prepared.

1.4 Solvent-Free Synthesis of Zeolites

1.4.1 Solvent-Free Synthesis of Nanoporous Materials

Solvent-free synthesis has proven to be an efficient method in organic synthesis since the 1980s, due to the environmental impact, safety, energy consumption, and economic cost associated with traditional solvent-intensive chemical processes [113–116]. Later, solvent-free thermal synthesis has also been applied in the preparation of ceramics, hydrides, and nitrides, which often require very high temperatures and repeat firings to ensure that the bond making/breaking and organization processes have enough energy for the formation of crystalline phases [113].

Porous metal–organic framework (MOF) materials are an intensely researched area, and mechanochemical synthesis of such phases had been demonstrated by Pichon et al. using an acid–base reaction between copper acetate and isonicotinic acid to give $Cu(ina)_2$ [117]. Neat grinding gave the porous framework quantitatively in a few minutes, with the acetic acid and water by-products partially lost and partially included in the pores. The porosity of mechanochemically prepared MOFs has been investigated. Yuan et al. found that the BET surface area of $Cu_3(btc)_2$ (HKUST-1) obtained by neat grinding or liquid-assisted grinding (LAG) of copper(II) acetate monohydrate with benzene-1,3,5-tricarboxylic acid was comparable to that of samples obtained by conventional solution-based routes [118]. Similar observations have been made by Klimakow et al. [119] who also extended the synthesis to MOF-14 $Cu_3(btb)_2$ (btb is the larger tricarboxylate 4,40,400-benzenetribenzoate).

In 2011, Zhang et al. have reported the synthesis of nanoporous aluminophosphates (SF-APOs) and metal-substituted aluminophosphates (SF-MAPOs, M = Co, Fe, Cr) from a mixture of raw materials for the first time via simple grinding and heating in the absence of solvent [120]. Characterization results showed that these mesoporous aluminophosphates had a hierarchically microporous/mesoporous structure. To understand the mechanism on "solvent-free" synthesis of hierarchically porous aluminophosphates and heteroatom-substituted aluminophosphates, in situ XRD patterns for synthesizing $SF\text{-}APO_{RT}$ were measured (Fig. 1.12). When the raw materials of $Al(O^iPr)_3$, $NH_4H_2PO_4$, $TMAOH.5H_2O$, and CTAB were mixed, the mixture exhibited typical XRD peaks of each compounds. By increasing aging time to 105 min at room temperature, the wide-angle XRD peak intensity associated with raw materials gradually reduced, suggesting that the crystalline structure of raw materials gradually lost, giving amorphous nature of $AlPO_4$, which was in good agreement with spontaneous dispersion of solid inorganic salts due to the increase of entropy (ΔS) [121]. In the meanwhile, small-angle XRD patterns of the samples aged at 60 min showed a small and new peak at about 2.3°, and this peak intensity increased with aging time. The appearance of peak at 2.3° was well consistent with

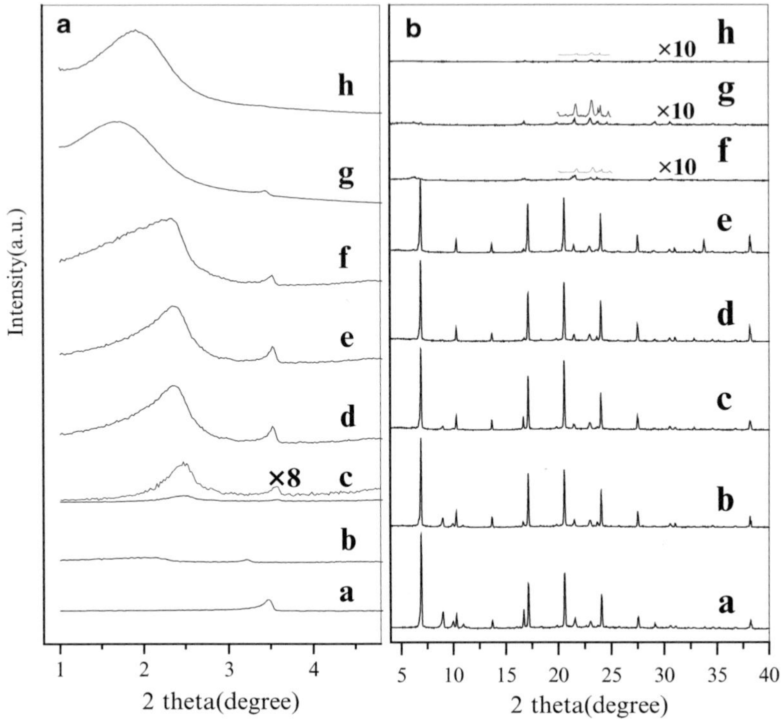

Fig. 1.12 (**A**) Small-angle and (**B**) wide-angle XRD patterns for (*a*) SF-APO$_{RT\text{-}0min}$, (*b*) SF-APO$_{RT\text{-}45\,min}$, (*c*) SF-APO$_{RT\text{-}60\,min}$, (*d*) SF-APO$_{RT\text{-}70\,min}$, (*e*) SF-APO$_{RT\text{-}90\,min}$, (*f*) SF-APO$_{RT\text{-}105\,min}$, (*g*) SF-APO$_{RT\text{-}2h}$, and (*h*) SF-APO$_{RT\text{-}27\,h}$, respectively (Reprinted with permission from Ref. [120], Copyright 2011 Royal Society of Chemistry)

CTAB micelle in the amorphous silicas and aluminophosphates. It was difficult to observe the peak at 3.4° assigned to CTAB raw material in small-angle XRD patterns at 27 h, suggesting complete dispersion of CTAB in the samples. All above results suggested that CTAB existed in the samples in two forms: one was monodispersed, and the other was aggregated micelle. After calcination at 500 °C for 6 h, the monodispersed CTAB molecule created the microporosity (0.60–0.74 nm), while the aggregated CTAB molecules templated the mesoporosity (1.7–2.6 nm).

Although solvent-free synthesis has been well studied in the preparation of inorganic materials, synthesis of zeolites in the absence of water is still scarce. In 1990, Xu et al. reported first example for the synthesis of zeolites from dry gel conversion (DGC) or vapor phase transport (VPT) technique, in which a prepared damp or dried sodium aluminosilicate gel was suspended above liquid in an autoclave and subjected to the mixed vapor of amine and water at elevated temperature and pressure [122]. However, solvents (e.g., water and alcohols) are still necessary for preparation of the starting gels, indicating that these methods are not "real" solvent-free synthesis.

1.4.2 Solvent-Free Synthesis of Aluminosilicate Zeolites

More recently, Ren et al. reported solvent-free synthesis of various zeolites from grinding of dry raw materials followed by heating to 180 °C [123]. Notably, zeolite products cannot be obtained if it is absent of a small amount of water (hydrated form of sodium silicate or hydrated form of silica) in the solid synthesis system, suggesting that a small amount of water is a critical parameter and might be favorable for facilitating hydrolysis and condensation of Si–O–Si bonds during the synthesis. Using this approach, Ren and coworkers managed to prepare a variety of some of the most industrially important zeolites, those with the MFI, SOD, MOR, BEA*, and FAU framework types. In addition, the process also allowed the incorporation of several different heteroatoms (Al, Fe, B, Ga) into the structures, opening up potential uses of the zeolites in catalysis.

To understand the mechanism of the solvent-free route, the crystallization process of ZSM-5 has been carefully investigated via XRD, UV-Raman, and ^{29}Si NMR techniques (Fig. 1.13). Before crystallization, the sample exhibits XRD patterns of each raw materials. After the treatment at 180 °C for 2 h, the peaks associated with the raw solids disappear, and a peak related to cubic NaCl phase is observed (interaction between Na_2SiO_3 and NH_4Cl). The disappearance of the XRD peaks is attributed to the spontaneous dispersion of solid salts on the amorphous support due to the increase of entropy (ΔS). At the same time, the bonds assigned to TPA^+ species

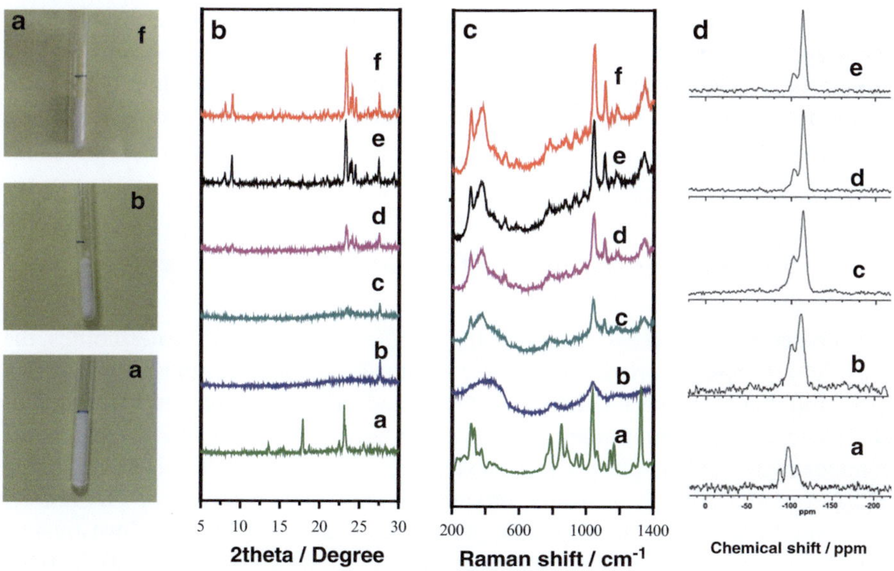

Fig. 1.13 (**A**) Photographs, (**B**) XRD patterns, (**C**) UV-Raman spectra, (**D**) ^{29}Si NMR spectra of the samples crystallized at (*a*) 0, (*b*) 2, (*c*) 10, (*d*) 12, (*e*) 18, and (*f*) 24 h for synthesizing silicalite-1 zeolite via solvent-free route (Reprinted with permission from Ref. [123], Copyright 2012 American Chemical Society)

in UV-Raman spectra are greatly reduced after the treatment at 180 °C for 2 h, which is in good agreement with the high disordering of TPA^+ species with weak Raman signals by the high dispersion of solid salts on the amorphous support. As observed from ^{29}Si NMR spectroscopy, the treatment of the sample at 180 °C for 2 h results in a significant condensation of silica species, giving that Q4 silica species $[Si(SiO)_4]$ are dominant. When the crystallization time reaches 10 h, XRD patterns and UV-Raman spectra confirmed that a small amount of S-Si-ZSM-5 crystals is formed. With the crystallization time increasing from 10 to 18 h, the intensities of XRD peaks and Raman bonds strongly increase, indicating the successful transformation from amorphous silica to zeolite crystals. When the crystallization time is over 18 h, there is no obvious change in XRD patterns and UV-Raman spectra of the samples, indicating that the crystallization of S-Si-ZSM-5 zeolite is basically finished. The photographs of the samples crystallized at various times confirmed that samples were always in a solid phase, and the sample volume was reduced remarkably after the treatment, attributed to the condensation of silica species due to the crystallization.

Compared with conventional hydrothermal synthesis of zeolites, solvent-free synthesis has obvious advantages: (1) high yields of zeolites, (2) better utilization of autoclaves, (3) a significant reduction of pollutants, (4) saving energy and simplifying synthetic procedures, and (5) a remarkable reduction of reaction pressure. As a consequence, Morris and James highlight the importance of solventless synthesis of zeolites shortly after its publication [124]. Particularly, they propose several questions on this new approach (Is the initial grinding itself important in inducing the reaction? How does the organic SDA interact with the silica in order to direct the synthesis toward one particular product? Can such processes be scaled up? Are the properties of these materials at least as good as those of the hydrothermally synthesized zeolites?) Thus, new progresses have been moved toward answering these questions.

1.4.3 Solvent-Free Synthesis of Aluminophosphate Zeolites

Encouraged by the success in solvent-free synthesis of aluminosilicate zeolites, the same group has reported solvent-free synthesis of aluminophosphate (AEL), silicoaluminophosphate (AEL, CHA, GIS), and heteroatom-substituted aluminophosphate zeolites from mixing, grinding, and heating the raw materials in a recent communication [125]. Chosen as a model, the solvent-free synthesis of SAPO-34 (S-SAPO-34) from mechanically mixing of solid raw materials of $NH_4H_2PO_4$, boehmite, fumed silica, and template (morpholine) has been carefully investigated. Very interestingly, a hysteresis loop occurred at a relative pressure of 0.50–0.98, indicating the presence of mesoporosity and macroporosity in the samples. SEM and TEM images demonstrate the presence of hierarchical macroporosity. S-SAPO-34 sample prepared in a solvent-free manner has a unique micro–meso–macroporous structure, which is very favorable for designing and preparing efficient catalysts.

The crystallization process of S-SAPO-34 has also been carefully studied via a series of techniques (Fig. 1.14). Similar to solvent-free synthesis of ZSM-5, the crystallization of S-SAPO-34 is also a solid process. The sample volume is also remarkably reduced after the treatment, due to the condensation for synthesizing the S-SAPO-34 zeolite. XRD patterns of samples show that the starting materials have sharp peaks associated with $NH_4H_2PO_4$. After treatment at 200 °C for 1 h, the sharp peaks completely disappear. After heating at 200 °C for 2 h, the sample shows weak peaks associated with the CHA structure, indicating that a small amount of SAPO-34 crystals has been formed. A cubic crystal can also be observed in the SEM image. However, N_2 sorption isotherms of the sample show only a small amount of microporosity, thus confirming that there is only a small amount of SAPO-34 crystals formed. As the crystallization time was increased from 3 h to 8 h, the intensities of XRD peaks gradually increased, indicating the successful solid transformation of S-SAPO-34 from the amorphous phase. As observed in the SEM image and N_2 sorption isotherms, the cubic crystals are dominant and the micropore

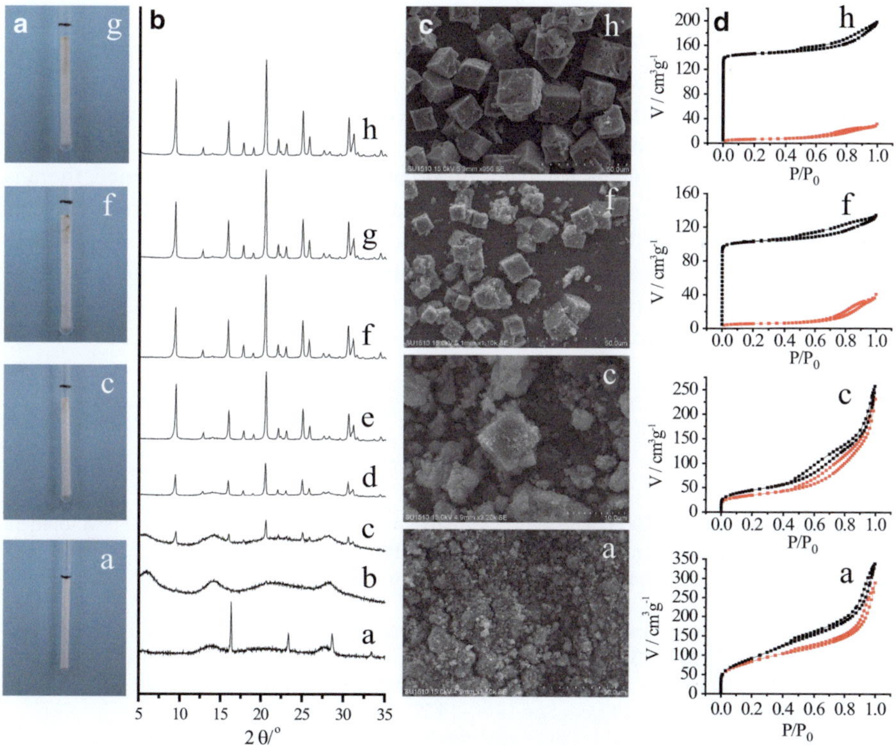

Fig. 1.14 Investigation on S-SAPO-34 crystallization. (**A**) Photographs, (**B**) XRD patterns, (**C**) SEM images, and (**D**) N_2 sorption isotherms (■ as-synthesized; ■ calcined) of the samples crystallized at (*a*) 0, (*b*) 1, (*c*) 2, (*d*) 3, (*e*) 4, (*f*) 8, (*g*) 24 h, and (*h*) 36 h for synthesizing S-SAPO-34 zeolite (Reprinted with permission from Ref. [125], Copyright 2013 American Chemical Society)

surface area is as high as 310 $m^2 g^{-1}$ at 8 h. A further increase in crystallization time from 8 h to 24 h resulted in the complete disappearance of the amorphous phase. When the crystallization time is over 24 h, there is no obvious change in the XRD pattern, N_2 sorption isotherms, or SEM image, which indicates that the crystallization of S-SAPO-34 is almost finished.

More interestingly, S-SAPO-34 zeolite gives full conversion and high selectivities for light olefins including ethylene, propylene, and butylenes in methanol-to-olefin (MTO) reaction. The total yield is as high as 88.9 %, which is comparable with that of conventional hydrothermally synthesized SAPO-34 under the same condition. Moreover, the S-SAPO-34 catalyst shows slightly lower selectivity for ethylene, but slightly higher selectivity for propylene and butylene, attributed to the presence of hierarchically porosity, which is favorable for enhancing propylene and butylene selectivities. The feature of solvent-free synthesis for combining the advantages such as low cost and good catalytic performance could be potentially important for industrial applications.

1.4.4 Summary and Perspective

In this chapter, sustainable routes for the preparation of zeolites have been described, in particular the organotemplate-free and solvent-free synthesis approaches. Notably, these sustainable approaches are separated, and as a consequence, there is always a balance of competing aspects. For example, organotemplate-free synthesis is mostly performed in hydrothermal process; many kinds of zeolites can be prepared in the absence of solvent, but still require the presence of organic templates. Thus, the combination of various sustainable routes should have a good future for synthesizing zeolites from an industrial overview.

References

1. Cundy CS, Cox PA (2003) The hydrothermal synthesis of zeolites: history and development from the earliest days to the present time. Chem Rev 103:663–701
2. Xu R, Pang W et al (eds) (2007) Chemistry of zeolites and related porous materials. Wiley, Singapore
3. Barrer RM (1948) Synthesis of a zeolitic mineral with chabazite-like sorptive properties. J Chem Soc 2:127–132
4. Barrer RM, Hinds L et al (1953) The hydrothermal chemistry of silicates. Part III. Reactions of analcite and leucite. J Chem Soc 1466–1475
5. Milton RM (1959) Molecular sieve adsorbents. US Patent 2,882,243, 14 Apr 1959
6. Milton RM (1959) Molecular sieve adsorbents. US Patent 2,882,244, 14 Apr 1959
7. Barrer RM, Denny PJ (1961) Hydrothermal chemistry of the silicates. Part IX. Nitrogenous aluminosilicates. J Chem Soc 971–982
8. Kerr GT, Kokotail GT (1961) Sodium zeolite ZK-4, a new synthetic crystalline aluminosilicate. J Am Chem Soc 83:4675

9. Meng X, Xiao FS (2014) Green routes for synthesis of zeolites. Chem Rev 114:1521–1543
10. Corma A (1995) Inorganic solid acids and their use in acid-catalyzed hydrocarbon reactions. Chem Rev 95:559–614
11. Corma A (1997) From microporous to mesoporous molecular sieve materials and their use in catalysis. Chem Rev 97:2373–2419
12. Vaughan DEW, Barrett MG (1982) High silica faujasite polymorph – CSZ-3 and method of synthesizing. US Patent 4,333,859, 8 Jun 1982
13. Vaughan DEW (1989) Composition and process for preparing ECR-30. US Patent 4,879,103, 7 Nov 1989
14. Liu S, Li L et al (2008) Synthesis of EMT-rich faujasite in the presence of organic template of low-cost polyquaternium-6. J Porous Mater 15:295–301
15. Wang J, Song J et al (2009) Tetramethylguanidine-templated synthesis of aluminophosphate-based microporous crystals with AFI-type structure. Microporous Mesoporous Mater 117:561–569
16. Gobbi A, Frenking G (1993) Y-conjugated compounds-the equilibrium geometries and electronic-structures of guanidine, guanidinium cation, urea, and 1,1-diaminoethylene. J Am Chem Soc 115:2362–2372
17. Zones SI, Hwang S-J (2002) Synthesis of high silica zeolites using a mixed quaternary ammonium cation, amine approach: discovery of zeolite SSZ-47. Chem Mater 14:313–320
18. Zones SI (2008) Preparation of molecular sieves using a structure directing agent and an N, N, N-triakyl benzyl quaternary ammonium cation. US Patent 20,080,075,656, 27 Mar 2008
19. Moscoso JG, Lewis GJ et al (2004) Crystalline aluminosilicate zeolitic composition: UZM-9. US Patent 6,713,041, 30 Mar 2004
20. Lewis GJ, Miller MA et al (2004) Experimental charge density matching approach to zeolite synthesis. Stud Surf Sci Catal 154:364–372
21. Miller MA, Moscoso JG et al (2007) Synthesis and characterization of the 12-ring zeolites UZM-4 (BPH) and UZM-22 (MEI) via the charge density mismatch approach in the Choline-Li_2O-SrO-Al_2O_3-SiO_2 system. Stud Surf Sci Catal 170:347–354
22. Ren L, Zhu L et al (2011) Designed copper-amine complex as an efficient template for one-pot synthesis of Cu-SSZ-13 zeolite with excellent activity for selective catalytic reduction of NO_X by NH_3. Chem Commun 47:9789–9791
23. Ren L, Zhang Y et al (2012) Design and synthesis of a catalytically active Cu-SSZ-13 zeolite from a copper-amine complex template. Chin J Catal 33:92–105
24. Takewaki T, Beck LW et al (1999) Zincosilicate CIT-6: a precursor to a family of *BEA-type molecular sieves. J Phys Chem B 103:2674–2679
25. Takewaki T, Beck LW et al (1999) Synthesis of CIT-6, a zincosilicate with the (*)BEA topology. Top Catal 9:35–42
26. Takewaki T, Hwang SJ et al (1999) Synthesis of *BEA-type molecular sieves using mesoporous materials as reagents. Microporous Mesoporous Mater 32:265–278
27. Jones CW, Hwang SJ et al (2001) Synthesis of hydrophobic molecular sieves by hydrothermal treatment with acetic acid. Chem Mater 13:1041–1050
28. Jones CW, Tsuji K et al (2001) Tailoring molecular sieve properties during SDA removal via solvent extraction. Microporous Mesoporous Mater 48:57–64
29. Lee H, Zones SI et al (2003) A combustion-free methodology for synthesizing zeolites and zeolite-like materials. Nature 425:385–388
30. Meng X, Xie B et al (2009) Organotemplate-free routes for synthesizing zeolites. Chin J Catal 30:965–971
31. Argauer RJ, Landolt GR (1972) Crystalline zeolite ZSM-5 and method of preparing the same. US Patent 3,702,886, 14 Nov 1972
32. Grose, RW, Flanigen EM (1977) Novel zeolite compositions and processes for preparing and using same. Belgian Patent 851,066
33. Grose, RW, Flanigen EM (1980) Novel zeolite compositions and processes for preparing and using same. British Patent 574,840

34. Grose, RW, Flanigen EM (1981) Novel zeolite compositions and processes for preparing and using same. US Patent 4,257,885, 24 Mar 1981
35. Shiralkar VP, Clearfield A (1989) Synthesis of the molecular-sieve zsm-5 without the aid of templates. Zeolites 9:363–370
36. Wadlinger RL, Kerr GT et al (1967) Catalytic composition of a crystalline zeolite. US Patent 3,308,069, 7 Mar 1967
37. Xie B, Song J et al (2008) An organotemplate-free and fast route for synthesizing Beta zeolite. Chem Mater 20:4533–4535
38. Kamimura Y, Chaikittisilp W et al (2010) Critical factors in the seed-assisted synthesis of zeolite Beta and "green Beta" from OSDA-free Na^+-aluminosilicate gels. Chem Asian J 5:2182–2191
39. Zhang H, Xie B et al (2013) Rational synthesis of Beta zeolite with improved quality by decreasing crystallization temperature in organotemplate-free route. Microporous Mesoporous Mater 180:123–129
40. Xie X, Zhang H et al (2011) Seed-directed synthesis of zeolites with enhanced performance in the absence of organic templates. Chem Commun 47:3945–3947
41. Kamimura Y, Tanahashi S et al (2011) Crystallization behavior of zeolite Beta in OSDA-free, seed-assisted synthesis. J Phys Chem C 115:744–750
42. De Baerdemaeker T, Yilmaz B et al (2013) Catalytic applications of OSDA-free Beta zeolite. J Catal 308:73–81
43. Yilmaz B, Müller U et al (2013) A new catalyst platform: zeolite Beta from template-free synthesis. Catal Sci Technol 3:2580–2586
44. Zhang H, Chu L et al (2013) One-pot synthesis of Fe-Beta zeolite by an organotemplate-free and seed-directed route. J Mater Chem A 1:3254–3257
45. Dougnier F, Patarin J et al (1992) Synthesis, characterization, and catalytic properties of silica-rich Faujasite-type zeolite (FAU) and its hexagonal analog (EMT) prepared by using crown-ethers as templates. Zeolites 12:160–166
46. Ng E-P, Chateigner D et al (2012) Capturing ultrasmall EMT zeolite from template-free systems. Science 335:70–73
47. Rosinski EJ, Rubin MK (1974) Crystalline zeolite ZSM-12. US Patent 3,832,449, 27 Aug 1974
48. Trewella JC, Schlenker JL et al (1985) The Si-29 MAS-NMR spectrum of ZSM-12. Zeolites 5:130–131
49. Chiche BH, Dutartre R et al (1995) Study of the sorption and acidic properties of MTW-type zeolite. Catal Lett 31:359–366
50. Ritsch S, Ohnishi N et al (1998) High-resolution electron microscopy study of ZSM-12 (MTW). Chem Mater 10:3958–3965
51. Mitra A, Kirby CW et al (2002) Synthesis of pure-silica MTW powder and supported films. Microporous Mesoporous Mater 54:175–186
52. Iyoki K, Kamimura Y et al (2010) Synthesis of MTW-type zeolites in the absence of organic structure-directing agent. Chem Lett 39:730–731
53. Kamimura Y, Itabashi K et al (2012) Seed-assisted, OSDA-free synthesis of MTW-type zeolite and "Green MTW" from sodium aluminosilicate gel systems. Microporous Mesoporous Mater 147:149–156
54. Kamimura Y, Itabashi K et al (2012) OSDA-free synthesis of MTW-type zeolite from sodium aluminosilicate gels with zeolite beta seeds. Microporous Mesoporous Mater 163:282–290
55. Ernst S, Weitkamp J et al (1989) Synthesis and shape-selective properties of ZSM-22. Appl Catal 48:137–148
56. Marler B (1987) Silica-ZSM-22: synthesis and single-crystal structure refinement. Zeolites 7:393–397
57. Gunawardane RP, Gies H et al (1988) Long-chain polyamines and amine boric-acid pairs as templates for the synthesis of porous tectosilicates. Zeolites 8:127–131
58. Zones SI (1989) Synthesis of pentasil zeolites from sodium-silicate solutions in the presence of quaternary imidazole compounds. Zeolites 9:458–467

59. Rollmann LD, Schlenker JL et al (1999) On the role of small amines in zeolite synthesis. J Phys Chem B 103:7175–7183
60. Masih D, Kobayashi T et al (2007) Hydrothermal synthesis of pure ZSM-22 under mild conditions. Chem Commun 31:3303–3305
61. Wang Y, Wang X et al (2014) Seed-directed and organotemplate-free synthesis of TON zeolite. Catal Today 226:103–108
62. Plank CJ, Rosinski EJ ct al (1978) Crystalline zeolite ZSM-23 and synthesis thereof. US Patent 4,076,842, 28 Feb 1978
63. Valyocsik EW (1984) Synthesis of ZSM-23 zeolite. US Patent 4,490,342, 25 Dec 1984
64. Parker LM, Bibby DM et al (1983) Synthesis and some properties of 2 novel zeolites, KZ-1 and KZ-2. Zeolites 3:8–11
65. Araya A, Lowe BM (1987) Synthesis of zeolite EU-13 from a reaction mixture containing tetramethylammonium compound. US Patent 4,705,674, 10 Nov 1987
66. Zones SI (1991) Zeolite SSZ-32. US Patent 5,053,373, 1 Oct 1991
67. Nakagawa Y (1998) Process for preparing zeolites having MTT crystal structure using small, neutral amines. US Patent 5,707,601, 13 Jan 1998
68. Moini A, Schmitt KD et al (1994) The role of diquaternary cations as directing agents in zeolite synthesis. Zeolites 14:504–511
69. Rane SJ, Chakrabarty DK (1991) Shape selective catalysis by zeolite KZ-1 synthesis, characterization and isomerization of meta-xylene. Appl Catal 75:281–288
70. Wu Q, Wang X et al (2014) Organotemplate-free, seed-directed, and rapid synthesis of Al-rich zeolite MTT with improved catalytic performance in isomerization of *m*-xylene. Microporous Mesoporous Mater 186:106–112
71. Müller K, Bein T (2011) Crystallization and porosity of ZSM-23. Microporous Mesoporous Mater 143:253–262
72. Vortmann S, Marler B et al (1995) Synthesis and crystal-structure of the new borosilicate zeolite RUB-13. Microporous Mater 4:111–121
73. Lee GS, Zones SI (2002) Polymethylated [4.1.1] octanes leading to zeolite SSZ-50. J Solid State Chem 167:289–298
74. Yokoi T, Yoshioka M et al (2009) Diversification of RTH-type zeolite and its catalytic application. Angew Chem Int Ed 48:9884–9887
75. Gies H, Gunawardane RP (1987) One-step synthesis, properties and crystal-structure of aluminum-free Ferrierite. Zeolites 7:442–445
76. Morris RE, Weigel SJ et al (1994) A synchrotron X-ray-diffraction, neutron-diffraction, Si-29 MAS-NMR, and computational study of the siliceous form of zeolite Ferrierite. J Am Chem Soc 116:11849–11855
77. Asensi MA, Martínez A (1999) Selective isomerization of n-butenes to isobutene on high Si Al ratio ferrierite in the absence of coke deposits: implications on the reaction mechanism. Appl Catal A 183:155–165
78. Lee S, Shin C et al (2004) Investigations into the origin of the remarkable catalytic performance of aged H-ferrierite for the skeletal isomerization of 1-butene to isobutene. J Catal 223:200–211
79. Rakoczy RA, Breuninger M et al (2002) Template-free synthesis of zeolite ferrierite and characterization of its acid sites. Chem Eng Technol 25:273–275
80. Zhang H, Guo Q et al (2011) Organotemplate-free synthesis of high-silica ferrierite zeolite induced by CDO-structure zeolite building units. J Mater Chem 21:9494–9497
81. Baerlocher Ch, McCusker LB et al (2007) Atlas of zeolite framework types, sixth rev. edn. Elsevier, Amsterdam
82. Gottardi G, Galli E (1825) Natural zeolites. Springer, Berlin
83. Kerr GT (1969) Synthetic zeolite and method for preparing the same. US Patent 3,459,676, 5 Aug 1969
84. Short GD, Whittam TV (1983) Zeolite Nu-3. US Patent 4,372,930, 8 Feb 1983

85. Kuehl GH (1985) Process for making zeolite ZSM-45 with a dimethyldiethylammonium directing agent. US Patent 4,495,303, 22 Jan 1985
86. Han B, Lee S-H et al (2005) Zeolite synthesis using flexible diquaternary alkylammonium ions $(C_nH_{2n+1})_2HN^+(CH_2)_5N^+H(C_nH_{2n+1})_2$ with n=1-5 as structure-directing agents. Chem Mater 17:477–486
87. Xu H, Li J et al (2009) Synthesis and properties of a zeolite LEV analogue from the system-Na2O-Al2O3-SiO2-N, N-dimethylpiperidine chloride-H2O. Catal Today 148:6–11
88. Inoue T, Itakura M et al (2009) Synthesis of LEV zeolite by interzeolite conversion method and its catalytic performance in ethanol to olefins reaction. Microporous Mesoporous Mater 122:149–154
89. Lok BM, Messina CA et al (1984) Silicoaluminophosphate molecular-sieves – another new class of microporous crystalline inorganic solids. J Am Chem Soc 106:6092–6093
90. Grunewald A, Gies H (1994) Quinuclidine and 3-azabicyclo[3.2.2]nonane, 2 versatile templates in the synthesis of porous silicates in the SiO_2-B_2O_3-H_2O-template system. Microporous Mater 3:159–164
91. Zhu G, Xiao F-S et al (1997) Synthesis and characterization of a new microporous aluminophosphate with levyne structure in the presence of HF. Microporous Mater 11:269–273
92. Barrett PA, Jones RH (2000) Evidence for ordering of cobalt ions in the microporous solid acid catalyst CoDAF-4 by single crystal X-ray diffraction and resonant X-ray powder diffraction. Phys Chem Chem Phys 2:407–412
93. Zhang H, Yang C et al (2012) Organotemplate-free and seed-directed synthesis of levyne zeolite. Microporous Mesoporous Mater 155:1–7
94. Lawton L, Bennett JM et al (1993) Synthesis and proposed framework topology of zeolite SUZ-4. J Chem Soc Chem Commun 11:894–896
95. Paik WC, Shin CH et al (2000) Synthesis of zeolites P1 and SUZ-4 through a synergy of organic N,N,N,N′,N′,N′-hexaethylpentanediammonium and inorganic cations. Chem Commun 17:1609–1610
96. Zhang W, Wu Y et al (2011) Organotemplate-free route for synthesizing SUZ-4 zeolite under static hydrothermal condition. Mater Res Bull 46:1451–1454
97. Rubin MK, Rosinski EJ et al (1978) Hydrocarbon conversion with crystalline zeolite ZSM-34. US 4,116,813, 26 Sept 1978
98. Givens EN, Plank CJ et al (1978) Manufacture of light olefins. US 4,079,095, 14 Mar 1978
99. Givens EN, Plank CJ et al (1978) Manufacture of light olefins. US 4,079,096, 14 Mar 1978
100. Occelli ME, Innes RA et al (1987) Quaternary ammonium cation effects on the crystallization of offretite erionite type zeolites. 1. Synthesis and catalytic properties. Zeolites 7:265–271
101. Vartuli JC, Kennedy GJ et al (2000) Zeolite syntheses using diamines: evidence for in situ directing agent modification. Microporous Mesoporous Mater 38:247–254
102. Wu Z, Song J et al (2008) Organic template free synthesis of ZSM-34 zeolite from an assistance of zeolite L seeds solution. Chem Mater 20:357–359
103. Zhang L, Yang C et al (2010) Organotemplate-free syntheses of ZSM-34 zeolite and its heteroatom-substituted analogues with good catalytic performance. Chem Mater 22: 3099–3107
104. Yang C, Ren L et al (2012) Organotemplate-free and seed-directed synthesis of ZSM-34 zeolite with good performance in methanol-to-olefins. J Mater Chem 22:12238–12245
105. Vaughan DEW, Strohmaier KG (1987) Crystalline zeolite (ECR-1) and process for preparing it. US 4,657,748, 14 Apr 1987
106. Leonowicz ME, Vaughan DEW (1987) Proposed synthetic zeolite ECR-1 structure gives a new zeolite framework topology. Nature 329:819–821
107. Chen CSH, Schlenker JL et al (1996) Synthesis and characterization of synthetic zeolite ECR-1. Zeolites 17:393–400
108. Gualtieri AF, Ferrari S et al (2006) Rietveld structure refinement of zeolite ECR-1. Chem Mater 18:76–84

109. Song J, Dai L et al (2006) Organic template free synthesis of aluminosilicate zeolite ECR-1. Chem Mater 18:2775–2777
110. Ding H, Song J et al (2009) Synthesis of zeolite ECR-1 from hydrothermal phase transformation of zeolite Y. Chem J Chin Univ 30:255–257
111. Zhang L, Liu S et al (2012) Organic template-free synthesis of ZSM-5/ZSM-11 co-crystalline zeolite. Microporous Mesoporous Mater 147:117–126
112. Itabashi K, Kamimura Y et al (2012) A working hypothesis for broadening framework types of zeolites in seed-assisted synthesis without organic structure-directing agent. J Am Chem Soc 134:11542–11549
113. Tanaka K, Toda F (2000) Solvent-free organic synthesis. Chem Rev 100:1025–1074
114. Walsh PJ, Li H et al (2007) A green chemistry approach to asymmetric catalysis: solvent-free and highly concentrated reactions. Chem Rev 107:2503–2545
115. Martins MAP, Frizzo CP et al (2009) Solvent-free heterocyclic synthesis. Chem Rev 109:4140–4182
116. James SL, Adams CJ et al (2012) Mechanochemistry: opportunities for new and cleaner synthesis. Chem Soc Rev 41:413–447
117. Pichon A, Lazuen-Garay A et al (2006) Solvent-free synthesis of a microporous metal-organic framework. CrystEngComm 8:211–214
118. Yuan W, Lazuen-Garay A et al (2010) Study of the mechanochemical formation and resulting properties of an archetypal MOF: Cu-3(BTC)(2) (BTC = 1,3,5-benzenetricarboxylate). CrystEngComm 12:4063–4065
119. Klimakow M, Klobes P et al (2010) Mechanochemical synthesis of metal-organic frameworks: a fast and facile approach toward quantitative yields and high specific surface areas. Chem Mater 22:5216–5221
120. Zhang P, Wang L et al (2011) "Solvent-free" synthesis of thermally stable and hierarchically porous aluminophosphates (SF-APOs) and heteroatom-substituted aluminophosphates (SF-MAPOs). J Mater Chem 21:12026–12033
121. Xie Y, Tang Y (1990) Spontaneous monolayer dispersion of oxides and salts onto surfaces of supports: applications to heterogeneous catalysis. Adv Catal 37:1–43
122. Xu W, Dong J et al (1990) A novel method for the preparation of zeolite ZSM-5. J Chem Soc Chem Commun :755–756
123. Ren L, Wu Q et al (2012) Solvent-free synthesis of zeolites from solid raw materials. J Am Chem Soc 134:15173–15176
124. Morris RE, James SL (2013) Solventless synthesis of zeolites. Angew Chem Int Ed 52:2163–2165
125. Jin Y, Sun Q et al (2013) Solvent-free synthesis of silicoaluminophosphate zeolites. Angew Chem Int Ed 52:9172–9175

Chapter 2
Ionothermal Synthesis of Molecular Sieves

Zhi-Jian Tian and Hao Liu

Abstract The synthesis of molecular sieves in ionic liquid or deep eutectic solvent is termed ionothermal synthesis. It involves an ionic reaction environment with extremely low volatility, and thus, can avoid the safety risks of hydro/solvothermal reactions, and bring great convenience and flexibility to both the synthesis of molecular sieves and the study on their formation mechanisms. In this chapter, first introduced are the special physicochemical properties of ionic liquids and deep eutectic solvents, and the distinctive opportunities brought by them to synthesize molecular sieves. A detailed methodological illustration of ionothermal synthesis technique is provided subsequently. The molecular sieves ionothermally synthesized are summarized by their features such as composition, structure, porosity and morphology. Then follows a review on the fascinating structure-directing effect of various guest species in ionothermal synthesis. Finally, a prospect is given on the future development of ionothermal synthesis.

Keywords Ionothermal synthesis • Molecular sieves • Structure-directing effect • Crystallization • Zeolite membranes • Microwave chemistry • Ionic liquids • Deep eutectic solvents

2.1 Introduction

Because of their negligible volatility and favourable physicochemical properties, ionic liquid (IL) and its analogue deep eutectic solvent (DES) are considered as *green* solvents for various synthesis and catalysis processes. In 2004, Prof. R. E. Morris

Z.-J. Tian (✉)
Dalian National Laboratory for Clean Energy, Dalian Institute of Chemical Physics, Chinese Academy of Sciences, Dalian 116023, China

State Key Laboratory of Catalysis, Dalian Institute of Chemical Physics, Chinese Academy of Sciences, Dalian 116023, China
e-mail: tianz@dicp.ac.cn

H. Liu
Dalian National Laboratory for Clean Energy, Dalian Institute of Chemical Physics, Chinese Academy of Sciences, Dalian 116023, China

F.-S. Xiao, X. Meng (eds.), *Zeolites in Sustainable Chemistry*,
Green Chemistry and Sustainable Technology,
DOI 10.1007/978-3-662-47395-5_2

(University of St Andrews, UK) and his colleagues reported their pioneering work on the synthesis of AlPOs by using IL/DES as both solvent and structure-directing agent [1] and termed this new process *ionothermal synthesis*. Compared with hydro/solvothermal routes, the most significant property of ionothermal synthesis is that it can take place at ambient pressure with a few safety risks. Because of its convenience and flexibility, ionothermal synthesis has attracted great research interest in recent years. Up until now, hundreds of papers in this area, including several excellent reviews [2–7], have been published. These papers mainly focus on following topics: (i) technique improvement brought by ionothermal synthesis, (ii) molecular sieves ionothermally synthesized, and (iii) formation mechanisms of these molecular sieves. Herein, a practical guide will be provided on ionothermal synthesis of molecular sieves, in particular the zeolitic ones. According to this criterion, progress achieved in above fields will be surveyed, accompanied by detailed descriptions and discussions on how the relevant experiments were carried out.

2.2 Reaction Medium in Ionothermal Synthesis

2.2.1 IL and DES

In recent literature, the term *ionic liquid* is traditionally defined as a salt with melting point below or near the boiling point of water. A detailed definition and a historical introduction of IL can be found in Wilkes et al.'s work [8]. In general, IL is composed of large organic cations and organic/inorganic anions (Fig. 2.1). Its physicochemical property is strongly determined by the composition and structure of these ions. Considering the wide variety of both cations and anions, the potential number of ILs is extremely large.

DES is an alternative solvent system that has similar physicochemical properties of IL. It is composed of several components, which associate with each other in some way and result in a considerably lower freezing point of the mixture compared with each component [9, 10]. Among the variety of DES, the mixture of quaternary ammonium salt and hydrogen bond donor is the most widely used type of DES in ionothermal synthesis of molecular sieves. In the original paper on this type DES, Abbott et al. reported the interesting solvent properties and the temperature-dependent high conductivities of some binary mixtures of quaternary ammonium salts and amides [11]. Using NMR spectroscopy, they observed the intense cross-correlation between fluoride anions and urea protons in DES choline fluoride–urea (1/2, [molar ratio of the components, the same hereinafter]). They also observed the presence of chloride anions with one or two urea molecules ($M^{-}=95$ or 155, respectively) from MS of DES choline chloride–urea (1/2). According to these results, as well as other phenomena, Abbott et al. suggested that the decrease in the freezing point of this type DES should be attributed to the formation of hydrogen bonds between the anions of the quaternary ammonium salts and the amides (as hydrogen bond donors).

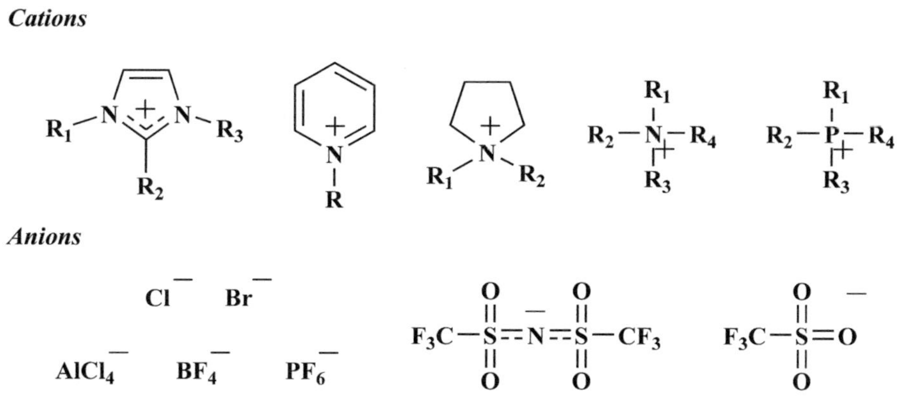

Fig. 2.1 Commonly used ions for ionic liquids

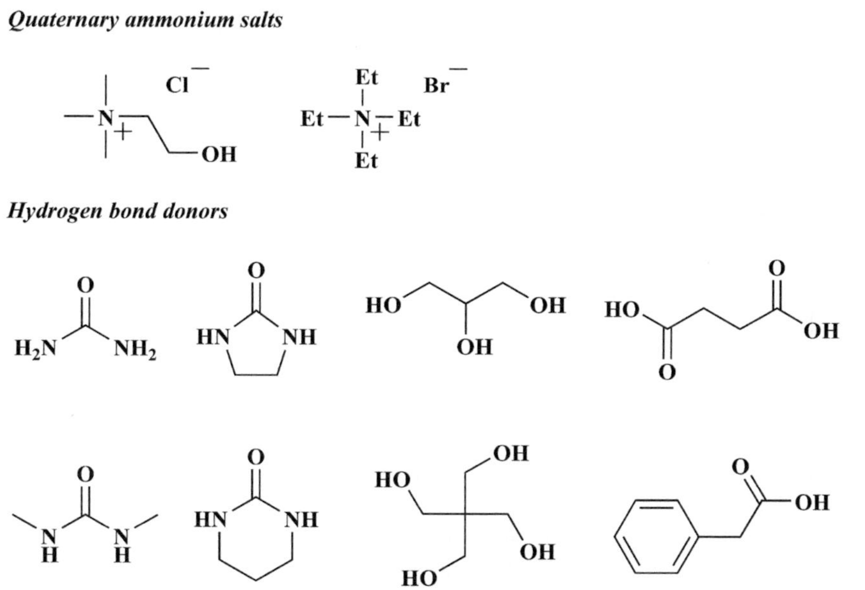

Fig. 2.2 Commonly used components for the quaternary ammonium salt–hydrogen bond donor type deep eutectic solvent

Abbott et al.'s study shows a facile way to prepare solvents with similar physicochemical properties of IL. The quaternary ammonium salt–hydrogen bond donor type DES can be conveniently prepared by simply heating and mixing the components, which are mainly the naturally occurring, renewable and well-characterized bulk chemicals (Fig. 2.2). Thus, various DESs can be obtained and used in ionothermal synthesis of molecular sieves.

As well established, hydrogen bonding system X–H···Y extensively exists between the hydrogen bond donor, *i.e.*, a relatively polarized X–H bond (X=N, O, F, or even C), and the hydrogen bond acceptor, which could be a highly electronegative atom or an electron-rich aromatic system. Thus, a more universal description of the formation process of the above-mentioned DES can be given as following equilibrium [12]:

$$\text{Organic salt} + \text{Uncharged molecule} \rightleftharpoons \text{Cation} + \text{Hydrogen-bonded complex anion}$$

or

$$\text{Organic salt} + \text{Uncharged molecule} \rightleftharpoons \text{Hydrogen-bonded complex cation} + \text{Anion}$$

Accompanying this process, the initial ordered structures of the components are disrupted. The stability difference between these two kinds of hydrogen-bonded complex ions determines which of these two equilibrium processes is preferred during the formation of DES. In any cases, ions constitute the majority of the obtained mixture, which exhibits extremely high electrical conductivity as well as other IL-like properties.

2.2.2 Hydrogen Bonding Interactions in Ionothermal Synthesis

Reactant molecules may also form hydrogen bonds with IL/DES. For instance, the formation of hydrogen bonds between water molecules and the anions of ILs has been confirmed by both theoretical studies and spectroscopic experiments [13–15]. As these studies show, when diluted in IL, water molecules tend to be isolated from each other rather than form network. The resultant deactivation of water can stabilize the bonds that are sensitive to hydrolysis [16, 17]. Furthermore, Weber et al. reported that hydrolysis reaction rate could be controlled by tuning the strength of hydrogen bonds between water molecules and the IL cations [18]. They suggested that compared with 1-butyl-3-methylimidazolium (BMIm) cations, 1-butyl-2,3-dimethylimidazolium (BDMIm) cations form weaker hydrogen bonds with water molecules, and therefore lead to an increase of the hydrolysis reaction rate.

Another interesting phenomenon caused by the formation of hydrogen bonds is that the volatility of water may decrease to some degree when it is mixed with hydrophilic IL/DES. By monitoring the pressure changes of the mixtures containing 5 g of [BMIm]Br and different amount of water in a 10 mL sealed vessel at 175 °C under microwave radiation, Cai et al. found that there is no autogenic pressure for the mixture with water/[BMIm]Br molar ratio = 0.5/1 [19]. This water concentration is much higher than that required for an efficient crystallization of AlPOs under similar condition [20].

Table 2.1 ^{1}H NMR spectral data (chemical shifts δ relative to D_2O, 400 MHz, 80 °C) of 1-butyl-3-methylimidazolium bromide and its mixtures with organic amines [21]

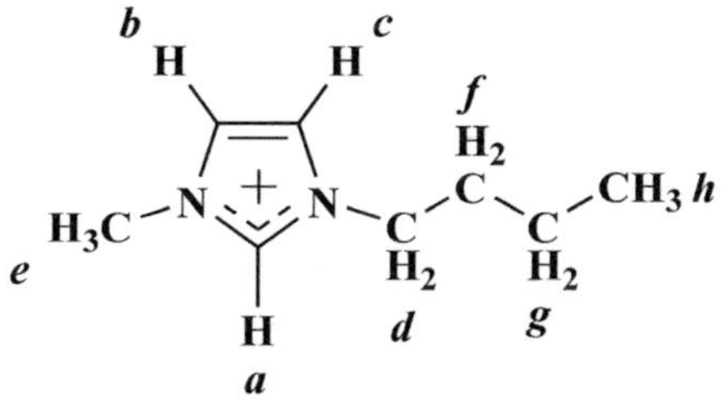

	δ (ppm)							
Sample[a]	H(*a*)	H(*b*)	H(*c*)	H(*d*)	H(*e*)	H(*f*)	H(*g*)	H(*h*)
[BMIm]Br	11.04	9.33	9.24	5.43	5.14	2.81	2.19	1.73
[BMIm]Br–1-MIm (2/1)	11.01	9.22	9.11	5.41	5.10	2.85	2.65	1.82
[BMIm]Br–*n*-DPA (2/1)	10.79	9.17	9.07	5.42	5.11	2.83	2.23	1.77

[a]*BMIm* 1-butyl-3-methylimidazolium, *1-MIm* 1-methylimidazole, *n-DPA* *n*-dipropylamine

The addition of various organic amines/ammonium salts into IL/DES has shown great advantage in flexibly adjusting the crystallization environment for molecular sieves. The influence of these additive organic species is strongly related to the hydrogen bonding interaction between organic species and the IL/DES cations. In the very first literature on the structure-directing effect of organic amines in ionothermal synthesis, Wang et al. pointed out that the hydrogen-bonded complex cations may stabilize the 12-membered ring channels and facilitate the formation of AFI structure [21]. The presence of the hydrogen bonds between organic amines and BMIm cations can be indicated by the downfield shift of the signals corresponding to the protons of imidazolium ring in ^{1}H NMR spectra (Table 2.1, hydrogen atoms at positions *a*, *b*, and *c*). Further discussion on this structure-directing effect will be given in Sect. 2.5.3.2.

2.2.3 Synthesis Opportunities

In Table 2.2 are listed selected physicochemical data of some ILs/DESs and molecular solvents, both of which are commonly used in the synthesis of molecular sieves. It can be found that the physicochemical properties of IL/DES are strongly determined by the composition and structure of the components. Similar to the selected molecular solvents, these ILs/DESs have wide liquid range and medium to strong polarity. Therefore, in theory, they can successfully replace the molecular solvents used for hydro/solvothermal synthesis. The most significant differences of ILs/DESs compared with molecular solvents are their extremely low volatility and high

Table 2.2 Selected physicochemical data of some ionic liquids, deep eutectic solvents and molecular solvents[a]

Solvent	T_m (°C)	T_f (°C)	T_g (°C)	T_b (°C)	κ (mS/cm)	ε	E_T^N	$\Delta_v H_m$ (kJ/mol)
Ionic liquid[b]								
[EMIm]Br	79 [22]	30 [22]			32.2 (100 °C) [23]			
[EMIm][BF_4]	11 [22]	−63 [22]			69 (100 °C) [23]	12.8 (25 °C) [24]	0.71 [25]	135.5 (25 °C) [26]
[EMIm][PF_6]	62 [22]	5 [22]			37.0 (100 °C) [23]			143.6 (25 °C) [26]
[BMIm]Br	73 [27]		−50 [28]					152.2 (25 °C) [29]
[BMIm][BF_4]			−85 [28]		21.80 (80 °C) [30]	11.7 (25 °C) [24]	0.68 [25]	
[BMIm][PF_6]	11 [28]		−76 [28]		12.14 (80 °C) [30]	11.4 (25 °C) [24]	0.68 [25]	
[EDMIm]Br	141 [22]	92 [22]						
Deep eutectic solvent[c]								
ChCl–urea (1/2)		12 [11]			1.8 (40 °C) [31]		0.84 [32]	
ChCl–EG (1/2)		−66 [33]			7.61 (20 °C) [34]		0.80 [32]	
Molecular solvent								
Water	0			100	4×10^{-5} (18 °C) [35]	80.20 (20 °C) [35]	1.000 [36]	40.66 (T_b) [35]
Ethanol	−114 [35]			78.3 [35]	1.35×10^{-6} (25 °C) [35]	25.3 (20 °C) [35]	0.654 [36]	38.6 (T_b) [35]
Ethylene glycol				196–198 [35]		41.4 (20 °C) [35]	0.790 [36]	50.5 (T_b) [35]
Pyridine	−41.6 [35]			115.2 [35]	5.3×10^{-5} (18 °C) [35]	13.26 (20 °C) [35]	0.302 [36]	35.1 (T_b) [35]

[a] T_m melting point, T_f freezing point, T_g glass transition point, T_b boiling point, κ electrical conductivity, ε dielectric constant, E_T^N normalized polarity value at room temperature, $\Delta_v H_m$ vaporization enthalpy

[b] *EMIm* 1-ethyl-3-methylimidazolium, *BMIm* 1-butyl-3-methylimidazolium, *EDMIm* 1-ethyl-2,3-dimethylimidazolium

[c] *ChCl* choline chloride, *EG* ethylene glycol

electronic conductivity. Owing to these properties, ILs/DESs bring some distinctive opportunities for the synthesis of molecular sieves.

2.2.3.1 Extremely Low Volatility

One of the most significant properties of IL is its extremely low volatility, which is attributed to its entire ionic character. The vaporization enthalpies of ILs are almost an order of magnitude larger than that of molecular solvents. As a result, in common literature, IL is always described as *involatile* solvent, and the upper limit of its liquid range is usually its decomposition temperature. For DES, the boiling point (or decomposition temperature) of the constituents are always high. Therefore, similar to IL, DES also has long liquid range and low volatility. At the crystallization temperature for molecular sieves (usually $\leq$200 °C), no autogenic pressure of the solvent IL/DES can be detected in principle. The resultant dramatic decrease of the system pressure makes ionothermal synthesis of molecular sieves much safer and controllable compared with hydro/solvothermal routes. Furthermore, the extremely low volatility of IL/DES makes the ionothermal crystallization process visible and may simplify the *in situ* experiment.

2.2.3.2 High Microwave-Adsorption Capacity

Because of their high electrical conductivity and dielectric constant, IL and DES are the appropriate medium for microwave heating [37]. In 2006, Prof. Y.-P. Xu and other colleagues in our group reported the microwave-enhanced ionothermal synthesis of molecular sieves for the first time [38]. The microwave reaction system they used is illustrated in Fig. 2.3. The static crystallization process took place in ordinary glass equipment at 150 °C and no special devices were required. Xu et al.'s work shows that similar to the results obtained in hydro/solvothermal synthesis, microwave-enhanced ionothermal synthesis leads to a more rapid crystallization rate compared with that using conventional heating method.

2.2.3.3 Stability

In ionothermal synthesis, it is occasionally observed the decomposition of IL/DES. Herein, imidazolium salts, which are the most widely used ILs in ionothermal synthesis of molecular sieves, are taken as examples to illustrate the decomposition mechanism of IL/DES.

As Chambreau et al.'s work shows, the thermal decomposition of 1-ethyl-3-methylimidazolium (EMIm) bromide prefers to take place *via* the evolution of alkylbromide and alkylimidazole through an S_N2 type mechanism (Scheme 2.1) [39]. The difference between the experimentally evaluated activation enthalpies of the two possible processes indicates that the thermal decomposition of [EMIm]Br

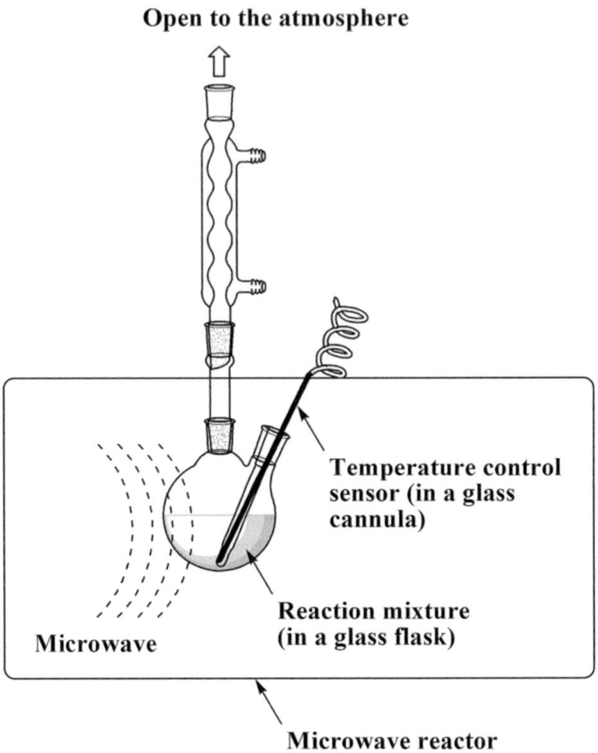

Fig. 2.3 Schematic illustration of the microwave reaction system Xu et al. used [38]

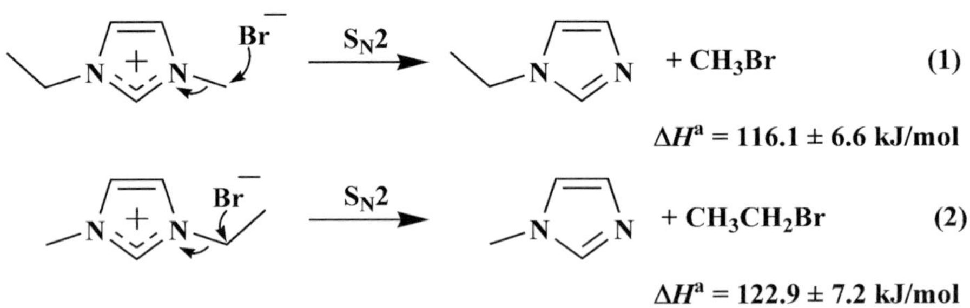

Scheme 2.1 Thermal decomposition mechanism of 1-ethyl-3-methylimidazolium bromide [39]

favours the formation of 1-ethylimidazole and bromomethane (Process 1), rather than the formation of 1-methylimidazole and bromoethane (Process 2). At 181 °C, the branching ratio for CH_3Br/C_2H_5Br analyzed *via* tunable vacuum ultraviolet photoionization time-of-flight MS is 0.76/0.24. Based on *ab initio* quantum chemical

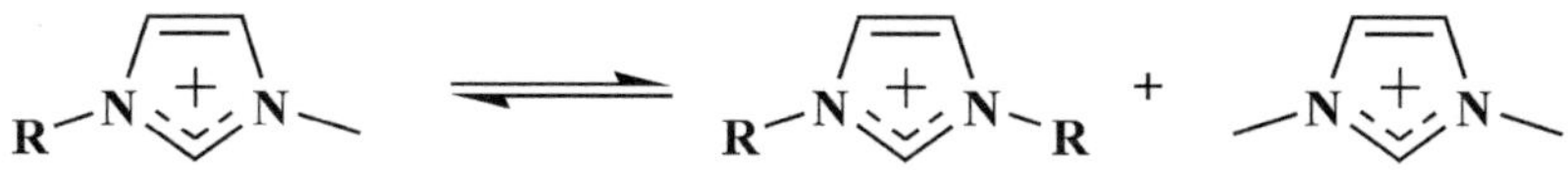

Scheme 2.2 Transalkylation reaction of 1-alkyl-3-methylimidazolium cations

calculation study on the possible thermal decomposition ways of 1-alkyl-3-methylimidazolium chloride, Kroon et al. suggested that the effect of the alkyl chain length on both the activation energy and the thermal decomposition temperature is very small [40].

The alkylbromides and alkylimidazoles formed *via* thermal decomposition may reconstitute and lead to transalkylation (Scheme 2.2). In some cases, the resultant new imidazolium cations could play the role of structure-directing agent during the crystallization of molecular sieves [41, 42]. Parnham et al. found that in sealed system with the presence of fluoride anions, 1-alkyl-3-methylimidazolium bromides with alkyl chains longer than ethyl would decompose and form 1,3-dimethylimidazolium (DMIm) cations *via* transalkylation [41]. The resultant DMIm cations can direct the formation of fluorinated AlPOs with CHA topology, of which the *cha* cages are unfavourable for the presence of cations larger than EMIm cation. In a later study, Griffin et al. reported that a similar transalkylation reaction can also take place for [EMIm]Cl, which has a more nucleophilic anion compared with bromide anion [42].

Another decomposition pathway for imidazolium salts relates to the high acidity at C2 position of imidazolium ring. Imidazolium cations would deprotonate in basic environment, followed by complexation and/or decomposition side reactions [43]. Therefore, commonly used imidazolium salts are not suitable for reactions involving strong base, which is often required in the synthesis of zeolites (see also Sect. 2.4.2.2).

Similar to imidazolium salts, other varieties of ILs have also been found to break down during ionothermal synthesis [44]. Besides these, the quaternary ammonium cations composing DES are also instable at high temperature or in high basic environment. Oil-like liquid is often found in the autoclaves after the crystallization using DES as reaction medium. The existence of these organic species may influence the state of the reaction mixture and the nature of the final product. As will be discussed in Sect. 2.4.4.2, the decomposition product of DES embedded into molecular sieves has been employed as templates for mesopores [45].

The decomposed IL can also supply framework species for the synthesis of molecular sieves. Yang et al. prepared a series of boron-containing ILs 1-alkyl-3-methylimidazolium bis(oxalato)borates (Fig. 2.4) and used them as B–O source to synthesize metal borophosphates [46, 47]. Though strictly speaking, the synthesis route they employed is not ionothermal, Yang et al. illustrated an inspirational example of the application of designed ILs to synthesize molecular sieves.

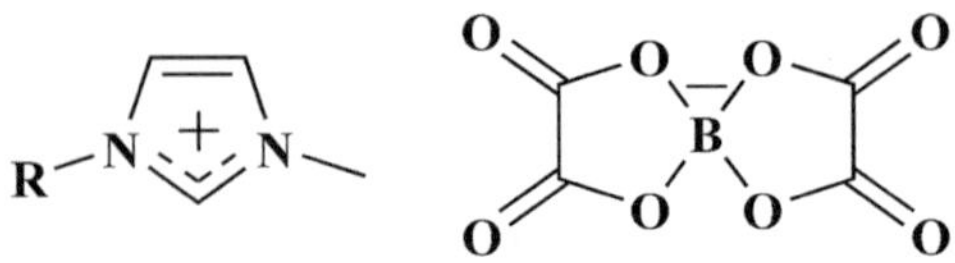

Fig. 2.4 Structure of 1-alkyl-3-methylimidazolium bis(oxalato)borate

2.3 Ionothermal Synthesis Technique

2.3.1 General Steps of Ionothermal Synthesis

The typical procedure for ionothermal synthesis of molecular sieves follows the sequence shown in Scheme 2.3. Compared with hydro/solvothermal routes, one of the significant differences in ionothermal routes is that the preparation and recovery of the solvent, *i.e.*, IL/DES, are important to be considered. Another difference in ionothermal synthesis is that the ionothermal crystallization can take place at ambient pressure with fewer safety risks but more flexibility.

2.3.1.1 Preparation of IL/DES

Although huge varieties of ILs are commercially available now, most of the ILs used in routine laboratory experiments are self-synthesized. Considering their easy synthesis from cheap raw materials, convenient physicochemical properties and relatively low toxicity, imidazolium salts are the most preferred ILs in ionothermal synthesis. A typical synthesis route of imidazolium-based IL is shown in Scheme 2.4.

Compared with the substitution and anion-exchange processes shown in Scheme 2.4, the preparation route of DES is much simpler. Commonly, the homogeneous liquid formed by mixing and heating the components is just the desired DES. It seems that more and more research interest has been attracted to the use of DES in ionothermal synthesis because of its simple preparation method and changeable composition.

2.3.1.2 Mixing and Ageing

Unlike hydro/solvothermal synthesis, the mixing and ageing processes of ionothermal synthesis always take place at a temperature higher than room temperature (usually 70–100 °C). Therefore, it is unclear the temperature boundary between the mixing–ageing and the crystallization processes in ionothermal synthesis compared with those in hydro/solvothermal methods. Thus, there is only little literature that focuses on the pre-crystallization condition of the reaction mixture for ionothermal synthesis.

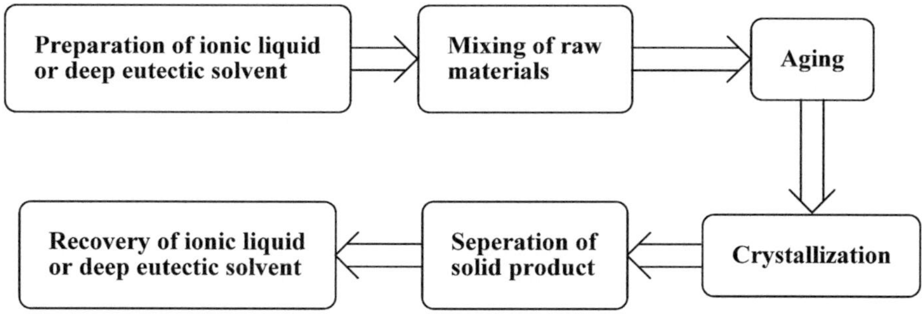

Scheme 2.3 Typical procedure for ionothermal synthesis of molecular sieves

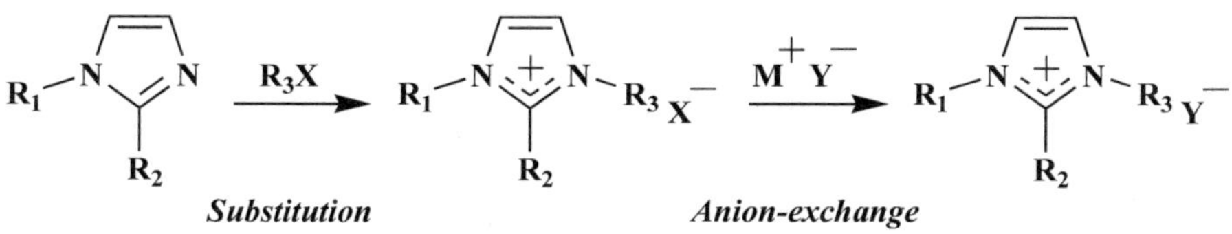

Scheme 2.4 Typical synthesis route of imidazolium-based ionic liquid

Using *in situ* ^{1}H–^{1}H NMR rotating-frame nuclear Overhauser effect spectroscopy, Xu et al. provided an elaborate illustration of the pre-crystallization reaction mixture with molar composition

$$Al_2O_3 / P_2O_5 / HF / [BMIm]Br / morpholine = 1/3/0.4/40/(2-5)$$

for synthesizing AlPO-AFI [48]. In the initial mixture of [BMIm]Br and morpholine exists a complex structure of BMIm cations and morpholine molecules through intermolecular hydrogen bonds. When H_3PO_4 is added into the reaction system, this complex structure is stable at 80 °C but invisible at 140 °C. Xu et al. attributed this change to the strong interaction of H_3PO_4 with morpholine, as well as the charge repulsion between protonated morpholine molecules and BMIm cations. After the further addition of aluminium isopropoxide ($Al[OCH(CH_3)_2]_3$) at 140 °C, the complex structure of BMIm cations and morpholine molecules recovers. This indicates the formation of Al–O–P species and the release of morpholine molecules. Further research shows that the concentration of organic amine determines both the existence and the structure-directing ability of the IL/DES–organic amine complex cations (see Sect. 2.5.3.2).

2.3.1.3 Recovery and Reuse of IL/DES

Considering the toxicity and cost of ILs/DESs, the recovery and reuse of them should be comprehensively studied. However, this topic is rarely discussed so far, though hundreds of papers have been published on ionothermal synthesis in the past decade.

A practical procedure to recover [EMIm]Br after ionothermal synthesis of AlPO was illustrated by Cooper et al. [1]. They announced the successful synthesis of two types of AlPOs by using the recovered IL. Similar result was achieved by Han et al. [49]. In Han et al.'s work, both ^{1}H NMR spectra and TGA curves show no significant differences between the fresh and recovered ILs. However, there is a little decrease of the thermal decomposition temperature of the recovered ILs compared with the fresh one.

Although water and other molecular species can be removed from IL, the raw materials, especially the phosphorus-containing species, which is always added in excess amount, may remain in the recovered ILs and disturb the stoichiometry for further synthesis. As Hou's work shows, AlPO-AEL can be successfully synthesized by using the first-time recovered [EMIm]Br as solvent [50]. Unfortunately, products obtained from [EMIm]Br recovered for the second and third times are the mixtures of AlPO-AEL and dense phase $AlPO_4$. The decrease in phase purity should be mainly attributed to the accumulation of phosphate anions during the recycling of IL.

2.3.2 Examples of Ionothermal Synthesis

In this section, taking the synthesis of AlPOs with –CLO topology as example, the typical procedure for ionothermal synthesis of molecular sieves is presented. Brief descriptions of the preparations of the IL [EMIm]Br and the DES tetraethylammonium chloride–pentaerythritol are also provided, respectively.

2.3.2.1 [EMIm]Br System [51]

Preparation of [EMIm]Br Under vigorous stirring, bromoethane is added dropwise to 1-methylimidazole in a three-necked round-bottomed flask with reflux condenser. The molar ratio of 1-methylimidazole/bromoethane is 1/1.5. After refluxing for 6 h, the mixture is distilled under reduced pressure at 85 °C. The residual liquid, which would crystallizes into white solid by cooling, is collected.

Synthesis of AlPO-CLO A glass beaker is charged with 276 mmol of [EMIm]Br, 7 mmol of H_3PO_4 (85 wt% in H_2O), 7 mmol of aluminium isopropoxide and 7 mmol of HF (40 wt% in H_2O). These reactants are stirred at 100 °C for 1 h. Then, 4 mmol of 1,6-hexanediamine is added into above mixture. After further stirring for 5 min, the final mixture is transferred into a PTFE-lined autoclave (volume = 100 mL) and crystallizes at 210 °C for 2 h. The product is cooled, filtered, washed with water and ethanol and then dried at 110 °C overnight.

2.3.2.2 Tetraethylammonium Chloride–Pentaerythritol System [52]

AlPO-CLO can also be synthesized by using tetraethylammonium chloride–pentaerythritol as solvent and 2-methylimidazole as co-structure-directing agent. This procedure can be performed thoroughly in an open vessel.

Preparation of DES Tetraethylammonium Chloride–Pentaerythritol This DES is prepared by mixing tetraethylammonium chloride and pentaerythritol with molar ratio 1/1.22, heating and stirring until a homogenous liquid forms.

Synthesis of AlPO-CLO Aluminium isopropoxide is added into above-mentioned DES. The mixture is stirred at 85 °C for 30 min. Then, HF (40 wt% in H_2O), H_3PO_4 (85 wt% in H_2O) and 2-methylimidazole are added successively. The molar ratio of the reaction mixture is

$$\text{tetraethylammonium chloride} / \text{pentaerythritol} / \text{Al} / \text{P} / \text{F} / 2-\text{methylimidazole} = 5.32 / 6.47 / 1.00 / 1.42 / 1.07 / 0.50$$

After sufficient stirring, the mixture is programmatically heated to 180 °C in 30 min and then crystallizes at this temperature for 4 h. The product is cooled, washed with water and acetone and then dried at 110 °C.

2.3.3 *Reactant Selection*

2.3.3.1 IL/DES and Additive Organic Species

Although the varieties of both ILs and DESs are extremely huge, only several kinds of ILs/DESs with low cost, convenient preparation and a few toxicity risk are commonly used in ionothermal synthesis. Considering their medium hydrogen bonding ability, chloride and bromide anions are preferred to be used as anions of IL/DES for ionothermal synthesis of molecular sieves. Organic species is always introduced into the reaction mixture to adjust the physicochemical properties of the solvent and play the role of co-structure-directing agent. A more comprehensive discussion on the structure-directing effect of IL/DES and additive organic species is given in Sect. 2.5.3.

2.3.3.2 Framework Sources

Similar to hydro/solvothermal synthesis, the nature of the products ionothermally synthesized is strongly determined by the kinds and amounts of the framework sources. As Cooper et al. reported in the initial paper on the ionothermal synthesis of AlPOs, different crystalline phases can be obtained by slightly alternating the composition of reaction mixture [1]. The later work of Wang et al. showed that both the kind of aluminium sources and the P/Al ratio of the reaction mixture can influence the obtained crystalline phase significantly [53]. When basic aluminium acetate ($Al_2O(CH_3COO)_4 \cdot 4H_2O$) is used as aluminium source, the best P/Al ratio for the synthesis of AlPO-AEL is 1.5/1. If the P/Al ratio is larger than this optimum value, both the phase purity and the yield of AlPO-AEL decrease. However, when aluminium isopropoxide is used as aluminium source, the optimum P/Al ratio increases to 3/1. Wang et al. also found that compared with aluminium isopropoxide, basic aluminium acetate disperses and dissolves into the reaction mixture more easily. They suggested that this phenomenon should be attributed to the hydrolysis mechanism difference between these two kinds of aluminium sources.

The addition of framework sources into the reaction system is often accompanied with the introduction of anions and water, both of which may strongly influence the physicochemical properties of IL/DES. Furthermore, because the commonly used IL/DES in ionothermal synthesis usually has a poorer coordination ability compared with water, the ionothermal crystallization of molecular sieves always involve a more unexpected competition between the potential ligands of metal cations (anions and water molecules). Thus, when added into the ionothermal reaction mixture as framework source, metal salts with different anions or different amount of crystallization water may show different reaction behaviour and result in different final products, some of which are possibly unachievable in traditional conditions.

2.3.3.3 Water

It is always difficult to catch the trace of water in the crystallization process of molecular sieves because it seems to be everywhere in conventional synthesis conditions. By carefully avoiding introducing water into the reaction mixture, Ma et al. provided a kinetic study on the effect of water on the ionothermal synthesis of AlPOs [20]. They found that when no water is added into the reaction mixture, the crystallization process of AlPO undergoes a quite long induction period to accumulate a certain amount of the *in situ* generated water. Then, the accumulated water improves the formation rate of solution active species, which crystallize and release more water into the reaction mixture. Thus, the overall ionothermal synthesis of molecular sieves appears to be an autocatalytic process. Ma et al. also found that the induction time could be reduced by introducing a reactant quantity of water into the reaction mixture—the amount of water accompanying the raw materials is always enough for this aim.

Water is indispensable in some cases, especially the ionothermal synthesis of zeolites [19, 54]. Therefore, the extra-addition of water is sometimes necessary. As mentioned in Sect. 2.2.2, water added into IL/DES with a low concentration can form hydrogen bonds with IL/DES and consequently be isolated and deactivated. However, when too much water is added into the reaction mixture, the crystallization system would change from ionothermal into hydrothermal. Wragg et al. carried out an extensive study to determine the influence of the water/IL ratio of reaction mixture [44]. Their study confirmed Ma et al.'s result [20] that a very small amount of added water is advantageous for the ionothermal synthesis of AlPOs. However, water added into the reaction mixture with a larger amount (as Wragg et al. suggested, the water/IL molar ratio > 0.7) would disrupt the structure-directing ability of IL and lead to the formation of dense phases.

2.3.4 Key Parameters of Crystallization

2.3.4.1 Reaction Vessel: Open or Sealed?

The low volatility of IL/DES brings convenience for synthesis of molecular sieves but also a question: *Does the ionothermal crystallization of molecular sieves taken place in an open vessel follow the same pathway as that in a sealed vessel?*

As mentioned in Sect. 2.2.3.3, Parnham et al. found that in sealed system with the presence of fluoride anions, 1-alkyl-3-methylimidazolium cations would decompose and subsequently form DMIm cations, which can direct the formation of AlPO-CHA [41]. However, only AlPOs with AEL and AFI topologies could be obtained in open vessels. They attributed this difference to the escape of bromoalkane molecules and thus the inhibition of the transalkylation reaction.

Furthermore, in an open vessel, the uncontrollable release of reactants such as H_2O, HF and other molecules from the reaction mixture may result a change in the reaction environment and even significantly prevent the crystallization process [19]. Therefore, to ensure the reproducibility of the synthesis, it is suggested to carry out the crystallization process in a sealed system.

2.3.4.2 Heating Method

Because of the advantages stated in Sect. 2.2.3.2, microwave irradiation is often employed instead of conventional heating method to ionothermally synthesize molecular sieves with a much shorter crystallization time. However, the change of heating method is doubtless accompanied by the change of crystallization equipment, and in some cases, may lead to a different final product.

By microwave irradiating the mixture of dry gel precursor, [BMIm]Br and extra-added water, Cai et al. synthesized pure silica zeolite with MFI topology [19]. However, no crystalline phase could be detected after heating the mixture for 24 h in conventional heating method. They attributed above different results to the different water-holding abilities of the crystallization equipments employed in the two synthesis processes. As Cai et al. described, by using conventional heating method, 'the excess amount of water could not be retained in an open vessel even before the temperature of [BMIm]Br reached 175 °C'. However, a *reflux* process was observed under microwave radiation and as a result, 'water can be supplied back to [BMIm]Br for zeolite crystallization'.

Although no further details of their crystallization equipments were provided, Cai et al.'s study reminds us to be careful when making comparison between different synthesis conditions, especially on the crystallization mechanism and kinetics. What the use of IL/DES provides is not only flexibility in the synthesis of molecular sieves, but also much more variables that may influence the product formation.

2.4 Molecular Sieves Ionothermally Synthesized

2.4.1 Zeolitic Metal Phosphates

Until the time of writing this chapter, to the best of our knowledge, nearly 25 zeolite framework types have been ionothermally synthesized. Most of these zeolitic molecular sieves are metal phosphates, of which the synthesis details are summarized in Table 2.3. The synthesis procedures of these zeolitic metal phosphates are always similar to that of AlPO-CLOs described in Sect. 2.3.2.

2.4.1.1 New Zeolitic Materials

As Table 2.3 shows, besides the zeolitic molecular sieves that can be synthesized *via* conventional routes, several metal phosphates with novel structures and compositions have been obtained by ionothermal method.

SIZ-7 ($Co_{12.8}Al_{19.2}(PO_4)_{32}$) is an ionothermally synthesized CoAPO which has a novel zeolite framework type SIV (Fig. 2.5) [56]. The SIV-type framework displays a three-dimensional eight-membered ring channel system. SIZ-7 has a monoclinic space group *C2/c* (No. 15) with $a=10.2959(4)$ Å, $b=14.3715(5)$ Å, $c=28.599(1)$ Å and $\beta=91.094(1)°$. Figure 2.6 shows the simulated XRD pattern of SIZ-7.

–CLO is an interrupted zeolite framework type with 20-membered pore openings. In conventional synthesis conditions, only GaPO-CLO (cloverite) and its metal substituted analogues have been synthesized. In 2010, Wei et al. reported the successful synthesis of AlPO-CLO (DNL-1) with formula

$$\left[(C_6N_2H_{18})_{104}(C_6N_2H_{11})_{80}(H_2O)_{910}\right]\left[Al_{768}P_{768}O_{2976}(OH)_{192}F_{288}\right]$$

in [EMIm]Br with 1,6-hexanediamine as co-structure-directing agent [81]. DNL-1, which is the first AlPO with 20-membered pore openings, shows higher thermal stability compared with cloverite. The *in situ* XRD patterns of DNL-1 show that no significant loss of crystallinity occurs until the temperature reaches approximately 950 °C. Moreover, the ordered crystalline structure of DNL-1 can remain after calcination at 850 °C and subsequent exposure at room temperature with 60 % humidity for 6 d.

Table 2.3 Summary of zeolitic metal phosphates ionothermally synthesized

Framework type	Solvent	T atoms	Additive organic species	References
ABW	Choline chloride–N,N′-dimethylurea	Co+P	(None)	[55]
AEI	1-Ethyl-3-methylimidazolium bromide	Co/Al+P	(None)	[56]
AEL	1-Butyl-3-methylimidazolium bromide	Al+P	(None)	[20, 21, 38, 57–60]
	1-Butyl-3-methylimidazolium bromide	Al+P	Morpholine	[61]
	1-Butyl-3-methylimidazolium bromide	Al+P	*n*-Dipropylamine	[21, 57]
	1-Butyl-3-methylimidazolium bromide	Al+P	Tetrabutylammonium bromide	[58]
	1-Butyl-3-methylimidazolium bromide	Al+P	Choline chloride	[58]
	1-Butyl-3-methylimidazolium bromide	Al+P	Tetrapropylammonium bromide	[58]
	1-Butyl-3-methylimidazolium bromide	Mg/Al+P	(None)	[62]
	1-Butyl-3-methylimidazolium bromide	Mg/Al+P	*n*-Dipropylamine	[62]
	1-Ethyl-3-methylimidazolium bromide	Al+P	(None)	[1, 20, 38, 44, 53, 63–67]
	1-Ethyl-3-methylimidazolium bromide	Co/Al+P	(None)	[68]
	1-Ethyl-3-methylimidazolium bromide	Co/Al+P	1,6-Hexanediamine	[68]
	1-Ethyl-3-methylimidazolium bromide	Al+Si+P	(None)	[38, 63]
	1-Ethyl-3-methylimidazolium chloride	Al+P	(None)	[60]
	1-Ethyl-2,3-dimethylimidazolium bromide	Al+P	(None)	[60]
	1-Ethylpyridinium bromide	Al+P	(None)	[44]
	1-Ethyl-3-methylimidazolium bromide–1-ethyl-2,3-dimethylimidazolium bromide	Al+P	(None)	[67]
	1-Ethyl-3-methylimidazolium bromide–1-hexyl-2,3-dimethylimidazolium bromide	Al+P	(None)	[67]
	Choline chloride–urea	Mg/Al+P	(None)	[69]
AEN	Choline chloride–urea	Mg/Al+P	(None)	[69]
AFI	1-Allyl-2,3-dimethylimidazolium bromide	Al+P	(None)	[60]
	1-Benzyl-3-methylimidazolium chloride	Al+P	(None)	[70]

(continued)

Table 2.3 (continued)

Framework type	Solvent	T atoms	Additive organic species	References
	1,3-Dibutylimidazolium chloride	Al+P	(None)	[60]
	1-Butyl-3-methylimidazolium bromide	Al+P	(None)	[20, 21, 38]
	1-Butyl-3-methylimidazolium bromide	Al+P	*n*-Butylamine	[21]
	1-Butyl-3-methylimidazolium bromide	Al+P	Diethylamine	[21]
	1-Butyl-3-methylimidazolium bromide	Al+P	Triethylamine	[21]
	1-Butyl-3-methylimidazolium bromide	Al+P	1-Methylimidazole	[21]
	1-Butyl-3-methylimidazolium bromide	Al+P	Morpholine	[48, 61]
	1-Butyl-3-methylimidazolium bromide	Al+P	*n*-Dipropylamine	[21, 57, 58]
	1-Butyl-3-methylimidazolium bromide	Al+P	*i*-Dipropylamine	[21]
	1-Butyl-3-methylimidazolium bromide	Al+P	Pyrrolidine	[21]
	1-Butyl-3-methylimidazolium bromide	Al+P	Choline chloride	[58]
	1-Butyl-3-methylimidazolium bromide	Al+P	Tetraethylammonium chloride	[58]
	1-Butyl-3-methylimidazolium bromide	Mg/Al+P	(None)	[62]
	1-Butyl-2,3-dimethylimidazolium bromide	Al+P	(None)	[60, 67]
	1-Ethyl-2,3-dimethylimidazolium bromide	Al+P	(None)	[60, 67, 71]
	1-Ethyl-2,3-dimethylimidazolium bromide	Mn/Al+P	(None)	[72, 73]
	1-Hexyl-2,3-dimethylimidazolium bromide	Al+P	(None)	[67]
	1-Butyl-1-methylpyrrolidinium chloride	Al+P	(None)	[60]
	1-Ethyl-3-methylimidazolium bromide–1-ethyl-2,3-dimethylimidazolium bromide	Al+P	(None)	[67]
	Choline chloride–pentaerythritol	Al+Si+P	(None)	[74]
	Choline chloride–succinic acid	Fe/Al+P	Pyridine	[75]
	Tetraethylammonium bromide–pentaerythritol	Al+P	(None)	[76]
	Choline chloride–tetraethylammonium bromide–succinic acid	Fe/Al+P	(None)	[45, 77, 78]

AST	Choline chloride–pentaerythritol	Al + Si + P	(None)	[74]
	Choline chloride–succinic acid	Fe/Al + P	Pyridine	[75]
ATV	1-Butyl-3-methylimidazolium bromide	Al + P	*n*-Dipropylamine	[21]
AWW	1,1-Dimethylpiperidinium chloride–pentaerythritol	Al + P	(None)	[76]
CHA	1-Butyl-3-methylimidazolium bromide	Al + P	(None)	[41]
	1-Butyl-3-methylimidazolium bromide	Al + P	1-Methylimidazole	[79]
	1-Butyl-3-methylimidazolium bromide	Al + P	2-Methylimidazole	[79]
	1-Butyl-3-methylimidazolium bromide	Al + P	4-Methylpyridine	[79]
	1-Ethyl-3-methylimidazolium bromide	Al + P	(None)	[1, 41, 42, 44, 65]
	1-Ethyl-3-methylimidazolium bromide	Al + P	Tetraethylammonium hydroxide	[80]
	1-Ethyl-3-methylimidazolium bromide	Co/Al + P	1,6-Hexanediamine	[68]
	1-Ethyl-3-methylimidazolium chloride	Al + P	(None)	[42]
	1-Methyl-3-pentylimidazolium bromide	Al + P	(None)	[41]
	1-Methyl-3-propylimidazolium bromide	Al + P	(None)	[41]
	1-Methyl-3-isopropylimidazolium bromide	Al + P	(None)	[41]
	1,1′-Dimethyl-3,3′-hexamethylene-diimidazolium dibromide	Al + P	(None)	[41]
	1-Ethylpyridinium bromide	Al + P	(None)	[44]
	1,4-Diethyl-1,4-dimethylpiperazinium dibromide–pentaerythritol	Al + P	(None)	[76]
–CLO	1-Ethyl-3-methylimidazolium bromide	Al + P	1,6-Hexanediamine	[81]
	1-Ethyl-3-methylimidazolium bromide	Co/Al + P	1,6-Hexanediamine	[68]
	1-Ethyl-3-methylimidazolium bromide	Ga + P	(None)	[82]
DFT	Choline chloride–2-imidazolidone	Mn + P	(None)	[55]
	Choline chloride–2-imidazolidone	Fe + P	(None)	[55]
	Choline chloride–2-imidazolidone	Co + P	(None)	[55]
	Choline chloride–2-imidazolidone	Zn + P	(None)	[83]

(continued)

Table 2.3 (continued)

Framework type	Solvent	T atoms	Additive organic species	References
ERI	1,1-Dimethylpiperidinium chloride–pentaerythritol	Al + P	(None)	[76]
GIS	Choline chloride–urea	Mg/Al + P	(None)	[69]
JBW	Tetraethylammonium bromide–1,3-dimethylurea	Zn + P	(None)	[84]
	N,N-Dimethylpiperidinium chloride–1,3-dimethylurea	Zn + P	(None)	[84]
LAU	1-Ethyl-3-methylimidazolium tosylate	Mn/Al + P	Imidazole	[85]
LEV	Choline chloride–succinic acid	Co/Al + P	Cyclam	[16]
	Choline chloride–succinic acid	Co/Al + P	Pyridine	[16]
LTA	1-Benzyl-3-methylimidazolium chloride	Al + P	(None)	[70]
	1-Benzyl-3-methylimidazolium chloride	Al + P	Tetramethylammonium bromide	[70]
	1-Butyl-3-methylimidazolium bromide	Al + P	*n*-Butylamine	[57, 86]
	1-Butyl-3-methylimidazolium bromide	Al + P	Tributylamine	[57, 86]
	1-Butyl-3-methylimidazolium bromide	Al + P	Cyclohexanamine	[57, 86]
	1-Butyl-3-methylimidazolium bromide	Al + P	Diethylamine	[57, 86]
	1-Butyl-3-methylimidazolium bromide	Al + P	Triethylenetetramine	[86]
	1-Butyl-3-methylimidazolium bromide	Al + P	Hexamethylenetetramine	[57, 86]
	1-Butyl-3-methylimidazolium bromide	Al + P	Imidazole	[57, 86]
	1-Butyl-3-methylimidazolium bromide	Al + P	Morpholine	[61]
	1-Butyl-3-methylimidazolium bromide	Al + P	*n*-Dipropylamine	[57, 86]
	1-Butyl-3-methylimidazolium bromide	Al + P	Pyridine	[57, 86]
	1-Butyl-3-methylimidazolium bromide	Al + P	Tetramethylammonium hydroxide	[58]
	1-Butyl-3-methylimidazolium bromide	Ga + P	(None)	[82]
	1-Butyl-3-methylimidazolium chloride	Al + P	(None)	[60]
	1-Ethyl-3-methylimidazolium bromide	Al + P	(None)	[87]

	1-Ethyl-3-methylimidazolium bromide	Ga + P	(None)	[82, 87]
	1-Hexyl-3-methylimidazolium bromide	Ga + P	(None)	[82]
	1-Methyl-3-pentylimidazolium bromide	Ga + P	(None)	[82]
	1-Methyl-3-propylimidazolium bromide	Ga + P	(None)	[82]
	1-Butylpyridinium chloride	Al + P	(None)	[60]
	Choline chloride–tetrahydro-2-pyrimidione	Ga + P	(None)	[88]
SIV	1-Ethyl-3-methylimidazolium bromide	Co/Al + P	(None)	[56]
SOD	1-Ethyl-3-methylimidazolium bromide	Al + P	(None)	[49]
	1-Ethyl-3-methylimidazolium bromide	Co/Al + P	(None)	[49, 56]
	1-Ethyl-3-methylimidazolium bromide	Fe/Al + P	(None)	[49]
	1-Butyl-3-methylimidazolium bromide	Al + P	(None)	[49]
	1-Butyl-3-methylimidazolium tetrafluoroborate	Co/Al + P	(None)	[49]
THO	Tetraethylammonium bromide–1,3-dimethylurea	Zn + P	(None)	[84]
ZON	Choline chloride–pentaerythritol	Al + P	(None)	[76]

Only the formally published results are included. In some cases, the products contain more than one crystalline phase

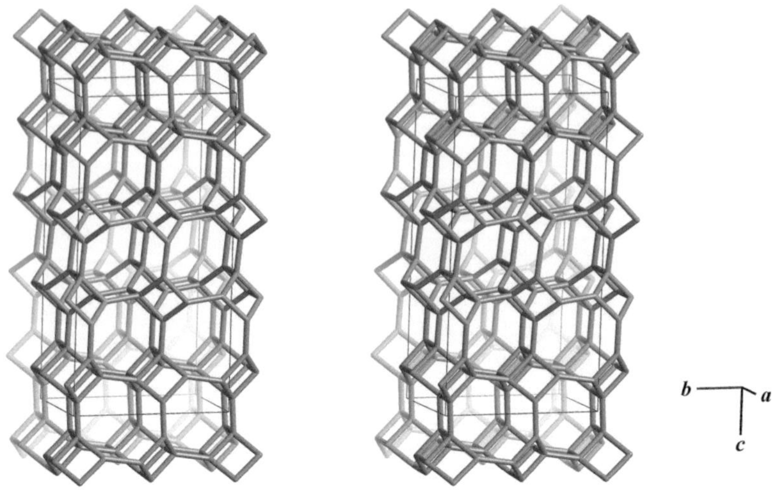

Fig. 2.5 Stereographic drawing of SIZ-7 framework [56]

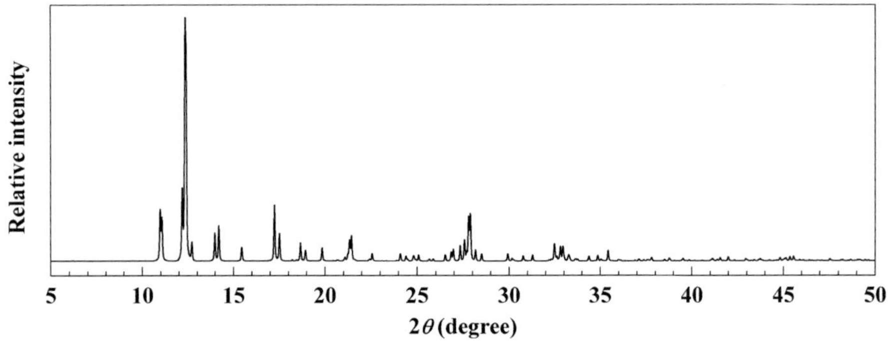

Fig. 2.6 Simulated XRD pattern of SIZ-7 [89]. Wavelength: 1.54180 Å (Cu K_α)

Besides the report of Wei et al., other several ionothermal routes have also been developed to synthesize AlPO-CLO. Pei synthesized AlPO-CLOs in a variety of IL/DES-organic amine systems (Table 2.4) [52]. Furthermore, Lohmeier prepared another AlPO-CLO sample (LUH-2) from a mixed-anion IL system [EMIm]$Br_{0.88}$(*L*-lactate)$_{0.12}$ by microwave heating the reaction mixture

$$\mathrm{Al}\left(L-\mathrm{lactate}\right)_3 / \mathrm{H_3PO_4} / \mathrm{HF} / \left[\mathrm{EMIm}\right]\mathrm{Br} / \left[\mathrm{EMIm}\right]\left(L-\mathrm{lactate}\right) = 1.00/3.00/0.84/18.70/2.67$$

at 160 °C for 60 min [90]. The flexibility in the ionothermal synthesis of AlPO-CLOs, as shown by above examples, indicates the great potential of ionothermal method for the synthesis of novel molecular sieves.

Table 2.4 Synthesis details of AlPO-CLOs reported by Pei [52]

Synthesis system		Molar ratio of reactants	Crystallization condition	
Solvent[a]	Amine[b]	(Solvent/Al/P/F/amine)	Temperature (°C)	Time (h)
[MPIm]Br	*n*-DPA	7.89/1.00/1.20/1.07/1.19	180	4
[EMIm]Br	*n*-DPA	8.43/1.00/1.20/1.07/1.20	180	2
TEAC–Pent (1/1.22)	2-MIm	5.32/1.00/1.42/1.07/0.50	180	4

[a]*MPIm* 1-methyl-3-propylimidazolium, *EMIm* 1-ethyl-3-methylimidazolium, *TEAC* tetraethylammonium chloride, *Pent* pentaerythritol
[b]*n-DPA* *n*-dipropylamine, *2-MIm* 2-methylimidazole

2.4.1.2 SAPOs

Silicon-containing zeolitic molecular sieves, such as zeolites and SAPOs, are widely used as heterogeneous acid catalysts. Since the very beginning of the study on ionothermal synthesis, researchers have tried to introduce silicon into the framework of zeolitic molecular sieves [1, 63]. However, only limited results have been achieved previously because of the poor reaction activity of silicon sources in ionothermal condition.

In the study carried out by Xu et al., SAPOs were synthesized by adding tetraethylorthosilicate into the synthesis procedure for AlPOs [38, 63]. Combining the analysis results of ^{29}Si MAS NMR, NH_3-TPD and EDX, they declared the successful incorporation of silicon into the framework of AEL type molecular sieves [38]. Later, Zhao et al. reported their synthesis of SAPO-AFI by using DES choline chloride–pentaerythritol as reaction medium and fumed silica as silicon source [74]. Unfortunately, according to the EDX analysis results, some of the products they obtained may contain silicon-rich phases as well as SAPO molecular sieves.

2.4.2 Zeolites

2.4.2.1 Zeolites Synthesized from Sol/Gel–IL Mixtures

The main obstacle to ionothermal synthesis of silicon-containing molecular sieves, as illustrated in Sect. 2.4.1.2, is the extremely low hydrolysis rate of silicon sources in ionothermal environment. By carrying out the hydrolysis of raw materials and crystallization under different conditions, *i.e.*, by ionothermally treating the prehydrolyzed silicon-containing reaction mixtures, several zeolites have been synthesized successfully.

By heating the mixture of [EMIm]Br and an aqueous sol with following composition

$$SiO_2 / Al_2O_3 / Na_2O / K_2O / H_2O = 2.2/1/4/1.4/80$$

in a temperature range between 50 and 150 °C with ambient pressure, Ma et al. obtained a series of zeolite products [91]. They found that the product changes from sodalite to zeolite X when the ratio of [BMIm]Br to the aqueous sol decreases. Obviously, this process cannot be regarded as ionothermal synthesis wherein IL/DES should play the role of the main solvent [44]. Subsequently, Cai et al. reported their synthesis of pure silica zeolite with MFI topology by microwave-heating the mixture of [BMIm]Br, extra-added water and a dry gel precursor [19]. Their study showed that the addition of water is essential for the crystallization of zeolite.

2.4.2.2 Zeolites Synthesized in Basic IL [BMIm]$OH_{0.65}Br_{0.35}$

Mineralizers, such as H_2O, OH^- and F^-, are always essential for the hydrolysis of silicon sources. However, in some senses, the addition of water into IL/DES means the loss of properties of ionothermal synthesis. This contradiction prompted an attempt to use task specific IL for ionothermal synthesis. By using basic IL [BMIm]$OH_{0.65}Br_{0.35}$ as solvent, structure-directing agent and mineralizer, Wheatley et al. synthesized pure silica zeolites with MFI and TON topologies [54]. [BMIm]$OH_{0.65}Br_{0.35}$, the use of which is the key to the successful ionothermal synthesis of these zeolites, was prepared by anion exchange of [BMIm]Br. However, it should be noted that the anion exchange process was accompanied with the degradation of BMIm cations. In fact, the poor basic stability of IL/DES has strongly restricted their application [92].

2.4.3 Other Crystalline Microporous Materials

Besides the above-mentioned zeolitic materials, a huge variety of other crystalline molecular sieves, such as inorganic open-framework materials and metal-organic frameworks, has also been ionothermally synthesized. For instance, a novel open-framework zirconium phosphate, $[(C_2H_7NH)_8(H_2O)_8][Zr_{32}P_{48}O_{176}F_8(OH)_{16}]$ (ZrPOF-EA), was synthesized by Liu et al. [93]. After removing the guest organic species by calcination in air at 410 °C for 9 h, the framework of ZrPOF-EA remains and can display a high CO_2/CH_4 separation capacity. Some ionothermally synthesized metal-organic frameworks with good thermal stability and porosity also exhibit excellent performance in adsorption and separation processes [94–96].

2.4.4 Mesoporous and Hierarchical Porous Materials

2.4.4.1 Mesoporous γ-Al_2O_3

In conventional solvent environment, ILs are commonly used as surfactants to direct the formation of mesoporous materials. Therefore, in theory, a similar structure-directing process can also take place in ionothermal condition. An example of

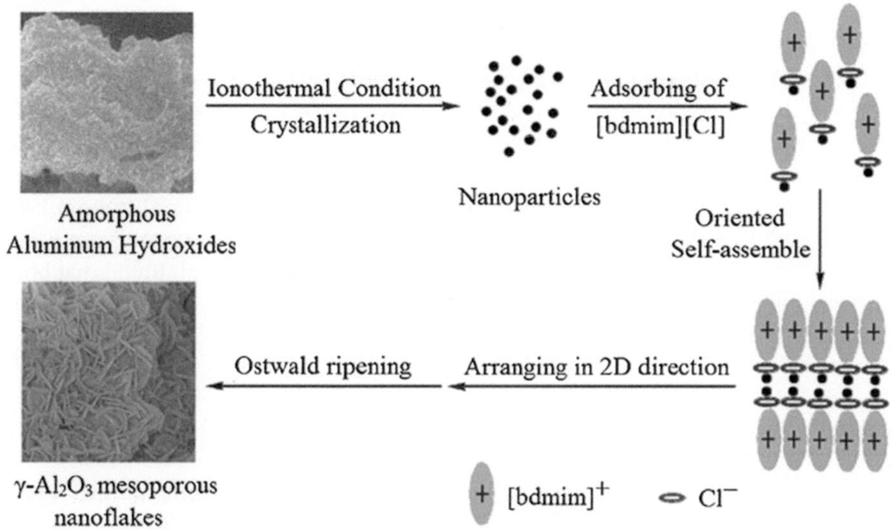

Fig. 2.7 Schematic illustration of the formation mechanism of the mesoporous γ-Al_2O_3 synthesized by Lian et al. (Reproduced from Ref. [97] with permission of The Royal Society of Chemistry)

ionothermal synthesis of mesoporous materials was reported by Lian et al., who obtained mesoporous γ-Al_2O_3 by heating a mixture of $AlCl_3$, $NaNH_2$ and [BDMIm]Cl at 150 °C for 10 h [97]. The authors supposed that the formation of the mesoporous structure follows the process shown in Fig. 2.7.

2.4.4.2 Hierarchical Porous FeAPOs

Using microwave-assisted ionothermal method, Zhao et al. synthesized several hierarchical porous FeAPO-AFI samples [45]. The authors suggested that the formation of mesopores should be attributed to both the agglomeration of particles and the removal of embedded organic species. An interesting phenomenon reported by Zhao et al. is the significantly unusual TGA curves of their hierarchical porous samples. For example, as one of the TGA curves (Figure 10 in Ref. [45]) shows, the weight loss of the sample S1.1–1.5 is about 45 % at 550 °C. This value is almost two times as large as that of the MnAPO-AFI sample ionothermally synthesized by Ng et al. [72]. Obviously, this difference indicates that the sample S1.1–1.5 is a hybrid of organic species and crystalline molecular sieves. However, no further details of the organic species were provided in Zhao et al.'s paper.

2.4.5 *Molecular Sieves with Special Morphologies*

2.4.5.1 Morphology-Controlled Single Crystals

The control of the crystal morphology has attracted great research interest because morphology details can always show the traces of crystallization mechanism. Ma et al. synthesized GaPO-LTA single crystals in 1-alkyl-3-methylimidazolium bromides and found that the sizes and morphologies of these single crystals are influenced by the alkyl chain length of these IL cations [82]. Wang et al. found that both the reaction temperature and the cooling rate can strongly affect the sizes and morphologies of the obtained ZIF-8 crystals (Scheme 2.5) [98].

2.4.5.2 Zeolite Films and Membranes

Ionothermal method has also been used in the preparation of zeolite films and membranes. Cai et al. reported their ionothermal synthesis of AlPO-AEL and SAPO-AEL films on aluminium alloy AA 2024-T3 [99]. Both the XRD patterns and SEM images (Fig. 2.8) show that crystals of the SAPO-AEL film are packed more densely and more highly oriented compared with that of the AlPO-AEL film. The authors attributed the orientation difference of the AlPO-AEL and SAPO-AEL films to their

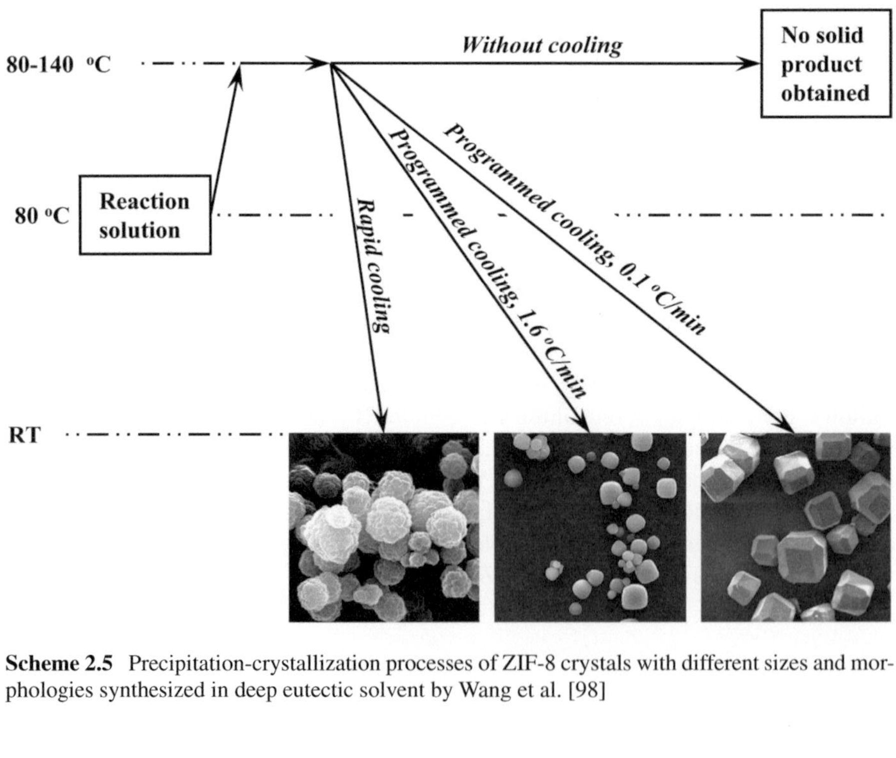

Scheme 2.5 Precipitation-crystallization processes of ZIF-8 crystals with different sizes and morphologies synthesized in deep eutectic solvent by Wang et al. [98]

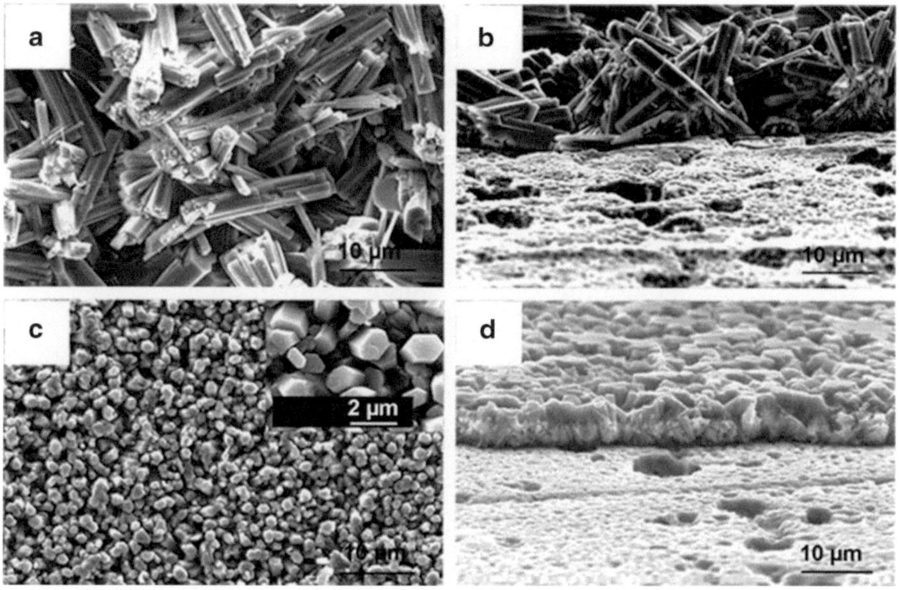

Fig. 2.8 SEM images of the AlPO-AEL and SAPO-AEL films synthesized by Cai et al. (**a**) AlPO-AEL (surface). (**b**) AlPO-AEL (cross section). (**c**) SAPO-AEL (surface, inset is higher magnification). (**d**) SAPO-AEL (cross section, mildly polished surface) (Reproduced in part from Ref. [99] by permission of John Wiley and Sons)

different crystallization rates. For SAPO-AEL, a dense nuclei layer forms on the substrate surface first because of its slow crystallization rate. Then, in subsequent crystal growth process, the limited space between the formed growth centres makes the in-plane growth suppressed. Therefore, it is preferred the formation of highly *c*-axis oriented SAPO-AEL film. However, for AlPO-AEL, the fast crystallization rate results in a single-step isotropic crystallization and thus the obtained film should not exhibit any preferred orientation.

Later, Li et al. reported their synthesis of AlPO membranes through ionothermal conversion of the δ-alumina substrate surface [100, 101]. Cross-sectional SEM and EDX results of the AlPO-CHA membrane show that an intermediate layer (~20 μm) exists between the molecular sieve layer (~100 μm) and the substrate (Fig. 2.9) [100]. Along the three layers of molecular sieves, intermediate and substrate, the P/Al ratio drops from unity to zero. According to this phenomenon, Li et al. proposed the formation mechanism of the AlPO-CHA membrane they synthesized as following [100, 101]:

1. An aluminophosphate precursor layer forms on the substrate surface in the reaction between the surface δ-alumina and the reactants in solution.
2. Nucleation of molecular sieves occurs at the solution–precursor layer interface and subsequently crystals grow to form a continuous membrane. The nucleation and crystal growth processes follow the solid-phase transformation mechanism.

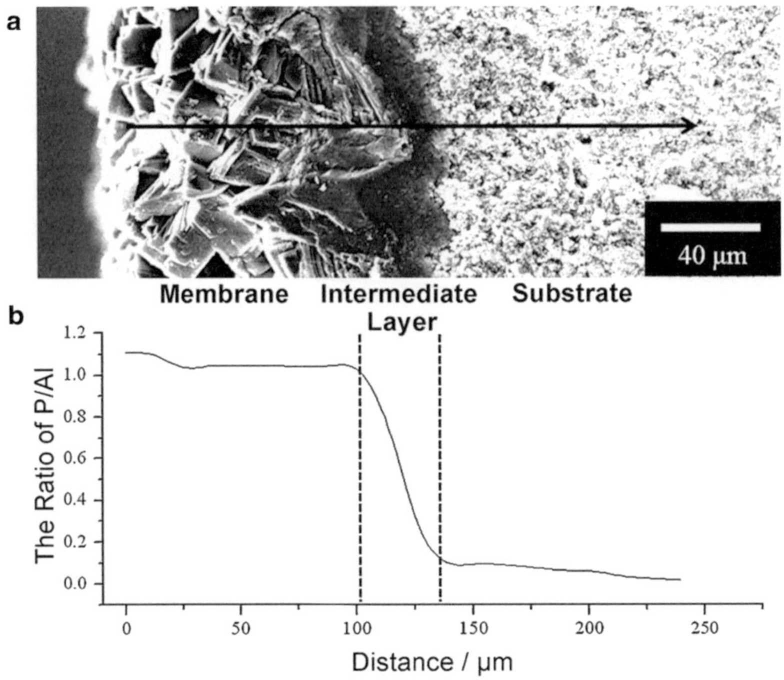

Fig. 2.9 (**a**) Cross-sectional SEM image of the AlPO-CHA membrane synthesized by Li et al. The *arrow* represents the path of the EDX line scan. (**b**) P/Al ratio along the path of the EDX line scan (Reproduced from Ref. [100] by permission of John Wiley and Sons)

3. The formation of the precursor layer continues until the interaction between the substrate and the solution is obstructed by the formation of continuous membrane. Finally, the crystallization process finishes when the nutrient in the precursor layer is depleted.

2.5 Structure-Directing Effect in Ionothermal Synthesis

2.5.1 IL/DES

In the review entitled *The Synthesis of Molecular Sieves from Non-Aqueous Solvents*, Morris and Weigel classified organic solvents into four categories according to their tendency to form hydrogen bonds [102]. The authors pointed out that the interaction between structure-directing agent and solvent should be intermediate. Weak interaction is adverse to the solvation and isolation of the molecules of structure-directing agent. However, strong interaction would prevent these molecules from interacting with framework species. As a result, successful structure-directing process requires a balance of the interactions between solvent, framework species and the *potential* structure-directing agent.

In their paper published in 2004, Cooper et al. announced that the competition between framework–structure-directing agent interaction and framework–solvent interaction can be subtly removed by using IL/DES as both the solvent and the structure-directing agent [1]. Since then, consequent studies have proven that the cations of IL/DES, which are also commonly used as structure-directing agents in conventional synthesis routes, can play the role of structure-directing agent in ionothermal synthesis.

The physicochemical properties of IL are strongly related to not only the cation but also the anion. The study of Parnham and Morris showed that by replacing the commonly used hydrophilic IL [EMIm]Br with the hydrophobic one [EMIm][Tf_2N] (Tf_2N = bis(trifluoromethylsulfonyl)amide), an aluminophosphate chain structure ($Al(H_2PO_4)_2F$) was synthesized [103]. The authors attributed the formation of this low dimensional structure to the use of hydrophobic IL. They suggested that the IL [EMIm][Tf_2N] with low polarity prevents the solvation of framework species. Therefore, the crystallization would be a solid-phase transformation without the participation of EMIm cations. Although the use of ILs with anions rather than halide anions seems not very successful in ionothermal synthesis of zeolitic molecular sieves, these ILs have shown great potential in the synthesis of metal-organic frameworks. Discussion on the structure induction effect of the anions in the ionothermal synthesis of metal-organic frameworks, especially the chiral inducting effect of chiral ILs, can be found in some elaborate reviews such as Refs. [4] and [104].

2.5.2 Decomposition Species of ILs

As discussed in Sect. 2.2.3.3, the decomposition of IL/DES occasionally occurs in ionothermal synthesis and the decomposition product can strongly influence the nature of the molecular sieves ionothermally synthesized. An interesting example of the decomposition–transalkylation process of imidazolium ILs has also been discussed in that section [41].

2.5.3 Additive Organic Species

As shown in Table 2.3, considering their cost, toxicity and physicochemical properties, only the imidazolium ILs and the chlorine chloride-based DESs have been widely selected for ionothermal synthesis previously. Undoubtedly, compared with the preparation of IL/DES with novel structures, the employment of organic species as co-structure-directing agent is a more facile and efficient way to adjust the structure-directing process during ionothermal synthesis.

2.5.3.1 Co-Structure-Directing Effect

The term *co-structure-directing effect* is always used to describe the crystallization process that influenced by more than one kind of potential structure-directing agent. For instance, Xing et al. synthesized a new open-framework aluminophosphate (JIS-1) in [EMIm]Br by using 1-methylimidazole as co-structure-directing agent [105]. The EMIm cation and the protonated 1-methylimidazolium cation coexist in the intersection of the tridirectional channels of JIS-1. Despite the structure similarity between these two kinds of cations, JIS-1 could not be synthesized in the absence of 1-methylimidazole. Theoretical calculation showed that the replace of 1-methylimidazolium cations by EMIm cations is energetically unfavourable.

Structure analysis of the ionothermally synthesized AlPO-LTA [70] and DNL-1 [81, 106] shows an interesting cooperating mode of multiple structure-directing agents with significantly different shapes and sizes. Based on Rietveld refinement and NMR spectroscopy, Fayad et al. suggested that the structure-directing agents 1-benzyl-3-methylimidazolium cations, tetramethylammonium cations and fluoride anions fill in the *lta* cages, *sod* cages and *d4r* units of AlPO-LTA respectively (Fig. 2.10) [70]. In the structure of DNL-1, it was found that 1,6-hexanediamine molecules prefer to fill in the *lta* cages with random orientation, while EMIm cations present in both the *lta* cages and the empty space at the intersection of the 20-membered ring channels [106].

2.5.3.2 Structure-Directing Effect of Hydrogen Bonded Complex Cations

Besides directing the structure formation as the process discussed in above section, the additive organic species can also form large complex cations with the IL/DES cations *via* hydrogen bonds. These complex cations may adjust the physicochemical properties of reaction mixture and play the role of structure-directing agent during the crystallization of molecular sieves.

Pei et al. found that the structure-directing effect of the IL cations is adjusted by the concentration of organic amine added into the reaction mixture for the synthesis of AlPOs [57]. By increasing the concentration of *n*-dipropylamine in [BMIm]Br, the product transforms in the sequence

$$\text{AEL} \rightarrow \text{AFI} \rightarrow \text{LTA}$$

gradually. The same transformation sequence was also observed in the reaction system

$$[\text{BMIm}]\text{Br} / \text{Al}_2\text{O}_3 / \text{P}_2\text{O}_5 / \text{HF} / \text{morpholine} = 40 / 1.0 / 3.0 / 0.5 / x$$

by Xu et al. (Fig. 2.11) [61]. The *in situ* ^{1}H–^{1}H NMR rotating-frame nuclear Overhauser effect spectroscopy of these reaction mixtures (Fig. 2.12) clearly indicates the existence of the hydrogen-bonded BMIm–morpholine hybrid in the

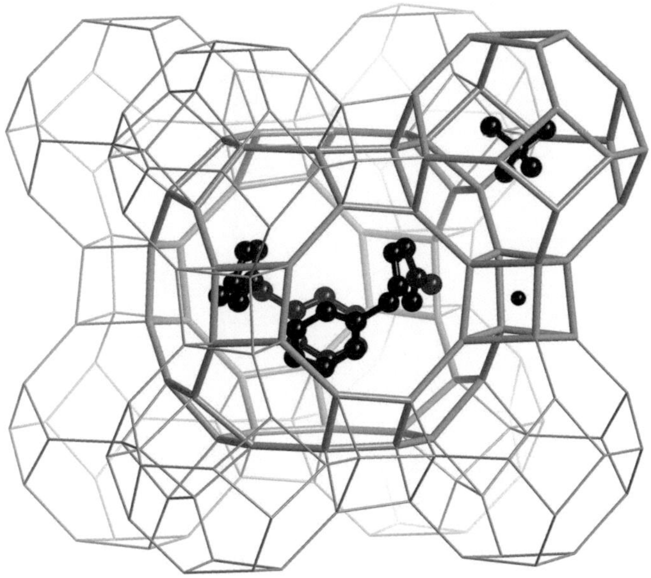

Fig. 2.10 1-Benzyl-3-methylimidazolium cations, tetramethylammonium cations and fluoride anions fill in the *lta* cages, *sod* cages and *d4r* units respectively and play the role of co-structure-directing agents in the ionothermal synthesis of AlPO-LTA [70]

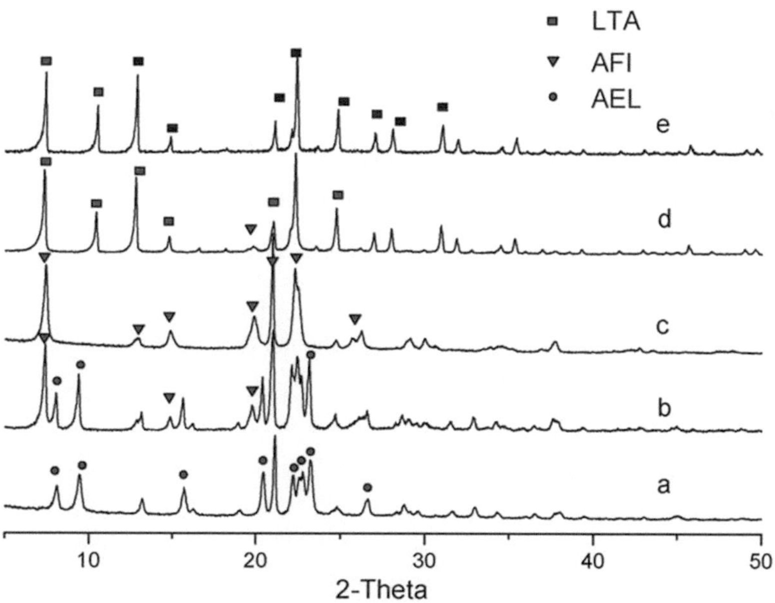

Fig. 2.11 XRD patterns of the solid samples synthesized from the reaction mixtures with different molar ratios of morpholine and 1-butyl-3-methylimidazolium bromide. (**a**) 1/60. (**b**) 1/30. (**c**) 1/23. (**d**) 1/7.5. (**e**) 1/7. The crystallization is carried out at 150 °C for 6 h (Reproduced from Ref. [61] with permission of the PCCP Owner Societies)

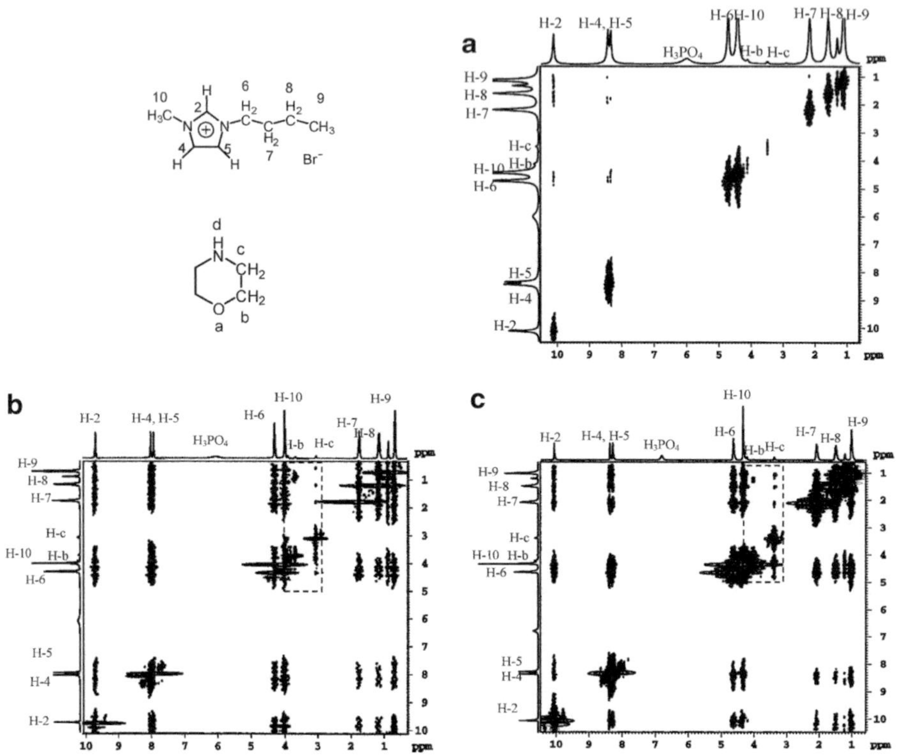

Fig. 2.12 *In situ* ^{1}H–^{1}H NMR rotating-frame nuclear Overhauser effect spectroscopy of the reaction mixture with different molar ratios of morpholine and 1-butyl-3-methylimidazolium bromide. (**a**) 1/60. (**b**) 1/23. (**c**) 1/7. Each spectrum was acquired at 150 °C with a mixing time of 600 ms. The C and H positions of morpholine and 1-butyl-3-methylimidazolium bromide are marked (Adapted from Ref. [61] with permission of the PCCP Owner Societies)

crystallization system with high morpholine concentration. The authors assumed that with the increase of morpholine concentration larger hydrogen-bonded complex cations should form in the reaction mixture and direct the formation of structures with larger channels or cavities.

2.5.3.3 Other Influences

Besides above-mentioned structure-directing effects, it is also observed the competition of structure-directing process between the additive organic species and the IL/DES cations [107]. In this case, the product is always a mixture of different crystalline phases. Furthermore, the basicity enhancement of the reaction mixture induced by adding organic amine may also influence the crystallization and result a product transformation [68].

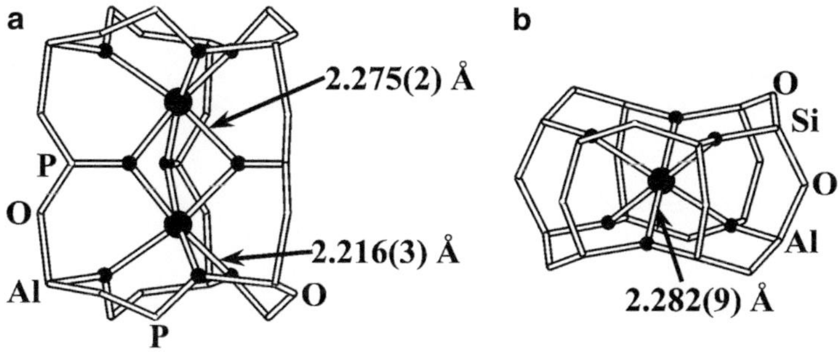

Fig. 2.13 Comparison of the coordination modes of Mn^{2+} cations in the [6^28^3] cage of DNL-2 (**a**) and in the *d6r* cages of Mn^{2+}-exchanged zeolite X (**b**). Mn and the coordinating O atoms are shown as *large and small black balls* respectively (Reproduced from Ref. [17] with permission of The Royal Society of Chemistry)

2.5.4 Transition Metal Cations

In the ionothermal synthesis of both CoAPO [56] and MgAPO [62], it was found that the hetero-metal species added into the reaction mixture contributes a structure-directing effect to the formation of the final product. However, no details of these phenomena were elaborated in these papers. Recently, Liu et al. reported their ionothermal synthesis of an open-framework fluorinated aluminium phosphite-phosphate (DNL-2) [17]. The authors found that the *in situ* released Mn^{2+} cations are employed as structure-directing agent in the ionothermal synthesis of DNL-2. As Fig. 2.13 shows, the coordination mode of Mn^{2+} cations in DNL-2 is similar to that of the site I Mn^{2+} cations in Mn^{2+}-exchanged zeolite X. However, in conventional conditions, only the structure-directing ability of alkali/alkaline-earth metal cations has been observed. Liu et al. suggested that the structure-directing effect of Mn^{2+} cations in the ionothermal synthesis of DNL-2 depends on the unique property of [EMIm]Br, which has intermediate coordination ability and can isolate and deactivate H_2O or other strong ligands.

2.6 Prospect and Conclusion

With the development in the last decade, ionothermal synthesis has become a widely used method, which can provide a changeable crystallization environment to prepare molecular sieves with fascinating structures and various compositions. Considering its great advantages as well as its obvious limitations, we think that in the future study, more attention should be paid to following topics.

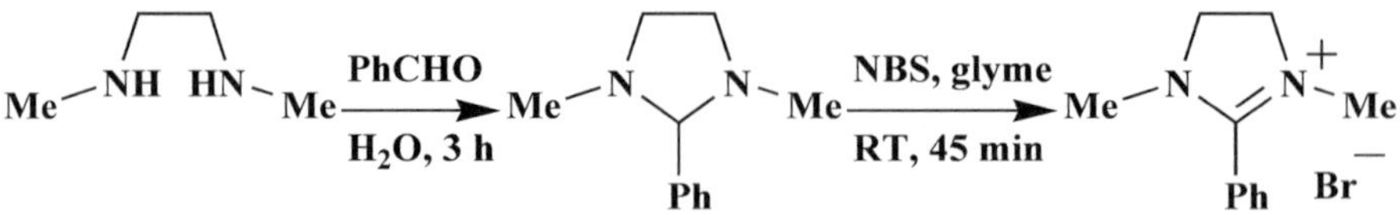

Scheme 2.6 Preparation route of a basically stable ionic liquid reported by Jurčík and Wilhelm [108]

Expand the Selection of ILs/DESs Though it is believed that the variety of ILs/DESs is hundreds of thousands times more than that of molecular solvents, only a few kinds of ILs/DESs, *i.e.*, the imidazolium ILs and the chlorine chloride-based DESs, have been widely used in ionothermal synthesis of molecular sieves. As mentioned in Sect. 2.4.2.2, $[BMIm]OH_{0.65}Br_{0.35}$ is the only task specific IL used in ionothermal synthesis. Because of its poor basic stability, $[BMIm]OH_{0.65}Br_{0.35}$ is not the perfect choice to satisfy the demand for the synthesis of zeolites. However, as early as in 2005, Jurčík and Wilhelm had reported their preparation of an imidazolinium salt with phenyl substituent at C2 position (Scheme 2.6). They announced that this IL can be used as solvent for reactions involving medium to strong bases such as quinuclidine and Grignard reagents [108]. Unfortunately, to our best knowledge, there are no reports on whether this variety of ILs fits the basic environment required for the synthesis of zeolites.

Emphasize the Complementarity Between Ionothermal Synthesis and Conventional Methods, Rather Than Their Differences As discussed in this chapter, water is indispensable in some ionothermal cases. However, when designing experiments, researchers always try to introduce water into the reaction mixture as little as possible, despite the fact that a too low concentration of water may result in a poor crystallization of the product. We always overemphasize the differences between ionothermal synthesis and hydrothermal synthesis, but overlook the hydrogen bonding interactions (Sect. 2.2.2) accompanying the dispersion of water molecules in IL/DES. Previous study has shown that various organic solvents and even water can have significant structure-directing effects. Thus, except their characteristic physicochemical properties, *i.e.*, the extremely low volatility and the high microwave-adsorption capacity, IL/DES shows no special differences compared with pyridine or other organic amines that can be used as reaction medium for the synthesis of molecular sieves. Therefore, the fruitful results accumulated in the study on synthesis of molecular sieves by other routes are beneficial to the understanding of the reaction process in ionothermal condition. For instance, the successful synthesis of a series of zeolites from solvothermal [109] and solvent-free [110–112] routes, both of which can be carried out in non-aqueous condition, suggests a possible access to the ionothermal synthesis of zeolites, which is still one of the greatest challenges in this area.

In conclusion, ionothermal synthesis has shown excellent advantages for the synthesis of molecular sieves with high efficiency and low safety risk. Because of the cost and other factors, it seems that ILs/DESs would never be the substitute for water in industrial synthesis process of molecular sieves. However, ionothermal synthesis is an essential complement to conventional synthesis routes in mechanism study and preparation of novel molecular sieves.

Acknowledgement The authors gratefully acknowledge National Natural Science Foundation of China for its funding support (Grant No. 21373214).

References

1. Cooper ER, Andrews CD, Wheatley PS, Webb PB, Wormald P, Morris RE (2004) Ionic liquids and eutectic mixtures as solvent and template in synthesis of zeolite analogues. Nature 430:1012–1016
2. Taubert A, Li Z (2007) Inorganic materials from ionic liquids. Dalton Trans (7):723–727
3. Parnham ER, Morris RE (2007) Ionothermal synthesis of zeolites, metal-organic frameworks, and inorganic–organic hybrids. Acc Chem Res 40:1005–1013
4. Morris RE (2009) Ionothermal synthesis – ionic liquids as functional solvents in the preparation of crystalline materials. Chem Commun (21):2990–2998
5. Ma Z, Yu J, Dai S (2010) Preparation of inorganic materials using ionic liquids. Adv Mater 22:261–285
6. Freudenmann D, Wolf S, Wolff M, Feldmann C (2011) Ionic liquids: new perspectives for inorganic synthesis? Angew Chem Int Ed 50:11050–11060
7. Wang Y, Xu Y, Tian Z, Lin L (2012) Research progress in ionothermal synthesis of molecular sieves. Chin J Catal 33:39–50
8. Wilkes JS, Wasserscheid P, Welton T (2008) Introduction. In: Wasserscheid P, Welton T (eds) Ionic liquids in synthesis, vol 1, 2nd edn. Wiley, Weinheim, pp 1–6
9. Zhang Q, De Oliveira VK, Royer S, Jérôme F (2012) Deep eutectic solvents: syntheses, properties and applications. Chem Soc Rev 41:7108–7146
10. Ruß C, König B (2012) Low melting mixtures in organic synthesis – an alternative to ionic liquids? Green Chem 14:2969–2982
11. Abbott AP, Capper G, Davies DL, Rasheed RK, Tambyrajah V (2003) Novel solvent properties of choline chloride/urea mixtures. Chem Commun (1):70–71
12. D'Agostino C, Harris RC, Abbott AP, Gladden LF, Mantle MD (2011) Molecular motion and ion diffusion in choline chloride based deep eutectic solvents studied by ^{1}H pulsed field gradient NMR spectroscopy. Phys Chem Chem Phys 13:21383–21391
13. Cammarata L, Kazarian SG, Salter PA, Welton T (2001) Molecular states of water in room temperature ionic liquids. Phys Chem Chem Phys 3:5192–5200
14. Hanke CG, Lynden-Bell RM (2003) A simulation study of water-dialkylimidazolium ionic liquid mixtures. J Phys Chem B 107:10873–10878
15. Danten Y, Cabaço MI, Besnard M (2009) Interaction of water highly diluted in 1-alkyl-3-methyl imidazolium ionic liquids with the PF_6^- and BF_4^- anions. J Phys Chem A 113: 2873–2889
16. Drylie EA, Wragg DS, Parnham ER, Wheatley PS, Slawin AMZ, Warren JE, Morris RE (2007) Ionothermal synthesis of unusual choline-templated cobalt aluminophosphates. Angew Chem Int Ed 46:7839–7843
17. Liu H, Tian Z-J, Gies H, Wei Y, Marler B, Wang L, Wang Y-S, Li D-W (2014) Mn^{2+} cation-directed ionothermal synthesis of an open-framework fluorinated aluminium phosphite-phosphate. RSC Adv 4:29310–29313
18. Weber CC, Masters AF, Maschmeyer T (2012) Controlling hydrolysis reaction rates with binary ionic liquid mixtures by tuning hydrogen-bonding interactions. J Phys Chem B 116: 1858–1864

19. Cai R, Liu Y, Gu S, Yan Y (2010) Ambient pressure dry-gel conversion method for zeolite MFI synthesis using ionic liquid and microwave heating. J Am Chem Soc 132: 12776–12777
20. Ma H, Tian Z, Xu R, Wang B, Wei Y, Wang L, Xu Y, Zhang W, Lin L (2008) Effect of water on the ionothermal synthesis of molecular sieves. J Am Chem Soc 130:8120–8121
21. Wang L, Xu Y, Wei Y, Duan J, Chen A, Wang B, Ma H, Tian Z, Lin L (2006) Structure-directing role of amines in the ionothermal synthesis. J Am Chem Soc 128:7432–7433
22. Ngo HL, LeCompte K, Hargens L, McEwen AB (2000) Thermal properties of imidazolium ionic liquids. Thermochim Acta 357–358:97–102
23. Vila J, Fernández-Castro B, Rilo E, Carrete J, Domínguez-Pérez M, Rodríguez JR, García M, Varela LM, Cabeza O (2012) Liquid-solid-liquid phase transition hysteresis loops in the ionic conductivity of ten imidazolium-based ionic liquids. Fluid Phase Equilib 320:1–10
24. Wakai C, Oleinikova A, Ott M, Weingärtner H (2005) How polar are ionic liquids? Determination of the static dielectric constant of an imidazolium-based ionic liquid by microwave dielectric spectroscopy. J Phys Chem B 109:17028–17030
25. Park S, Kazlauskas RJ (2001) Improved preparation and use of room-temperature ionic liquids in lipase-catalyzed enantio- and regioselective acylations. J Org Chem 66:8395–8401
26. Zaitsau DH, Fumino K, Emel'yanenko VN, Yermalayeu AV, Ludwig R, Verevkin SP (2012) Structure-property relationships in ionic liquids: a study of the anion dependence in vaporization enthalpies of imidazolium-based ionic liquids. ChemPhysChem 13:1868–1876
27. Hu H-C, Soriano AN, Leron RB, Li M-H (2011) Molar heat capacity of four aqueous ionic liquid mixtures. Thermochim Acta 519:44–49
28. Fredlake CP, Crosthwaite JM, Hert DG, Aki SNVK, Brennecke JF (2004) Thermophysical properties of imidazolium-based ionic liquids. J Chem Eng Data 49:954–964
29. Verevkin SP, Emel'yanenko VN, Zaitsau DH, Ralys RV, Schick C (2012) Ionic liquids: differential scanning calorimetry as a new indirect method for determination of vaporization enthalpies. J Phys Chem B 116:4276–4285
30. Yu Y-H, Soriano AN, Li M-H (2009) Heat capacities and electrical conductivities of 1-n-butyl-3-methylimidazolium-based ionic liquids. Thermochim Acta 482:42–48
31. Abbott AP, Capper G, McKenzie KJ, Ryder KS (2007) Electrodeposition of zinc-tin alloys from deep eutectic solvents based on choline chloride. J Electroanal Chem 599:288–294
32. Gorke JT, Srienc F, Kazlauskas RJ (2008) Hydrolase-catalyzed biotransformations in deep eutectic solvents. Chem Commun (10):1235–1237
33. Shahbaz K, Mjalli FS, Hashim MA, Al Nashef IM (2010) Using deep eutectic solvents for the removal of glycerol from palm oil-based biodiesel. J Appl Sci 10:3349–3354
34. Abbott AP, Harris RC, Ryder KS (2007) Application of hole theory to define ionic liquids by their transport properties. J Phys Chem B 111:4910–4913
35. Speight JG (ed) (2005) Lange's handbook of chemistry, 16th edn. McGraw-Hill, New York
36. Reichardt C (1994) Solvatochromic dyes as solvent polarity indicators. Chem Rev 94:2319–2358
37. Hoffmann J, Nüchter M, Ondruschka B, Wasserscheid P (2003) Ionic liquids and their heating behaviour during microwave irradiation – a state of the art report and challenge to assessment. Green Chem 5:296–299
38. Xu Y-P, Tian Z-J, Wang S-J, Hu Y, Wang L, Wang B-C, Ma Y-C, Hou L, Yu J-Y, Lin L-W (2006) Microwave-enhanced ionothermal synthesis of aluminophosphate molecular sieves. Angew Chem Int Ed 45:3965–3970
39. Chambreau SD, Boatz JA, Vaghjiani GL, Koh C, Kostko O, Golan A, Leone SR (2011) Thermal decomposition mechanism of 1-ethyl-3-methylimidazolium bromide ionic liquid. J Phys Chem A 116:5867–5876
40. Kroon MC, Buijs W, Peters CJ, Witkamp G-J (2007) Quantum chemical aided prediction of the thermal decomposition mechanisms and temperatures of ionic liquids. Thermochim Acta 465:40–47
41. Parnham ER, Morris RE (2006) 1-Alkyl-3-methyl imidazolium bromide ionic liquids in the ionothermal synthesis of aluminium phosphate molecular sieves. Chem Mater 18:4882–4887

42. Griffin JM, Clark L, Seymour VR, Aldous DW, Dawson DM, Iuga D, Morris RE, Ashbrook SE (2012) Ionothermal ^{17}O enrichment of oxides using microlitre quantities of labelled water. Chem Sci 3:2293–2300
43. Sowmiah S, Srinivasadesikan V, Tseng M-C, Chu Y-H (2009) On the chemical stabilities of ionic liquids. Molecules 14:3780–3813
44. Wragg DS, Slawin AMZ, Morris RE (2009) The role of added water in the ionothermal synthesis of microporous aluminium phosphates. Solid State Sci 11:411–416
45. Zhao X, Chen J, Sun Z, Li A, Li G, Wang X (2013) Formation mechanism and catalytic application of hierarchical structured FeAlPO-5 molecular sieve by microwave-assisted ionothermal synthesis. Microporous Mesoporous Mater 182:8–15
46. Yang M, Xu F, Liu Q, Yan P, Liu X, Wang C, Welz-Biermann U (2010) Chelated orthoborate ionic liquid as a reactant for the synthesis of a new cobalt borophosphate containing extra-large 16-ring channels. Dalton Trans 39:10571–10573
47. Yang M, Yan P, Xu F, Ma J, Welz-Biermann U (2012) Role of boron-containing ionic liquid in the synthesis of manganese borophosphate with extra-large 16-ring pore openings. Microporous Mesoporous Mater 147:73–78
48. Xu R, Zhang W, Guan J, Xu Y, Wang L, Ma H, Tian Z, Han X, Lin L, Bao X (2009) New insights into the role of amines in the synthesis of molecular sieves in ionic liquids. Chem Eur J 15:5348–5354
49. Han L, Wang Y, Li C, Zhang S, Lu X, Cao M (2008) Simple and safe synthesis of microporous aluminophosphate molecular sieves by inothermal approach. AIChE J 54:280–288
50. Hou L (2007) Study on synthesis of aluminophosphate and hetero-atom substituted aluminophosphate molecular sieves in ionic liquid. Master's thesis, Dalian Institute of Light Industry, Dalian
51. Wei Y (2010) Studies on the ionothermal synthesis of phosphate molecular sieves. Doctoral dissertation, Dalian Institute of Chemical Physics, Chinese Academy of Sciences, Dalian
52. Pei R (2010) The structure directing effect in the ionothermal synthesis of aluminophosphate molecular sieves. Doctoral dissertation, Dalian Institute of Chemical Physics, Chinese Academy of Sciences, Dalian
53. Wang S-J, Hou L, Xu Y-P, Tian Z-J, Yu J-Y, Lin L-W (2008) Effect of Al-containing precursors on ionothermal synthesis of aluminophosphate molecular sieve. Chin J Process Eng 8:93–96
54. Wheatley PS, Allan PK, Teat SJ, Ashbrook SE, Morris RE (2010) Task specific ionic liquids for the ionothermal synthesis of siliceous zeolites. Chem Sci 1:483–487
55. Yonemoto BT, Lin Z, Jiao F (2012) A general synthetic method for MPO_4 (M = Co, Fe, Mn) frameworks using deep-eutectic solvents. Chem Commun 48:9132–9134
56. Parnham ER, Morris RE (2006) The ionothermal synthesis of cobalt aluminophosphate zeolite frameworks. J Am Chem Soc 128:2204–2205
57. Pei R, Wei Y, Li K, Wen G, Xu R, Xu Y, Wang L, Ma H, Wang B, Tian Z, Zhang W, Lin L (2010) Mixed template effect adjusted by amine concentration in ionothermal synthesis of molecular sieves. Dalton Trans 39:1441–1443
58. Pei R, Tian Z, Wei Y, Li K, Xu Y, Wang L, Ma H (2010) Ionothermal synthesis of $AlPO_4$ molecular sieves in the presence of quaternary ammonium cation. Mater Lett 64:2118–2121
59. Zhao Z, Zhang W, Xu R, Han X, Tian Z, Bao X (2012) Ionothermal synthesis process for aluminophosphate molecular sieves in the mixed water/ionic liquid system. Dalton Trans 41:990–994
60. Stark A (2014) Shaping micro- and macroscopic properties of ionic liquid-solute systems: multi-functional task-specific agents. J Mol Liq 192:144–152
61. Xu R, Shi X, Zhang W, Xu Y, Tian Z, Lu X, Han X, Bao X (2010) Cooperative structure-directing effect in the synthesis of aluminophosphate molecular sieves in ionic liquids. Phys Chem Chem Phys 12:2443–2449
62. Wang L, Xu Y-P, Wang B-C, Wang S-J, Yu J-Y, Tian Z-J, Lin L-W (2008) Ionothermal synthesis of magnesium-containing aluminophosphate molecular sieves and their catalytic performance. Chem Eur J 14:10551–10555

63. Xu Y, Tian Z, Xu Z, Wang B, Li P, Wang S, Hu Y, Ma Y, Li K, Liu Y, Yu J, Lin L (2005) Ionothermal synthesis of silicoaluminophosphate molecular sieve in N-alkyl imidazolium bromide. Chin J Catal 26:446–448
64. Hu Y, Liu Y-J, Yu J-Y, Xu Y-P, Tian Z-J, Lin L-W (2006) Ionothermal synthesis of aluminophosphate molecular sieve. Chin J Inorg Chem 22:753–756
65. Wragg DS, Byrne PJ, Giriat G, Le Ouay B, Gyepes R, Harrison A, Whittaker AG, Morris RE (2009) In situ comparison of ionothermal kinetics under microwave and conventional heating. J Phys Chem C 113:20553–20558
66. Xu R, Zhang W, Xu J, Tian Z, Deng F, Han X, Bao X (2013) Multinuclear solid-state NMR studies on the formation mechanism of aluminophosphate molecular sieves in ionic liquids. J Phys Chem C 117:5848–5854
67. Shi Y, Liu G, Wang L, Zhang X (2014) Ionothermal synthesis of phase pure $AlPO_4$-5 using a series of tri-substituted imidazolium bromides. Microporous Mesoporous Mater 193:1–6
68. Li D, Xu Y, Wang Y, Liu H, Wang B, Ma H, Xu R, Tian Z (2014) Ionothermal syntheses and characterizations of cobalt-substituted extra-large pore aluminophosphate molecular sieves with –CLO topology. Microporous Mesoporous Mater 198:153–160
69. Carvalho MM, Ruotolo LAM, Fernandez-Felisbino R (2013) Synthesis of aluminophosphate by the ionothermal method using factorial design. Microporous Mesoporous Mater 165:163–167
70. Fayad EJ, Bats N, Kirschhock CEA, Rebours B, Quoineaud A-A, Martens JA (2010) A rational approach to the ionothermal synthesis of an $AlPO_4$ molecular sieve with an LTA-type framework. Angew Chem Int Ed 49:4585–4588
71. Khoo DY, Kok W-M, Mukti RR, Mintova S, Ng E-P (2013) Ionothermal approach for synthesizing AlPO-5 with hexagonal thin-plate morphology influenced by various parameters at ambient pressure. Solid State Sci 25:63–69
72. Ng E-P, Sekhon SS, Mintova S (2009) Discrete MnAlPO-5 nanocrystals synthesized by an ionothermal approach. Chem Commun (13):1661–1663
73. Ng E-P, Itani L, Sekhon SS, Mintova S (2010) Micro- to macroscopic observations of MnAlPO-5 nanocrystal growth in ionic-liquid media. Chem Eur J 16:12890–12897
74. Zhao X, Wang H, Kang C, Sun Z, Li G, Wang X (2012) Ionothermal synthesis of mesoporous SAPO-5 molecular sieves by microwave heating and using eutectic solvent as structure-directing agent. Microporous Mesoporous Mater 151:501–505
75. Zhao X, Kang C, Wang H, Luo C, Li G, Wang X (2011) Ionothermal synthesis of FeAlPO-16 molecular sieve by microwave irradiation in eutectic mixture. J Porous Mater 18:615–621
76. Liu L, Li X, Xu H, Li J, Lin Z, Dong J (2009) Template control in ionothermal synthesis of aluminophosphate microporous materials. Dalton Trans (47):10418–10421
77. Zhao X, Wang H, Dong B, Sun Z, Li G, Wang X (2012) Facile synthesis of FeAlPO-5 molecular sieve in eutectic mixture via a microwave-assisted process. Microporous Mesoporous Mater 151:56–63
78. Zhao X, Sun Z, Zhu Z, Li A, Li G, Wang X (2013) Evaluation of iron-containing aluminophosphate molecular sieve catalysts prepared by different methods for phenol hydroxylation. Catal Lett 143:657–665
79. Pei R, Tian Z, Wei Y, Li K, Xu Y, Wang L, Ma H (2010) Ionothermal synthesis of $AlPO_4$-34 molecular sieves using heterocyclic aromatic amine as the structure directing agent. Mater Lett 64:2384–2387
80. Wragg DS, Le Ouay B, Beale AM, O'Brien MG, Slawin AMZ, Warren JE, Prior TJ, Morris RE (2010) Ionothermal synthesis and crystal structures of metal phosphate chains. J Solid State Chem 183:1625–1631
81. Wei Y, Tian Z, Gies H, Xu R, Ma H, Pei R, Zhang W, Xu Y, Wang L, Li K, Wang B, Wen G, Lin L (2010) Ionothermal synthesis of an aluminophosphate molecular sieve with 20-ring pore openings. Angew Chem Int Ed 49:5367–5370
82. Ma H, Xu R, You W, Wen G, Wang S, Xu Y, Wang B, Wang L, Wei Y, Xu Y, Zhang W, Tian Z, Lin L (2009) Ionothermal synthesis of gallophosphate molecular sieves in 1-alkyl-3-methyl imidazolium bromide ionic liquids. Microporous Mesoporous Mater 120:278–284

83. Liu L, Kong Y, Xu H, Li JP, Dong JX, Lin Z (2008) Ionothermal synthesis of a three-dimensional zinc phosphate with DFT topology using unstable deep-eutectic solvent as template-delivery agent. Microporous Mesoporous Mater 115:624–628
84. Liu L, Wragg DS, Zhang H, Kong Y, Byrne PJ, Prior TJ, Warren JE, Lin Z, Dong J, Morris RE (2009) Ionothermal synthesis, structure and characterization of three-dimensional zinc phosphates. Dalton Trans (34):6715–6718
85. Bieniok A, Brendel U, Sereni P, Musso M (2013) Raman spectroscopy and crystal structure investigation of solvo- and ionothermally prepared microporous metal-aluminophosphates with the laumontite framework structure. Z Krist 228:374–381
86. Pei R, Xu Y, Wei Y, Wen G, Li K, Wang L, Ma H, Tian Z, Lin L (2010) The cooperative templating effect of organic amine in the ionothermal synthesis of LTA type aluminophosphate molecular sieves. Chin J Catal 31:1083–1089
87. Han L, Wang Y, Zhang S, Lu XM (2008) Ionothermal synthesis of microporous aluminum and gallium phosphates. J Cryst Growth 311:167–171
88. Parnham ER, Drylie EA, Wheatley PS, Slawin AMZ, Morris RE (2006) Ionothermal materials synthesis using unstable deep-eutectic solvents as template-delivery agents. Angew Chem Int Ed 45:4962–4966
89. Baerlocher C, McCusker LB (2007) Database of zeolite structures. http://www.iza-structure.org/databases/
90. Lohmeier SJ (2011) Synthese und Charakterisierung von in ionischen Flüssigkeiten hergestellten Aluminiumphosphaten. Doctoral dissertation, Gottfried Wilhelm Leibniz Universität Hannover, Hannover
91. Ma Y-C, Xu Y-P, Wang S-J, Wang B-C, Tian Z-J, Yu J-Y, Lin L-W (2006) Synthesis of sodalite in room-temperature ionic liquid. Chem J Chin Univ 27:739–741
92. Yuen AKL, Masters AF, Maschmeyer T (2013) 1,3-Disubstituted imidazolium hydroxides: dry salts or wet carbenes? Catal Today 200:9–16
93. Liu L, Yang J, Li J, Dong J, Šišak D, Luzzatto M, McCusker LB (2011) Ionothermal synthesis and structure analysis of an open-framework zirconium phosphate with a high CO_2/CH_4 adsorption ratio. Angew Chem Int Ed 50:8139–8142
94. Zhang J, Wu T, Chen S, Feng P, Bu X (2009) Versatile structure-directing roles of deep-eutectic solvents and their implication in the generation of porosity and open metal sites for gas storage. Angew Chem Int Ed 48:3486–3490
95. Fu H, Qin C, Lu Y, Zhang Z-M, Li Y-G, Su Z-M, Li W-L, Wang E-B (2012) An ionothermal synthetic approach to porous polyoxometalate-based metal-organic frameworks. Angew Chem Int Ed 51:7985–7989
96. Liu J, Zhang F, Zou X, Yu G, Zhao N, Fan S, Zhu G (2013) Environmentally friendly synthesis of highly hydrophobic and stable MIL-53 MOF nanomaterials. Chem Commun 49:7430–7432
97. Lian J, Ma J, Duan X, Kim T, Li H, Zheng W (2010) One-step ionothermal synthesis of γ-Al_2O_3 mesoporous nanoflakes at low temperature. Chem Commun 46:2650–2652
98. Wang Y, Xu Y, Ma H, Xu R, Liu H, Li D, Tian Z (2014) Synthesis of ZIF-8 in a deep eutectic solvent using cooling-induced crystallisation. Microporous Mesoporous Mater 195:50–59
99. Cai R, Sun M, Chen Z, Munoz R, O'Neill C, Beving DE, Yan Y (2008) Ionothermal synthesis of oriented zeolite AEL films and their application as corrosion-resistant coatings. Angew Chem Int Ed 47:525–528
100. Li K, Tian Z, Li X, Xu R, Xu Y, Wang L, Ma H, Wang B, Lin L (2012) Ionothermal synthesis of aluminophosphate molecular sieve membranes through substrate surface conversion. Angew Chem Int Ed 51:4397–4400
101. Li K, Li X, Wang Y, Li D, Liu H, Xu R, Tian Z (2013) Ionothermal synthesis of AEL-type aluminophosphate molecular sieve membrane and its formation mechanism. Acta Chim Sin 71:573–578
102. Morris RE, Weigel SJ (1997) The synthesis of molecular sieves from non-aqueous solvents. Chem Soc Rev 26:309–317

103. Parnham ER, Morris RE (2006) Ionothermal synthesis using a hydrophobic ionic liquid as solvent in the preparation of a novel aluminophosphate chain structure. J Mater Chem 16:3682–3684
104. Morris RE, Bu X (2010) Induction of chiral porous solids containing only achiral building blocks. Nat Chem 2:353–361
105. Xing H, Li J, Yan W, Chen P, Jin Z, Yu J, Dai S, Xu R (2008) Cotemplating ionothermal synthesis of a new open-framework aluminophosphate with unique Al/P ratio of 6/7. Chem Mater 20:4179–4181
106. Martineau C, Bouchevreau B, Tian Z, Lohmeier S-J, Behrens P, Taulelle F (2011) Beyond the limits of X-ray powder diffraction: description of the nonperiodic subnetworks in aluminophosphate-cloverite by NMR crystallography. Chem Mater 23:4799–4809
107. Wei Y, Marler B, Zhang L, Tian Z, Graetsch H, Gies H (2012) Co-templating ionothermal synthesis and structure characterization of two new 2D layered aluminophosphates. Dalton Trans 41:12408–12415
108. Jurčík V, Wilhelm R (2005) An imidazolinium salt as ionic liquid for medium and strong bases. Green Chem 7:844–848
109. Kuperman A, Nadimi S, Oliver S, Ozin GA, Garcés JM, Olken MM (1993) Non-aqueous synthesis of giant crystals of zeolites and molecular sieves. Nature 365:239–242
110. Ren L, Wu Q, Yang C, Zhu L, Li C, Zhang P, Zhang H, Meng X, Xiao F-S (2012) Solvent-free synthesis of zeolites from solid raw materials. J Am Chem Soc 134:15173–15176
111. Jin Y, Sun Q, Qi G, Yang C, Xu J, Chen F, Meng X, Deng F, Xiao F-S (2013) Solvent-free synthesis of silicoaluminophosphate zeolites. Angew Chem Int Ed 52:9172–9175
112. Tian Y, McPherson MJ, Wheatley PS, Morris RE (2014) Ionic liquid assisted synthesis of zeolite-TON. Z Anorg Allg Chem 640:1177–1181

Chapter 3
Interlayer Expansion of the Layered Zeolites

Toshiyuki Yokoi and Takashi Tatsumi

Abstract Zeolites with three-dimensional (3D) open-framework structures are generally crystallized under hydrothermal conditions. In addition to this conventional route, the formation of the 3D structures can be achieved by the structural conversion of zeolitic hydrous layer silicates (HLSs) through topotactic dehydration–condensation of silanols on the HLSs. The interlayer spacings can be expanded by the interlayer silylation of zeolitic HLSs, forming interlayer-expanded zeolite (IEZ) materials. The IEZ materials are crystalline and show similar physical and chemical properties to the conventional 3D zeolites. Creating larger interlayer space will provide more open entrance for reactants and decrease the diffusion constrains in catalytic reaction. In this chapter, recent developments of the IEZ materials, in particular interlayer-expanded MWW-, FER-, and CDO-type zeolites, are featured.

Keywords Hydrous layer silicates • Silylation • Pillaring • Interlayer expansion • Interlayer-expanded zeolite • Ferrierite • PLS-1 • MCM-22

3.1 Introduction

Microporous crystalline zeolites have been utilized in many industrial technologies, including gas adsorption, ion exchange, separation, and catalysis for their unique porosity and high surface area [1–3]. Until now, more than 200 different framework topological structures have been known [4]; however, only some of which have important commercial values [5]. Nevertheless, zeolites are still attracting a wide research interest because novel frameworks may have new physicochemical properties and potential applications. So far, a variety of successful methods and strategies have been employed to synthesize zeolites with new topological structures, summarized in excellent papers [2, 6].

T. Yokoi (✉) • T. Tatsumi
Chemical Resources Laboratory, Tokyo Institute of Technology,
4259 Nagatsuta, Midori-ku, Yokohama 226-8503, Japan
e-mail: yokoi.t.ab@m.titech.ac.jp

F.-S. Xiao, X. Meng (eds.), *Zeolites in Sustainable Chemistry*,
Green Chemistry and Sustainable Technology,
DOI 10.1007/978-3-662-47395-5_3

Zeolites with three-dimensional (3D) open-framework structures are generally crystallized under hydrothermal conditions. In addition to this conventional route, the formation of 3D structures can be achieved by the structural conversion of zeolitic hydrous layer silicates (HLSs); zeolitic HLS is converted into a zeolite with 3D structures through topotactic dehydration–condensation of silanols on the HLS. For example, the layered precursors PREFER and MCM-22(P) have been converted into ferrierite (FER) and MCM-22 (MWW), respectively. To date, various topotactic relationships between HLSs and the corresponding 3-dimensional zeolite crystals such as PLS-1 (PLS-4, RUB-36, MCM-47, MCM-65, UZM-13, UZM-17, and UZM-19)-CDS-1 (CDO), NU-6(1)-Nu-6(2) (NSI), EU-19(P)-EU-19 (CAS-NSI), RUB-39-RUB-41 (RRO), RUB-18-RUB-24 (RWR), and RUB-15-sodalite (SOD) have been reported and summarized in excellent reviews [7, 8].

In heterogeneous catalysis, the accessibility to the active sites strongly affects a catalytic performance. This is rendered possible by introducing a second larger porosity level, i.e., by creating hierarchically porous structures. Pillaring between interlayer spacings is a useful way of intercalating guest species between interlayer spacings of layered materials [9]. The interlayer spacings can be supported by pillars after being expanded through a suitable intercalation process. Furthermore, pillaring can provide new pores in addition to original micropores.

Based on the concept of "pillaring," the periodic silylation using, e.g., diethoxydimethylsilane (DEDMS, $SiMe_2(OEt)_2$) and dichlorodimethylsilane (DCDMS, $SiMe_2Cl_2$), on the surface of HLS followed by connection between the interlayers leads to hierarchical structures consisting of two types of pores, namely, the original intralayer micropores and newly formed interlayer pores [10–14]. The pioneering work has been done by Kuroda and his co-workers; the silylation on well-known layered silicates (octosilicate, magadiite, and kenyaite) with alkoxychlorosilanes have been investigated; large flexible micropores were formed in the interlayers by pillaring with bulky organic molecules [15–18].

In 2007, a novel methodology of preparing an interlayer-expanded zeolite (IEZ) material was first developed through the interlayer silylation of zeolitic hydrous layer silicates, pure silica PLS-1, with DCDMS [19]. When PLS-1 was thermally treated, a zeolite denoted CDS-1 (IZA code: CDO) was obtained as a result of dehydration–condensation of the PLS layers [20]. The interlayer micropores of CDS-1 consist of two-dimensional straight 8-membered-ring (8MR) channels. Since such a small microporosity of CDS-1 inhibits typical organic molecules from adsorbing in their micropores, we have attempted to prepare an interlayer-expanded microporous material being capable of accommodating larger organic molecules through the interlayer silylation of pure silica PLS-1 with DCDMS and have found remarkable adsorption properties of the organic–inorganic hybrid microporous material IEZ-1 (Interlayer-Expanded Zeolite 1) in contrast to its purely inorganic analogue, IEZ-2, which was obtained after calcination of IEZ-1 at 500 °C. After this report, in 2008, we successfully reported a versatile method based on molecular alkoxysilylation of lamellar precursors such as MWW(P), PREFER, PLS-1, and MCM-47 for constructing new 3D zeolitic structures with expanded pore windows [21]. The IEZ products are crystalline and show physical and chemical properties comparable with microporous silicate zeolites and rather different from the pioneering work of Kuroda and his co-workers who applied similar strategies to layered silicates to obtain more or less amorphous products. Figure 3.1 depicts the strategy for converting zeolite layered precursors to new zeolite structure with the interlayer pores expanded, forming the IEZ materials.

3.2 Interlayer-Expanded MWW-Type Zeolite

The MWW-type zeolite has a structure which combines large cavities formed by 12-membered rings with a system of independent 10-membered-ring channels [22–25]. The MWW-type aluminosilicate, Al-MWW (so-called MCM-22), has found important industrial applications to the selective production of bulk petrochemicals such as cumene and ethylbenzene and has exhibited unique catalytic properties due to its unique pore structure and acidity [26–31]. Although potential advantages of the MWW-type materials are expected from their supercages and side pockets, access to the supercages is seriously restricted by the openings of the elliptical 10-MR pores. Therefore, the delamination and expansion of interlayer space are effective way to make the active sites in the supercages available for catalysis involving bulky substrate molecules as described below.

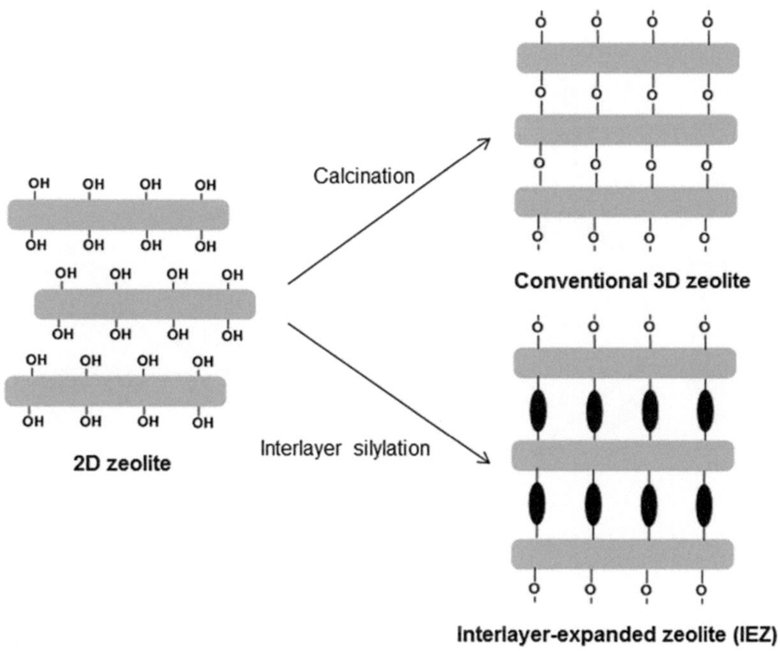

Fig. 3.1 Strategy for converting zeolite layered precursors to new zeolite structure with the interlayer pores expanded, forming the IEZ materials

3.2.1 Structural Diversity of MWW-Type Zeolite

The major structural interests in the MWW zeolites were its layered precursor and reversible structural conversion. The layered precursor of MWW zeolite (MWW(P)) can be swollen, exfoliated, and pillared under suitable conditions. In 1993, Kresge and Roth successfully synthesized MCM-36, a catalytically active material with a high surface area, by swelling the MWW (P) with tetrapropylammonium hydroxide (TPAOH) and cetyltrimethylammonium chloride (CTMACl) in base conditions followed by pillaring with silica species, e.g., TEOS [32, 33].

3.2.2 ITQ-2

Corma et al. successfully synthesized delaminated zeolite material, ITQ-2, by swelling the Al-MWW (P) with cetyltrimethylammonium bromide (CTMABr) [34, 35]. Figure 3.2 shows a scheme for the preparation of the different materials obtained from the MWW precursor. In this way, the solid thin sheets (approximately 2.5 nm high) with an extremely high external surface area (>700 m^2/g) were produced. These sheets consist of a hexagonal array of cups that penetrate into the sheet from both sides. These cups have an aperture of approximately 0.7 nm, formed by a 12-membered ring (12 MR). ITQ-2 showed greatly enhanced activities in acid-catalyzed reactions of large molecules [36–41]. This is due to the much higher well-defined external surface area, i.e., a larger number of cups present in the ITQ-2 structure, which in turn gives a larger amount of structurally accessible acid sites.

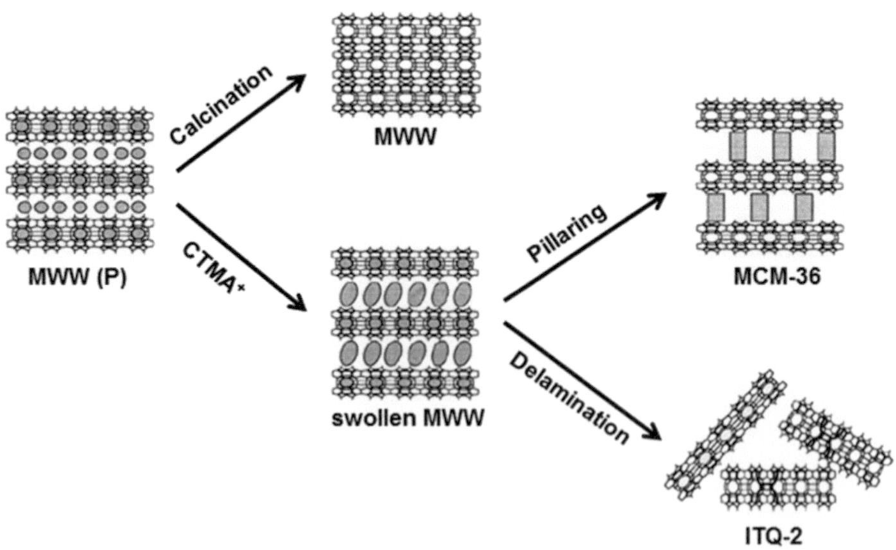

Fig. 3.2 Different materials obtained from the MWW-type zeolite precursor (Reprinted from Ref. [35], Copyright 1999, with permission from Elsevier)

3.2.3 *Al-IEZ-MWW*

The periodic silylation using diethoxydimethylsilane (DEDMS, $SiMe_2(OEt)_2$) or dichlorodimethylsilane (DCDMS, $SiMe_2Cl_2$) on the surface of layered silicates followed by connection between the interlayers leads to hierarchical structures consisting of two types of pores, namely, the original intralayer micropores and newly formed interlayer spacings. The treatment of Al-MWW(P) with DEDMS in acidic media leads to the formation of an aluminosilicate-type interlayer-expanded zeolite MWW (Al-IEZ-MWW) with expanded 12-membered-ring (12-MR) micropores (Fig. 3.3) [21]. By the interlayer silylation of Al-MWW(P), the micropore diameter of interlayers with 12-MR supercages enlarged from ca. 7.0 to 8.0 Å [21].

In the silylation, acid, e.g., HNO_3, enhances the extraction of the structure-directing agent (SDA) in the interlayer as well as the hydrolysis of the silylating agent followed by the condensation reaction of the silylating agent and the interlayer-silanol groups. The silylation process under acidic conditions simultaneously causes a dealumination from the MWW framework, resulting in the decrease in the acid amount; the Si/Al atomic ratios of Al-MWW(P) and Al-IEZ-MWW were 15 and 35, respectively [42, 43]. To suppress the dealumination, the vapor-phase silylation of Al-MWW(P) with dichlorodimethylsilane [42] and the preparation of Al-YNU-1 from deboronated MWW have been investigated [43]. We have developed the "two-step" silylation treatment with DEDMS via the first silylation in 0.1 M HNO_3 and the following silylation in 1.0 M NH_3 or water. This two-step silylation treatment successfully gave Al-IEZ-MWW that retains almost all the framework Al atoms [44]. Thus, prepared Al-IEZ-MWW serves as a useful acid catalyst for large molecules, e.g., the Friedel-Crafts acylation of anisole with acetic anhydride [42–44].

Very recently, we also found a different method for preparing the Al-IEZ-MWW without leaching of the Al species [45]. The strategy is to conduct the silylation under weakly acidic conditions; the interlayer silylation of Al-MWW(P) was performed by using DEDMS under reflux conditions in aqueous solution of 1.0 M ammonium salts such as NH_4NO_3, NH_4Cl, and CH_3COONH_4, instead of 1.0 M HNO_3. Furthermore, the combination of the silylation under weakly acidic conditions and the following acid treatment led to the formation of Al-IEZ-MWW with a

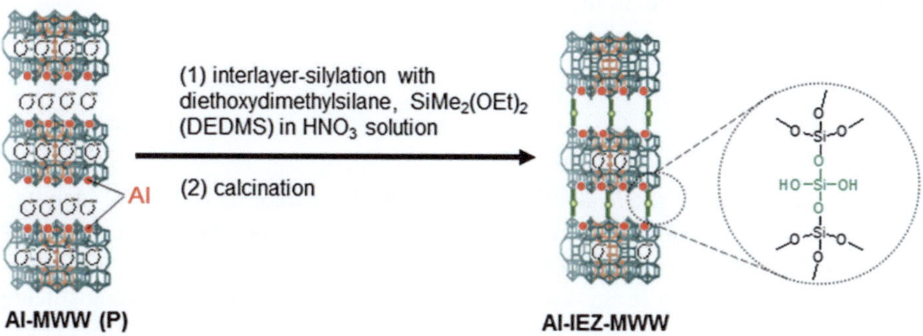

Fig. 3.3 Formation scheme of interlayer-expanded zeolite (IEZ)

high catalytic performance in the acylation of anisole compared to typical Al-MWW and Al-IEZ-MWW directly prepared under normal acidic conditions.

3.2.4 MWW-Type Titanosilicate

Since the discovery of TS-1 [46], "titanosilicate" zeolite catalysts have been of scientific and technological interest because of their ability to promote environmentally friendly catalytic oxidation of a variety of organic compounds [47]. TS-1, titanium-substituted silicalite-1, is one of the most efficient catalysts for the oxidation of a variety of organic molecules with hydrogen peroxide [48–53]. However, the pore opening of TS-1 limits its applications to small molecules, typically linear alkanes or alkenes, or monosubstituted aromatic compounds such as phenol, anisole, and aniline. One of the strategies to overcome the disadvantage of TS-1 is the development of large-pore titanosilicates to meet increasing demands for synthesizing fine chemicals in an environmentally friendly way. For this purpose, Ti-Beta, Ti-MTW, Ti-MOR, Ti-ITQ-7, Ti-MCM-68, and Ti-YNU-2, all with 12-membered-ring (12-MR) pores, as well as Ti-containing mesoporous materials have been prepared, and these efforts are summarized in recent review [54]. However, none of these catalysts are efficient enough in terms of intrinsic activity and stability except for Ti-MCM-68 and Ti-YNU-2 that require a very complex structure-directing agent.

Ti-MWW has proved to be much more active than TS-1 and Ti-Beta in the epoxidation of linear alkenes with H_2O_2 [55–61], but it is still inferior to Ti-Beta in the oxidation of bulky molecules owing to severe steric restrictions. Actually, many efforts based on the unique structural properties of the MWW zeolite have been made to improve the catalytic performance of Ti-MWW.

Corma et al. successfully prepared Ti-ITQ-2 by grafting titanocene on ITQ-2 material [62]. This material contains highly accessible and well-ordered titanium sites. Ti-ITQ-2 can be used as an oxidation catalyst for the epoxidation of cyclohexene with tert-butyl hydroperoxide (TBHP). Unfortunately, Ti-ITQ-2 only shows the catalytic performance when organic hydroperoxide oxidants such as TBHP are used as oxidant under extremely harsh conditions (i.e., the absence of water in the system). This is probably due to the presence of a large number of hydrophilic silanol groups that reside on the exterior surface (Fig. 3.4) [62].

3.2.5 Ti-YNU-1

Direct condensation of the as-synthesized lamellar precursor of Ti-MWW (denoted by Ti-MWW(P)) by calcination results in the formation of the MWW structure having a 10-MR interlayer pore (3D Ti-MWW). Interestingly, when Ti-MWW (P) was calcined after refluxing in an aqueous solution of HNO_3, the novel titanosilicate structurally analogous to the MWW precursor (designated as Ti-YNU-1) was

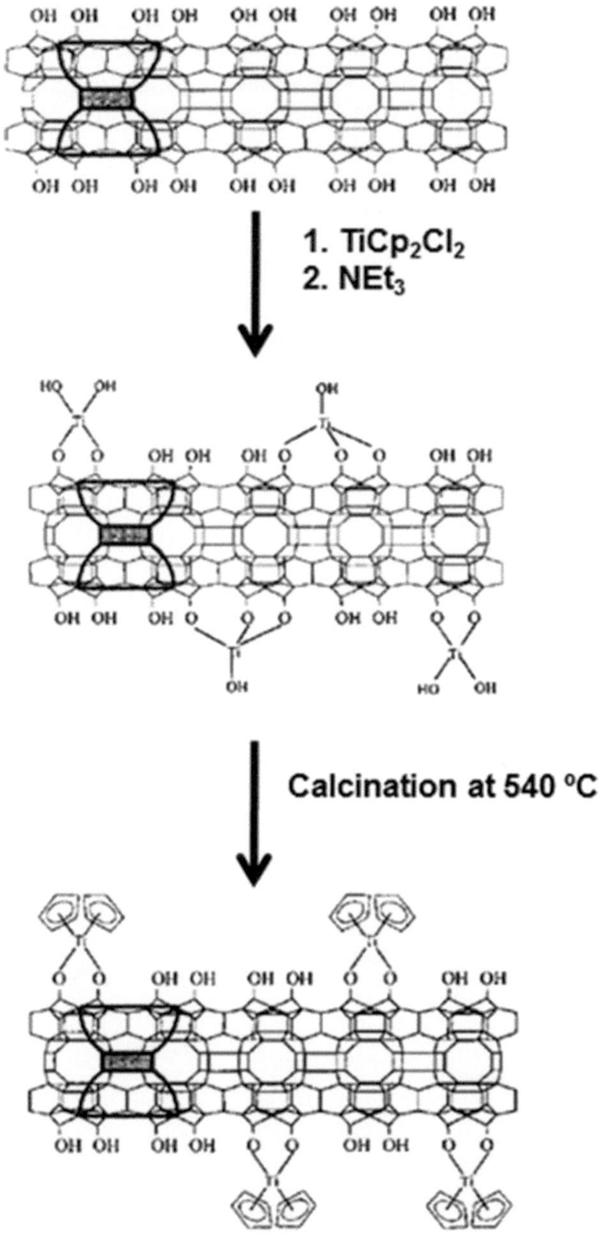

Fig. 3.4 Schematic grafting of titanocene on the ITQ-2 surface (Reproduced from Ref. [62] by permission of The Royal Society of Chemistry)

obtained (Fig. 3.5) [63–66]. Ti-YNU-1 shows much higher oxidation ability, epoxide selectivity, and stability than Ti-Beta in the oxidation of bulky cycloalkenes. TEM studies showed that Ti-YNU-1 has a large interlayer pore space corresponding to the 12-MR zeolite [65]. Since no Si source to expand the interlayer spacings has been added, it is assumed that silica "debris" formed by the decomposition of a part of the MWW layer acted as the Si source. Although Ti-YNU-1 exhibits a high catalytic performance, its preparation requires elaborated technical skill; a subtle difference in the conditions for acid treatment strongly affects the structural properties and catalytic performances, and its formation depends greatly on the Ti content of Ti-MWW (P). Recently, Moliner and Corma reported that the titanosilicate form of the expanded MWW-related material was synthesized starting from a pure silica precursor, ITQ-1, and that it showed a better activity than regular Ti-MWW in the selective epoxidation reaction of cyclohexene using H_2O_2 as oxidant [67].

3.2.6 *Ti-IEZ-MWW*

In 2009, Wu et al. have successfully developed interlayer-expanded MWW titanosilicate, designated as "Ti-IEZ-MWW," by a direct silylation of Ti-MWW(P) with diethoxydimethylsilane ($Si(OEt)_2Me_2$, DEDMS) under mild acidic conditions using 0.1 M HNO_3[68]. In acid solution, the hexamethyleneimine (HMI) as SDA molecule was extracted from the interlayer space, making the silane molecules

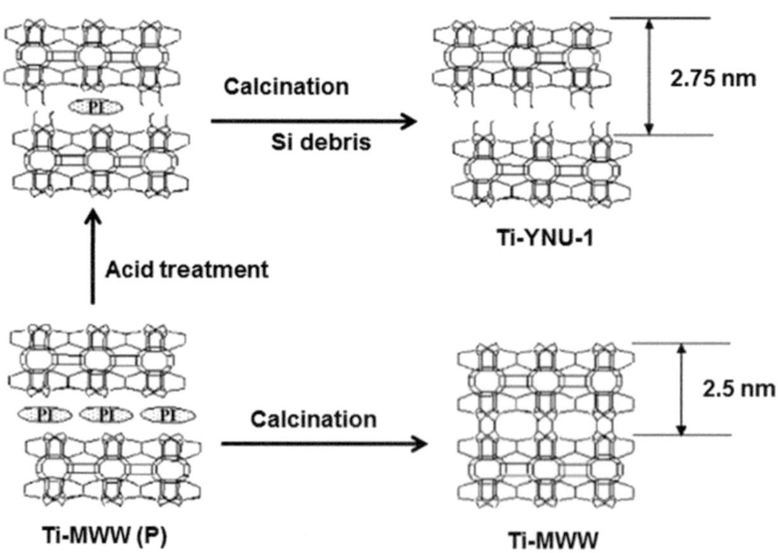

Fig. 3.5 Possible scheme for the formation of Ti-YNU-1 and 3D Ti-MWW, as well as their structures

accessible to the silanol groups on the layer surface to induce silylation. Ti-IEZ-MWW with a wide range of Si/Ti ratios (20–100) were prepared by changing the Ti content in Ti-MWW(P). Note that the structure of Ti-IEZ-MWW was very similar to that of previously reported Ti-YNU-1 [64, 65]. However, the present silylation method is independent of the Ti content of precursors and proves to be more feasible and useful for the preparation of large-pore titanosilicates.

They also investigated the catalytic properties of Ti-IEZ-MWW in the epoxidation of 1-hexene and cyclohexene with H_2O_2. 3D Ti-MWW and Ti-IEZ-MWW did not show significant differences in the epoxidation of 1-hexene because the molecules of linear 1-hexene are small enough to enter into both the interlayer and intralayer 10-MR channels of the 3D MWW structure. In contrast, there is a marked difference in the performance for the epoxidation of cyclohexene between the two catalysts; Ti-IEZ-MWW showed a much higher conversion than 3D Ti-MWW (Fig. 3.6) [68]. This is because cyclohexene molecules with a larger kinetic diameter hardly penetrate the distorted intralayer 10-MR channels.

It is an important characteristic of the MWW zeolite that structural interchange between 3D crystalline structure and the layered precursors is reversible. Recently, we found an effectiveness of reversible structural conversion (RSC) of the MWW zeolite in preparing Ti-IEZ-MWW with a high catalytic performance [69]. In our strategy, first, Ti-MWW(P) was treated in an aqueous solution of HNO_3 to remove octahedrally coordinated Ti species in Ti-MWW(P). The whole formation process of Ti-IEZ-MWW-RSC is depicted in Fig. 3.7. This acid treatment also led to the partial removal of piperidine molecules, which served as pillaring species between the MWW layers as well as the structure-directing agent for the MWW-type zeolite, forming the 3D MWW structure. Next, to cause the structural change from the 3D MWW structure to the original layered precursor, Ti-MWW(P)-AT was hydrothermally treated in an aqueous solution containing PI, forming a layered material

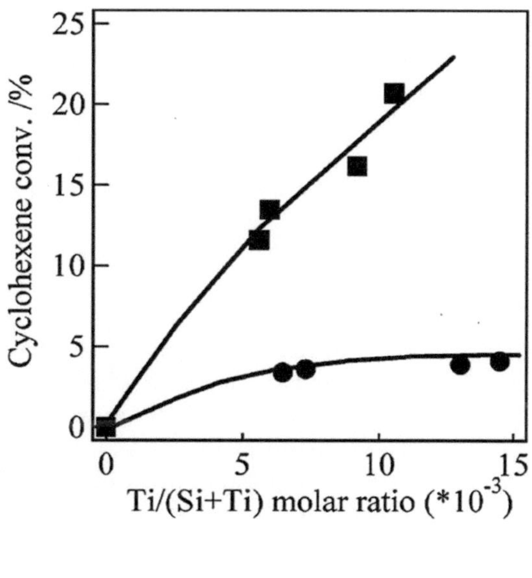

Fig. 3.6 Comparison of the catalytic activity between 3D Ti-MWW and IEZ-Ti-MWW with different Si/Ti mole ratios in the epoxidation of cyclohexene. Reaction conditions: cat., 0.05 g; cyclohexene, 10 mmol; H_2O_2, 10 mmol; MeCN, 10 mL; temp., 333 K; time, 2 h (Reproduced from Ref. [68] by permission of The Royal Society of Chemistry)

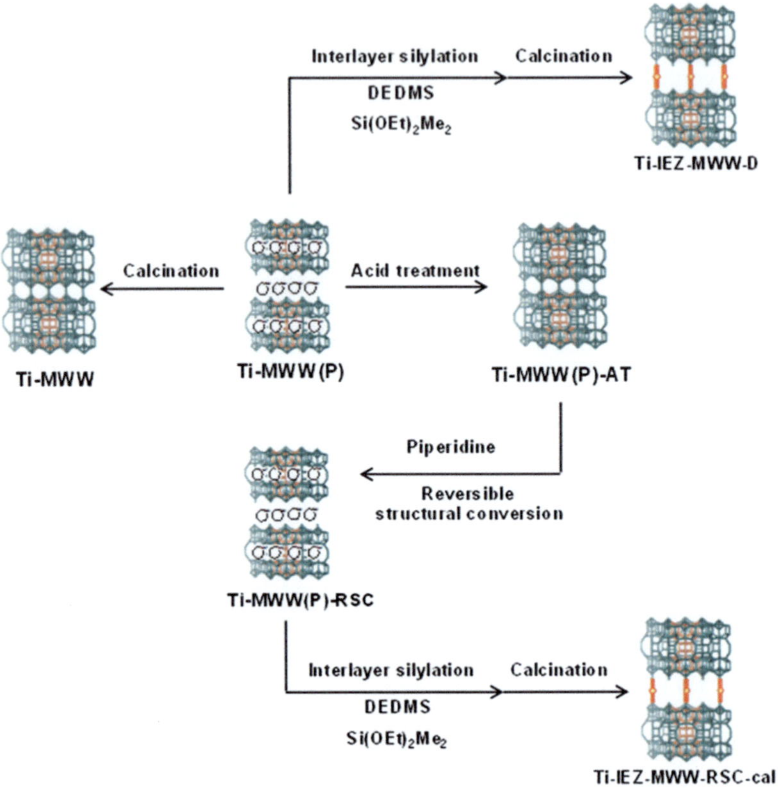

Fig. 3.7 The whole formation process of Ti-IEZ-MWW-RSC from Ti-MWW(P) (Reprinted from Ref. [69], Copyright 2014, with permission from Elsevier)

(Ti-MWW(P)-RSC), which has virtually the same structure as Ti-MWW(P). Finally, the interlayer silylation of Ti-MWW(P)-RSC was performed using DEDMS under acidic conditions (Ti-IEZ-MWW-RSC). Ti-IEZ-MWW-RSC contained unique framework Ti species as confirmed by UV-vis. spectroscopy; unlike typical Ti-MWW and Ti-IEZ-MWW, Ti-IEZ-MWW-RSC-cal contains the unique framework Ti atoms (TifA) as evidenced by the band at 210 nm in addition to typical ones (TifB) as evidenced by the band at 230 nm. The UV-vis. spectrum of Ti-IEZ-MWW prepared by Wu et al. [68] did not show a significant band at 210 nm (TifA), being consistent with our findings. TifA is completely linked with four T atoms, i.e., Ti(OSi)4, while TifB contains Ti-OH species, i.e., Ti(OSi)n(OH)4-n [70]. Ti-IEZ-MWW-RSC-cal proved to be the best in term of the turnover number (TON) for epoxidations of both cyclohexene and cyclooctene compared to Ti-IEZ-MWW directly prepared from Ti-MWW(P) as well as the conventional 3D Ti-MWW. We speculate that the existence of "TifA" in the MWW framework would be a key to the success in the preparation of Ti-IEZ-MWW catalyst with a unique catalytic performance and that the RSC process is efficient for introducing the framework Ti atoms that are linked with four T atoms, "TifA" (Fig. 3.8).

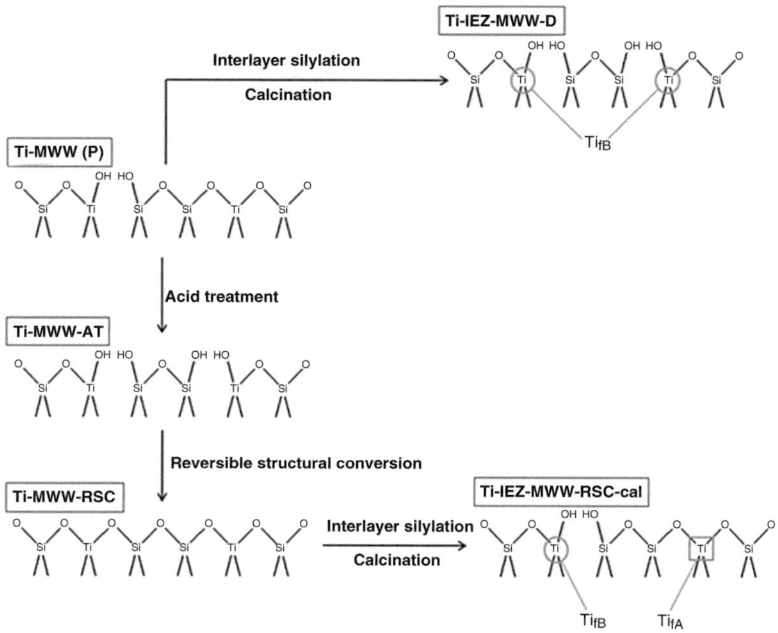

Fig. 3.8 Possible structures of Ti species in Ti-IEZ-MWW-RSC and related materials (Reprinted from Ref. [69], Copyright 2014, with permission from Elsevier)

3.2.7 Layered Zeolitic Organic–Inorganic Material Based on the MWW-Type Structure

When 1,4-bis(triethoxysilyl)benzene (BTEB) is employed as the organic agent, novel layered zeolitic organic–inorganic materials (MWW-BTEB) have been synthesized by intercalation and stabilization of acrylic silsesquioxane molecules between inorganic zeolitic MWW layers [71]. The preparation of MWW-BTEB with layered precursors of the MWW zeolite and intercalated aryl-bridged silsesquioxane molecules was performed after the swelling and intercalation of the layered precursors with cetyltrimethylammonium hydroxide solution (CTMA) and BTEB, respectively. The organic linkers are covalently bonded to the surface of the zeolitic layers by reaction of terminal alkoxide groups from disilanes with silanol groups from the inorganic layers (Fig. 3.9). Note that in this case, two condensed BTEB molecules were acted as pillars, and then the basal space was increased from 27 to 40.1 Å compared to the typical 3D-MWW structure.

Furthermore, post-synthesis treatments of the zeolitic hybrids have allowed the functionalization of the organic counterpart; amino groups can be introduced via nitration of the aryl groups derived from BTEB, resulting in bifunctional acid-base catalysts together with the acid sites generated by the presence of framework Al atoms in the individual MWW layers. Thus, obtained amino groups containing MWW-BTEB can be applied in the one-pot synthesis of benzylidene malononitrile

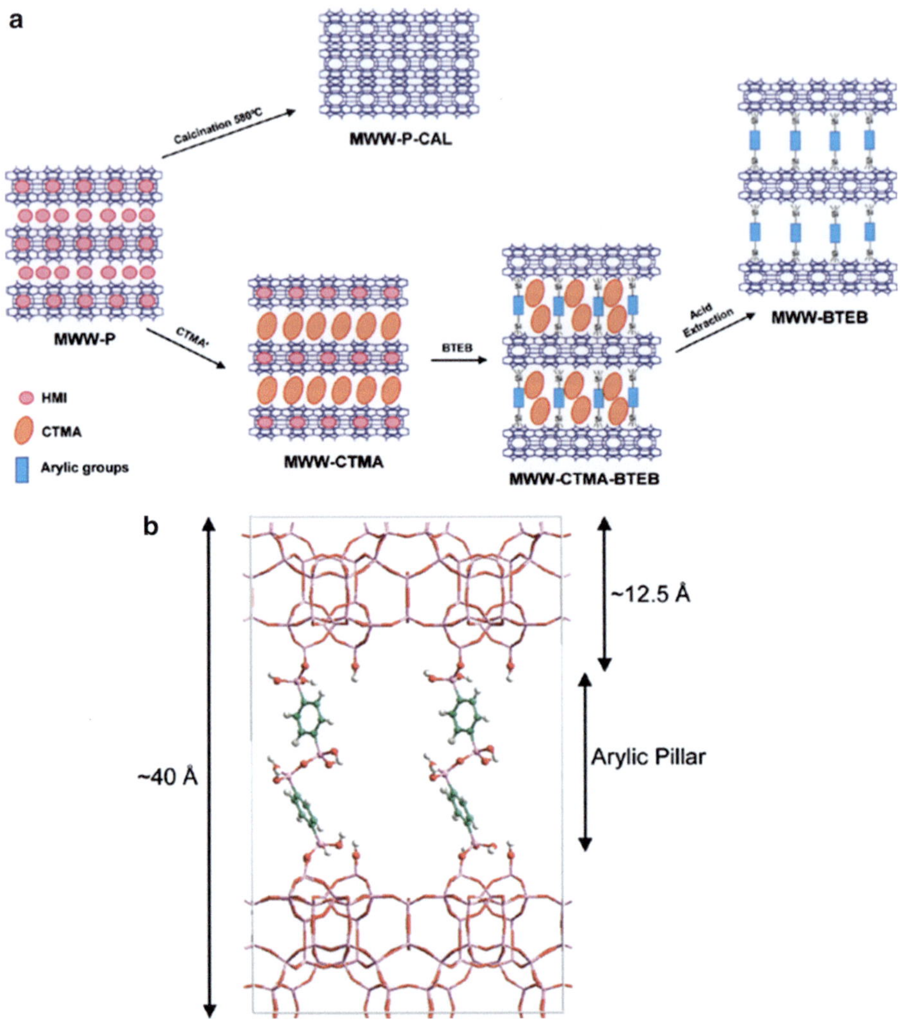

Fig. 3.9 Artistic representation of (**a**) methodology employed to obtain pillared hybrid zeolitic materials from MWW precursors and (**b**) layered hybrid material obtained by pillaring with BTEB silsesquioxane molecules (MWW-BTEB) (Reprinted with the permission from Ref. [71]. Copyright 2010 American Chemical Society)

from malononitrile and benzaldehyde dimethylacetal; the process involves the hydrolysis of the acetal catalyzed by the Brønsted acid sites present in the inorganic building blocks, followed by a Knoevenagel condensation reaction catalyzed by the basic sites located in the organic fragments which form the mesoporous gallery [71].

3.3 Interlayer-Expanded FER- and CDO-Type Zeolites

3.3.1 Introduction of IEZ-FER and IEZ-CDO

As described above, in 2007, we first reported a novel methodology of preparing the IEZ material (IEZ-1/-2) through the interlayer silylation of PLS-1 (CDO topology) with DCDMS [19]. Ruan et al. elucidated the structure of the IEZ-FER material obtained after the interlayer silylation of PREFER, which is a lamellar precursor of ferrierite (FER), with DEDMS [72]. IEZ-FER preserves the pentasil layers corresponding to that found in the well-known 3D FER zeolite, but it also shows a distinct expansion of the layer spacing (~5 Å). The newly formed channel system is distinct from that of the 3D FER structure, which has 8-membered ring (MR) and 10MR pores along the [010] and [001] directions, respectively; IEZ-FER exhibits 10MR and 12MR pores along the [010] and [001] directions, respectively, in the interlayer space. The formation of the large pores could be achieved by the insertion of monolayer of Si species between the layers.

Thus, the concept of the preparation of IEZ materials have been expanded to other zeolites obtained from zeolitic hydrous layer silicates (HLSs). The IEZ materials derived from the HLSs with ferrierite (FER) layers by silylation reaction have been called interlayer-expanded FER (IEZ-FER), or interlayer-expanded CDO (IEZ-CDO), referring to the FER or CDO zeolite framework structure, which has been obtained by topotactic condensation of FER layers from the HLSs with different stacking modes; for example, PREFER and RUB-36 lead to IEZ-FER and IEZ-CDO, respectively. The IEZ materials through the interlayer silylation of the zeolitic HLSs with FER-layers including PLS-1, PLS-3, PLS-4, RUB-36, and PREFER are described below.

3.3.2 COE-3/4

Inagaki et al. [19] and Wu et al. [21] reported on the preparation of the IEZ-CDO materials that have been obtained through an interlayer pore expansion reaction of PLS-1 [20] with dichlorodimethylsilane (DCDMS). In 2012, Gies et al. also reported the IEZ-CDO materials through an interlayer pore expansion reaction of a RUB-36 precursor with DCDMS [73]. Thus, obtained products have been named COE-3/-4 as reference to the International Network of Centers of Excellence, INCOE coordinated by BASF. The linker group (-O-Si(CH_3)2-O-) still has the two methyl groups in the as-synthesized form (material name COE-3 [$Si_{20}O_{38}(CH_3)_4$] for the silicate framework) rendering hydrophobic properties. COE-3 is thermally stable and can be calcined at 550 °C to yield a hydrophilic material COE-4 [$Si_{20}O_{38}(OH)_4$]. The COE-3/-4 materials have the expanded, two-dimensional 10-membered-ring (10-MR) pore system (Fig. 3.10).

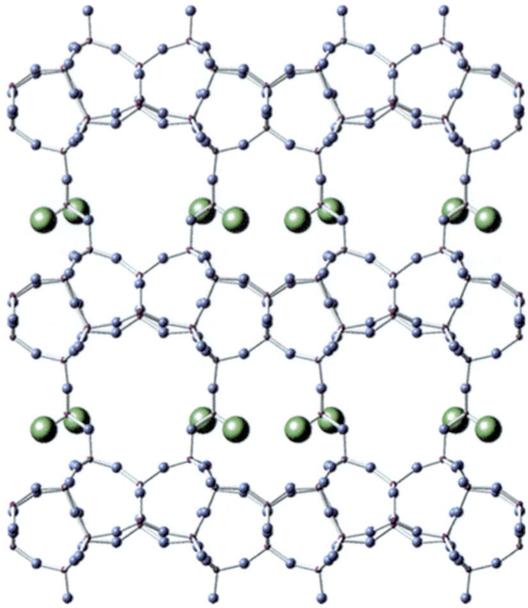

Fig. 3.10 Structure of IEZ-CDO (named "COE-3/-4"), interlayer-expanded zeolite obtained from RUB-36 (Reprinted with the permission from Ref. [73]. Copyright 2012 American Chemical Society)

3.3.3 Al-COE-3/4

INCOE group succeeded in preparing a ferrierite-type layered aluminosilicate (Al-RUB-36), and its IEZ materials denoted Al-COE-3 and Al-COE-4, which are solid acid catalysts with unique shape-selective nature [74]. They found that Pt-loaded Al-COE-4 exhibited a high catalytic performance in the decane hydroconversion compared to Pt/MgAPO-11 (AEL) and Pt/ZSM-22 (TON).

3.3.4 Ti-COE-3/-4

Furthermore, INCOE group has succeeded in preparing a ferrierite-type layered titanosilicate (Ti-RUB-36) using dimethyldiethylammonium (DMDEA) hydroxide as the organic structure-directing agent (OSDA) [75]. They also successfully prepared its IEZ materials named "Ti-COE-3" and "Ti-COE-4" (Fig. 3.11). Ti-RUB-36 is a lamellar titanosilicate with a structure similar to that of lamellar PLS-1 [20]. The Si/Ti ratios of Ti-COE- 3 and Ti-COE-4 are 149 and 150, respectively. The UV-vis. spectrum of Ti-RUB-36 showed one peak at 230 nm, indicating the presence of tetrahedrally coordinated framework Ti species, while both Ti-COE-3 and Ti-COE-4 showed an additional broad band centered at 360 nm, which was ascribed to TiO_2 species, but framework Ti species appearing at 230 nm were still predominant.

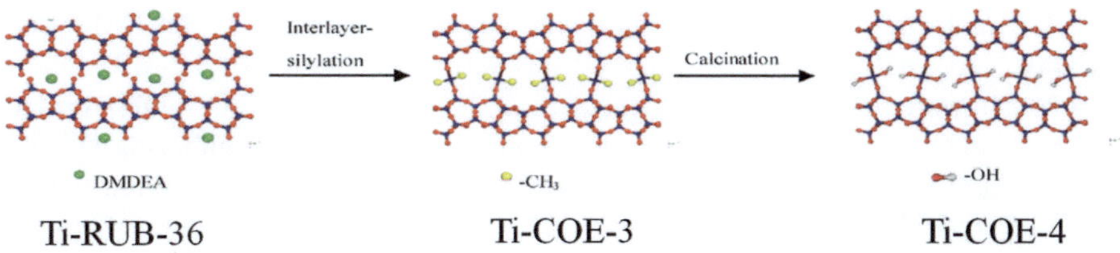

Fig. 3.11 Scheme of the transformation of Ti-RUB-36 into Ti-COE-3 and Ti-COE-4 (Reproduced from Ref. [75] by permission of John Wiley & Sons Ltd)

Ti-COE-3 and Ti-COE-4 exhibited a higher catalytic performance in the oxidation of cyclohexene with H_2O_2 compared to TS-1 [75]. Notably, Ti-COE-3 and Ti-COE-4 have the same structure but contain differently functionalized silicate bridging units. Ti-COE-4 is much more active than Ti-COE-3. They considered that the hydroxyl groups in Ti-COE-4 would support the adsorption of H_2O_2 during the reaction, resulting in a high catalytic performance in the cyclohexene conversion. This phenomenon is in good agreement with the results published previously in that hydrophilicity/hydrophobicity of titanosilicates strongly influences their catalytic activities [76, 77].

Very recently, the density functional theory (DFT) has been employed to study the framework stability and Brønsted acidity of the zeolite T-COE-4, in which the tetravalent Si is isomorphously substituted by a trivalent Fe, B, Ga, or Al heteroatom at the linker position. The influences of substitution energy and equilibrium geometry parameters on the stability of T-COE-4 have been investigated in detail [78]. There was no significant change in the interlayer distance of the Fe-, Al-, and Ga-substituted zeolites; however, insertion of B atoms led to the collapse of the layers. The substitution energies at the linker position increased in the order Al-COE-4 < Ga-COE-4 < Fe-COE-4 < B-COE-4. The relative Brønsted acid strength of the interlayer-expanded T-COE-4 zeolite decreases in the order of Al-COE-4 > Ga-COE-4 > Fe-COE-4 > B-COE-4. These findings may be helpful for the structural design and functional modification of interlayer-expanded zeolites.

3.3.5 *APZ-1–APZ-4*

In 2010, Ikeda et al. reported that layered silicates PLS-1, PLS-3, PLS-4, and PREFER, which consist of FER layer sheets, were successfully converted into new open-framework microporous materials by pillaring the interlayers with SiO_2(–OH)2 fragments, designated as APZ-1, APZ-2, APZ-3, and APZ-4, respectively [79]. They claimed that the fragments probably originated from the collapse of the crystalline structure and interlayer silylation occurred with the migration of the fragments (dissolution-condensation). Note that APLs can be prepared by the thermal acid (HCl) treatment of layered silicates at 170 °C for 24 h. The simple preparation method without using alkoxysilane has great merit in comparison with that for the IEZ materials prepared via interlayer silylation using, e.g., DCDMS and DCDMS. All APZs have 2D microporous structures with 10×10-MR and/or 12×10-MR pore openings, whose structures are constructed of a $SiO_2(-OH)_2$ bridge (Q2-structure) between terminal oxygen atoms (Q3 sites) of neighboring silicate layers. The stacking sequence of the ferrierite layer for each APZ is similar to that of the layered silicates used.

The APZs obtained exhibited high gas adsorption properties and narrow pore size distributions although the micropore structure is formed incompletely owing to the lack of Q2 pillars [79]. Furthermore, it has been clarified that APZs have a high hydrophilicity due to the existence of a large number of hydroxyl groups in micropores and that APZs showed a high thermal stability up to 600 °C without losing porosity [79].

3.3.6 Interlayer-Expanded PLS-4 Zeolite

Although APZ-n (n = 1–4) materials have the interlayer micropore structures, they cannot be always constructed completely because the Si species as a result of framework dissolution under the acid treatment are probably not enough in amount for creating the highly crystalline IEZ structure [79]. Thus, the introduction of additional Si sources to modify the interlayer micropore structures would be an alternative.

In 2013, Wu and his co-workers reported that "PLS-4" lamellar precursors, which is comprised of the FER layers and organic structure-directing agent of diethyldimethylammonium cations, were interlayer expanded by silylation with DEDMS (Me2Si(OEt)2) [80]. Thus, obtained interlayer-expanded material, named "PLS-4-sil" (as-made form), has a larger porosity than directly calcined PLS-4 with the CDO topology. The "PLS-4-sil" material can be calcined to change the pore openness and hydrophilicity/hydrophobicity. With the increase of calcination temperature, the silylated materials tend to adsorb hydrophilic molecules as well as bulky organic molecules.

3.4 Other Examples of Interlayer-Expanded Zeolites

RUB-39 was originally used as a case study to show that the condensation product "RUB-41" with the RRO topology has unique sorption properties and that the introduction of Al leads to a selective microporous acid catalyst [81]. INCOE group reported that the hydrous layer silicate RUB-39 has been subjected to interlayer expansion reaction using DCDMS at 180 °C, yielding new, crystalline microporous frameworks, IEZ-RRO named "COE-1" (Fig. 3.12). [82]. After calcination at 500 °C, the organic groups in COE-1 were completely removed, yielding the siliceous IEZ-RRO (COE-2).

The Nu-6(1) lamellar precursor with the NSI topology and its calcined derivative Nu-6(2) have been reported by Whittam [83]. Very recently, the inter-

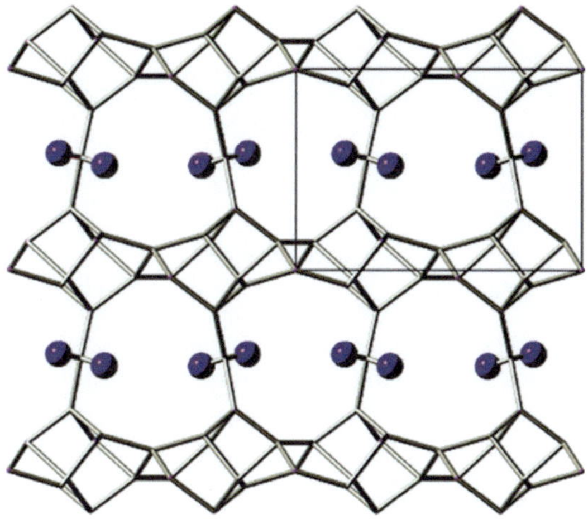

Fig. 3.12 Structure of IEZ-RRO (named "COE-1/-2"), interlayer-expanded zeolite obtained from RUB-39(Reprinted with the permission from Ref. [82]. Copyright 2011 American Chemical Society)

layer expansion of the NSI zeolite, "IEZ-Nu-6(1)" has been developed by the acid treating the Nu-6(1) in ethanol without an additional silane source by Wu and his co-workers [84]. The treatment of the as-made Nu-6(1) lamellar precursor with the HCl (2 M)-EtOH solution at 200 °C for only 30 min led to the partially delaminated analogue, denoted as "ECNU-4." It was then calcined at 550 °C for 6 h yielding the "ECNU-4-Cal" zeolite. Prolonging the HCl–EtOH treatment for 24 h, the IEZ-Nu-6(1) was obtained (Fig. 3.13). A continuous interlayer structural deconstruction and reconstruction would occur during the treatment, and the removable silicon species were finally inserted between the layers, yielding IEZ-Nu-6(1).

Wu and his co-workers also found that ECNU-4 can be swollen in the mixture of tetrapropylammonium hydroxide (TPAOH) and cetyltrimethylammonium bromide (CTAB) at room temperature under vigorous stirring. The swollen material named "swollen ECNU-4" was further sonicated to yield a more deeply delaminated zeolite, "Del-Nu-6" (Fig. 3.14) [84]. ECNU-4 and Del-Nu-6 differed in delamination degree, structural disordering, and textural properties, especially surface area.

Rojas and Camblor have developed a new organosilicate layered material, HPM-2 [85]. Upon calcination, HPM-2 transforms into the pure silica MTF zeolite by topotactic condensation. The interlayer expansion reaction of HPM-2 with DMDCD was apparently successful, but the obtained material is non-microporous unlike typical IEZ materials. This is likely due to an unfavorable disposition of silanols in close couples within each layer. But the silanol condensation between the newly incorporated silicon species occurs across the main window producing narrower (6 MR) rather than wider (10MR) windows (Fig. 3.15) [85]. However, this structural change may contribute to the development of zeolitic materials with new pore structure.

Very recently, a top-down strategy that involves the disassembly of a parent layered zeolite, UTL, and its reassembly into two complex zeolites with targeted topologies, IPC-2 and IPC-4, has been developed [86].

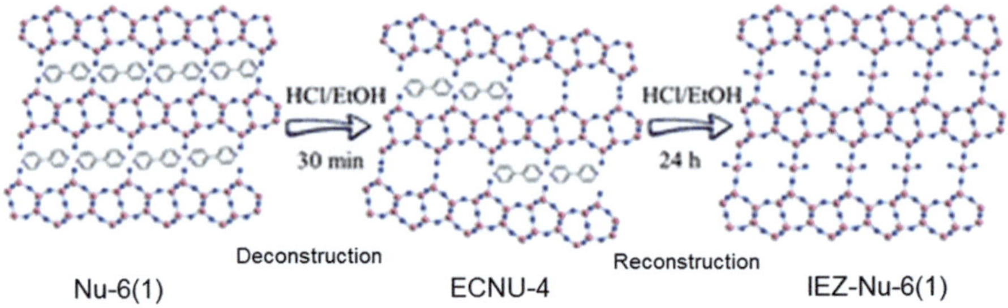

Fig. 3.13 Synthesis scheme of interlayer expansion of the **NSI** zeolite, "IEZ-Nu-6(1)" (Reproduced from Ref. [84] by permission of The Royal Society of Chemistry)

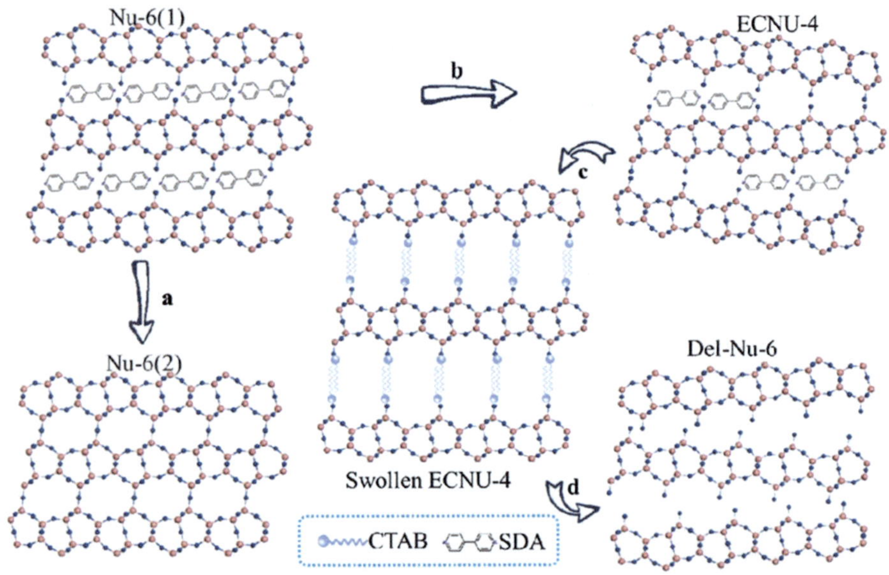

Fig. 3.14 Structural diversity of Nu-6(1) zeolite in posttreatment, including calcination (*a*), short-time treatment in HCl–EtOH solution (*b*), swelling (*c*), and delamination (*d*) (Reproduced from Ref. [84] by permission of The Royal Society of Chemistry)

3.5 Conclusions and Outlook

The synthesis of 3D microporous zeolite materials from layered zeolites opened the potential possibility to insert different heteroatoms into the layered zeolites at the linker positions. Different silylating agents as linker groups should lead to larger pore openings. In 2014, Wu and his co-workers reported that dimeric silanes, ClMe2Si-SiMe2Cl, were applied to serve as pillars between the MWW layers to construct larger interlayer space compared to that by using dichlorodimethylsilane

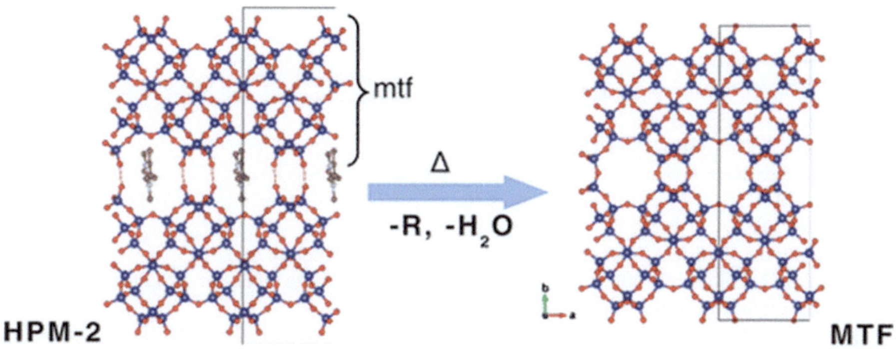

Fig. 3.15 HPM-2, a new organosilicate layered material synthesized by the fluoride route using 2-ethyl-1,3,4-trimethylimidazolium (Reprinted with the permission from Ref. [85]. Copyright 2014 American Chemical Society)

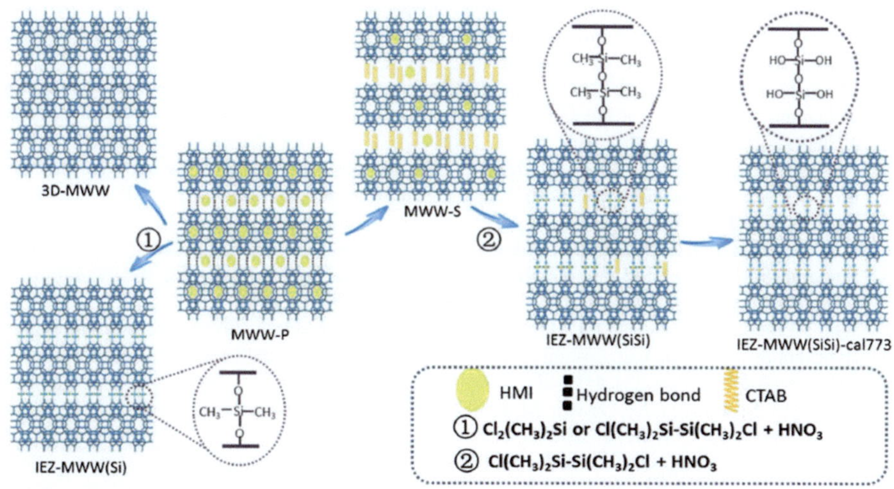

Fig. 3.16 Illustration of post-syntheses of interlayer-expanded MWW zeolites through pre-swelling-assisted silylation (Reprinted from Ref. [87], Copyright 2014, with permission from Elsevier)

(DCDMS, $SiMe_2Cl_2$) (Fig. 3.16) [87]. Thus, creating larger interlayer space will provide more open entrance for reactants and decrease the diffusion constrains in catalytic reaction. Replacing the methyl group with other functional groups at the linker position can add new functionalities accessible in the void space for chemical, catalytic reactions. Thus, the chemistry of interlayer-expanded zeolites will demonstrate the development of microporous materials with new physicochemical properties and functions.

References

1. Corma A (1995) Inorganic solid acids and their use in acid-catalyzed hydrocarbon reactions. Chem Rev 95:559–614
2. Cundy CS, Cox PA (2003) The hydrothermal synthesis of zeolites: history and development from the earliest days to the present time. Chem Rev 103:663–702
3. Corma A (2003) State of the art and future challenges of zeolites as catalysts. J Catal 216:298–312
4. International Zeolite Association. URL: http://www.iza-structure.org/
5. Bellussi G, Pollesel P (2005) Industrial applications of zeolite catalysis: production and uses of light olefins. Stud Surf Sci Catal 158:1201–1212
6. Bellussi G, Carati A, Rizzo C, Millini R (2013) New trends in the synthesis of crystalline microporous materials. Catal Sci Technol 3:833–857
7. Roth WJ, Dorset DL (2011) Expanded view of zeolite structures and their variability based on layered nature of 3-D frameworks. Microporous Mesoporous Mater 142:32–36
8. Roth WJ, Nachtigall P, Morris RE, Čejka J (2014) Chem Rev (in press) dx.doi.org/10.1021/cr400600f
9. Tsapatsis M, Maheshwari S (2008) Pores by pillaring: not always a maze. Angew Chem Int Ed 47:4262–4263
10. Mochizuki D, Shimojima A, Imagawa T, Kuroda K (2005) Molecular manipulation of two-and three-dimensional silica nanostructures by alkoxysilylation of a layered silicate octosilicate and subsequent hydrolysis of alkoxy groups. J Am Chem Soc 127:7183–7191
11. Mochizuki D, Kuroda K (2006) Design of silicate nanostructures by interlayer alkoxysilylation of layered silicates (magadiite and kenyaite) and subsequent hydrolysis of alkoxy groups. New J Chem 30:277–284
12. Mochizuki D, Kuroda K (2006) Synthesis of microporous inorganic-organic hybrids from layered octosilicate by silylation with 1, 4-bis (trichloro-and dichloromethyl-silyl) benzenes. Chem Mater 18:5223–5229
13. Ishii R, Ikeda T, Itoh T, Ebina T, Yokoyama T, Hanaoka T, Mizukami F (2006) Synthesis of new microporous layered organic–inorganic hybrid nanocomposites by alkoxysilylation of a crystalline layered silicate, ilerite. J Mater Chem 16:4035–4043
14. Ishii R, Ikeda T, Mizukami F (2009) Preparation of a microporous layered organic–inorganic hybrid nanocomposite using p-aminotrimethoxysilane and a crystalline layered silicate, ilerite. J Colloid Interface Sci 331:417–424
15. Okutomo S, Kuroda K, Ogawa M (1999) Preparation and characterization of silylated-magadiites. Appl Clay Sci 15:253–264
16. Shimojima A, Mochizuki D, Kuroda K (2001) Synthesis of silylated derivatives of a layered polysilicate kanemite with mono-, di-, and trichloro(alkyl)silanes. Chem Mater 13:3603–3609
17. Mochizuki D, Shimojima A, Kuroda K (2002) Formation of a new crystalline silicate structure by grafting dialkoxysilyl groups on layered octosilicate. J Am Chem Soc 124:12082–12083
18. Fujita I, Kuroda K, Ogawa M (2003) Synthesis of interlamellar silylated derivatives of magadiite and the adsorption behavior for aliphatic alcohols. Chem Mater 15:3134–3141
19. Inagaki S, Yokoi T, Kubota Y, Tatsumi T (2007) Unique adsorption properties of organic–inorganic hybrid zeolite IEZ-1 with dimethylsilylene moieties. Chem Commun 5188–5190
20. Ikeda T, Akiyama Y, Oumi Y, Kawai A, Mizukami F (2004) The topotactic conversion of a novel layered silicate into a new framework zeolite. Angew Chem Int Ed 43:4892–4896
21. Wu P, Ruan J, Wang L, Wu L, Wang Y, Liu Y, Fan W, He M, Terasaki O, Tatsumi T (2008) Methodology for synthesizing crystalline metallosilicates with expanded pore windows through molecular alkoxysilylation of zeolitic lamellar precursors. J Am Chem Soc 130:8178–8187
22. Rubin M, Chu P (1990) US Patent 4,954,325, to Mobil Oil Corporation

23. Leonowicz M, Lawton J, Lawton S, Rubin M (1994) MCM-22: a molecular sieve with two independent multidimensional channel systems. Science 264:1910–1913
24. Lawton S, Leonowicz M, Partridge R, Chu P, Rubin M (1998) Twelve-ring pockets on the external surface of MCM-22 crystals. Microporous Mesoporous Mater 23:109–117
25. Bevilacqua M, Meloni D, Sini F, Monaci R, Montanari T (2008) A study of the nature, strength, and accessibility of acid sites of H-MCM-22 zeolite. J Phys Chem C 112:9023–9033
26. Corma A, Gonzàlez-Alfaro V, Orchillès AV (1995) Catalytic cracking of alkanes on MCM-22 zeolite: comparison with ZSM-5 and beta zeolite and its possibility as an FCC cracking additive. Appl Catal A Gen 129:203–215
27. Kumar N, Lindfors L (1996) Synthesis, characterization and application of H-MCM-22, Ga-MCM-22 and Zn-MCM-22 zeolite catalysts in the aromatization of n-butane. Appl Catal A Gen 147:175–187
28. Corma A, Martinez-Triguero J (1997) The use of MCM-22 as a cracking zeolitic additive for FCC. J Catal 165:102–120
29. Corma A, Martínez-Soria V, Schnoeveld E (2000) Alkylation of benzene with short-chain olefins over MCM-22 zeolite: catalytic behaviour and kinetic mechanism. J Catal 192:163–173
30. Degnan TF Jr, Morris Smith C, Venkat Chaya R (2001) Alkylation of aromatics with ethylene and propylene: recent developments in commercial processes. Appl Catal A Gen 221:283–294
31. Perego C, Ingallina P (2002) Recent advances in the industrial alkylation of aromatics: new catalysts and new processes. Catal Today 73:3–22
32. Kresge CT, Roth WJ (1993) Crystalline oxide material. US Patent 5,266,541, to Mobil Oil Corporation
33. Kresge CT, Roth WJ (1994) Composition of synthetic porous crystalline material, its synthesis. US Patent 5,278,115, to Mobil Oil Corporation
34. Corma A, Fornés V, Pergher SB, Maesen Th LM, Buglass JG (1998) Delaminated zeolite precursors as selective acidic catalysts. Nature 396:353–356
35. Corma A, Fornés V, Martinez-Triguero J, Pergher S, Maesen Th LM (1999) Delaminated zeolites: combining the benefits of zeolites and mesoporous materials for catalytic uses. J Catal 186:57–63
36. Maheshwari S, Martínez C, Portilla MT, Llopis FJ, Corma A, Tsapatsis M (2010) Influence of layer structure preservation on the catalytic properties of the pillared zeolite MCM-36. J Catal 272:298–308
37. Corma A, Diaz U, Fornés V, Guil JM, Martínez-Triguero J, Creyghton EJ (2000) Characterization and catalytic activity of MCM-22 and MCM-56 compared with ITQ-2. J Catal 191:218–224
38. Corma A, Fornés V, Guil JM, Pergher S, Maesen Th LM, Buglass JG (2000) Microporous Mesoporous Mater 38:301
39. Aguilar J, Pergher SBC, Detoni C, Corma A, Melo FV, Sastre E (2008) Alkylation of biphenyl with propylene using MCM-22 and ITQ-2 zeolites. Catal Today 133–135:667–672
40. Jung H, Park S, Shin C, Park Y, Hong S (2007) Comparative catalytic studies on the conversion of 1-butene and n-butane to isobutene over MCM-22 and ITQ-2 zeolites. J Catal 245:65–74
41. Inagaki S, Kamino K, Kikuchi E, Matsukata M (2007) Shape selectivity of MWW-type aluminosilicate zeolites in the alkylation of toluene with methanol. Appl Catal A Gen 318:22–27
42. Inagaki S, Tatsumi T (2009) Vapor-phase silylation for the construction of monomeric silica puncheons in the interlayer micropores of Al-MWW layered precursor. Chem Commun 2583–2585
43. Fan W, Wei S, Yokoi T, Inagaki S, Li J, Wang J, Kondo JN, Tatsumi T (2009) Synthesis, characterization, and catalytic properties of H-Al-YNU-1 and H-Al-MWW with different Si/Al ratios. J Catal 266:268–278
44. Inagaki S, Imai H, Tsujiuchi S, Yakushiji H, Yokoi T, Tatsumi T (2011) Enhancement of catalytic properties of interlayer-expanded zeolite Al-MWW via the control of interlayer silylation conditions. Microporous Mesoporous Mater 142:354–362

45. Yokoi T, Mizuno S, Imai H, Tatsumi T (in press) Dalton Trans. doi:10.1039/C4DT00352G
46. Taramasso M, Perego G, Notari B (1983) Preparation of porous crystalline synthetic material comprised of silicon and titanium oxides. US Patent 4,410,501
47. Notari B (1996) Microporous crystalline titanium silicates. Adv Catal 41:253–334
48. Tatsumi T, Nakamura M, Negishi S, Tominaga H (1990) Shape-selective oxidation of alkanes with H_2O_2 catalysed by titanosilicate. Chem Commun 476–477
49. Thangraj A, Kumar R, Ratnasamy P (1991) Catalytic properties of crystalline titanium silicalites III. Ammoximation of cyclohexanone. J Catal 131:294–297
50. Tatsumi T, Yako M, Nakamura M, Yuhara Y, Tominaga H (1993) Effect of alkene structure on selectivity in the oxidation of unsaturated alcohols with titanium silicalite-1 catalyst. J Mol Catal 78:L41–L45
51. Sonawane HR, Pol AV, Moghe PP, Biswas SS (1994) Selective catalytic oxidation of arylamines to azoxybenzenes with H_2O_2 over Zeolites. J Chem Soc Chem Commun 1215–1216
52. Kumar SB, Mirajkar SP, Paris GCG, Kumar P, Kumar R (1995) Epoxidation of styrene over a titanium silicate molecular sieve TS1 using dilute H_2O_2 as oxidizing agent. J Catal 156:163–166
53. Hulea V, Moreau P (1996) The solvent effect in the sulfoxidation of thioethers by hydrogen peroxide using Ti-containing zeolites as catalysts. J Mol Catal A 113:499–505
54. Moliner M, Corma A (2014) Advances in the synthesis of titanosilicates: from the medium pore TS-1 zeolite to highly-accessible ordered materials. Microporous Mesoporous Mater 189:31–40
55. Blasco T, Corma A, Navarro MT, Perez Pariente J (1995) Synthesis, characterization, and catalytic activity of Ti-MCM-41 structures. J Catal 156:65–74
56. Newalker BL, Olanrewaju J, Komarneni S (2001) Direct synthesis of titanium-substituted mesoporous SBA15 molecular sieve under microwave hydrothermal conditions. Chem Mater 13:552–557
57. Koyano K, Tatsumi T (1996) Synthesis of titanium-containing mesoporous molecular sieves with a cubic structure. Chem Commun 145–146
58. Wu P, Tatsumi T, Komatsu T, Yashima T (2000) A novel titanosilicate with MWW structure: II. Catalytic properties in the selective oxidation of alkenes. Chem Lett 202:774–775
59. Wu P, Tatsumi T (2001) Extremely high trans selectivity of Ti-MWW in epoxidation of alkenes with hydrogen peroxide. Chem Commun 897–898
60. Wu P, Tatsumi T, Komatsu T, Yashima T (2001) A novel titanosilicate with MWW structure: II. Catalytic properties in the selective oxidation of alkenes. J Catal 202:245–255
61. Wu P, Tatsumi T (2002) Unique trans-selectivity of Ti-MWW in epoxidation of cis/trans-alkenes with hydrogen peroxide. J Phys Chem B 106:748–753
62. Corma A, Díaz U, Fornés V, Jordá JL, Domine M, Rey F (1999) Ti/ITQ-2, a new material highly active and selective for the epoxidation of olefins with organic hydroperoxides. Chem Commun 779–780
63. Nuntasri D, Wu P, Tatsumi T (2003) Highly active delaminated Ti-MWW for epoxidation of bulky cycloalkenes with hydrogen peroxide. Chem Lett 32:326–327
64. Wu P, Nuntasri D, Ruan J, Liu Y, He M, Fan W, Terasaki O, Tatsumi T (2004) Delamination of Ti-MWW and high efficiency in epoxidation of alkenes with various molecular sizes. J Phys Chem B 108:19126–19131
65. Ruan J, Wu P, Slater B, Terasaki O (2005) Structure elucidation of the highly active titanosilicate catalyst Ti-YNU-1. Angew Chem Int Ed 44:6719–6723
66. Fan W, Wu P, Namba S, Tatsumi T (2006) Synthesis and catalytic properties of a new titanosilicate molecular sieve with the structure analogous to MWW-type lamellar precursor. J Catal 243:183–191

Chapter 4
Mesostructured Zeolites

Ryong Ryoo, Kanghee Cho, and Filipe Marques Mota

Abstract Mesoporous materials constructed with microporous zeolitic frameworks (i.e., mesoporous zeolites) are of great interest owing to the very short diffusion path lengths across thin zeolite layers and the presence of large external surfaces containing strong Brønsted acid sites. These characteristics of mesoporous zeolites are highly advantageous for a wide range of applications, particularly in heterogeneous catalysis. The mesoporous materials show unprecedentedly high catalytic performances (e.g., high catalytic conversion and catalytic longevity) as zeolites in various petrochemical reactions and fine-chemical organic reactions and especially in reactions involving bulky molecules. In this chapter, we describe the various methods currently available for the synthesis of mesoporous zeolites.

Keywords Hierarchical zeolite • Mesoporous zeolite • Nanosheets • Nanosponge • Synthetic methods • Demetallation • Hard-templating • Soft-templating • Heterogeneous catalysis

4.1 Hierarchically Porous Structure: Efficient System for Facile Diffusion (or Transportation)

Conventionally, zeolites are synthesized in the form of micrometer-size particles. Even in such small crystallites, billions of micropores (usually <1 nm) exist in every dimension. Because the intracrystalline diffusion of guest molecules into the zeolite

R. Ryoo (✉)
Center for Nanomaterials and Chemical Reactions,
Institute for Basic Science (IBS), Daejeon 305-701, Republic of Korea

Department of Chemistry, Korea Advanced Institute of Science and Technology (KAIST), Daejeon 305-701, Republic of Korea
e-mail: rryoo@kaist.ac.kr

K. Cho • F.M. Mota
Center for Nanomaterials and Chemical Reactions,
Institute for Basic Science (IBS), Daejeon 305-701, Republic of Korea

F.-S. Xiao, X. Meng (eds.), *Zeolites in Sustainable Chemistry*,
Green Chemistry and Sustainable Technology,
DOI 10.1007/978-3-662-47395-5_4

particle often results in severe limitations, the full application of a zeolite is rarely achieved [1–3]. The slow diffusion inside a zeolite is responsible for diffusion limitations in many catalytic applications, with reaction rates limited by the diffusion rate. Some catalytic reactions involving molecules larger than the zeolite pore apertures are furthermore restricted to the external surface of the solid. In some cases, the diffusion of reactants and desired products is sufficiently fast, but the zeolite catalyst produces by-products that are polymerized into bulky coke species [3–5] that accumulate on the catalytic site and decrease the catalytic lifetime. In these cases, downsizing the zeolite particle size is helpful for improving by-product diffusion and enhancing the catalytic lifetime. To overcome this limitation, continuous exploratory work has led to the synthesis of zeolites possessing large micropores [6–17] and zeolites with nanoscale crystal size [18–21] and the generation of mesopores in zeolites [22–82]. Of these, the mesopore-generating synthesis approach has been considered the most promising as a versatile and inexpensive strategy. The nanometer-scale hierarchical pore network provides highly efficient diffusional pathways between zeolitic micropores and mesopores in a confined space. Mehlhorn et al. [83] reported the impact of the presence of mesopores in intracrystalline diffusion using a pulsed field gradient NMR technique. In this context, many research groups have proposed a wide number of synthesis methods for such mesoporous or mesostructured zeolites. To optimize the design of hierarchical zeolites, a quantitative correlation between pore topology (i.e., the amount, size, connectivity, and distribution of each porosity level) and the properties of the acid solid is imperative [84]. Successful synthesis methods are introduced in the following sections.

4.2 Post-synthetic Chemical Treatment to Generate Mesostructure in the Zeolite Crystals

With this method, the framework constituents (i.e., Al and Si) of a pre-synthesized and calcined zeolite with a bulk crystal morphology are partially dissolved by acids and bases. The dissolution condition is carefully controlled so that the dissolution can excavate intracrystalline mesoporous channels or void spaces into the zeolite crystal. The texture (e.g., mesopore volume and pore diameter) of the mesopores in the resultant zeolite samples varies considerably according to the detailed parameters, such as the chemical composition and structure of the pristine zeolite, solution temperature, acid or base concentration, and contact time of zeolite in the solution.

Post-synthetic demetallation can be classified into dealumination and desilication, depending on the treatment condition. Particularly, introduction of mesoporosity by dealumination has been industrialized as an economic post-synthesis modification procedure. Aluminum dissolution in high-Al zeolites can result in the generation of mesopores. The steam-thermal dealumination of Y zeolite has been well documented in the fields of fluid catalytic cracking and hydrocracking [25–31]. The procedure consists of sequential high-temperature steaming and subsequent acid leaching treatments. The resultant Y zeolite, referred to as an "ultrastable Y zeolite" (USY), displays high hydrothermal stability, strong acidity, and moderate mesoporosity. The treatment causes the generation of mesoporous channels (5–100 nm in pore diameter) in the zeolite crystal. As high intracrystalline mesoporosity is attained, samples often exhibit inaccessible isolated cavities that render part of the obtained mesoporosity ineffective for improving diffusion. This has been revealed by 3D-computer tomography using transmission electron micrograph (TEM) images and by N_2 adsorption and mercury porosimetry (Fig. 4.1). With the dissolution of Al species, the number of Brønsted acid sites significantly decreases. The dealuminated zeolite often exhibits extra-framework octahedral Al that is ineffectively removed during the acid leaching treatment. The dealumination method has been further applied to other types of zeolites including mazzite [32, 33], mordenite [34, 35], and ZSM-5 [36–38] under various hydrothermal treatment conditions. Similar to the hydrothermal dealumination, zeolites can be dealuminated using aqueous solutions containing ammonium hexafluorosilicate [39, 40] and $SiCl_4$ [40–43]. In this case, extracted framework Al species may be further replaced by Si atoms to obtain high-silica zeolites.

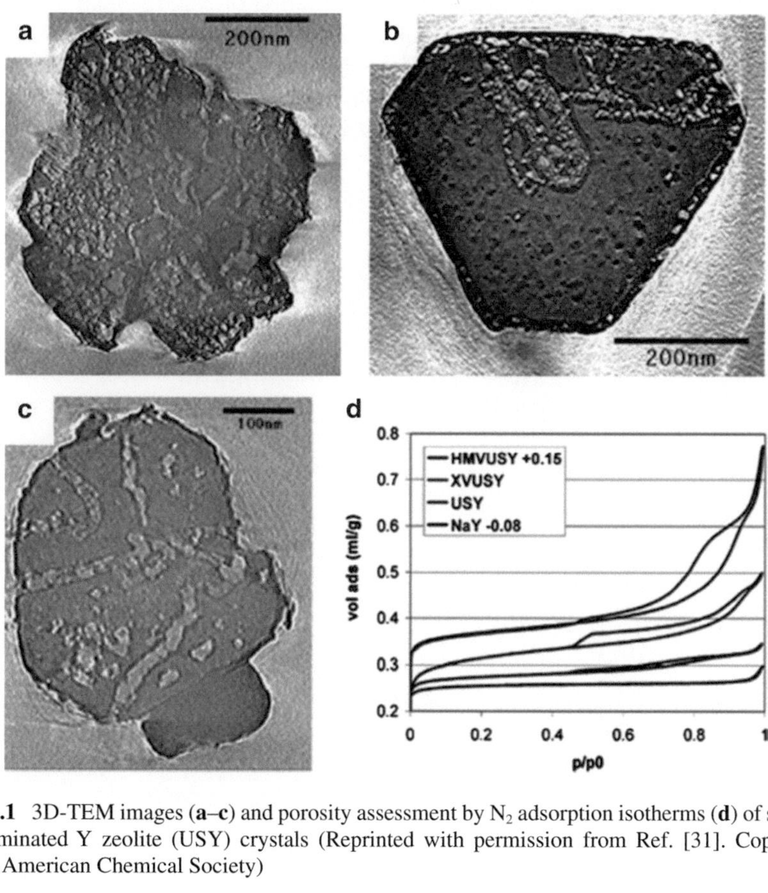

Fig. 4.1 3D-TEM images (**a–c**) and porosity assessment by N_2 adsorption isotherms (**d**) of steam-dealuminated Y zeolite (USY) crystals (Reprinted with permission from Ref. [31]. Copyright 2002, American Chemical Society)

In recent years, the generation of mesopores by alkaline-mediated leaching of framework Si atoms has become a widely applied approach. Post-synthesis desilication has demonstrated noteworthy versatility in the treatment of a wide number of zeolite frameworks, generating accessible mesopores and rendering sites in the crystal interior active. Highly mesoporous hierarchical MFI zeolites prepared by post-synthesis desilication have been obtained with an interconnected mesopore distribution in the range of 30–100 nm. The interconnected architecture is advantageous in reducing the limited access and diffusion within the microporosity of the zeolite. Desilication is performed with an alkali metal hydroxide solution and is particularly effective for high-silica zeolites (usually Si/Al > 20) [44–51]. Several reports have focused on the concentration of the alkaline solution, temperature, and time of the post-synthesis treatment or on the Si-to-Al molar ratio of the parent solid.

The degree of desilication depends on the content and distribution of Al in the pristine zeolite crystals [45, 48, 49]. To better illustrate the impact of the Si/Al ratio of the bulk zeolite, Pérez-Ramírez et al. treated a series of ZSM-5 zeolite samples with an Al content in the range of Si/Al = 20 ~ 100, using an aqueous solution containing 0.2 M NaOH [48]. As framework SiO_4^- sites bonded to Al atoms are resistant to the attack by bases, the concentration of NaOH failed to generate mesopores for Si-to-Al ratios lower than 20. In contrast, for Si/Al > 50, almost the entire portion of the zeolite was lost into the solution, whereas the framework silica quickly dissolved in an uncontrollable manner (Fig. 4.2). The intermediate between these two extreme cases, i.e., Si/Al = 25 ~ 50, was thus considered the

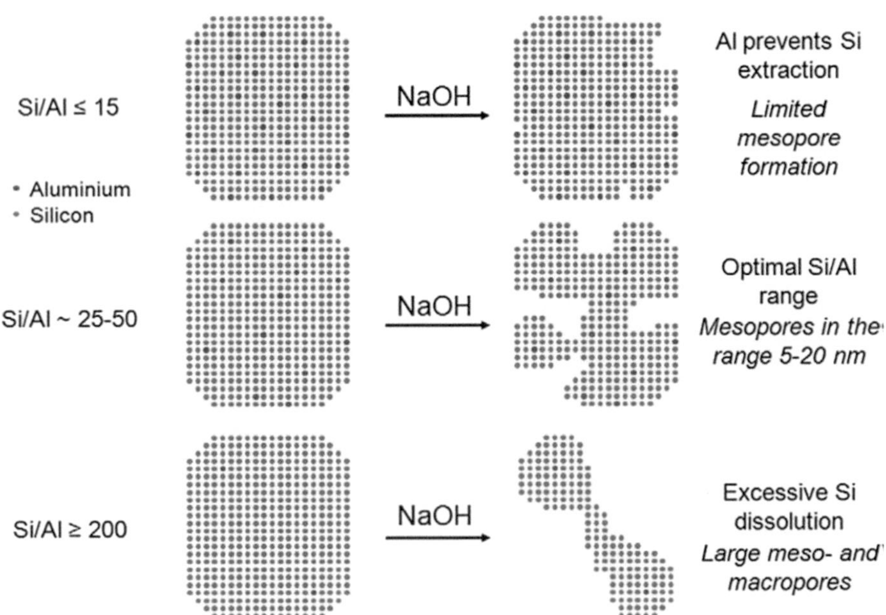

Fig. 4.2 Schematic representation of the influence of Al content on the desilication treatment of MFI zeolites in alkali solution (Reprinted with permission from Ref. [48]. Copyright 2004, American Chemical Society)

optimum range for the generation of mesopores through desilication under the selected conditions.

In addition, Al spatial distribution within a zeolite particle is a critical factor. When large ZSM-5 crystals with a single crystal platelike morphology are treated in a basic solution, the desilication treatment often results in the selective dissolution of the internal core of each zeolite particle. This selectivity is the result of the comparatively higher solubility of the silica-rich core compared to the Al-rich outer region. As the crystal size decreases, the Al distribution becomes progressively more homogeneous, and the dissolution accordingly can occur more homogeneously throughout the entire crystal. Based on this principle, Groen et al. prepared MFI zeolites possessing uniform mesoporosity, using submicron zeolite crystals instead of large bulk zeolite crystals [49]. Later, Chal et al. added ammonium surfactants such as cetyltrimethylammonium bromide (CTAB) in the desilication of the ZSM-5 [52]. The addition of CTAB was very effective in generating mesopores with a uniform diameter. CTAB is presumed to stabilize the zeolite frameworks by the surfaces of surfactant micelles that can form along the desilicated mesoporous channels. That is, the formation of surfactant micelles inside a freshly generated mesopore is believed to protect the surrounding zeolite frameworks from further desilication, which may otherwise occur in an uncontrollable manner.

Both post-synthetic demetallation treatments can be applied in sequential steps for the generation of mesopores, as de Jong et al. demonstrated in treating a zeolite Y [53]. The mesopore size distribution of the resultant zeolite was bimodal, with two peaks centered at 3 and 27 nm (Fig. 4.3). The smaller mesopores were interpreted to result from desilication while the larger mesopores were generated by dealumination. Thus, the zeolite product exhibited two kinds of mesopores with different pore diameters, in addition to the intrinsic zeolitic micropores. According to de Jong et al., this zeolite would be suitable as a catalyst for hydrocracking of bulky molecules (e.g., with a heavy oil component).

Desilication and dealumination are top-down approaches, by principle deforming part of the initial zeolite crystals during post-synthesis treatments. Both methods intrinsically change the framework composition of the zeolite, reduce the thermal and hydrothermal stabilities of the zeolites, and lose a significant amount of Brønsted acid sites. In addition, it is difficult to control the mesopore volume and diameters. An amorphous mesoporous silica phase can be generated from the dissolved silicate species, making it difficult to separate from the zeolite phase. Nevertheless, the dealumination and/or desilication method is currently most widely adopted for industrial applications because of its suitability for low-cost mass production [53].

4.3 Hard-Templating Strategy

Tailoring the shape size and connectivity of a mesoporous structure within a high-crystalline zeolite can be achieved by the addition of a hard template in the zeolite synthesis composition [54–80]. Originally, the principle behind the hard-templating approach relied on a confined zeolite crystallization within the pores or voids of a

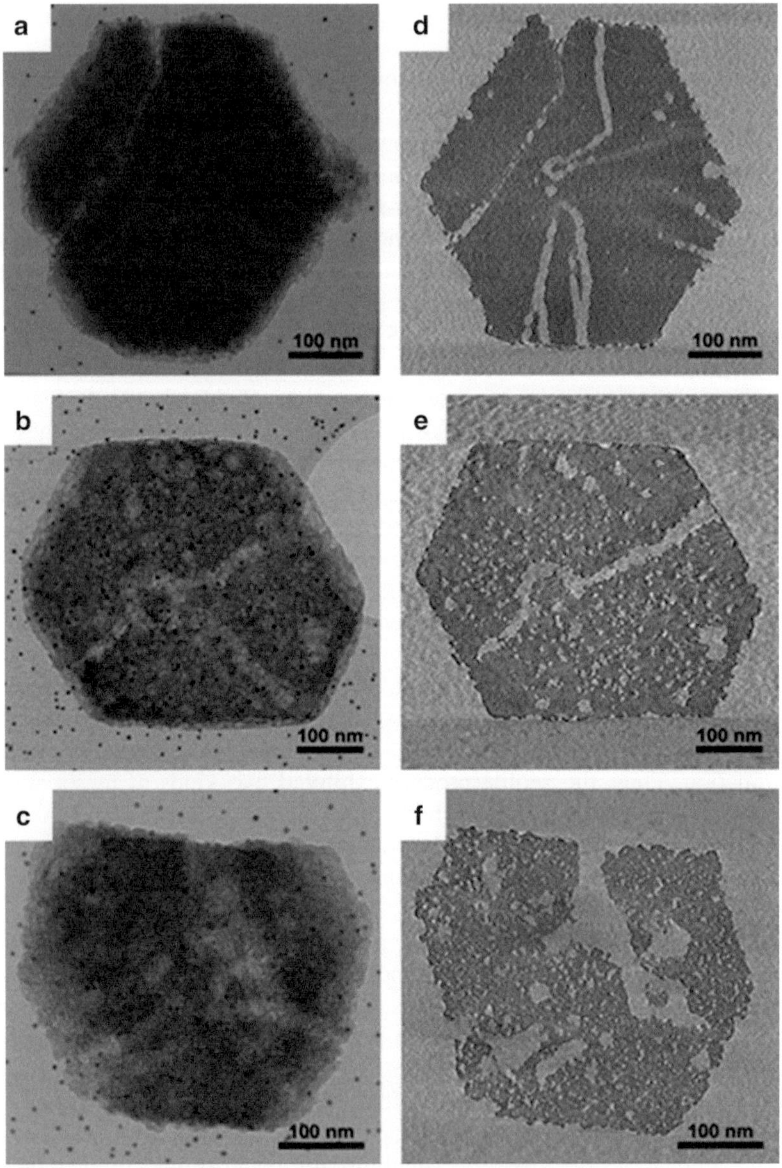

Fig. 4.3 Electron microscopy and electron tomography study of parent HY-30, base-leached HY-A (0.05m NaOH) and HY-B (0.10m NaOH) samples. The TEM micrographs show that base leaching of the parent HY-30 (**a**) leads to generation of more-porous structures as can be seen from HY-A (**b**) and HY-B (**c**), yet without revealing the true nature of the mesopore network. The numerical cross-sections through 3D reconstructions of the particles provided by electron tomography clearly depict the presence of both small (ca. 3 nm) and large (ca. 30 nm) mesopores, as well as their interconnectivity and shape: (**d**) 0.56 nm thick slice of HY-30; (**e**) 0.82 nm thick slice of HY-A; (**f**) 0.56 nm thick slice of HY-B sample (Reproduced from Ref. [53] by permission of John Wiley & Sons Ltd)

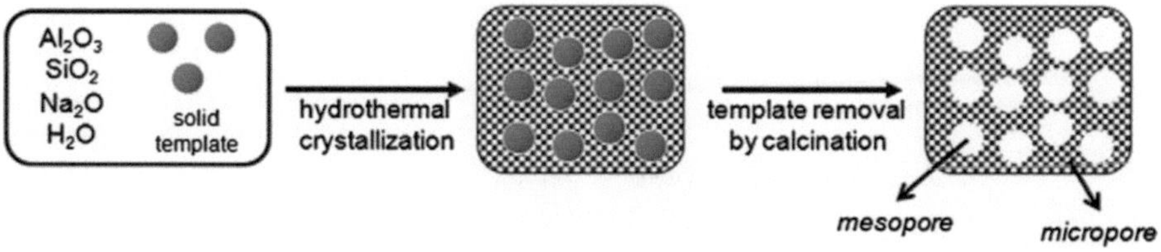

Fig. 4.4 Schematic representation of solid-templating synthesis strategy (Reprinted with permission from Ref. [54]. Copyright 2000, American Chemical Society)

nano-template. Depending on the selected synthesis conditions and rates of nucleation and crystal growth, hard-templating has further proved to generate mesoporous zeolite single crystals. The resulting mesostructure can be tailored through careful selection of hard templates with a suitable texture. Template incorporation within a zeolite and the zeolite crystallization within a nano-template are, however, thermodynamically unfavorable processes. Accordingly, crystallization typically occurs via a dry-gel synthesis route, inducing pseudomorphic crystallization of zeolites in a confined space. A high affinity between the hard templates and the zeolite framework and the appropriate synthesis conditions are required to prevent the isolation of the silicate gel with consequent formation of a non-mesoporous zeolite.

Most representative hard templates are nano-sized solids composed of carbonaceous species. Nanoporous materials, such as mesoporous or macroporous carbons and nonporous nanobeads or nanofibers, are generally employed. These mesopore-generating agents are able to retain their nanostructures during zeolite synthesis and be removed afterward via thermal decomposition (Fig. 4.4). Jacobsen et al. synthesized single crystal MFI zeolites using carbon nanoparticles as a hard template [54]. The sample exhibited an intracrystalline mesopore volume exceeding 1.0 $cm^3 g^{-1}$ and a uniform pore size comparable to the particle sizes of the carbon templates (Fig. 4.5). Similarly, Tao et al. [61] used a carbon aerogel monolith, obtained through the pyrolysis of resorcinol–formaldehyde resin [62–66, 77], to generate a uniform mesoporous MFI zeolite. The authors confirmed that mesoporous textural properties could be tuned according to the particle size of the carbon aerogel. The use of uniform carbon nano-templates additionally proved successful in the synthesis of various hierarchical zeolites (MEL [67], MTW [68], and BEA [69]) and aluminophosphate zeotypes (CHA and AFI [70]).

Zeolites containing ordered mesoporosity generated in the presence of ordered mesoporous carbons have been extensively reported in the literature. Tsapatsis et al. used a three-dimensionally ordered mesoporous (3DOM) carbon synthesized by replication of colloidal crystals composed of size-tunable silica nanoparticles (Fig. 4.6) [76]. The zeolite exhibited low-angle XRD peaks, suggesting the presence of an ordered mesoporous system. In agreement with the size of the particles, the framework thickness could be adjusted within a range of 10–40 nm. Holland et al. previously reported confined growth within the voids created by assembled polystyrene spheres [78]. The resulting zeolite contained macropores with a pore diameter (250 nm) determined by the bead size (Fig. 4.7). In recent years, CMK-n-type mesoporous carbons

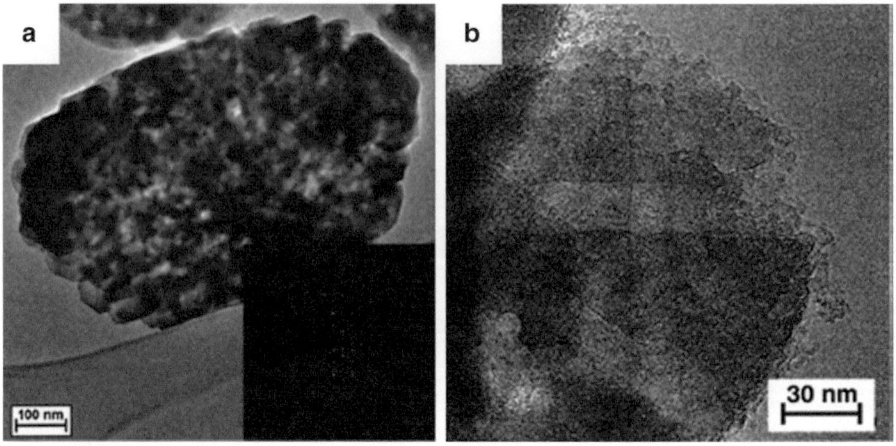

Fig. 4.5 (**a**) TEM images of single-crystalline mesoporous zeolites synthesized using carbon nanoparticles as a hard template (Reprinted with permission from Ref. [54]. Copyright 2000, American Chemical Society). (**b**) TEM images of single-crystalline mesoporous zeolites synthesized using carbon nanotubes as a hard template (Reprinted with permission from Ref. [56]. Copyright 2001, American Chemical Society)

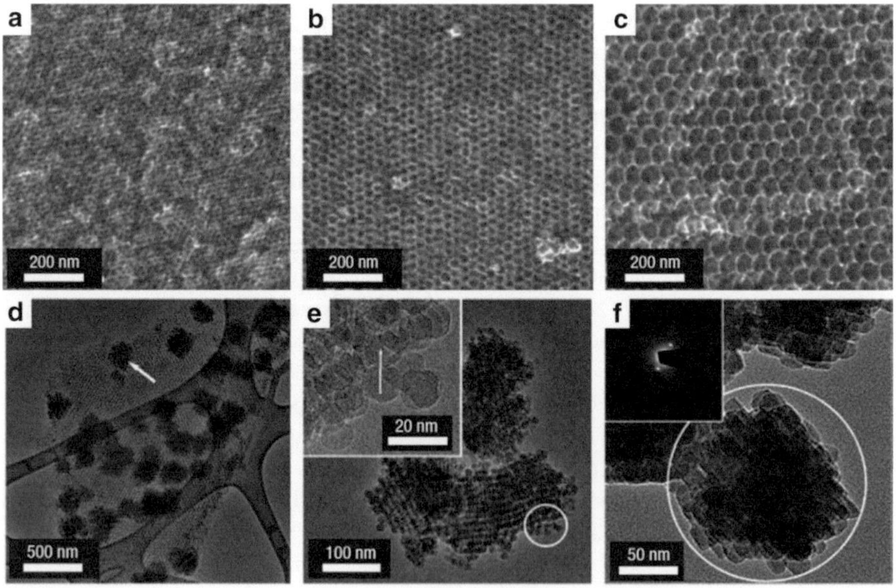

Fig. 4.6 SEM images (**a–c**) of empty 3DOM carbon replicas formed from colloidal silica with the size of ca. 10 nm (**a**), 20 nm (**b**), and 40 nm (**c**). TEM images (**d–f**) of 3DOM-imprinted silicalite-1 crystals. TEM (**d**) shows isolated crystalline domains (*arrow*) dispersed throughout the 3DOM carbon template, and TEM images (**e, f**) show representative particulate domains after the 3DOM carbon template was removed (Reprinted by permission from Macmillan Publishers Ltd: Ref. [76], Copyright 2008)

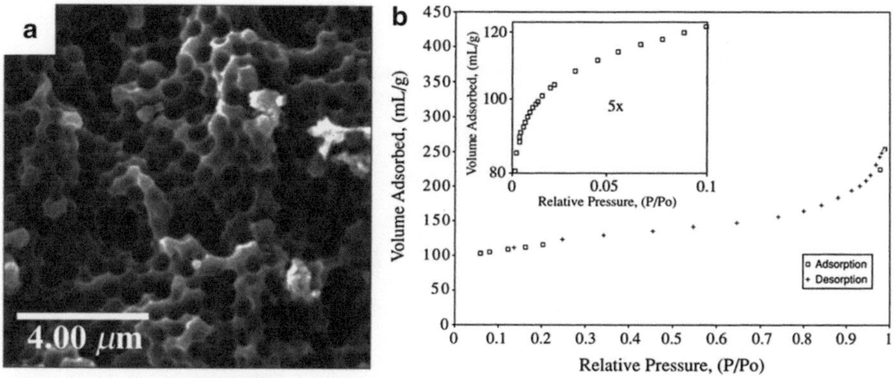

Fig. 4.7 SEM image (**a**) and N_2 adsorption–desorption isotherms (**b**) of mesoporous MFI zeolite synthesized using polystyrene beads as a solid template (Reprinted with permission from Ref. [78]. Copyright 1999, American Chemical Society)

(with a sufficiently wide pore diameter >10 nm) were equally demonstrated to generate ordered mesoporous zeolites. In the synthesis of ordered mesoporous MFI zeolites, Sakthivel et al. [71] and Yang et al. [72] employed ordered CMK-1 and CMK-3 mesoporous carbons, respectively [73–75]. Notwithstanding the presence of uniform mesopores with large pore volumes (0.8–0.9 cm^3 g^{-1}), zeolites exhibited low-intensity low-angle XRD peaks. Cho et al. recently followed a similar approach in the synthesis of silicate zeolite (MFI topology) (Fig. 4.8) [79]. Surprisingly, they successfully achieved a higher mesoporosity degree upon control of the humidity conditions during the synthesis procedure. Faithful replication of the carbon template into the zeolite was further claimed to equally depend on the humidity of the synthetic gel.

4.4 Soft-Templating Strategy: Mesopore Generation via Supramolecular Self-Assembly

4.4.1 Discovery of Ordered Mesoporous Molecular Sieves Using a Soft-Templating Strategy

Soft-templating strategies imply the use of nanometer-sized supramolecular micelles assembled by surfactant molecules and capable of inducing the formation of a mesostructure. Ideally, the surfactant molecules are assembled to form mesostructured surfactant micelles while the head groups can generate zeolite frameworks. If the soft-templating occurs effectively, surfactant templates have the advantage of a facile molecular tuning of their functional groups and chain lengths and of geometrical packing parameters of the selected surfactant molecule [2, 81, 82, 85–88]. The latter can be further controlled by the addition of auxiliary organic additives such as swelling agents, e.g., trimethylbenzene, cosurfactants, and inorganic

Fig. 4.8 SEM images of mesoporous carbon materials synthesized using SBA-15 silica as a hard template under 85 % (**a**) and 100 % (**b**) humid condition and TEM images (**c** and **d**) of the carbon materials synthesized under the condition of 100 % humidity (Reprinted from Ref. [79], Copyright 2011, with permission from Elsevier)

salts [89, 90]. Yanagisawa et al. originally reported the soft-templating approach for the synthesis of alkyltrimethylammonium–kanemite silica complexes and their conversion to mesoporous FSM-16 materials [91, 92]. In 1992, Kresge et al. reported a generalized synthesis approach to generate ordered mesoporous molecular sieves (OMMSs), later referred to as the M41S family [93]. By using surfactants as a mesopore-generating agent, the authors proposed a liquid-crystal-templating mechanism to explain the formation of the OMMSs. The mechanism was later extended to a broader concept suggesting a surfactant–silicate cooperative assembly into a composite micelle, presumably involving Coulomb interactions, hydrogen bonding, and van der Waals interactions [87–97].

The presumption of promising crystalline zeolitic mesopore walls generated in the presence of ammonium-type surfactants was incorrect, however, as the obtained mesoporous materials would instead display amorphous silicate frameworks. In addition to a considerably weaker Brønsted acidity, the resulting amorphous mesoporous aluminosilicates exhibited low thermal and hydrothermal framework stability. The M41S family could extend the concept of molecular sieves to a wider 2–50 nm regime, offering high surface areas and a narrow, uniform distribution of pores in the mesopore region. Further exploratory work led to the synthesis of OMMSs such as MCM-41, MCM-48 [93], and SBA-15 [94].

OMMSs have been extensively applied in acid catalysis and as a support for transition metal catalysts and organometallic complexes, adsorbent for bulky molecules, drug delivery vehicle, and hard template for mesoporous carbon [98–102]. One particular motivation for the development of new OMMSs was the challenge to create uniform mesopores for the catalytic conversion of bulky molecules inaccessible in microporous architectures. Notwithstanding the remarkable features of the OMMS mesostructure, the application of these materials would be confirmed to be ineffective in catalytic systems requiring a strong acid functionality.

4.4.2 Combination of Single-Molecular and Supramolecular Templating Strategies

Following the successful synthesis of OMMSs, ammonium surfactants were tentatively added to a zeolite synthesis composition, based on the assumption that the supramolecular micelles and the conventional zeolite SDA could cooperatively generate zeolite structures within mesopore walls (Fig. 4.9). Among the reported works,

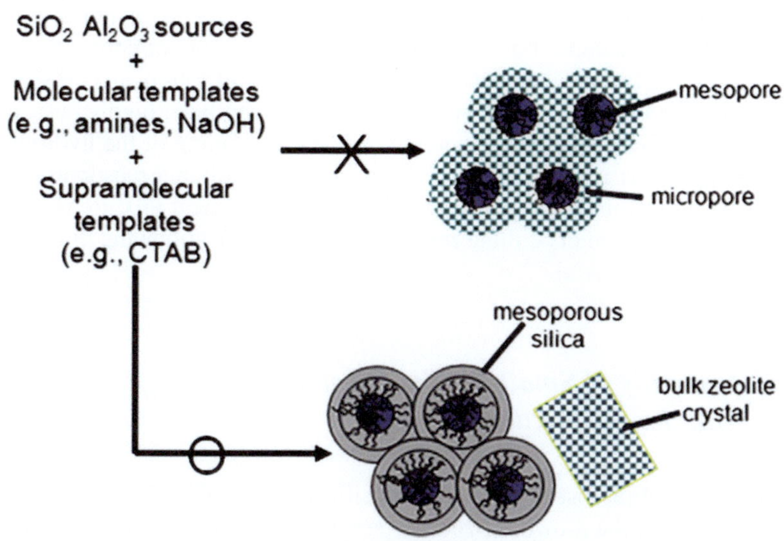

Fig. 4.9 Schematic representation of dual-templating synthesis approach by single-molecular and supramolecular templating methods, which eventually induces a separate formation of amorphous mesoporous materials and bulk zeolite crystals (Reprinted from Ref. [80], Copyright 2012, with permission from Elsevier)

Fig. 4.10 SEM image showing a physical mixture of MFI zeolite crystal and amorphous mesoporous silica material resultant, following a dual-templating synthesis (Reprinted from Ref. [103], Copyright 1998, with permission from Elsevier)

CTAB was commonly preferred as the supramolecular template. Karlsson et al. described a representative example in which they added the ammonium surfactant to a gel composition containing tetrapropylammonium for the synthesis of MFI zeolite [103]. Surprisingly, however, the two SDAs did not act cooperatively but rather in a competitive mechanism. Prior to zeolite crystallization, an OMMS phase was generated, but its dissolution would eventually occur under highly basic conditions, as in the described synthesis conditions. Further tuning of the hydrothermal time and temperature resulted in each case in a solely amorphous mesoporous material, a solely crystalline microporous zeolite, or their physical mixture (Fig. 4.10). To date, the synthesis of crystalline zeolitic mesopore walls following this approach has remained unsuccessful.

4.4.3 Assembly of Small Zeolite Crystallites with the Surfactant Micelles

To obtain an ideal mesostructured aluminosilicate material with crystalline walls, Liu et al. proposed the growth of zeolite "seeds" (proto-zeolitic entities) within an organized mesoporous matrix [104–107]. The strategy was intended to hinder phase separation between surfactant micelles and the zeolite crystals, as reported in the previous section. Following cautious control of the synthesis conditions (e.g., pH and temperature), formerly aged small zeolite crystallites were reported to be assembled around the mesostructured micelles of cationic surfactants. Synthesized samples exhibited well-defined low-angle XRD patterns in good agreement with those of the OMMS structure. Under the required milder conditions, the synthesis of highly crystalline zeolite frameworks was nonetheless unsuccessful. Representative high-angle XRD patterns confirmed the absence of atomic order characteristics of a crystalline zeolite framework. The resulting zeolites were thus described as "pseudo-zeolites" or "proto-zeolites" according to their framework nature. The acidity and hydrothermal stability were significantly improved compared to the MCM-41-type mesoporous aluminosilicate. The pseudo-mesoporous zeolites were successfully applied as catalysts in various catalytic reactions requiring weak Brønsted acidity or Lewis acidity, e.g., Friedel–Crafts alkylations [108]. Regardless, the evidenced acidity was inadequate for many catalytic applications requiring strong acid functionality [109].

At the expense of the mesoporous quality, enhanced crystallinity was achieved when the assembly of the zeolite crystallites into the mesostructured material was performed under steam-assisted crystallization (i.e., dry-gel synthesis) [110]. Even the modified synthesis strategy could not, however, effectively prevent the formation of a physical mixture of bulk zeolite and an OMMS phase. Several mesostructured/zeolitic composite materials, including ZSM-5/MCM-41 [111], ZSM-5/MCM-48 [112], Beta/MCM-41 [113, 114], and Beta/MCM-48 [115], have been synthesized. Improved catalytic activities were witnessed compared to the respective binary physical mixtures of zeolites and amorphous aluminosilicates [116].

4.4.4 Use of Ammonium Surfactant as a Capping Agent of Nanocrystalline Zeolites

Despite strong electrostatic interaction with silicate frameworks, CTAB has been the subject of continuous reports substantiating its effectiveness in generating a mesoporous structure within the highly crystalline domains of a zeolite. Recently, however, one exceptional case was the synthesis of LTA zeolite nanocrystals at low crystallization and growth rate at room temperature in the presence of surfactant micelles [117]. After the removal of CTAB, the sample showed large intercrystalline mesopores attributed to the voids between the aggregated zeolite nanocrystallites. The surfactant was concluded not to act as an effective mesopore-generating agent but rather as a capping agent, hindering crystal growth by covering the external surfaces of zeolite nanoparticles. Potential competition between CTA^+ ions and other cationic species, e.g., Na^+ and zeolite-structure-directing ammonium ions, was considered to impact the effectiveness of zeolite capping. Recently, Jo et al. reported that multivalent surfactants containing two or more ammonium head groups could be used as effective capping agents in the synthesis of nanocrystalline zeolites with various micropore structures (e.g., MOR, FAU, CHA, and MFI) [118]. Owing to a stronger electrostatic interaction, the multivalent surfactants were presumed to take on a more effective role as capping agents, hindering the zeolite nanocrystal surfaces from further growth. Accordingly, much higher external surfaces were attained, compared to their conventional counterparts.

4.4.5 *Use of Organosilane Molecules as Mesopore-Generating Agents*

Regardless of continuous exploratory work, phase-separation phenomena between amorphous silica and bulk zeolite have made it extremely challenging for years to synthesize crystalline zeolites exhibiting a uniform mesoporous architecture. In 2006, Choi et al. reported a direct-synthesis route to generating mesoporous hierarchical zeolites using tunable amphiphilic organosilanes acting simultaneously as a silica source and a supramolecular mesopore-generating agent [119, 120]. These rationally designed surfactant molecules contain a hydrophobic long-chain alkylammonium moiety and a hydrolysable methoxysilyl group, linked together through a chemically stable Si–C bond (Fig. 4.11). [3-(trimethoxysilyl)propyl]hexadecyldimethylammonium chloride ($[(CH_3O)_3SiC_3H_6\text{-}N^+(CH_3)_2C_{16}H_{33}]Cl^-$, "TPHAC" in short) and their structural analogues were reported as typical examples in the synthesis procedure [121]. The underlying idea is that the hydrolysable methoxysilyl moiety can strongly interact with growing crystal domains through the formation of covalent bonds with other SiO_2 and Al_2O_3 sources. TPHAC surfactants are able to be well incorporated within the zeolite product. The TPHAC surfactant was added to the synthesis composition of ordinary crystalline aluminosilicates, which contained molecular templates for zeolites (e.g., tetrapropylammonium) and sodium silicate (or TEOS) as the silica source (Fig. 4.12).

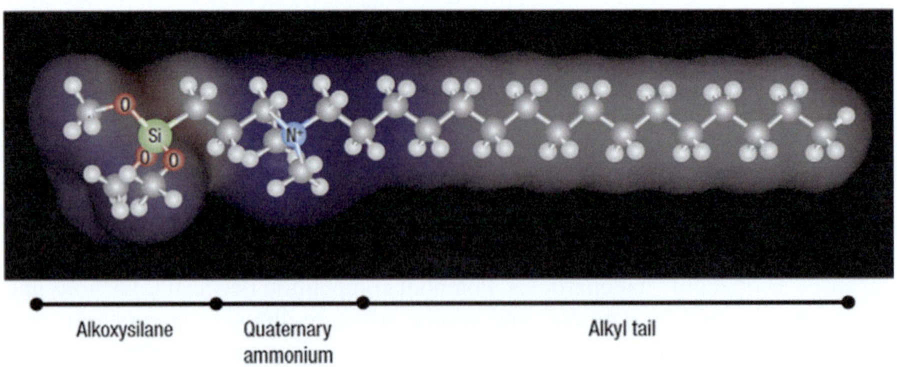

Fig. 4.11 Molecular structure of typical organosilane surfactant (Reprinted by permission from Macmillan Publishers Ltd: Ref. [120], Copyright 2006)

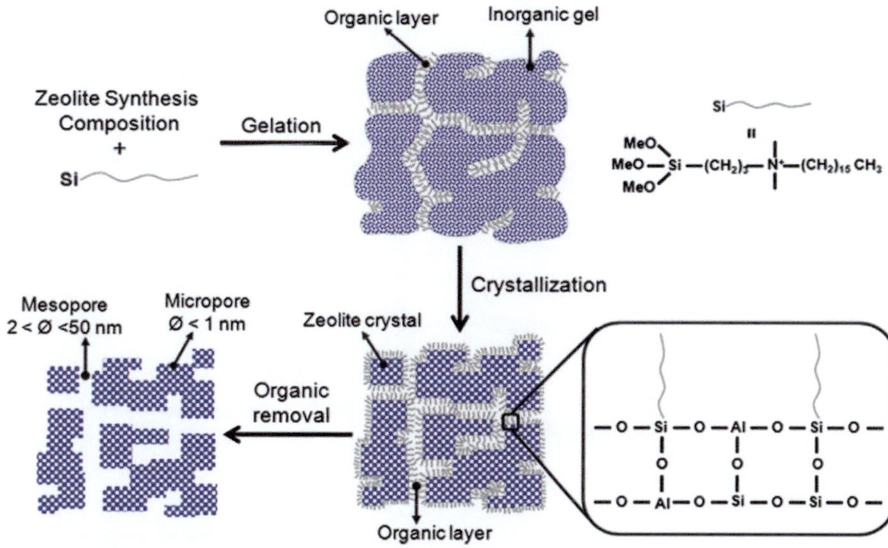

Fig. 4.12 Schematic representation of crystallization process of mesoporous zeolite using organosilane surfactant as a mesopore structure-directing agent (Reprinted from Ref. [80], Copyright 2012, with permission from Elsevier)

The resulting zeolites possessed high mesopore volume, uniform mesopore diameter distribution, and high BET-specific surface area. In addition, the mesopore diameters could be controlled according to the tail length of the organosilane surfactant or by tuning the hydrothermal temperature (Fig. 4.13) [119]. The mesoporous architecture was assessed by the generation of Pt nanowires within the structure of the hierarchical zeolite [122–124]. Representative TEM images revealed cylindrical Pt nanowires interconnected in a wormhole-like manner. The nanowires exhibited cylindrical uniform mesopores in the 3–4 nm range, in agreement with porosity assessment through N_2 isotherm analysis, similar to the KIT-1 mesoporous silica [122]. The presence of intracrystalline mesopores was further confirmed in high-resolution scanning electron microscope (SEM) images of ion-beam cross-sectioned crystals (Fig. 4.14) [125].

The organosilane-directed synthesis route was extended to a wide range of hierarchical zeolite and zeotype materials. Successful reports include the synthesis of mesoporous LTA [119, 124], SOD [126], and BEA [127] zeolite architectures and the synthesis of mesoporous aluminophosphates constructed with crystalline microporous AFI and AEL topologies [128]. Recent reports claimed the synthesis of X zeolite (FAU topology) with high Al content (Si/Al < 1.5), exhibiting a pore system in a micro–meso–macro hierarchical interconnection [129]. Notwithstanding its simplicity, the method had limited success at times, depending on the selected synthesis conditions and zeolite architecture. High organosilane content would, for

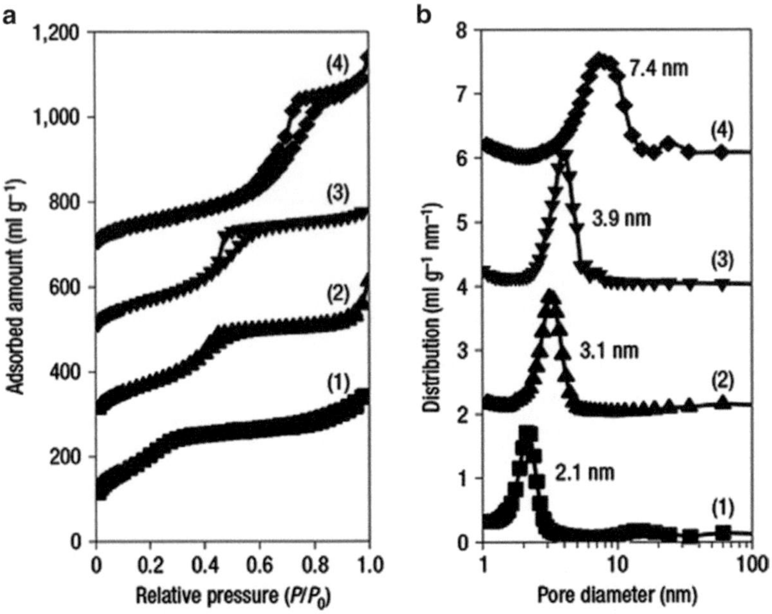

Fig. 4.13 N_2 adsorption isotherms (**a**) and their corresponding pore size distributions (**b**) of MFI zeolite crystals with tunable mesoporosity. According to the tail length of organosilane surfactant and/or the hydrothermal temperature, the average mesopore diameter can be uniformly controlled (Reprinted by permission from Macmillan Publishers Ltd: Ref. [119], Copyright 2006)

instance, generate the presence of impurity phases such as zeolite P (GIS topology) and amorphous aluminosilicates [124, 125]. Kim et al. recently reported that the presence of bulk zeolite seeds in the synthesis conditions enabled a reduction in the required hydrothermal time, extending the application of the strategy [130]. Representative TEM images revealed the complete absence of the bulk zeolite additives, confirming its dissolution into tiny fragments. The synthesis was successfully achieved for mesoporous CHA, FAU, and MOR architecture-type zeolites.

The use of organosilane molecules as a mesopore-directing agent has been successfully used to improve the efficiency and selectivity of hierarchical zeolites in various catalytic reactions. Mesoporous zeolites possessing strong Brønsted acidity displayed superior catalytic performances in acid-catalyzed reactions involving diffusion-limited steps for bulky molecules, compared to conventional and desilicated hierarchical zeolites [119, 131, 132]. The cost of catalyst regeneration or replacement, particularly in fixed-bed industrially relevant processes, often compromises the practicality of zeolites as heterogeneous catalysts. Remarkably improved resistance to deactivation in a wide range of acid-catalyzed reactions was demonstrated to be a noteworthy feature of the hierarchical materials [112]. Facile diffusion of coke precursors out of the zeolitic structure is believed to be notably enhanced by the presence of mesopores. Whereas solely microporous zeolites could

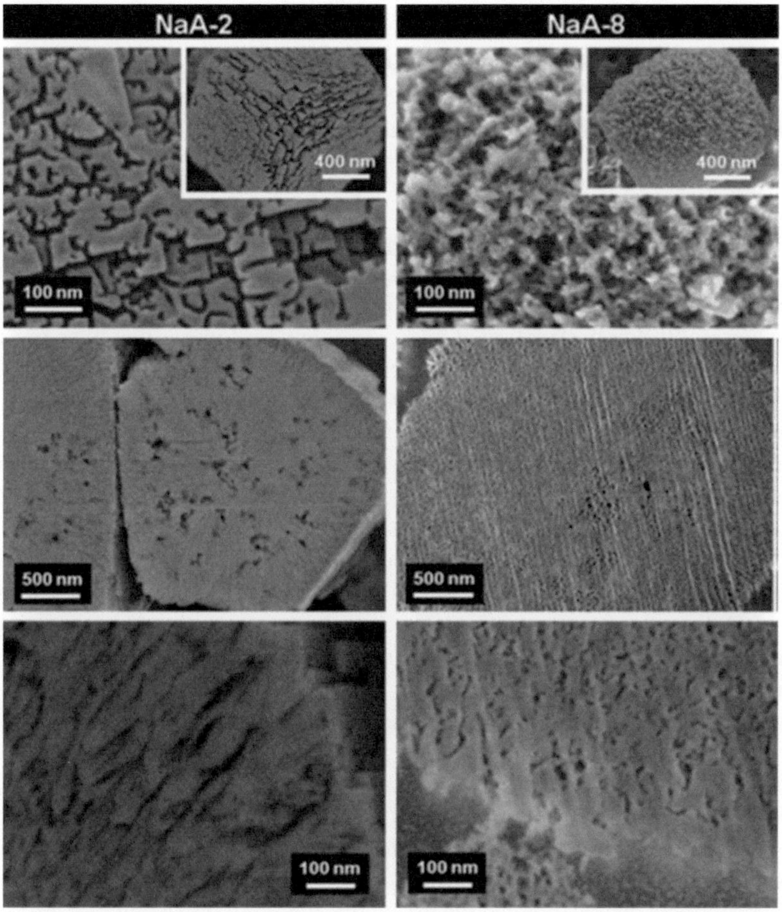

Fig. 4.14 Cross-sectional SEM image of mesoporous LTA zeolite, displaying intercrystalline mesopores inside the zeolite crystals as well as on the external surface. The *top images* were taken from the external surfaces, whereas the others were taken after cross-sectioning by Ar–ion beam (Reprinted from Ref. [125], Copyright 2010, with permission from Elsevier)

not be functionalized due to the lack of free silanol groups inside micropores, the hierarchical zeolites were suitable to be functionalized at the mesopore surface by using a high density of silanol groups [126, 133]. Organic-functionalized hierarchical zeolites exhibited further hydrothermal stability and reusability in catalytic applications, as compared with organic-functionalized mesoporous silica samples. The former were considered to increase the number of opportunities in the design of catalyst, sensor, adsorbent, and ion-exchange materials. Similarly, other synthesis routes were developed using nonionic organosilane [134], silylated polymer [135], and cationic polymers [136] as a mesopore SDA.

4.5 Generation of Mesostructured Zeolites by the Zeolite-Structure-Directing Surfactants

4.5.1 *Synthesis of MFI Nanosheets Using Diquaternary Ammonium Surfactant*

Despite numerous attempts to find appropriate surfactant molecules, the synthesis of uniform mesoporous materials with highly crystalline zeolitic walls remained a dream for years in the zeolite science field. In principle, hierarchical zeolites exhibiting a hierarchical pore structure of micropores and mesopores would be presumed to be synthesized in the presence of a zeolite SDA with a hydrophobic alkyl tail. Notwithstanding the ability to generate the mesoporous structure, ammonium-type surfactants were consistently ineffective in generating zeolitic mesopore walls, displaying instead amorphous silicate frameworks. Despite the success of the organosilane-directed synthesis route applied to a wide range of hierarchical zeolite and zeotype materials, the surfactant exhibited no zeolite-structure-directing role. Furthermore, none of these approaches could generate "ultrathin" zeolites with thicknesses below the typical values between 5 and 100 nm. Nanowires (one-dimensional) and nanosheets (two-dimensional), expected when confining the crystal thickness to, respectively, two and one spatial dimension, remained a remarkable challenge.

In 2009, Ryoo et al. introduced an effective strategy for synthesizing extremely thin nanosheets of MFI-type zeolite using multiammonium surfactants [137, 138]. Particularly, diquaternary ammonium surfactants such as $C_{22}H_{45}$–$N^{+}(CH_3)_2$–C_6H_{12}–$N^{+}(CH_3)_2$–C_6H_{13} (designated hereafter as C_{22-6-6}) contain a long alkyl tail (e.g., –$C_{22}H_{45}$) and two quaternary ammonium groups, connected by a –C_6H_{12}– spacer. The ammonium head parts directed the MFI zeolite framework, whereas self-assembled long hydrophobic chains directed the crystal morphology into a lamellar mesostructure. The hydrophobic alkyl tails were confirmed to block the excessive growth of the zeolite crystallites beyond the ammonium region, limiting the zeolite crystallization to the head part of the surfactants. No other conventional MFI zeolite SDA was required, confirming that a zeolitic microporous architecture can be simply synthesized using the porogenic activity of a functional group of the template molecule. The resultant zeolite took the form of nanosheets with a single-unit-cell thickness. Straight 10-MR channels along the *b*-crystallographic axis intersected the 10-MR *zigzag* channel system extended into the *a* and *c* directions. The 2.5-nm-thick framework was terminated in the *b* direction resulting in 10-MR openings on the surface. The thickness of the MFI zeolite nanosheets could be systematically controlled by changing the number of zeolite-structure-directing units in the head groups of the surfactants (Fig. 4.15) [139]. Ultrathin nanosheets (~1.5 nm), with only ¾ of the *b* lattice parameter, were even synthesized by using a Gemini-type tri-ammonium surfactant with the formula of $C_{18}H_{37}$-$N^{+}(CH_3)_2$-$C_6H_{12}N^{+}(CH_3)_{2\text{-C6H12-}}$ $_{N}{}^{+}(CH_3)_2$-C_6H_{13} [140].

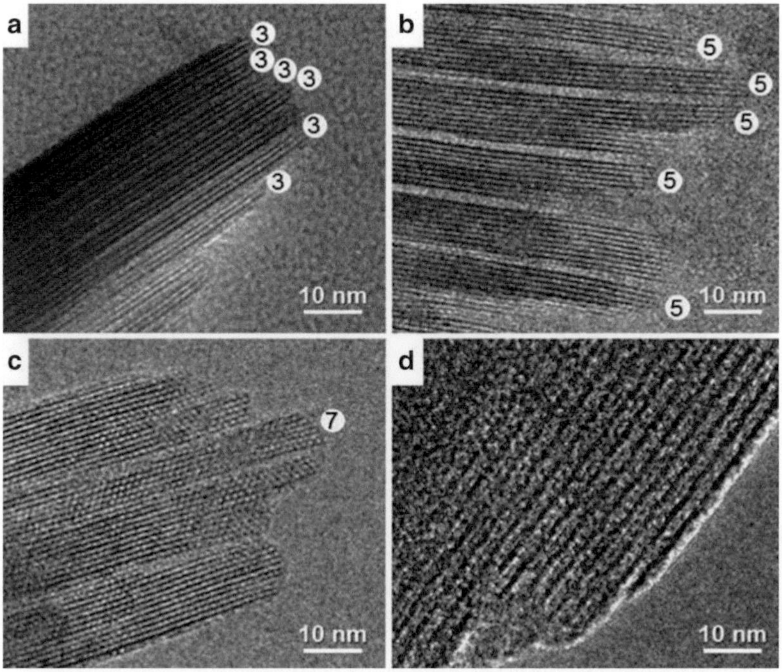

Fig. 4.15 High-resolution TEM images for as-synthesized products obtained with surfactants containing 2 (**a**), 3 (**b**), 4 (**c**), and 1 (**d**) ammonium group. The value in *white circles* indicates the number of pentasil sheets in a nanosheet (Reprinted with the permission from Ref. [139]. Copyright 2011, American Chemical Society)

The MFI zeolite nanosheets could be synthesized under a wide range of synthesis conditions and using any ordinary silica sources, such as sodium silicate, TEOS, silica gel, and fumed silica. Depending on the synthesis conditions, the MFI zeolite nanosheets could be synthesized as either a fully disordered assembly or through regular stacking into an ordered multilamellar mesostructure (Figs. 4.16 and 4.17). A uniform interlamellar distance was witnessed between alternating multilamellar MFI nanosheet layers (Fig. 4.17b). Small-angle XRD peaks up to the fourth-order reflections were in agreement with the corresponding interlayer distance because of the interlamellar structural coherence [130]. The interlamellar spacing could be tuned according to the surfactant tail lengths [134]. Partial condensation of the multilamellar structure was observed upon surfactant removal during calcination. Much of the intersheet space was however retained as disordered mesopores after calcination, owing to partially intergrown mesostructures presumed to act as "self-pillars." There were slight deviations in the crystal orientation in the *a*–*c* plane preventing complete topotactic condensation between nanosheets. The calcined product exhibited quite a high BET surface area (520 $m^2\ g^{-1}$) and large pore volume

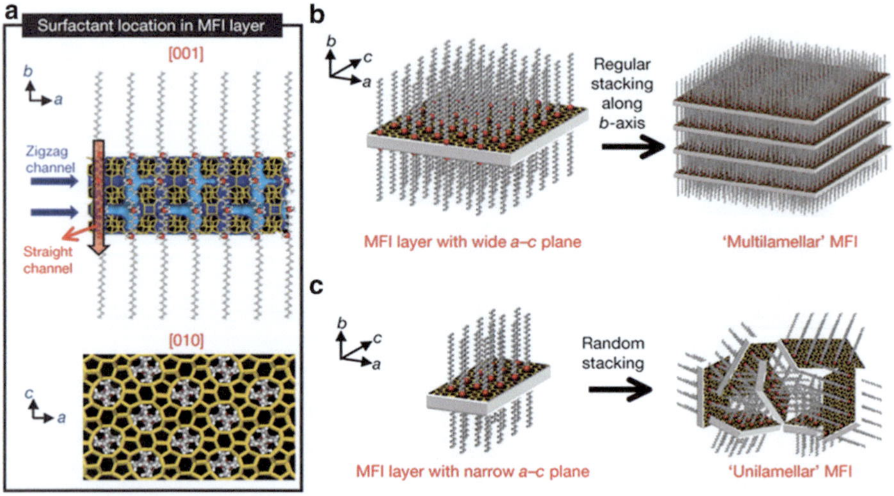

Fig. 4.16 Schematic representation of crystallization of MFI nanosheets. Proposed structure model for the single MFI nanosheet (**a**), where the surfactants are aligned along the straight channel of MFI framework and di-quaternary ammonium head groups are embedded inside the zeolite framework. Depending on the synthesis condition, the MFI nanosheet can be regularly assembled into multilamellar mesostructure (**b**) or randomly stacked into unilamellar (**c**) (Reprinted by permission from Macmillan Publishers Ltd: Ref. [137], Copyright 2009)

(0.5 $cm^3\ g^{-1}$), compared to conventional MFI (400 $m^2\ g^{-1}$, 0.3 $cm^3\ g^{-1}$). It is also remarkable that the ordered multilamellar structure could be maintained by generating silica pillars between zeolite layers (Fig. 4.18) [141]. Unilamellar MFI nanosheets similarly revealed very narrow *a–c* planes. These nanosheets supported each other in a random pile, exhibiting no mesostructure orders (Fig. 4.17c, d). The intersheet mesopores were not uniform in diameter but were well retained even after the complete removal of the surfactant by calcination. The calcined sample equally possessed a high BET surface area (700 $m^2\ g^{-1}$) and large pore volume (1.2 $cm^3\ g^{-1}$). Comparable textural characterization was observed with random zeolite nanosheets assembly, generated through exfoliation of layered zeolite precursors [142–145].

Careful selection of the crystallization conditions, particularly the hydrothermal time, proved decisive for the synthesis of unilamellar or multilamellar hierarchical MFI zeolite. Exploratory work revealed that under Na^+-free conditions, unilamellar MFI nanosheets could be slowly generated (11 d) at 140 °C [146]. Surprisingly, however, the unilamellar product maintained for 15 d was afterward converted into a multilamellar mesophase. The phenomenon was tentatively attributed to crystal ripening in the *a–c* plane and subsequent reorganization of randomly oriented MFI nanosheets into regular stacking. The unilamellar MFI nanosheets could still be collected before the conversion to multilamellar. The synthesis was successful within a

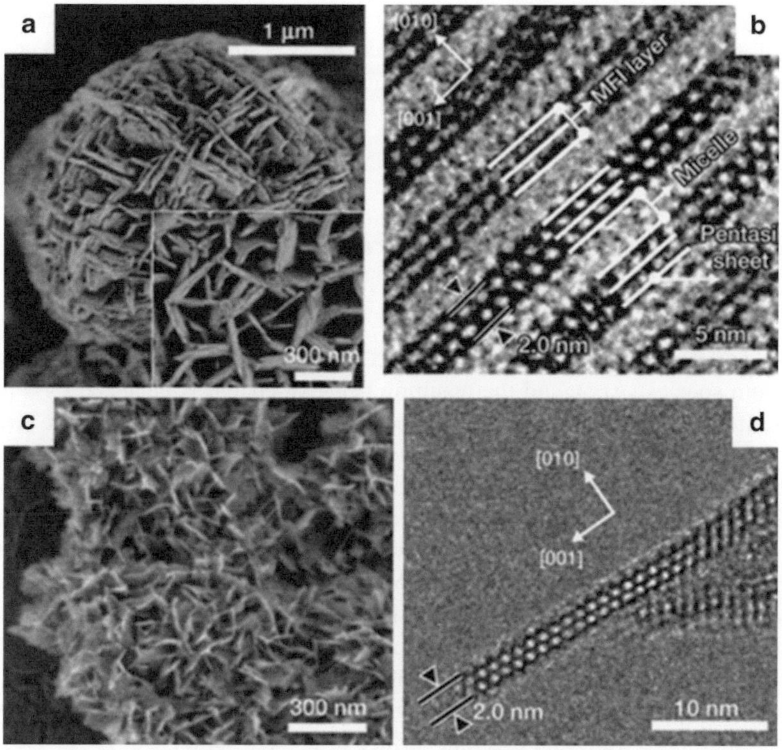

Fig. 4.17 SEM (**a**, **c**) and TEM (**b**, **d**) images of multilamellar (**a**, **b**) and unilamellar (**c**, **d**) MFI zeolite nanosheets (Reprinted by permission from Macmillan Publishers Ltd: Ref. [137], Copyright 2009)

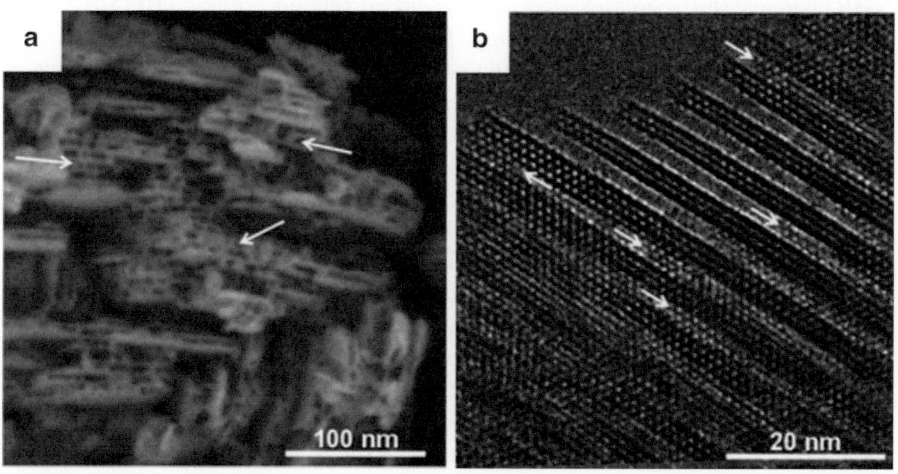

Fig. 4.18 High-resolution SEM (**a**) and TEM (**b**) images of the multilamellar MFI zeolite nanosheets, pillared with silica and subsequently calcined to remove the surfactants (Reprinted with the permission from Ref. [141]. Copyright 2010, American Chemical Society)

relatively short synthesis time (5 d at 140 °C) under Na^+-rich conditions using a sodium silicate water glass. The presence of Na^+ and high pH was however considered to allow for a quick transformation of the initial unilamellar into multilamellar zeolite.

4.5.2 Large Amount of Strong Brønsted Acid Sites at the External Surfaces of MFI Nanosheets

Brønsted acid sites in zeolites are developed when Si^{4+} is isomorphically replaced by aluminum species (Al^{3+}), and the resulting negative charge is balanced by a proton attached to an oxygen atom. In this case, both Al–O and Si–O bonds become virtually equivalent following a global rearrangement in the structure. Aluminum distribution and the Si–O–Al bond angle in the zeolite framework remarkably impact the strength of a Brønsted acid site [146]. The MFI zeolite exhibiting strong Brønsted acidity has been widely used as a heterogeneous acid catalyst in numerous catalytic applications [2, 3, 147, 148]. Brønsted acidity of a MFI zeolite nanosheet with a thickness of 2.5 nm was successfully characterized using ^{31}P MAS NMR with adsorbed trimethylphosphine oxide (TMPO) and tributylphosphine oxide (TBPO) [149]. Phosphine oxides are well known to be strongly adsorbed on Brønsted acid sites. TMPO has been reportedly used as a probe molecule to detect total Brønsted acid sites in both the internal micropore walls and external surface of an acid solid [150–153]. According to the ^{31}P NMR spectroscopic method, the phosphorus probe has partially negatively charged oxygen atoms that strongly interact with the bridging hydroxyl groups on the zeolite surface, thereby inducing the ^{31}P resonance to shift downfield [153]. The ^{31}P NMR chemical shift increases according to acid site binding affinity. TBPO (0.82 nm) is too large to penetrate into the intrinsic MFI microporous architecture (0.53×0.56 and 0.51×0.55 nm) and thus is limited to acid sites located on the external surface of the crystal. The acid strength and concentration of acid sites were assessed by chemical shifts and peak areas of the adsorbed molecules appearing in NMR spectra (Fig. 4.19). The ^{31}P NMR investigation identified Brønsted acid sites with three different strengths on external surfaces; four types were identified inside the microporous architecture. According to the ^{31}P NMR spectroscopic method, 32 % were located on the external surfaces among the total Brønsted acid sites. This result is quite impressive considering that in typical MFI crystal sizes, almost all Brønsted acid sites are located inside the microporous architecture of the zeolite rather than in the external surface. Full utilization of a catalyst is in this case scarcely achieved in bulk MFI zeolites. Among the external acid sites, 5.8 % of the total Al content was evaluated as strong acid sites. The MFI nanosheets were considered to be a valuable addition for catalytic reactions requiring strong acidity and involving bulk molecules. Nonetheless, the acid strength of the external Brønsted acid sites was apparently weaker than the internal acid sites because the former were less strained from perfect tetrahedral geometry.

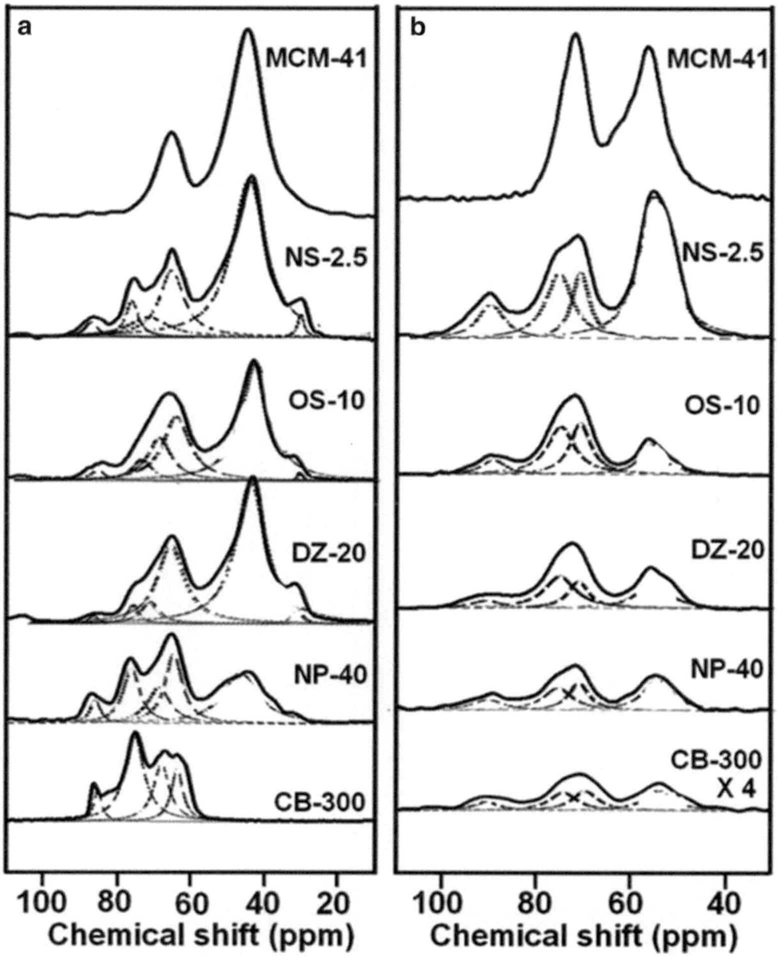

Fig. 4.19 ^{31}P MAS NMR spectra of (**a**) TMPO and (**b**) TBPO adsorbed on the MFI zeolite samples with various framework thicknesses and amorphous silica MCM-41. The *dotted curves* indicate results of spectral analyses by Gaussian deconvolution. All spectra are plotted on the absolute intensity scale after collecting with the same number of acquisitions and being reduced to the same sample mass (Reprinted with the permission from Ref. [149]. Copyright 2013, American Chemical Society)

4.5.3 Synthesis of MFI Nanosponges in the Presence of MFI Zeolite Seeds

Notwithstanding the remarkable features of the surfactant-directed MFI nanosheets, a long hydrothermal reaction time is required for synthesis, especially for materials with high Al content. Broader mesopore size distribution and eventual partial collapse of the achieved mesoporosity by an Ostwald-ripening process can be expected to occur. Additionally, the multiammonium surfactant can decompose by Hoffman elimination under strong basic synthesis conditions. In this context, the synthesis of hierarchical MFI zeolite was recently performed by adding small amounts of bulk MFI zeolite seeds [154]. Seeding was confirmed to remarkably reduce the zeolite crystallization time, similarly to seed-assisted synthesis of conventional MFI zeolites. As a result, the mesoporous zeolites could be rapidly generated even at high Al content. More important, the crystal seeding gave rise to the formation of an unexpected morphology of the hierarchical zeolite (Fig. 4.20).

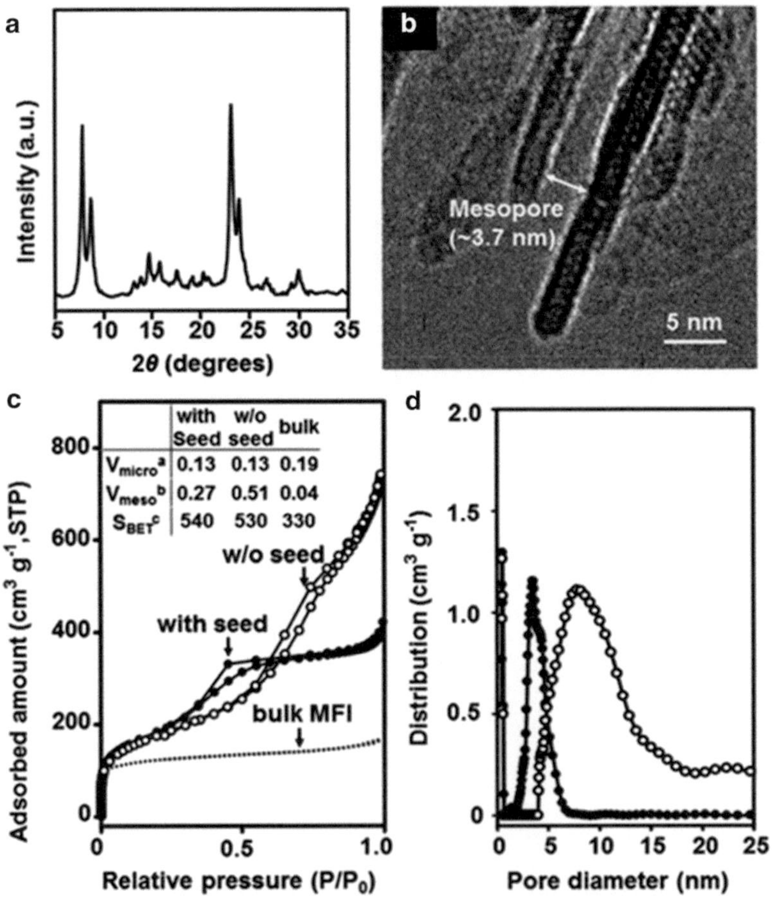

Fig. 4.20 Structure characterization data for a representative calcined zeolite nanosponge with Si/Al = 20 synthesized using diquaternary ammonium surfactant and bulk MFI seeds at 150 °C for 2.5 d. (**a**) XRD pattern, (**b**) TEM image, (**c**) Ar adsorption–desorption isotherm, and (**d**) pore size distribution. In (**c**) and (**d**), the zeolite nanosponge (*black circle*) is compared with the sample synthesized without seeding (*white circle*) and with the bulk MFI zeolite (*dotted line*) (Reproduced from Ref. [154] with permission from the Royal Society of Chemistry)

Individual nanosheets, revealing a narrow mesopore diameter distribution centered at 4 nm, were arranged in a nanosponge-like morphology. The self-supporting structure was witnessed even after the removal of the surfactant by calcination.

4.5.4 Generalizing a Zeolite Synthesis Method Using SDA Surfactant to Other Hierarchically Nanoporous Zeolites

Following the synthesis of hierarchical zeolites using multiammonium surfactants, Ryoo et al. generalized the approach to other zeolite architectures. A group of Gemini-type multiammonium surfactants (Fig. 4.21) was prepared to serve as structure-directing agents for the synthesis of mesostructured materials displaying crystalline zeolitic walls [155–157]. When a tri-quaternary ammonium surfactant with the formula of $C_{18}H_{37}$-$N^{+}(CH_3)_2$-C_6H_{12}-$N^{+}(CH_3)_2$-C_6H_{12}–$N^{+}(CH_3)_2$–$C_{18}H_{37}$ ($C_{18-6-6-18}$, Fig. 4.22a) was applied, hexagonally ordered MMS nanocrystals with an MFI-like microporous framework within the mesopore walls were generated (Fig. 4.22d). Representative SEM images revealed needle-shaped nanocrystals that were uniform in size and morphology. The zeolite-structure-directing role of the Gemini-type $C_{18-6-6-18}$ surfactant was elucidated by using 2D heteronuclear correlation (HETCOR) NMR, which resolves the molecular proximities of the surfactant heads to a solid framework (Fig. 4.23). The strong interaction between the ammonium groups and the aluminosilicate frameworks evidenced the microporogenic role of the ammonium region and their inclusion within the micropores. Structural coherence along the microscopic and mesoscopic arrangements was evidenced by powder XRD analysis. Because the mesopore walls were composed of only a single layer of zeolitic micropores, less than a single-unit-cell dimension of a bulk zeolite, structural information was insufficient for the precise determination of the microporous framework topology.

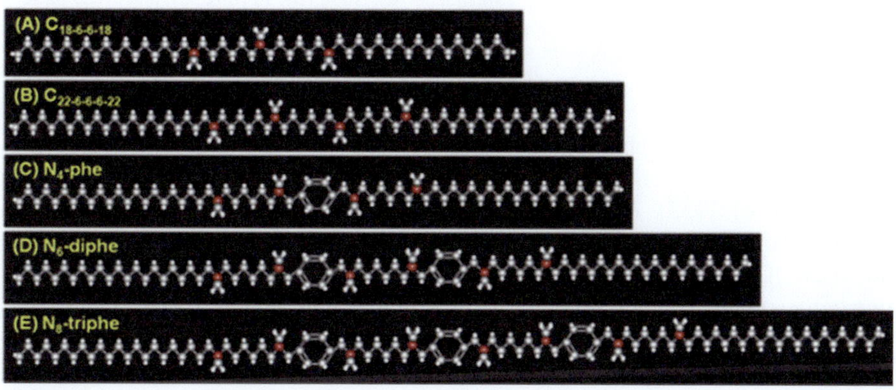

Fig. 4.21 Family of Gemini-type surfactant that can act as dual-porogenic structure-directing agent for mesoporous molecular sieves built with crystalline microporous zeolitic framework (From [155]. Reprinted with permission from AAAS)

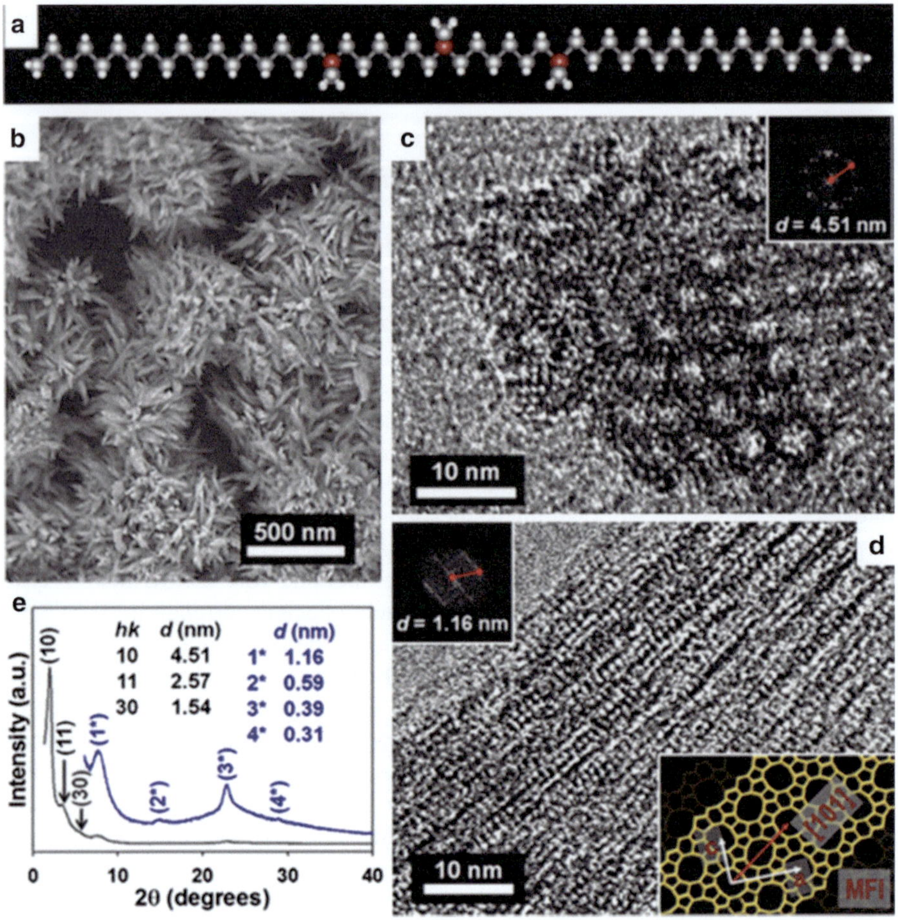

Fig. 4.22 Molecular structure of $C_{18-6-6-18}$ surfactant (**a**), SEM (**b**), TEM (**c, d**), and XRD pattern of hexagonally ordered crystalline mesoporous molecular sieve after surfactant removal (**e**) (From [155]. Reprinted with permission from AAAS)

The introduction of phenyl rings into the zeolite porogenic head of the surfactant was reported to generate a nanocrystalline zeolite with beta-type architecture (Fig. 4.21c–e) [155]. The beta nanocrystals were interconnected to form a nanosponge-like disordered mesoporous network with a uniform zeolite framework thickness. The disordered mesostructure exhibited in each case a narrow distribution of mesopore diameter. A notable feature of the dual-porogenic surfactant-driven synthesis route regarded the thickness of the crystalline microporous walls, uniformly tailored by the length of the microporogenic part of the surfactant. Wall thicknesses could be increased from 2.9 to 3.9 and 5.1 nm as the number of ammonium groups successively increased from 4 to 6 and 8, respectively. Such refined control driven by molecular design is so far unprecedented. The disordered network assembled by the zeolite framework had mesopores with a quite narrow distribution

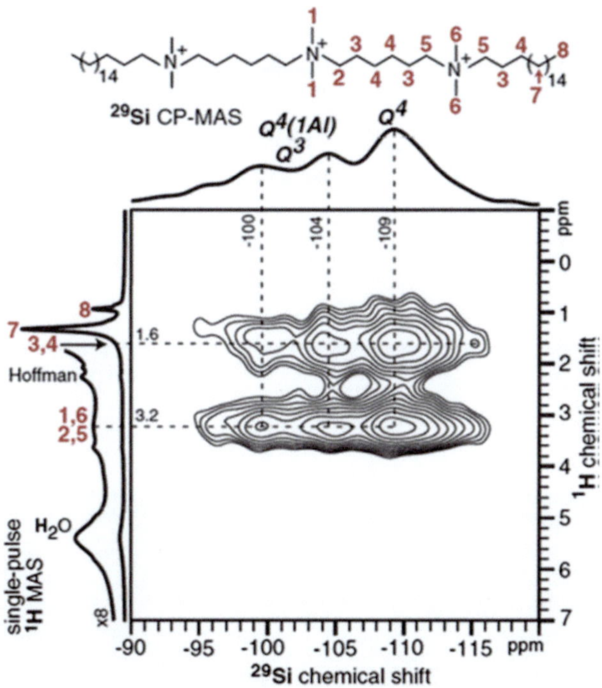

Fig. 4.23 Solid-state $^{29}Si\{1H\}$ 2D HETCOR NMR spectrum of the hexagonally ordered crystalline mesoporous molecular sieve in the as-synthesized form containing $C_{18\text{-}6\text{-}6\text{-}18}$ surfactant (From [155]. Reprinted with permission from AAAS)

of the pore diameter. In addition, the mesopore diameters could be tailored over a wide range of 3.8–21 nm by the addition of 1,3,5-trimethylbenzene (1,3,5-TMB) to the synthesis gel composition. Mesopore expansion by 1,3,5-TMB was later claimed to occur through reversible recrystallization of the zeolite frameworks in the presence of sufficiently high Na^+ content [158]. Mesopore expansion is described as occurring even after complete crystallization of the zeolite framework, whereas the phenomenon could be reversed following TMB removal by solvent extraction (Fig. 4.24). The thickness of the mesopore walls did not change during the consecutive pore expansion and shrinking processes.

Further exploratory work led to the synthesis of hierarchical MTW and MRE-like zeolites in the presence of surfactants with different head groups, including the presence of phenyl rings (Fig. 4.25) [159]. The framework thickness was tuned according to the number of ammonium groups in the surfactant, as previously reported with beta zeolite. As before, mesopore diameters could be conveniently increased by simple addition of 1,3,5-TMB [158].

The versatile synthesis strategy using multiammonium (or multiamine)-type surfactants as a zeolite SDA also proved to be efficient in the synthesis of hierarchical crystalline aluminophosphates [160]. Similarly, individual ammonium (or amine) head groups could direct the formation of microporous crystalline AEL.

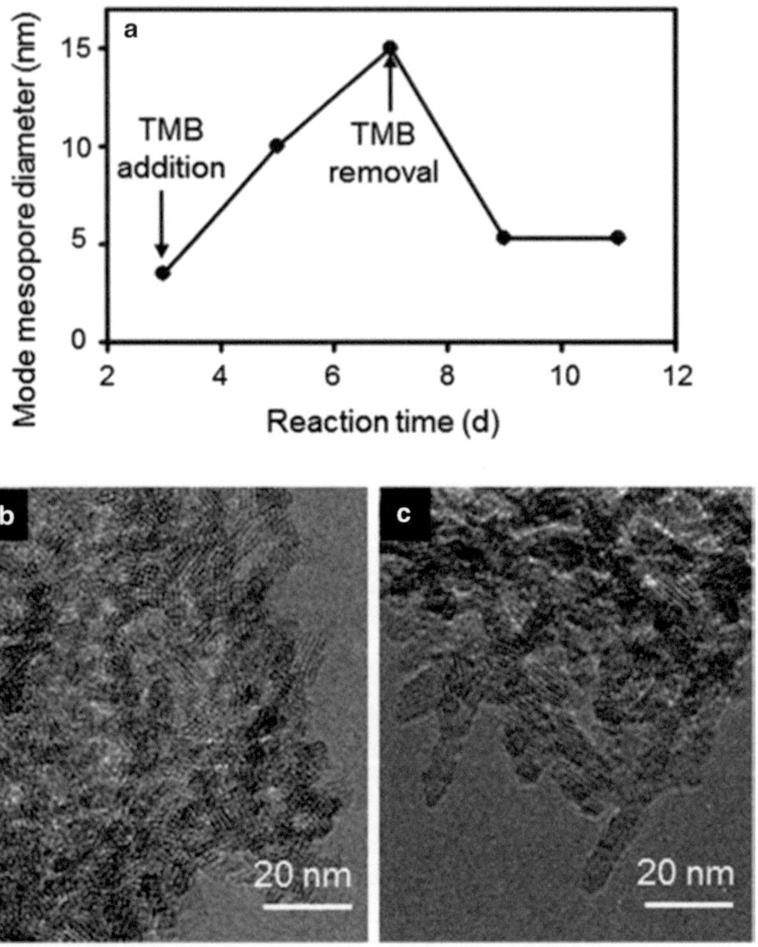

Fig. 4.24 (**a**) Mode mesopore diameters of solid precipitates, collected from the reaction mixture at various elapsed times during hydrothermal reaction at 413 K. High-resolution TEM images of nanosponge beta zeolites collected at (**b**) 3 d and (**c**) 7 d (Reprinted from Ref. [158], Copyright 2014, with permission from Elsevier)

4.5.5 Enhanced Properties of Surfactant-Directed Mesoporous Zeolites

Notwithstanding the remarkable features of the microporous structure, the narrow pores and dimensionality of the zeolitic porosity often induce severe diffusion limitations. The hierarchical zeolite, possessing a system of mesopores in addition to their intrinsic architecture, provides short diffusion path lengths inside the crystal framework and a high concentration of acid sites located on the resulting external

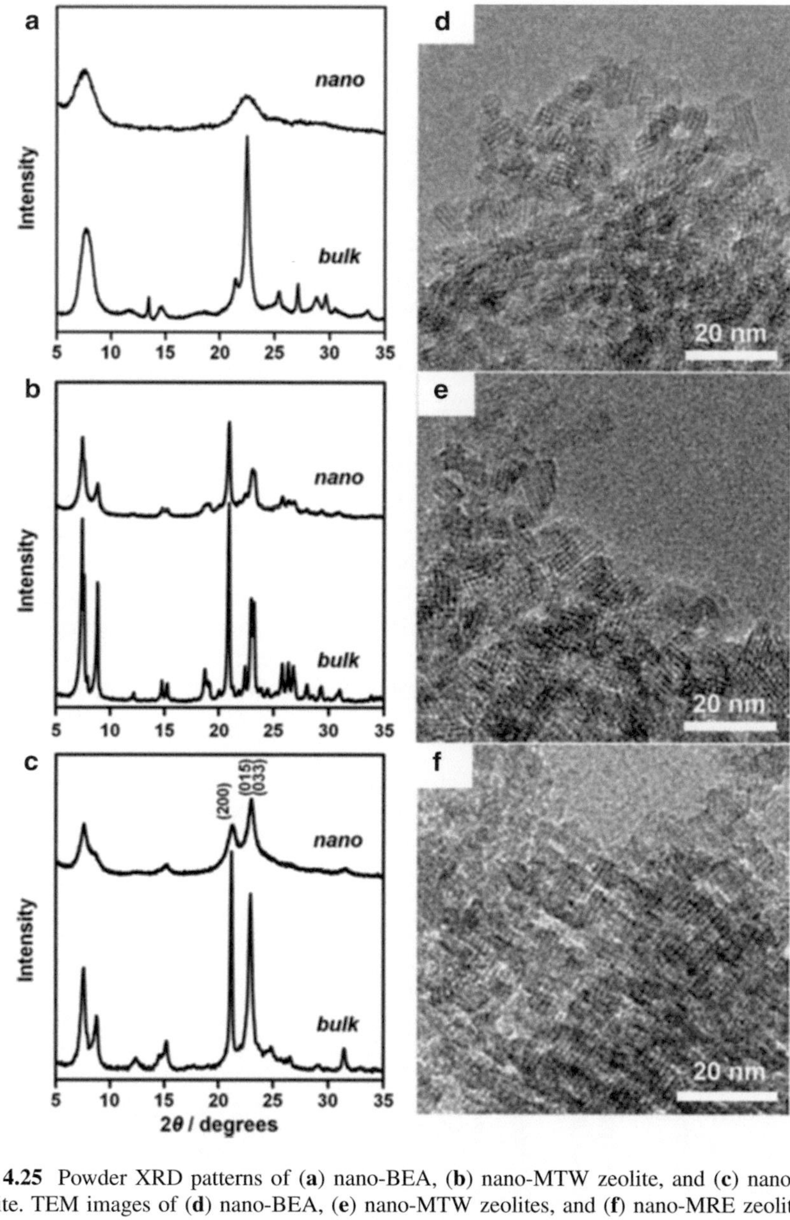

Fig. 4.25 Powder XRD patterns of (**a**) nano-BEA, (**b**) nano-MTW zeolite, and (**c**) nano-MRE zeolite. TEM images of (**d**) nano-BEA, (**e**) nano-MTW zeolites, and (**f**) nano-MRE zeolite. For comparison, XRD patterns of their bulk counterparts are also provided (Reprinted with the permission from Ref. [159]. Copyright 2013, American Chemical Society)

surface. Enhanced catalytic performances of surfactant-directed hierarchical zeolites have been assessed in a wide range of catalytic applications.

Within the tridimensional medium pore channel system of the MFI zeolite, the mechanism of deactivation by coke formation is widely accepted in the catalytic conversion of methanol-to-hydrocarbons. The rate of coke formation highly depends on zeolite topology, acid site density, and acid strength. MFI-based catalysts are deactivated during the catalytic reaction as the micropores are continuously filled with coke precursors, indirectly blocking diffusion into other micropores. In 2009, following the first reports of the synthesis of MFI nanosheets, Choi et al. described a fivefold higher catalytic lifetime observed with the hierarchical zeolite compared to its bulk counterpart (Fig. 4.26) [137]. The 2.5-nm thickness MFI framework facilitated the diffusion along the *b* direction. The authors proposed that because of the short diffusion of coke precursors across the nanosheets, slow deposition occurred exclusively at the external surface [161]. Likewise, MTW and MRE-type hierarchical zeolites exhibited higher catalytic longevity than conventional bulk counterparts in liquid-phase cumene reaction [159]. Slow catalytic deactivation was ascribed to the high number of strong Brønsted acid sites at the external surfaces permitting facile diffusion of reactant and product molecules. An extended catalytic lifetime has been further witnessed in catalytic reactions such as the Beckmann rearrangement (Fig. 4.27) [162], Friedel–Crafts alkylation [159, 163], and the oxidation of benzene with nitrous oxide [164]. High accessibility was also confirmed to generate higher catalytic activities, when Cu ion-exchanged ZSM-5 nanosheets were used in N_2O decomposition [165]. The chemical properties of the active Cu sites were also different. Higher reducibility of Cu^+ species (as evidenced by H_2-TPR data) on Cu-ZSM-5 nanosheets was inferred to lead to a higher activity. More facile desorption of oxygen from Cu-ZSM-5 nanosheets was further observed.

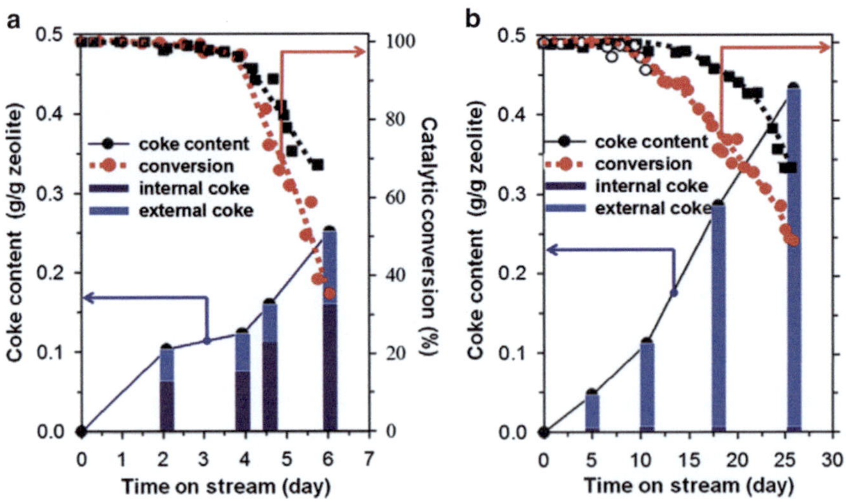

Fig. 4.26 Coke deposition in MFI zeolite nanosheet (**a**) and bulk MFI zeolite (**b**) during methanol-to-hydrocarbon conversion (Reprinted by permission from Macmillan Publishers Ltd: Ref. [137], Copyright 2009)

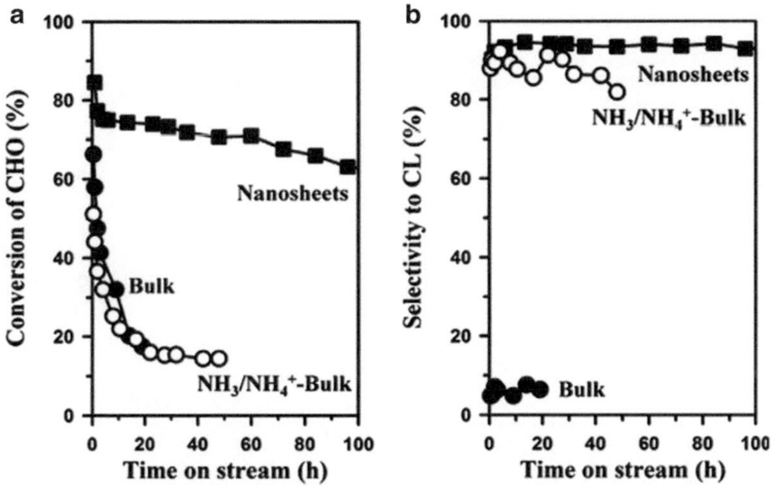

Fig. 4.27 Catalytic activity (**a**) and ε-caprolactam selectivity (**b**) as a function of the time on-stream during the gas-phase Beckmann rearrangement of cyclohexanone oxime into ε-caprolactam over MFI silica catalysts (Reprinted with the permission from Ref. [162]. Copyright 2011, American Chemical Society)

Improved catalytic stability and selectivity were also demonstrated in the selective formation of propylene from methanol. High-silica MFI nanosheets favored a higher molecular diffusion compared to the bulk counterpart, presumed to dramatically hinder secondary reactions like aromatization and hydrogen transfer [166].

Pt-loaded MFI zeolite nanosheets of different thicknesses were observed to display catalytic activity in *n*-heptane and *n*-decane hydroconversion [167–169]. Because of a short diffusion path length, catalytic performances revealed a higher isomerization yield compared to the bulk counterpart, as branched products were presumed to escape before cracking (Fig. 4.28). Molecular shape selectivity was equally observed with the hierarchical MFI zeolite. The *n*-decane test developed by Martens et al. [170, 171] has proved to be efficient for gaining information regarding the observed molecular shape selectivity with MFI nanosheets. Diffusion-based discrimination of reaction products in the MFI nanosheets was limited because of its remarkable ultrathin thickness [168]. The presence of molecular shape selectivity could still be clearly evidenced because of the framework termination in the *b* direction with 10-MR pore mouths.

The dimensionality and narrowness of the conventional porous system of zeolites often severely limits the surface accessibility and diffusion of bulky species. As heterogeneous acid catalysts, hierarchical zeolites have attracted increasing attention given the high concentration of external acid sites in the structure of the solids. The MFI nanosheets exhibited rapid catalytic conversion in acid-catalyzed reactions

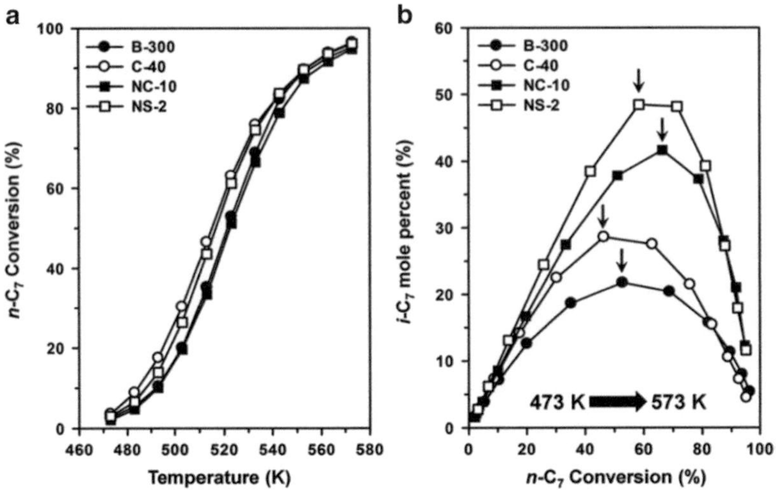

Fig. 4.28 (**a**) *n*-heptane conversion by Pt(IE)/MFI catalysts plotted as a function of reaction temperature (reaction conditions: WHSV = 6.8 h^{-1}, H_2/*n*-heptane molar ratio = 10, and atmospheric hydrogen pressure). (**b**) heptane isomer yield vs. total *n*-heptane conversion as a function of increasing reaction temperature. *Arrows* indicate the considered maximum isomerization yield (Reprinted from Ref. [167], Copyright 2013, with permission from Elsevier)

involving bulky molecular species (e.g., cracking of branched polyethylene, condensation of 2-hydroxyacetophenone with benzaldehyde, protection of benzaldehyde with pentaerythritol, cracking of decalin, and bio-oil) [137, 149, 172–174]. Furthermore, some of these reactions required strong Brønsted acidity at the external surface, which is a remarkable feature of the MFI nanosheets. With decalin cracking, a linear correlation was successfully established between the number of external strong Brønsted acid sites and observed catalytic activity [148]. Significant variations in TOF values were later observed over ZSM-5 nanocrystals and nanosheets in a gas-phase aromatic (chlorobenzene and nitrobenzene) chlorination reaction using TCCA, requiring mild and strong Brønsted acidity, respectively [175].

Valuable insights into the nature of external catalytic sites were equally provided in the cases of ultrathin MFI silica nanosheets during the gas-phase Beckmann rearrangement of cyclohexanone oxime to *e*-caprolactam (a precursor of Nylon-6) (Fig. 4.27) [162]. The reaction requires mild acidity of silanol groups in purely silicate frameworks. Nonetheless, whereas amorphous silicas exhibit poor selectivity values, pure MFI silica solids treated in ammonium solutions are known to be the most suitable catalyst in the gas-phase reaction. The ammonia treatment generates nest silanols by local desilication from the framework, resulting in high activity and product selectivity [176, 177]. On the MFI nanosheets, the high catalytic activity and selectivity were tentatively attributed to a zeolite confinement effect, termed the "nest effect," at the external pore openings (pore mouths) of the surface of the hierarchical zeolite (010 planes) [162, 178]. Under selected conditions, the solid displayed a tenfold catalytic longevity compared to the commercial catalyst, providing yet another example of the extended catalytic lifetime following the surfactant-directed synthesis approach.

Surfactant-directed hierarchical zeolites possessing large external surfaces represent a valuable addition as a supporting material for metal nanoparticles in a wide number of catalytic reactions, e.g., the gas-phase Fischer–Tropsch reaction. The MFI zeolite nanosponges exhibited remarkable dispersion of cobalt nanoparticles compared to the conventional bulk counterparts, resulting in higher catalytic conversion of CO to hydrocarbons [179]. Cobalt metal species display weak interaction with silanol groups in zeolite frameworks, tending to aggregate. The nanoparticles exhibited a diameter range between 4 and 7 nm, remarkably lower compared to those observed in conventional bulk MFI with small external surfaces (10 ~ 50 nm) (Fig. 4.29). More interestingly, the diameters of the cobalt nanoparticles are quite similar to the diameter of the mesopores of the MFI zeolite nanosponge, suggesting a confining effect for the generation of the cobalt nanoparticles.

Transition metals exhibiting redox properties were reported to be incorporated into the MFI ultrathin framework, using $C_{16-6-6}(OH)_2$ surfactant as a mesopore SDA [180]. Titanosilicate MFI nanosheets were suitable for the epoxidation of bulky cyclic olefins (e.g., cyclohexene and cyclooctene), displaying enhanced catalytic activity compared to TS-1. External Ti sites exhibited lower catalytic activity for the epoxidation of terminal alkenes, such as 1-hexene [180]. Higher catalytic activity was later obtained with a highly crystalline lamellar titanosilicate in the epoxidation of various bulky alkenes (e.g., 1-hexene, cyclohexene, cyclooctene, and cyclododecene) in the presence of tert-butyl hydroperoxide, cumene hydroperoxide, or aqueous H_2O_2 [181]. High-crystalline MFI-type stannosilicates (Sn–MFI) with nanosheet morphology and isolated framework Sn sites were also synthesized by another group [182]. The Sn source, Si/Sn ratio, crystallization time, and temperature were confirmed to affect framework Sn incorporation and nanosheet morphology. Compared to the bulk Sn–MFI counterpart, the 2.5-nm-thick nanosheets displayed high activity in the Baeyer–Villiger oxidation of bulky cyclic ketones using hydrogen peroxide (H_2O_2).

Aside from their successful application as heterogeneous catalysts, surfactant-directed mesoporous materials are expected to be of potential interest in future exploratory work on the advanced application of adsorption, separation, and sensor technologies. Positively charged MFI nanosheets containing the template prior to calcination have also been identified as suitable as anion exchangers (Fig. 4.30) [183]. The successful synthesis of silicalite-1 zeosil with a lamellar morphology was presumed to pave a new way for obtaining zeosil materials of controlled size and shape for molecular decontamination [184]. MFI nanosheets coated on porous α-alumina were further reported to be suitable as a selective membrane displaying high performance in xylene isomer separation, with a p-xylene/o-xylene separation

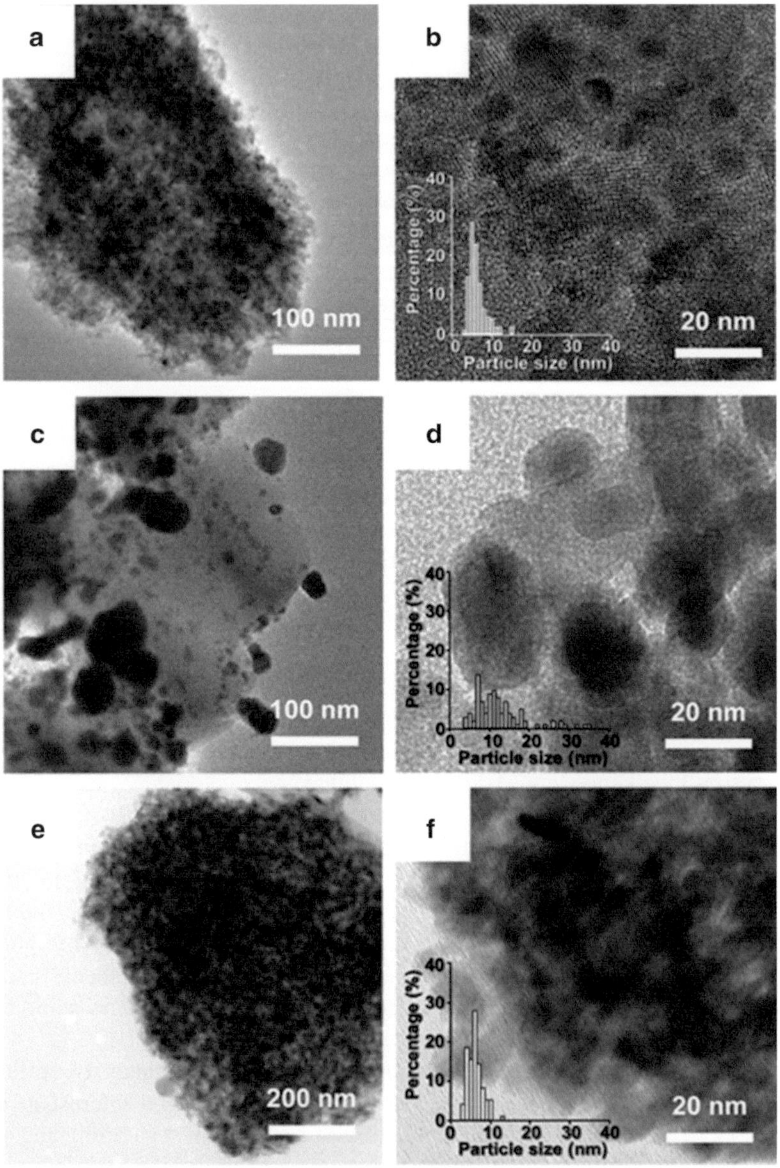

Fig. 4.29 TEM images of (**a**, **b**) Co/nanosponge MFI, (**c**, **d**) Co/bulk MFI, and (**e**, **f**) Co/γ-Al_2O_3 catalysts. Size distribution of cobalt nanoparticles supported on nanosponge MFI, bulk MFI, and γ-Al_2O_3 were derived from TEM images (Reprinted with the permission from Ref. [179]. Copyright 2014, American Chemical Society)

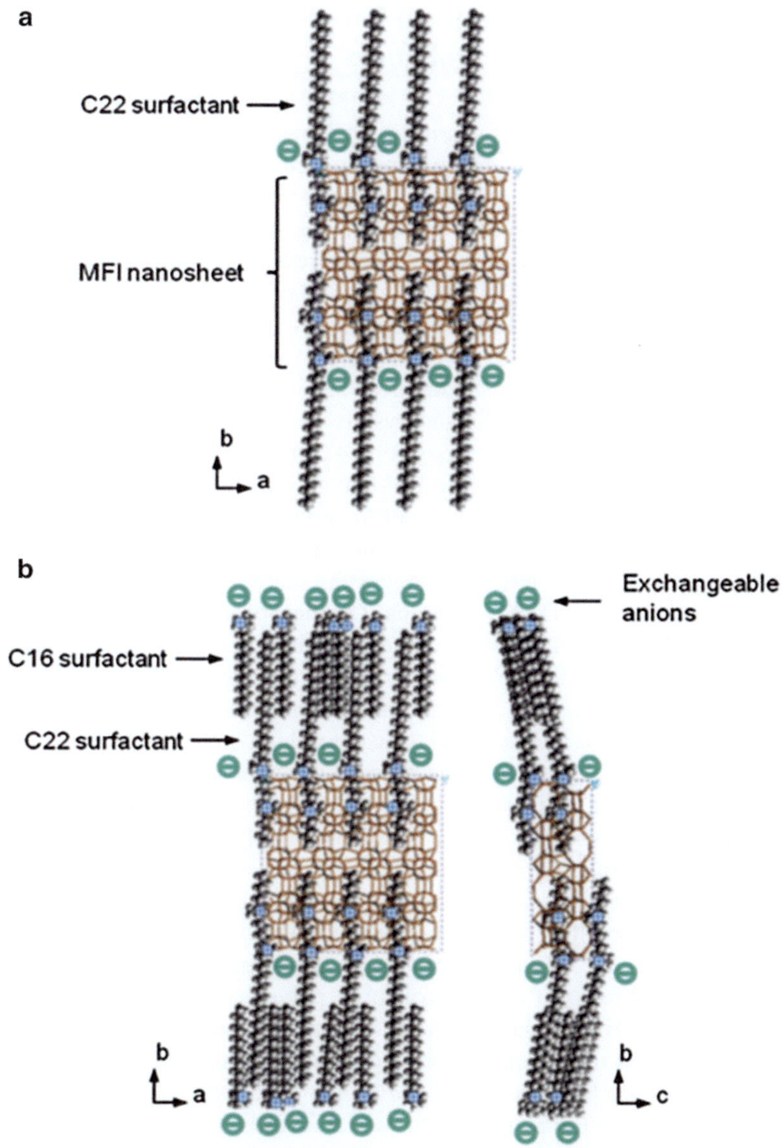

Fig. 4.30 Proposed structure models of positively charged MFI nanosheets prepared by **(a)** C_{16-6-6} and **(b)** C_{22-6-6} for the application of anion exchanger (Reproduced from Ref. [183] by permission of the Royal Society of Chemistry)

factor of 40–70 [185]. The nanosheets are in addition expected to be used as a membrane themselves, without any further treatments. The surfactant-directed mesoporous zeolites are equally suitable to be functionalized at the mesopore surface by using a high density of silanol groups, as previously reported for hierarchical zeolites synthesized in the presence of organosilane molecules.

4.6 Generation of Hierarchical Pore Structure in Zeolite by Random-Graft Copolymers

Following the synthesis of surfactant-directed hierarchical zeolites, Ryoo et al. reported a new type of dual micro- and mesoporous SDAs in which the multiammonium groups were functionalized to a linear polystyrene chain at a random interval [186]. Similar to previous works involving dual-porogenic surfactants, the multiammonium group was designed to direct the zeolite architecture. The strategy proved to be successful in the synthesis of a variety of zeolite architectures (e.g., MFI and BEA) and their analogues (e.g., aluminophosphate with structure type ATO).

Concerning the synthesis of hierarchical zeolites, e.g., ZSM-5, representative SEM images revealed thin MFI nanocrystals connected into a nanosponge-like three-dimensional network morphology (Fig. 4.31). Uniform nanocrystals, tailored to a thickness of 4.5 nm along the crystal *b*-axis, were further illustrated by high-resolution TEM analysis. The thickness of the crystals could be systematically tailored according to the size of multiammonium groups. The as-synthesized sample exhibited no micro- or mesoporosity, suggesting that the micropores in the zeolite domain were occupied by the SDA groups. The mesopores were likely occupied by a polystyrene backbone that was inaccessible in the zeolitic micropores. The nanosponge-type mesostructure could be retained after removal of polymer by calcination, with high mesopore volume and a large surface area being observed. Mesopores were, however, considered to be not as highly ordered as in mesoporous MCM-41 materials. The present strategy could generate a very narrow distribution of mesopore diameters, which were easily tailored according to the functionalization degree of multiammonium groups in the polymer [186].

To explain the formation of zeolite mesostructure, a polymer-fringed micellar crystallization mechanism was proposed (Fig. 4.32). Tight binding of the polymer side groups with inorganic species was believed to lead to an increase in inorganic concentration along the polymer chain. Hence, polymerization of inorganic species was promoted to form a mesostructured polymer–inorganic composite gel. Upon subsequent solvothermal treatment, the inorganic was transformed to a crystalline framework as tight bonding to the polymer was maintained. Following the crystal growth, steric hindrance imposed by the rigidly bound polymer-like capping agent was claimed to restrict the crystal thickness to a micelle-like region of only a few nanometers. Further crystal growth leading to Oswald ripening is therefore suppressed.

The synthesis of hierarchical materials capped by multiammonium-functionalized polymers was further reported to be successful for the synthesis of TiO_2 nanosheets of sub-nanometer thickness and mesoporous TiO_2, SnO_2, and ZrO_2. The simple and potentially mass-producible synthesis was concluded to be a valuable addition in gaining accessibility to unexplored mesostructured materials.

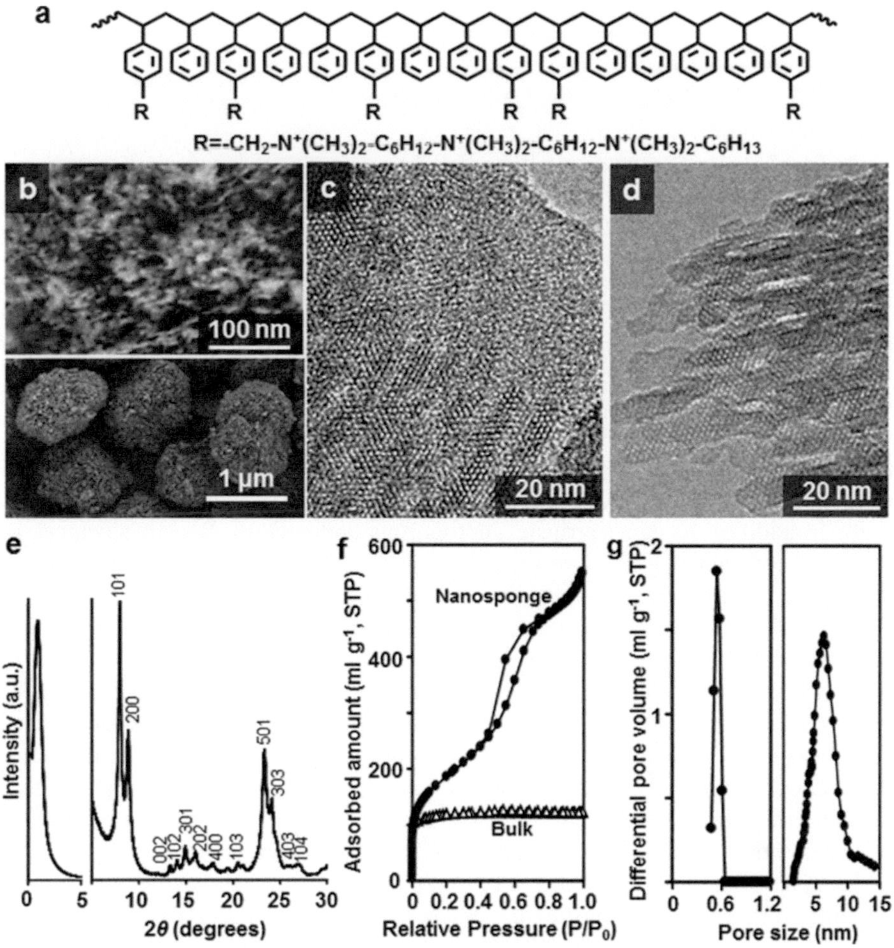

Fig. 4.31 MFI zeolite nanosponge synthesized using polymer-fringed crystallization mechanism. (**a**) Molecular structure of polymer used in the MFI zeolite synthesis. (**b**) and (**c**) SEM and TEM images of MFI zeolite nanosponge (as synthesized). (**d**) and (**e**) TEM image and XRD pattern of MFI zeolite after removal of polymer by calcination; (**f**) argon sorption isotherm of the calcined MFI zeolite at 87 K, in comparison with bulk MFI. (**g**) Pore size distribution derived from the adsorption isotherm, using nonlocal density-functional-theory method (Reproduced from Ref. [186] by permission of John Wiley & Sons Ltd)

4.7 Conclusion

The sole presence of the microporous architecture in zeolites often induces a severe intracrystalline diffusion limitation. Full utilization of a zeolite is, in this case, rarely achieved. Over the years, exploratory work has led to the synthesis of hierarchical

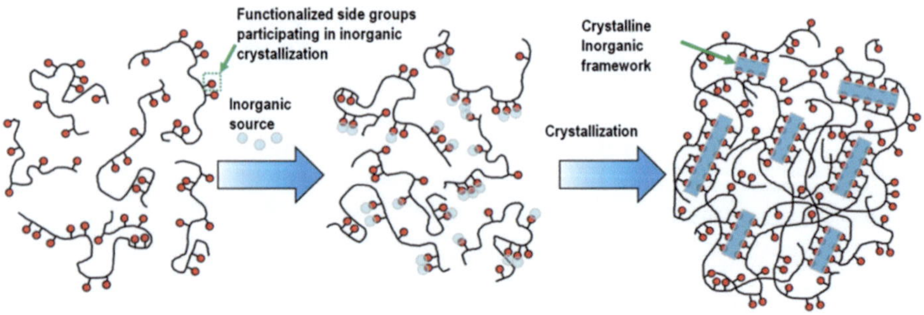

Fig. 4.32 Schematic description of polymer-fringed inorganic micellar crystallization (Reproduced from Ref. [186] by permission of John Wiley & Sons Ltd)

zeolites using top-down or bottom-up approaches. In this chapter, we have given an overview of various methods for synthesizing zeolites possessing a system of mesopores in addition to their intrinsic microporous architecture. Recent advances in strategies for enhancing zeolite accessibility and effectiveness for applications in adsorption, separation, and catalysis were carefully examined.

Of the top-down methods, post-synthetic demetallation is among the most commonly reported. This simple approach can effectively generate mesopores in zeolites, with practical applications at a mass-production scale. The treatment conditions, however, require further optimization for individual zeolites of particular structure types and, furthermore, of different Si-to-Al molar ratios. Furthermore, depending on the selected conditions, framework defect sites are often generated, leading to decreased crystallinity, changes in acidic properties, and poor hydrothermal or thermal stability when compared to the bulk counterpart.

Compared with demetallation, hard-templating approaches have the advantage of controlling the mesoporous texture and preserving zeolite crystallinity. Although a variety of solids have been used, the most general and versatile approaches apply different types of porous carbons. In addition to the use of expensive materials as hard templates, the synthesis conditions should be optimized to ensure that a confined zeolite crystallization occurs exclusively inside the pores of porous templates or the interstitial space between template particles.

Synthetic methods using zeolite-structure-directing surfactants are a valuable addition, suitable for the synthesis of a wide range of zeolitic architecture (e.g., MFI, BEA, MTW, and MRE). Various surfactants can be designed to control the zeolite framework types, framework thickness (typically, 1.5 ~ 6 nm), and mesopore diameters (typically, 3 ~ 15 nm) between frameworks in an active manner. The zeolite framework type is determined by a rational choice of the zeolite-structure-directing functional group. The thickness of the zeolite framework (i.e., mesopore wall) can be tailored by the number of the structure-directing modules in the function group. The mesopore diameter can be tuned further by the surfactant tails and by the addition of a pore-expanding agent. The surfactant-directed zeolites typically exhibit the morphologies of nanosheets or nanosponges. The mesopores are disordered but often have quite a narrow pore size distribution comparable to ordered mesoporous silicas. The resulting hierarchical zeolites (and zeotypes) have a high specific volume of mesopores and very high BET surface area. Another attractive approach is to use amorphous linear polymers as a two-level hierarchical structure-directing agent. In this method, the polymers are functionalized with zeolite-structure-directing groups as in the zeolite-structure-directing surfactants. The polymers can generate mesoporous zeolite nanosponges with controlled pore diameters and pore wall thickness.

Compared to their conventional bulk counterpart, hierarchical zeolites synthesized using the aforementioned methods show rapid molecular diffusion. In addition, the mesostructured zeolites have much larger external surfaces where the large molecules can be adsorbed. Such properties have significant merits and point to a high future potential of the mesostructured zeolites as heterogeneous catalysts and high-valued supporting materials for metal nanocatalysts.

References

1. Breck DW (1974) Zeolite molecular sieves: structure, chemistry, and use. Wiley, New York
2. Corma A (1997) From microporous to mesoporous molecular sieve materials and their use in catalysis. Chem Rev 97(6):2373–2419
3. Corma A (2003) State of the art and future challenges of zeolites as catalysts. J Catal 216(1–2):298–312
4. Perez-Ramirez J, Christensen CH, Egeblad K, Christensen CH, Groen JC (2008) Hierarchical zeolites: enhanced utilisation of microporous crystals in catalysis by advances in materials design. Chem Soc Rev 37(11):2530–2542
5. Rinaldi R, Schuth F (2009) Design of solid catalysts for the conversion of biomass. Energy Environ Sci 2(6):610–626
6. Davis ME, Saldarriaga C, Montes C, Garces J, Crowder C (1988) A molecular-sieve with 18-membered rings. Nature 331(6158):698–699
7. Davis ME (2002) Ordered porous materials for emerging applications. Nature 417(6891):813–821
8. Corma A, Diaz-Cabanas MJ, Jorda JL, Martinez C, Moliner M (2006) High-throughput synthesis and catalytic properties of a molecular sieve with 18-and 10-member rings. Nature 443(7113):842–845
9. Freyhardt CC, Tsapatsis M, Lobo RF, Balkus KJ, Davis ME (1996) A high-silica zeolite with a 14-tetrahedral-atom pore opening. Nature 381(6580):295–298
10. Strohmaier KG, Vaughan DEW (2003) Structure of the first silicate molecular sieve with 18-ring pore openings, ECR-34. J Am Chem Soc 125(51):16035–16039
11. Corma A, Diaz-Cabanas MJ, Rey F, Nicolooulas S, Boulahya K (2004) ITQ-15: the first ultralarge pore zeolite with a bi-directional pore system formed by intersecting 14- and 12-ring channels, and its catalytic implications. Chem Commun 12:1356–1357
12. Corma A, Diaz-Cabanas M, Martinez-Triguero J, Rey F, Rius J (2002) A large-cavity zeolite with wide pore windows and potential as an oil refining catalyst. Nature 418(6897):514–517
13. Jiang JX, Yu JH, Corma A (2010) Extra-large-pore zeolites: bridging the gap between micro and mesoporous structures. Angew Chem Int Ed 49(18):3120–3145
14. Blasco T, Corma A, Diaz-Cabanas MJ, Rey F, Vidal-Moya JA, Zicovich-Wilson CM (2002) Preferential location of Ge in the double four-membered ring units of ITQ-7 zeolite. J Phys Chem B 106(10):2634–2642

15. Corma A, Navarro MT, Rey F, Valencia S (2001) Synthesis of pure polymorph C of beta zeolite in a fluoride-free system. Chem Commun 16:1486–1487
16. Sastre G, Pulido A, Castaneda R, Corma A (2004) Effect of the germanium incorporation in the synthesis of EU-1, ITQ-13, ITQ-22, and ITQ-24 zeolites. J Phys Chem B 108(26):8830–8835
17. Sun JL, Bonneau C, Cantin A, Corma A, Diaz-Cabanas MJ, Moliner M, Zhang DL, Li MR, Zou XD (2009) The ITQ-37 mesoporous chiral zeolite. Nature 458(7242):1154–1157
18. Mintova S, Olson NH, Bein T (1999) Electron microscopy reveals the nucleation mechanism of zeolite Y from precursor colloids. Angew Chem Int Ed 38(21):3201–3204
19. Tosheva L, Valtchev VP (2005) Nanozeolites: synthesis, crystallization mechanism, and applications. Chem Mater 17(10):2494–2513
20. Larlus O, Mintova S, Bein T (2006) Environmental syntheses of nanosized zeolites with high yield and monomodal particle size distribution. Microporous Mesoporous Mater 96(1–3):405–412
21. Larsen SC (2007) Nanocrystalline zeolites and zeolite structures: synthesis, characterization, and applications. J Phys Chem C 111(50):18464–18474
22. Tao YS, Kanoh H, Abrams L, Kaneko K (2006) Mesopore-modified zeolites: preparation, characterization, and applications. Chem Rev 106(3):896–910
23. Egeblad K, Christensen CH, Kustova M, Christensen CH (2008) Templating mesoporous zeolites. Chem Mater 20(3):946–960
24. Schuth F (2003) Endo- and exotemplating to create high-surface-area inorganic materials. Angew Chem Int Ed 42(31):3604–3622
25. Scherzer J (1978) De-aluminated faujasite-type structures with SiO_2-Al_2O_3 ratios over 100. J Catal 54(2):285–288
26. Lynch J, Raatz F, Dufresne P (1987) Characterization of the textural properties of dealuminated HY forms. Zeolites 7(4):333–340
27. Cartlidge S, Nissen HU, Wessicken R (1989) Ternary mesoporous structure of ultrastable zeolite CSZ-1. Zeolites 9(4):346–349
28. Choifeng C, Hall JB, Huggins BJ, Beyerlein RA (1993) Electron-microscope investigation of mesopore formation and aluminum migration in USY catalysts. J Catal 140(2):395–405
29. Sasaki Y, Suzuki T, Takamura Y, Saji A, Saka H (1998) Structure analysis of the mesopore in dealuminated zeolite Y by high resolution TEM observation with slow scan CCD camera. J Catal 178(1):94–100
30. Janssen AH, Koster AJ, de Jong KP (2001) Three-dimensional transmission electron microscopic observations of mesopores in dealuminated zeolite Y. Angew Chem Int Ed 40(6):1102–1104
31. Janssen AH, Koster AJ, de Jong KP (2002) On the shape of the mesopores in zeolite Y: a three-dimensional transmission electron microscopy study combined with texture analysis. J Phys Chem B 106(46):11905–11909
32. Dutartre R, deMenorval LC, DiRenzo F, McQueen D, Fajula F, Schulz P (1996) Mesopore formation during steam dealumination of zeolites: influence of initial aluminum content and crystal size. Microporous Mater 6(5–6):311–320
33. McQueen D, Chiche BH, Fajula F, Auroux A, Guimon C, Fitoussi F, Schulz P (1996) A multitechnique characterization of the acidity of dealuminated mazzite. J Catal 161(2):587–596
34. Nesterenko NS, Thibault-Starzyk F, Montouillout V, Yuschenko VV, Fernandez C, Gilson JP, Fajula F, Ivanova II (2004) Accessibility of the acid sites in dealuminated small-port mordenites studied by FTIR of co-adsorbed alkylpyridines and CO. Microporous Mesoporous Mater 71(1–3):157–166
35. Lee KH, Ha BH (1998) Characterization of mordenites treated by HCl/steam or HF. Microporous Mesoporous Mater 23(3–4):211–219
36. Perez-Ramirez J, Kapteijn F, Groen JC, Domenech A, Mul G, Moulijn JA (2003) Steam-activated FeMFI zeolites: evolution of iron species and activity in direct N_2O decomposition. J Catal 214(1):33–45

37. Lago RM, Haag WO, Mikovsky RJ, Olson DH, Hellring SD, Schmitt KD, Kerr GT (1986) The nature of the catalytic sites in HZSM-5- activity enhancement. In: Murakami AI Y, Ward JW (eds) Studies in surface science and catalysis, vol 28. Elsevier, Amsterdam, pp 677–684
38. Rozwadowski M, Kornatowski J, Wloch J, Erdmann K, Golembiewski R (2002) Attempt to apply the fractal geometry for characterisation of dealuminated ZSM-5 zeolite. Appl Surf Sci 191(1–4):352 361
39. Levanmao R, Vo NTC, Sjiariel B, Lee L, Denes G (1992) Mesoporous aluminosilicates – preparation from Ca-a zeolite by treatment with ammonium fluorosilicate. J Mater Chem 2(6):595–599
40. Triantafyllidis KS, Vlessidis AG, Evmiridis NP (2000) Dealuminated H-Y zeolites: influence of the degree and the type of dealumination method on the structural and acidic characteristics of H-Y zeolites. Ind Eng Chem Res 39(2):307–319
41. Beyer HK, Belenykaja I (1980) A new method for the dealumination of faujasite-type zeolites. In: Imelik B, Naccache C, Ben Taarit Y, Vádrine JC, Condurier G, Praliaud H (eds) Studies in surface science and catalysis, vol 5. Elsevier, Amsterdam, pp 203–210
42. Scherzer J (1984) The preparation and characterization of aluminum-deficient zeolites. ACS Symp Ser 248:157–200
43. Goyvaerts D, Martens JA, Grobet PJ, Jacobs PA (1991) Factors affecting the formation of extra-framework species and mesopores during dealumination of zeolite Y. In: Poncelet G, Jacobs PA, Grange P, Delmon B (eds) Studies in surface science and catalysis, vol 63. Elsevier, Amsterdam, pp 381–395
44. Dessau RM, Valyocsik EW, Goeke NH (1992) Aluminum zoning in Zsm-5 as revealed by selective silica removal. Zeolites 12(7):776–779
45. Ogura M, Shinomiya SY, Tateno J, Nara Y, Nomura M, Kikuchi E, Matsukata M (2001) Alkali-treatment technique – new method for modification of structural and acid-catalytic properties of ZSM-5 zeolites. Appl Catal A Gen 219(1–2):33–43
46. Groen JC, Perez-Ramirez J, Peffer LAA (2002) Formation of uniform mesopores in ZSM-5 zeolite upon alkaline post-treatment? Chem Lett 1:94–95
47. Groen JC, Peffer LAA, Perez-Ramirez J (2003) Pore size determination in modified micro- and mesoporous materials: pitfalls and limitations in gas adsorption data analysis. Microporous Mesoporous Mater 60(1–3):1–17
48. Groen JC, Jansen JC, Moulijn JA, Perez-Ramirez J (2004) Optimal aluminum-assisted mesoporosity development in MFI zeolites by desilication. J Phys Chem B 108(35):13062–13065
49. Groen JC, Bach T, Ziese U, Donk AMPV, de Jong KP, Moulijn JA, Perez-Ramirez J (2005) Creation of hollow zeolite architectures by controlled desilication of Al-zoned ZSM-5 crystals. J Am Chem Soc 127(31):10792–10793
50. Groen JC, Moulijn JA, Perez-Ramirez J (2005) Decoupling mesoporosity formation and acidity modification in ZSM-5 zeolites by sequential desilication-dealumination. Microporous Mesoporous Mater 87(2):153–161; Groen JC, Moulijn JA, Perez-Ramirez J (2006) Desilication: on the controlled generation of mesoporosity in MFI zeolites. J Mater Chem 16(22):2121–2131; Verboekend D, Perez-Ramirez J (2011) Design of hierarchical zeolite catalysts by desilication. Catal Sci Technol 1(6):879–890; Milina M, Mitchell S, Crivelli P, Cooke D, Perez-Ramirez J (2014) Mesopore quality determines the lifetime of hierarchically structured zeolite catalysts. Nat Commun 5
51. Su LL, Liu L, Zhuang JQ, Wang HX, Li YG, Shen WJ, Xu YD, Bao XH (2003) Creating mesopores in ZSM-5 zeolite by alkali treatment: a new way to enhance the catalytic performance of methane dehydroaromatization on Mo/HZSM-5 catalysts. Catal Lett 91(3–4):155–167
52. Chal R, Cacciaguerra T, van Donk S, Gerardin C (2010) Pseudomorphic synthesis of mesoporous zeolite Y crystals. Chem Commun 46(41):7840–7842
53. de Jong KP, Zecevic J, Friedrich H, de Jongh PE, Bulut M, van Donk S, Kenmogne R, Finiels A, Hulea V, Fajula F (2010) Zeolite Y crystals with trimodal porosity as ideal hydrocracking

catalysts. Angew Chem Int Ed 49(52):10074–10078; Verboekend D, Vile G, Perez-Ramirez J (2012) Hierarchical Y and USY zeolites designed by post-synthetic strategies. Adv Funct Mater 22(5):916–928

54. Jacobsen CJH, Madsen C, Houzvicka J, Schmidt I, Carlsson A (2000) Mesoporous zeolite single crystals. J Am Chem Soc 122(29):7116–7117
55. Christensen CH, Johannsen K, Schmidt I, Christensen CH (2003) Catalytic benzene alkylation over mesoporous zeolite single crystals: improving activity and selectivity with a new family of porous materials. J Am Chem Soc 125(44):13370–13371
56. Schmidt I, Boisen A, Gustavsson E, Stahl K, Pehrson S, Dahl S, Carlsson A, Jacobsen CJH (2001) Carbon nanotube templated growth of mesoporous zeolite single crystals. Chem Mater 13(12):4416–4418
57. Boisen A, Schmidt I, Carlsson A, Dahl S, Brorson M, Jacobsen CJH (2003) TEM stereoimaging of mesoporous zeolite single crystals. Chem Commun 8:958–959
58. Janssen AH, Schmidt I, Jacobsen CJH, Koster AJ, de Jong KP (2003) Exploratory study of mesopore templating with carbon during zeolite synthesis. Microporous Mesoporous Mater 65(1):59–75
59. Zhu K, Egeblad K, Christensen CH (2007) Mesoporous carbon prepared from carbohydrate as hard template for hierarchical zeolites. Eur J Inorg Chem 25:3955–3960
60. Kustova M, Egeblad K, Zhu K, Christensen CH (2007) Versatile route to zeolite single crystals with controlled mesoporosity: in situ sugar decomposition for templating of hierarchical zeolites. Chem Mater 19(12):2915–2917
61. Tao YS, Kanoh H, Kaneko K (2003) ZSM-5 monolith of uniform mesoporous channels. J Am Chem Soc 125(20):6044–6045
62. Pekala RW, Alviso CT, Kong FM, Hulsey SS (1992) Aerogels derived from multifunctional organic monomers. J Non-Cryst Solids 145(1–3):90–98
63. Bekyarova E, Kaneko K (2000) Structure and physical properties of tailor-made Ce, Zr-doped carbon aerogels. Adv Mater 12(21):1625–1628
64. Hanzawa Y, Kaneko K, Yoshizawa N, Pekala RW, Dresselhaus MS (1998) The pore structure determination of carbon aerogels. Adsorption 4(3–4):187–195
65. Hanzawa Y, Hatori H, Yoshizawa N, Yamada Y (2002) Structural changes in carbon aerogels with high temperature treatment. Carbon 40(4):575–581
66. Tao YS, Kanoh H, Hanzawa Y, Kaneko K (2004) Template synthesis and characterization of mesoporous zeolites. Colloid Surf A 241(1–3):75–80
67. Kustova MY, Hasselriis P, Christensen CH (2004) Mesoporous MEL-type zeolite single crystal catalysts. Catal Lett 96(3–4):205–211
68. Wei XT, Smirniotis PG (2006) Synthesis and characterization of mesoporous ZSM-12 by using carbon particles. Microporous Mesoporous Mater 89(1–3):170–178
69. Pavlačková Z, Košová G, Žilková N, Zukal A, Čejka J (2006) Formation of mesopores in ZSM-5 by carbon templating. Stud Surf Sci Catal 162:905–912
70. Egeblad K, Kustova M, Klitgaard SK, Zhu KK, Christensen CH (2007) Mesoporous zeolite and zeotype single crystals synthesized in fluoride media. Microporous Mesoporous Mater 101(1–2):214–223
71. Sakthivel A, Huang SJ, Chen WH, Lan ZH, Chen KH, Kim TW, Ryoo R, Chiang AST, Liu SB (2004) Replication of mesoporous aluminosilicate molecular sieves (RMMs) with zeolite framework from mesoporous carbons (CMKs). Chem Mater 16(16):3168–3175
72. Yang ZX, Xia YD, Mokaya R (2004) Zeolite ZSM-5 with unique supermicropores synthesized using mesoporous carbon as a template. Adv Mater 16(8):727–732
73. Ryoo R, Joo SH, Jun S (1999) Synthesis of highly ordered carbon molecular sieves via template-mediated structural transformation. J Phys Chem B 103(37):7743–7746
74. Ryoo R, Joo SH, Kruk M, Jaroniec M (2001) Ordered mesoporous carbons. Adv Mater 13(9):677–681
75. Fang YM, Hu HQ (2006) An ordered mesoporous aluminosilicate with completely crystalline zeolite wall structure. J Am Chem Soc 128(33):10636–10637

76. Fan W, Snyder MA, Kumar S, Lee PS, Yoo WC, McCormick AV, Penn RL, Stein A, Tsapatsis M (2008) Hierarchical nanofabrication of microporous crystals with ordered mesoporosity. Nat Mater 7(12):984–991
77. Tao YS, Kanoh H, Kaneko K (2005) Synthesis of mesoporous zeolite a by resorcinol-formaldehyde aerogel templating. Langmuir 21(2):504–507
78. Holland BT, Abrams L, Stein A (1999) Dual templating of macroporous silicates with zeolitic microporous frameworks. J Am Chem Soc 121(17):4308–4309
79. Cho HS, Ryoo R (2012) Synthesis of ordered mesoporous MFI zeolite using CMK carbon templates. Microporous Mesoporous Mater 151:107–112
80. Na K, Choi M, Ryoo R (2013) Recent advances in the synthesis of hierarchically nanoporous zeolites. Microporous Mesoporous Mater 166:3–19
81. Barton TJ, Bull LM, Klemperer WG, Loy DA, McEnaney B, Misono M, Monson PA, Pez G, Scherer GW, Vartuli JC, Yaghi OM (1999) Tailored porous materials. Chem Mater 11(10):2633–2656
82. Ciesla U, Schuth F (1999) Ordered mesoporous materials. Microporous Mesoporous Mater 27(2–3):131–149
83. Mehlhorn D, Valiullin R, Karger J, Cho K, Ryoo R (2012) Exploring mass transfer in mesoporous zeolites by NMR diffusometry. Materials 5(4):699–720; Mehlhorn D, Valiullin R, Karger J, Cho K, Ryoo R (2012) Intracrystalline diffusion in mesoporous zeolites. Chemphyschem 13(6):1495–1499; Karger J, Valiullin R (2013) Mass transfer in mesoporous materials: the benefit of microscopic diffusion measurement. Chem Soc Rev 42(9):4172–4197
84. Milina M, Mitchell S, Cooke D, Crivelli P, Pérez-Ramírez J (2015) Impact of pore connectivity on the design of long-lived zeolite catalysts. Angew Chem Int Ed 54(5):1591–1594
85. Ying JY, Mehnert CP, Wong MS (1999) Synthesis and applications of supramolecular-templated mesoporous materials. Angew Chem Int Ed 38(1–2):56–77
86. Raman NK, Anderson MT, Brinker CJ (1996) Template-based approaches to the preparation of amorphous, nanoporous silicas. Chem Mater 8(8):1682–1701
87. Beck JS, Vartuli JC, Roth WJ, Leonowicz ME, Kresge CT, Schmitt KD, Chu CTW, Olson DH, Sheppard EW, Mccullen SB, Higgins JB, Schlenker JL (1992) A new family of mesoporous molecular-sieves prepared with liquid-crystal templates. J Am Chem Soc 114(27):10834–10843
88. Zhao DY, Huo QS, Feng JL, Chmelka BF, Stucky GD (1998) Nonionic triblock and star diblock copolymer and oligomeric surfactant syntheses of highly ordered, hydrothermally stable, mesoporous silica structures. J Am Chem Soc 120(24):6024–6036
89. Sayari A, Yang Y, Kruk M, Jaroniec M (1999) Expanding the pore size of MCM-41 silicas: use of amines as expanders in direct synthesis and postsynthesis procedures. J Phys Chem B 103(18):3651–3658
90. Schmidt-Winkel P, Lukens WW, Zhao DY, Yang PD, Chmelka BF, Stucky GD (1999) Mesocellular siliceous foams with uniformly sized cells and windows. J Am Chem Soc 121(1):254–255
91. Yanagisawa T, Shimizu T, Kuroda K, Kato C (1990) Trimethylsilyl derivatives of alkyltrimethylammonium-Kanemite complexes and their conversion to microporous SiO_2 materials. Bull Chem Soc Jpn 63(5):1535–1537
92. Inagaki S, Fukushima Y, Kuroda K (1994) Synthesis and characterization of highly ordered mesoporous material – FSM-16, from a layered polysilicate. Zeolites Relat Microporous Mater State Art 84:125–132
93. Kresge CT, Leonowicz ME, Roth WJ, Vartuli JC, Beck JS (1992) Ordered mesoporous molecular-sieves synthesized by a liquid-crystal template mechanism. Nature 359(6397):710–712
94. Zhao DY, Feng JL, Huo QS, Melosh N, Fredrickson GH, Chmelka BF, Stucky GD (1998) Triblock copolymer syntheses of mesoporous silica with periodic 50 to 300 angstrom pores. Science 279(5350):548–552

95. Monnier A, Schuth F, Huo Q, Kumar D, Margolese D, Maxwell RS, Stucky GD, Krishnamurty M, Petroff P, Firouzi A, Janicke M, Chmelka BF (1993) Cooperative formation of inorganic-organic interfaces in the synthesis of silicate mesostructures. Science 261(5126):1299–1303
96. Firouzi A, Kumar D, Bull LM, Besier T, Sieger P, Huo Q, Walker SA, Zasadzinski JA, Glinka C, Nicol J, Margolese D, Stucky GD, Chmelka BF (1995) Cooperative organization of inorganic-surfactant and biomimetic assemblies. Science 267(5201):1138–1143
97. Huo QS, Margolese DI, Ciesla U, Feng PY, Gier TE, Sieger P, Leon R, Petroff PM, Schuth F, Stucky GD (1994) Generalized synthesis of periodic surfactant inorganic composite-materials. Nature 368(6469):317–321
98. Moller K, Bein T (1998) Inclusion chemistry in periodic mesoporous hosts. Chem Mater 10(10):2950–2963
99. Hartmann M (2005) Ordered mesoporous materials for bioadsorption and biocatalysis. Chem Mater 17(18):4577–4593
100. Gusev VY, Feng XB, Bu Z, Haller GL, OBrien JA (1996) Mechanical stability of pure silica mesoporous MCM-41 by nitrogen adsorption and small-angle X-ray diffraction measurements. J Phys Chem 100(6):1985–1988
101. Jun S, Kim JM, Ryoo R, Ahn YS, Han MK (2000) Hydrothermal stability of MCM-48 improved by post-synthesis restructuring in salt solution. Microporous Mesoporous Mater 41(1–3):119–127
102. Cassiers K, Linssen T, Mathieu M, Benjelloun M, Schrijnemakers K, Van Der Voort P, Cool P, Vansant EF (2002) A detailed study of thermal, hydrothermal, and mechanical stabilities of a wide range of surfactant assembled mesoporous silicas. Chem Mater 14(5):2317–2324
103. Karlsson A, Stocker M, Schmidt R (1999) Composites of micro- and mesoporous materials: simultaneous syntheses of MFI/MCM-41 like phases by a mixed template approach. Microporous Mesoporous Mater 27(2–3):181–192
104. Liu Y, Zhang WZ, Pinnavaia TJ (2000) Steam-stable aluminosilicate mesostructures assembled from zeolite type Y seeds. J Am Chem Soc 122(36):8791–8792
105. Liu Y, Zhang WZ, Pinnavaia TJ (2001) Steam-stable MSU-S aluminosilicate mesostructures assembled from zeolite ZSM-5 and zeolite beta seeds. Angew Chem Int Ed 40(7):1255–1258
106. Zhang Z, Han Y, Xiao FS, Qiu S, Zhu L, Wang R, Yu Y, Zhang Z, Zou B, Wang Y, Sun H, Zhao D, Wei Y (2001) Mesoporous aluminosilicates with ordered hexagonal structure, strong acidity, and extraordinary hydrothermal stability at high temperatures. J Am Chem Soc 123(21):5014–5021
107. Liu Y, Pinnavaia TJ (2002) Assembly of hydrothermally stable aluminosilicate foams and large-pore hexagonal mesostructures from zeolite seeds under strongly acidic conditions. Chem Mater 14(1):3–5
108. Serrano DP, García RA, Vicente G, Linares M, Procházková D, Cejka J (2011) Acidic and catalytic properties of hierarchical zeolites and hybrid ordered mesoporous materials assembled from MFI protozeolitic units. J Catal 279(2):366–380
109. Xia YD, Mokaya R (2004) Are mesoporous silicas and aluminosilicas assembled from zeolite seeds inherently hydrothermally stable? Comparative evaluation of MCM-48 materials assembled from zeolite seeds. J Mater Chem 14(23):3427–3435
110. Naik SP, Chiang AST, Thompson RW (2003) Synthesis of zeolitic mesoporous materials by dry gel conversion under controlled humidity. J Phys Chem B 107(29):7006–7014
111. Huang L, Guo W, Deng P, Xue Z, Li Q (2000) Investigation of synthesizing MCM-41/ZSM-5 composites. J Phys Chem B 104(13):2817–2823
112. Xia Y, Mokaya R (2004) On the synthesis and characterization of ZSM-5/MCM-48 aluminosilicate composite materials. J Mater Chem 14(5):863–870
113. Guo W, Xiong C, Huang L, Li Q (2001) Synthesis and characterization of composite molecular sieves comprising zeolite Beta with MCM-41 structures. J Mater Chem 11(7):1886–1890

114. Guo W, Huang L, Deng P, Xue Z, Li Q (2001) Characterization of Beta/MCM-41 composite molecular sieve compared with the mechanical mixture. Microporous Mesoporous Mater 44–45(6):427–434
115. Prokešova P, Mintova S, Cejka J, Bein T (2003) Preparation of nanosized micro/mesoporous composites via simultaneous synthesis of Beta/MCM-48 phases. Microporous Mesoporous Mater 64(1–3):165–174
116. Naik SP, Chen JC, Chiang AST (2002) Synthesis of silicalite nanocrystals via the steaming of surfactant protected precursors. Microporous Mesoporous Mater 54(3):293–303
117. Hasan F, Singh R, Li G, Zhao D, Webley PA (2012) Direct synthesis of hierarchical LTA zeolite via a low crystallization and growth rate technique in presence of cetyltrimethylammonium bromide. J Colloid Interface Sci 382(1):1–12
118. Jo C, Jung J, Shin HS, Kim J, Ryoo R (2013) Capping with multivalent surfactants for zeolite nanocrystal synthesis. Angew Chem Int Ed 52(38):10014–10017
119. Choi M, Cho HS, Srivastava R, Venkatesan C, Choi DH, Ryoo R (2006) Amphiphilic organosilane-directed synthesis of crystalline zeolite with tunable mesoporosity. Nat Mater 5(9):718–723
120. Chmelka BF (2006) Zeolites – large molecules welcome. Nat Mater 5(9):681–682
121. Huttinger KJ, Jung MF (1989) Kinetics of the synthesis of trialkyl [–3-(trimethoxysilyl) propyl]-ammonium chloride and their antimicrobial action as fixed biocides. Chem Ing Tech 61(3):258–259
122. Ryoo R, Kim JM, Ko CH, Shin CH (1996) Disordered molecular sieve with branched mesoporous channel network. J Phys Chem 100(45):17718–17721
123. Galarneau A, Cambon H, Di Renzo F, Ryoo R, Choi M, Fajula F (2003) Microporosity and connections between pores in SBA-15 mesostructured silicas as a function of the temperature of synthesis. New J Chem 27(1):73–79
124. Cho K, Cho HS, de Menorval LC, Ryoo R (2009) Generation of mesoporosity in LTA zeolites by organosilane surfactant for rapid molecular transport in catalytic application. Chem Mater 21(23):5664–5673
125. Cho K, Ryoo R, Asahina S, Xiao C, Klingstedt M, Umemura A, Anderson MW, Terasaki O (2011) Mesopore generation by organosilane surfactant during LTA zeolite crystallization, investigated by high-resolution SEM and Monte Carlo simulation. Solid State Sci 13(4):750–756
126. Choi M, Lee DH, Na K, Yu BW, Ryoo R (2009) High catalytic activity of palladium(II)-exchanged mesoporous sodalite and NaA zeolite for bulky aryl coupling reactions: reusability under aerobic conditions. Angew Chem Int Ed 48(20):3673–3676
127. Valiullin R, Kärger J, Cho K, Choi M, Ryoo R (2011) Dynamics of water diffusion in mesoporous zeolites. Microporous Mesoporous Mater 142(1):236–244
128. Choi M, Srivastava R, Ryoo R (2006) Organosilane surfactant-directed synthesis of mesoporous aluminophosphates constructed with crystalline microporous frameworks. Chem Commun 42:4380–4382
129. Inayat A, Knoke I, Spiecker E, Schwieger W (2012) Assemblies of mesoporous FAU-type zeolite nanosheets. Angew Chem Int Ed 51(8):1962–1965
130. Kim J, Jo C, Lee S, Ryoo R (2014) Bulk crystal seeding in the generation of mesopores by organosilane surfactants in zeolite synthesis. J Mater Chem A 2(30):11905–11912
131. Srivastava R, Choi M, Ryoo R (2006) Mesoporous materials with zeolite framework: remarkable effect of the hierarchical structure for retardation of catalyst deactivation. Chem Commun 43:4489–4491
132. Shetti VN, Kim J, Srivastava R, Choi M, Ryoo R (2008) Assessment of the mesopore wall catalytic activities of MFI zeolite with mesoporous/microporous hierarchical structures. J Catal 254(2):296–303
133. Lee DH, Choi M, Yu BW, Ryoo R (2009) Organic functionalization of mesopore walls in hierarchically porous zeolites. Chem Commun 1:74–76

134. Vuong GT, Do TO (2007) A new route for the synthesis of uniform nanozeolites with hydrophobic external surface in organic solvent medium. J Am Chem Soc 129(13):3810–3811
135. Wang H, Pinnavaia TJ (2006) MFI zeolite with small and uniform intracrystal mesopores. Angew Chem Int Ed 45(45):7603–7606
136. Xiao FS, Wang L, Yin C, Lin K, Di Y, Li J, Xu R, Su DS, Schlögl R, Yokoi T, Tatsumi T (2006) Catalytic properties of hierarchical mesoporous zeolites templated with a mixture of small organic ammonium salts and mesoscale cationic polymers. Angew Chem Int Ed 45(19):3090–3093
137. Choi M, Na K, Kim J, Sakamoto Y, Terasaki O, Ryoo R (2009) Stable single-unit-cell nanosheets of zeolite MFI as active and long-lived catalysts. Nature 461(7261):246–249
138. Corma A (2009) Materials chemistry catalysts made thinner. Nature 461(7261):182–183
139. Park W, Yu D, Na K, Jelfs KE, Slater B, Sakamoto Y, Ryoo R (2011) Hierarchically structure-directing effect of multi-ammonium surfactants for the generation of MFI zeolite nanosheets. Chem Mater 23(23):5131–5137
140. Jung J, Jo C, Cho K, Ryoo R (2012) Zeolite nanosheet of a single-pore thickness generated by a zeolite-structure-directing surfactant. J Mater Chem 22(11):4637–4640
141. Na K, Choi M, Park W, Sakamoto Y, Terasaki O, Ryoo R (2010) Pillared MFI zeolite nanosheets of a single-unit-cell thickness. J Am Chem Soc 132(12):4169–4177
142. Maheshwari S, Jordan E, Kumar S, Bates FS, Penn RL, Shantz DF, Tsapatsis M (2008) Layer structure preservation during swelling, pillaring, and exfoliation of a zeolite precursor. J Am Chem Soc 130(4):1507–1516
143. Tsapatsis M, Maheshwari S (2008) Pores by pillaring: not always a maze. Angew Chem Int Ed 47(23):4262–4263
144. Roth WJ, Cejka J (2011) Two-dimensional zeolites: dream or reality? Catal Sci Technol 1(1):43–53
145. Schnell SK, Wu L, Koekkoek AJJ, Kjelstrup S, Hensen EJM, Vlugt TJH (2013) Adsorption of argon on MFI nanosheets: experiments and simulations. J Phys Chem C 117(46):24503–24510
146. Na K, Park W, Seo Y, Ryoo R (2011) Disordered assembly of MFI zeolite nanosheets with a large volume of intersheet mesopores. Chem Mater 23(5):1273–1279
147. Corma A (1995) Inorganic solid acids and their use in acid-catalyzed hydrocarbon reactions. Chem Rev 95(3):559–614
148. Cejka J, Centi G, Perez-Pariente J, Roth WJ (2012) Zeolite-based materials for novel catalytic applications: opportunities, perspectives and open problems. Catal Today 179(1):2–15
149. Seo Y, Cho K, Jung Y, Ryoo R (2013) Characterization of the surface acidity of MFI zeolite nanosheets by P-31 NMR of adsorbed phosphine oxides and catalytic cracking of decalin. ACS Catal 3(4):713–720
150. Lunsford JH, Rothwell WP, Shen W (1985) Acid sites in zeolite Y – a solid-state NMR and infrared study using trimethylphosphine as a probe molecule. J Am Chem Soc 107(6):1540–1547
151. Baltusis L, Frye JS, Maciel GE (1986) Phosphine oxide as NMR probes for adsorption sites on surfaces. J Am Chem Soc 108(22):7119–7120
152. Rakiewicz EF, Peters AW, Wormsbecher RF, Sutovich KJ, Müller KT (1998) Characterization of acid sites in zeolitic and other inorganic systems using solid-state P-31 NMR of the probe molecule trimethylphosphine oxide. J Phys Chem B 102(16):2890–2896
153. Zhao Q, Chen WH, Huang SJ, Wu YC, Lee HK, Liu SB (2002) Discernment and quantification of internal and external acid sites on zeolites. J Phys Chem B 106(17):4462–4469
154. Jo C, Cho K, Kim J, Ryoo R (2014) MFI zeolite nanosponges possessing uniform mesopores generated by bulk crystal seeding in the hierarchical surfactant-directed synthesis. Chem Commun 50(32):4175–4177
155. Na K, Jo C, Kim J, Cho K, Jung J, Seo Y, Messinger RJ, Chmelka BF, Ryoo R (2011) Directing zeolite structures into hierarchically nanoporous architectures. Science 333(6040):328–332

156. Möller K, Bein T (2011) Pores within pores-How to craft ordered hierarchical zeolites. Science 333(6040):297–298
157. Kore R, Srivastava R, Satpati B (2014) ZSM-5 zeolite nanosheets with improved catalytic activity synthesized using a new class of structure-directing agents. Chem Eur J 20(36):11511–11521
158. Jo C, Jung J, Ryoo R (2014) Mesopore expansion of surfactant-directed nanomorphic zeolites with trimethylbenzene. Microporous Mesoporous Mater 194:83–89
159. Kim W, Kim JC, Kim J, Seo Y, Ryoo R (2013) External surface catalytic sites of surfactant-tailored nanomorphic zeolites for benzene isopropylation to cumene. ACS Catal 3(2):192–195
160. Seo Y, Lee S, Jo C, Ryoo R (2013) Microporous aluminophosphate nanosheets and their nanomorphic zeolite analogues tailored by hierarchical structure-directing amines. J Am Chem Soc 135(24):8806–8809
161. Kim J, Choi M, Ryoo R (2010) Effect of mesoporosity against the deactivation of MFI zeolite catalyst during the methanol-to-hydrocarbon conversion process. J Catal 269(1):219–228
162. Kim J, Park W, Ryoo R (2011) Surfactant-directed zeolite nanosheets: a high-performance catalyst for gas-phase Beckmann rearrangement. ACS Catal 1(4):337–341
163. Kim JC, Cho K, Ryoo R (2014) High catalytic performance of surfactant-directed nanocrystalline zeolites for liquid-phase Friedel-Crafts alkylation of benzene due to external surfaces. Appl Catal A Gen 470:420–426
164. Koekkoek AJJ, Kim W, Degirmenci V, Xin H, Ryoo R, Hensen EJM (2013) Catalytic performance of sheet-like Fe/ZSM-5 zeolites for the selective oxidation of benzene with nitrous oxide. J Catal 299:81–89
165. Zou W, Xie P, Hua W, Wang Y, Kong D, Yue Y, Ma Z, Yang W, Gao Z (2014) Catalytic decomposition of N_2O over Cu-ZSM-5 nanosheets. J Mol Catal A Chem 394:83–88
166. Hu S, Shan J, Zhang Q, Wang Y, Liu Y, Gong Y, Wu Z, Dou T (2012) Selective formation of propylene from methanol over high-silica nanosheets of MFI zeolite. Appl Catal A Gen 445:215–220
167. Kim J, Kim W, Seo Y, Kim JC, Ryoo R (2013) n-Heptane hydroisomerization over Pt/MFI zeolite nanosheets: effects of zeolite crystal thickness and platinum location. J Catal 301:187–197
168. Verheyen E, Jo C, Kurttepeli M, Vanbutsele G, Gobechiya E, Korányi TI, Bals S, Tendeloo GV, Ryoo R (2013) Molecular shape-selectivity of MFI zeolite nanosheets in n-decane isomerization and hydrocracking. J Catal 300:70–80
169. Reddy JK, Motokura K, Koyama TR, Miyaji A, Baba T (2012) Effect of morphology and particle size of ZSM-5 on catalytic performance for ethylene conversion and heptane cracking. J Catal 289:53–61
170. Martens JA, Jacobs PA (1986) The potential and limitations of the normal-decane hydroconversion as a test reaction for characterization of the void space of molecular-sieve zeolites. Zeolites 6(5):334–348
171. Martens JA, Tielen M, Jacobs PA (1984) Estimation of the void structure and pore dimensions of molecular-sieve zeolites using the hydroconversion of normal-decane. Zeolites 4(2):98–107
172. Holm MS, Taarning E, Egeblad K, Christensen CH (2011) Catalysis with hierarchical zeolites. Catal Today 168(1):3–16
173. Lee HW, Park SH, Jeon JK, Ryoo R, Kim W, Suh DJ, Park YK (2014) Upgrading of bio-oil derived from biomass constituents over hierarchical unilamellar mesoporous MFI nanosheets. Catal Today 232:119–126
174. Liu F, Willhammar T, Wang L, Zhu L, Sun Q, Meng X, Carrillo-Cabrera W, Zou X, Xiao FS (2012) ZSM-5 zeolite single crystals with b-axis-aligned mesoporous channels as an efficient catalyst for conversion of bulky organic molecules. J Am Chem Soc 134(10):4557–4560

175. Boltz M, Losch P, Louis B, Rioland G, Tzanis L, Daou TJ (2014) MFI-type zeolite nanosheets for gas-phase aromatics chlorination: a strategy to overcome mass transfer limitations. RSC Adv 4(52):27242–27249
176. Heitmann GP, Dahlhoff G, Hölderich WF (1999) Catalytically active sites for the Beckmann rearrangement of cyclohexanone oxime to epsilon-caprolactam. J Catal 186(1):12–19
177. Degnan TF (2003) The implications of the fundamentals of shape selectivity for the development of catalysts for the petroleum and petrochemical industries. J Catal 216(1–2):32–46
178. Sastre G, Corma A (2009) The confinement effect in zeolites. J Mol Catal A Chem 305(1–2):3–7
179. Kim JC, Lee S, Cho K, Na K, Lee C, Ryoo R (2014) Mesoporous MFI zeolite nanosponge supporting cobalt nanoparticles as a Fischer-Tropsch catalyst with high yield of branched hydrocarbons in the gasoline range. ACS Catal 4(11):3919–3927
180. Na K, Jo C, Kim J, Ahn WS, Ryoo R (2011) MFI titanosilicate nanosheets with single-unit-cell thickness as an oxidation catalyst using peroxides. ACS Catal 1(8):901–907
181. Wang J, Xu L, Zhang K, Peng H, Wu H, Jiang JG, Liu Y, Wu P (2012) Multilayer structured MFI-type titanosilicate: synthesis and catalytic properties in selective epoxidation of bulky molecules. J Catal 288:16–23
182. Luo HY, Bui L, Gunther WR, Min E, Román-Leshkov Y (2012) Synthesis and catalytic activity of Sn-MFI nanosheets for the Baeyer-Villiger oxidation of cyclic ketones. ACS Catal 2(12):2695–2699
183. Schick J, Daou TJ, Caullet P, Paillaud J, Patarin J, Mangold-Callarec C (2011) Surfactant-modified MFI nanosheets: a high capacity anion-exchanger. Chem Commun 47(3):902–904
184. Kabalan I, Rioland G, Nouali H, Lebeau B, Rigolet S, Fadlallah MB, Toufaily J, Hamiyeh T, Daou TJ (2014) Synthesis of purely silica MFI-type nanosheets for molecular decontamination. RSC Adv 4(70):37353–37358
185. Varoon K, Zhang X, Elyassi B, Brewer DD, Gettel M, Kumar S, Lee JA, Maheshwari S, Mittal A, Sung CY, Cococcioni M, Francis LF, McCormick AV, Mkhoyan KA, Tsapatsis M (2011) Dispersible exfoliated zeolite nanosheets and their application as a selective membrane. Science 334(6052):72–75
186. Jo C, Seo Y, Cho K, Kim J, Shin HS, Lee M, Kim JC, Kim SO, Lee JY, Ihee H, Ryoo R (2014) Random-graft polymer-directed synthesis of inorganic mesostructures with ultrathin frameworks. Angew Chem Int Ed 53(20):5117–5121

Part II
New Developments in Characterization of Zeolites

Chapter 5
Structure Determination of Zeolites by Electron Crystallography

Tom Willhammar and Xiaodong Zou

Abstract Electron crystallography has shown to be an important method for structural characterization of zeolites. Electron crystallography is a method which comprises several important advantages over other characterization methods. With the electron as a probe, single-crystal diffraction data can be obtained from crystals million times smaller than what is possible with X-ray methods today. This is an important advantage especially for zeolites since they are often obtained as very small crystals. Electrons also enable the formation of images of a specimen with the atomic resolution. This is of essential importance when studying materials that are very complex or contain disorder. Over the years electron crystallography has been used for structure determination of zeolites. Through methodological advances during the last few years, it has evolved into an even more powerful method with crucial importance for structure determination. This chapter gives an introduction to electron crystallography and various electron crystallographic methods and their combinations with other methods used for structure determination of zeolite materials. Different routes for structure determination are described through examples from recently reported structure determinations.

Keywords Zeolites • Disordered materials • Stacking disorder • Electron crystallography • Structure determination • Electron microscopy • Electron diffraction • High-resolution transmission electron microscopy • 3D reconstruction • Powder X-ray diffraction

T. Willhammar • X. Zou (✉)
Berzelii Center EXSELENT on Porous Materials, and Inorganic and Structural Chemistry, Department of Materials and Environmental Chemistry, Stockholm University, Stockholm SE-106 91, Sweden
e-mail: xzou@mmk.su.se

F.-S. Xiao, X. Meng (eds.), *Zeolites in Sustainable Chemistry*,
Green Chemistry and Sustainable Technology,
DOI 10.1007/978-3-662-47395-5_5

5.1 Introduction

Knowledge about the structures of zeolites is essential for understanding their properties, for developing new applications, and for designing new materials. Structure determination of zeolites is challenging because zeolites are often polycrystalline, their structures are often complex with large unit cells, and the materials are sometimes disordered. The most common techniques used for structure determination of porous materials are single-crystal X-ray diffraction (SXRD), powder X-ray diffraction (PXRD), and transmission electron microscopy (TEM).

The standard technique for structure determination is single-crystal X-ray diffraction. The complete three-dimensional (3D) reciprocal lattice of a crystal can be collected from a crystal using either an in-house diffractometer or a synchrotron light source. The SXRD intensities are often kinematical and can be used to solve unknown structures of a large variety of complexity by a number of different methods. The structure solution is often straightforward and can be done in a few hours. The major limitation of structure determination by SXRD is the crystal size, which needs to be at least a few micrometers even using a brightest synchrotron light source. Another limitation is that it is difficult to determine structures of disordered crystals by SXRD. Only a limited number of zeolite materials were solved from SXRD, since it is relatively difficult to obtain large enough single crystals of zeolites. During the past 5 years, 11 new zeolite framework structures have been solved by single-crystal X-ray diffraction. These are aluminogermanates (PKU-9 (PUN)) [1], silicogermanates (SU-15 (SOF), SU-32 (STW)) [2], gallogermanates (CJ63 (JST), JU-64 (JSR)) [3, 4], aluminophosphates (CJ40 (JRY), CJ62 (JSW), CJ69 (JSN)) [3, 5–7], the beryllosilicate LSJ-10 (JOZ) [8], and the high-silica zeolite SSZ-57 (*SFN) with a disordered modulated structure [9].

For crystals too small for SXRD, powder X-ray diffraction has been an obvious choice. PXRD data is collected from millions of randomly oriented crystals in order to obtain strong enough diffraction intensities. The collection of PXRD data is very simple on in-house or synchrotron sources. The drawback of PXRD data is the peak overlap. Since intensities from crystals with different orientations are measured simultaneously, the 3D information of the reciprocal lattice is lost and projected onto one dimension. This results in overlap of reflections with similar *d*-spacings in PXRD patterns. This will lead to the fact that the intensity of some reflections cannot be determined unambiguously, which makes the structure determination by PXRD much more difficult than that by SXRD. Reflection intensities are extracted from PXRD data after a Le Bail fitting, which can give approximate single-crystal-like data for structure solution. Different methods have been developed to deal with the overlapping problem and to solve unknown structures by PXRD data [10–12]. During the past 5 years, PXRD has been the most important technique for structure determination of zeolites. More than 20 new zeolite structures have been solved from PXRD data. These are the ITQ-series (ITQ-32, ITQ-33 (ITT), ITQ-34 (ITR), ITQ-44 (IRR), ITQ-49) [13–17], SSZ-series (SSZ-52 (SFW), SSZ-56 (SFS), SSZ-65 (SSF), SSZ-77 (SVV), and SSZ-82 (SEW)) [18–22], IM-series (IM-16 (UOS) and IM-20 (UWY)) [23, 24], LZ-135 (LTF) [25], STA-15 (SAF) [26], MCM-70 (MVY) [27, 28], Linde type J (LTJ) [29], COK-14 (OKO) [30], ZnAlPO-57 (AFV) [31], ZnAlPO-59 (AVL) [31], ERS-18/SSZ-45 (EEI) [32], and the oxonitridophosphate-2 (NPT). Recently we demonstrated that zeolites and open-framework structures could be solved from PXRD data using a direct-space structure solution algorithm based on the building units identified by, for example, infrared spectroscopy [33].

Although structure solution by powder diffraction has been more and more powerful with new methods emerging, structures of many porous materials can still not be solved by PXRD alone. Among these are complex structures with large unit cells that cause severe reflection overlapping and crystals containing defects such as intergrowth and stacking disorders. For such materials, electron crystallography is the best alternative. Electron crystallography is an important tool for structure determination of materials where the crystals are too small to be studied by SXRD or the structures are too complex to be studied by PXRD. One important advantage of using electrons instead of X-rays is that electrons interact with matter much stronger than X-rays do and can be used for studying nano-sized crystals. This means that X-ray powder samples behave like single crystals under the electron beam. Another unique advantage is that high-resolution transmission electron microscopy (HRTEM) images can be obtained, which provide detailed information about atomic arrangements in a crystal. In addition, possible disorders can be directly seen by HRTEM. Electron crystallography has become more and more important for structure determination of unknown zeolites. Since 2006, 16 novel zeolite structures have been solved by electron crystallography alone or in combination with other techniques. These are TNU-9 (TUN) [34], IM-5 (IMF) [35], SSZ-74 (–SVR) [36], ITQ-26 (IWS) [37], ITQ-37 (–ITV) [38], ITQ-38 (ITG) [39], ITQ-39 (*-ITN) [40], ITQ-40 (–IRY) [41], ITQ-43 [42], ITQ-51 (IFO) [43], ITQ-53 [44], ITQ-54 [45], PKU-14 [46], PKU-16 (POS) [47], EMM-23 (*-EWT) [48], and SU-78 [49]. The power of electron microscopy for structural analysis of zeolites and other porous materials has been reviewed in several articles [50–55].

In this chapter we present the different electron crystallographic techniques and their applications on structure determination of zeolites. We will describe several advanced methods that are important for structure determination by electron crystallography: precession electron diffraction (PED) [56], 3D electron diffraction as represented by automated diffraction tomography (ADT) [57, 58] and rotation electron diffraction (RED) [59, 60], structure projection reconstruction by image processing [61–63], and 3D reconstruction from HRTEM images [64]. The recently developed techniques have made the structure determination by electron crystallography more feasible, even for non-TEM experts. We will demonstrate how the various electron crystallographic methods are chosen and used for solving different structural problems in porous materials. We will also show the benefits of combining electron crystallography and X-ray diffraction for studying complex zeolite structures. We give a number of examples where various electron crystallographic techniques are used and combined for structure determination of zeolites.

5.2 Recent Advances of Electron Crystallographic Methods

This chapter aims at describing recent advances in methods based on electron crystallography and electron microscopy. For a comprehensive general description of electron crystallography, we recommend the book "Electron Crystallography – Electron microscopy and electron diffraction" Oxford University Press, by Zou, Hovmöller, and Oleynikov [64].

Ab initio structure determination of porous materials can be done by electron diffraction and HRTEM. The advantage of using HRTEM images is that both the structure factor amplitudes and phases, which are needed for structure determination, can be obtained directly from HRTEM images. In diffraction, the structure factor phase information is lost and has to be determined by other methods. On the other hand, it is relatively easier to collect good electron diffraction patterns than to take high-quality HRTEM images.

5.2.1 Precession Electron Diffraction and 3D Electron Diffraction Techniques

The first publication of structure determination by electron diffraction (ED) was reported already in 1937 [65]. Although methods for quantifying electron diffraction patterns are developed [66] and examples of structure solution by electron diffraction were reported [67], the development of this method was hampered by the fear of dynamical effects. It was difficult to obtain complete 3D electron diffraction data because one needs to collect a large number of individual ED patterns along different zone axes. Furthermore, it was not easy to merge the 2D ED data in order to obtain the 3D data. Structure determination by electron diffraction was time consuming and required extensive expertise.

The developments on new electron diffraction techniques have made structure determination by electron diffraction more feasible. The precession electron diffraction (PED) technique invented by Vincent and Midgley 1994 [56] brought large attentions and interests in using electron diffraction data for structure solution. The electron precession is often controlled by dedicated PED hardware; the incident electron beam is tilted away by an angle from the optical axis and precessed around the optical axis. The PED pattern is an integration of all ED frames during the beam precession. We recently developed a software-based method for collecting PED data [68], where individual ED frames before the combination are collected and saved. The ED frames can be combined later to generate the PED pattern. Compared to conventional electron diffraction, the resolution of PED is higher and dynamical effects are reduced, due to the off-axis beam inclination where fewer reflections are excited simultaneously at the exact Bragg condition [69]. Precession electron diffraction has been used for solving unknown structures of inorganic compounds [70] and zeolites [71]. Intensities were extracted from PED patterns along different zone axes and merged to obtain 3D PED data for the structure solution.

Recently two automated methods have been developed for complete 3D electron diffraction data collection and processing: automated diffraction tomography (ADT) [57, 58] and rotation electron diffraction (RED) [59, 60]. In both ADT and RED methods, electron diffraction frames can be collected in either selected-area electron diffraction (SAED) mode or nano-electron diffraction (NED) mode. The ADT method is performed on a scanning transmission electron microscope (STEM) with or without precession. ED frames are collected sequentially in a fine step (usually 1°) by tilting around an arbitrary axis of a crystal. The sequential crystal tilt is computer controlled and crystal tracking during the tilt can be done in TEM or STEM mode. The software package ADT3D was developed for processing the ED frames and extracting the ED intensities, which includes the reconstruction of reciprocal space, determination of the unit cell parameters, reflection indexation, and intensity integration.

The rotation electron diffraction (RED) method runs in the TEM mode. The RED data collection combines the goniometer tilt with a fine electron beam tilt, which allows the collection of 3D electron diffraction data with fine sampling steps down to 0.001°; see Fig. 5.1 [59]. The data collection is controlled by the RED data collection software package [60]. Typically a beam tilt of 0.1–0.4° per step is combined with a goniometer tilt of 2–3° around a common tilt axis. More than 1000 ED frames can be collected in less than 1 h using this method. The ED patterns are analyzed using the RED data processing software package [60], which includes shift correction, peak search, unit cell determination, indexation of reflections, and intensity extraction.

The 3D ADT and RED methods have many advantages compared to the conventional 2D zone-axis electron diffraction methods. Firstly almost complete 3D data can be collected from one single crystal; the data within the missing cone may be generated from the symmetry-related reflections or collected from another crystal with different orientations. Secondly dynamical effects can be reduced because fewer reflections are present in each ED frame compared to a zone-axis ED pattern. More importantly, there is no need to have a crystal zone axis aligned in parallel to the electron beam and the data collection can start from any arbitrary orientation of the crystal. This means that the data collection no longer requires skilled operators and is feasible for non-TEM experts. The development of the ADT and RED methods has been revolutionary and turned a transmission electron microscope into a single-crystal electron diffractometer. Complete 3D single-crystal ED data can be collected from crystals of 50 nm in size, million times smaller than what is needed on a synchrotron source. As it will be shown later, the ADT and RED intensities have good quality and can be used for ab initio structure solution and refinement. A large and rapidly increasing number of new structures have been solved from the 3D electron diffraction data, among which are several zeolites.

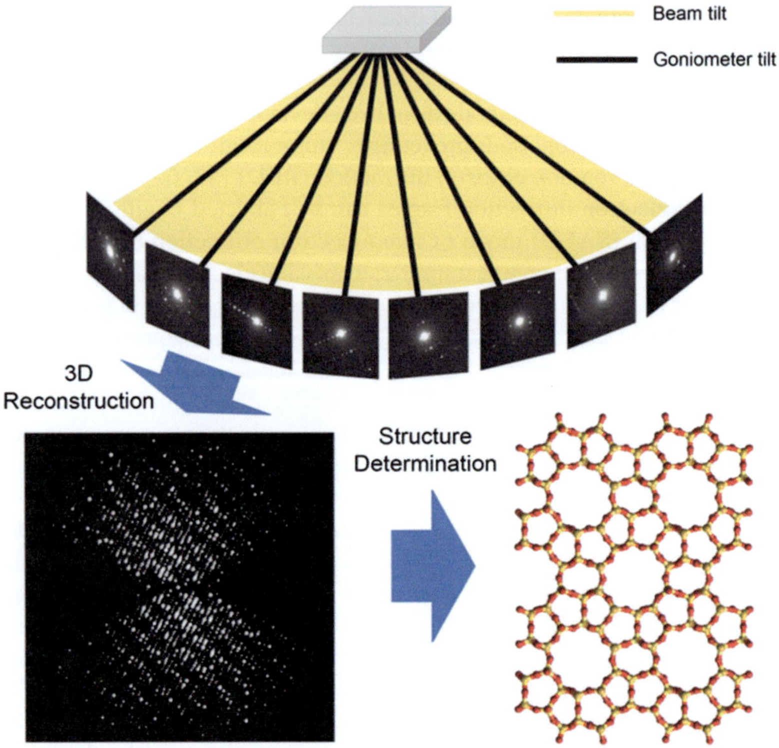

Fig. 5.1 Schematic representation of the concept of the rotation electron diffraction (*RED*) method. RED data (individual ED frames) are collected by combining beam tilt and goniometer tilt. The 3D reciprocal lattice is reconstructed from the RED data. Intensities and positions of the reflections in reciprocal space are obtained and can be visualized. After further data processing, the unit cell is determined and the reflections are indexed. A resultant *hkl* list is then used to determine the atomic structure of the crystals. The example used here is the zeolite silicalite-1 (Reprinted from Ref. [60]. Reproduced with permission of the International Union of Crystallography)

5.2.2 High-Resolution Transmission Electron Microscopy (HRTEM) and Structure Projection Reconstruction by Image Processing

High-resolution transmission electron microscopy (HRTEM) is the most common imaging technique used for studying porous materials at atomic scale [64]. Recent developments of transmission electron microscopes with image aberration correctors have pushed the resolution limit into the sub-angstrom level. However, for zeolites the image resolution is mainly limited by the materials due to the electron beam damage and not by the TEM. The stability of a porous material under electron beam can vary greatly, depending on the structure and composition of the material and the guest species in the pores.

For porous materials, low electron doses are often needed in order to minimize the radiation damage, and thus the signal to noise ratio in the obtained HRTEM images is usually very low. Working under low electron dose conditions requires high TEM skills and patience. More seriously, HRTEM image is usually not directly interpretable in terms of structure projection due to various reasons. One reason is dynamical effects, i.e., electrons are scattered more than once when passing through the sample. This should be minimized by working on very thin crystals. Another reason is the optical distortions caused by lens aberrations in the TEM. The image contrast changes with the imaging conditions, for example, the alignment of the electron optics and the focus of the objective lens; see Fig. 5.2a–d.

The HRTEM images show very different contrasts and do not always represent the structure projection (Fig. 5.2a–d). At different defocus values, the objective lens of a TEM transfers the structure to an HRTEM image differently, changing the contrast of the image. The effects of defocus and other optical parameters on an HRTEM image can be represented by a contrast transfer function (CTF) $T(\mathbf{u})$ [64]:

$$T(\mathbf{u}) = D(\mathbf{u})\sin\chi(\mathbf{u}) = D(\mathbf{u})\sin\left(\pi\varepsilon\lambda\mathbf{u}^2 + \tfrac{1}{2}Cs\lambda^3\mathbf{u}^4\right) \quad (5.1)$$

where λ is the electron wavelength, ε the defocus value, and C_s the spherical-aberration coefficient of the objective lens. $D(\mathbf{u})$ is an envelope function that slowly decreases from one at $u=0$ to zero at $u=$infinitive due to the partial coherence of the electron beam. Once the defocus value is determined, the CTF can be calculated according to Eq. (5.1).

The CTF transfers the diffracted electron beams in different ways. The diffraction amplitudes are directly proportional to the structure factors $F(\mathbf{u})$ of the crystal under kinematical approximation. The structure factors $F(\mathbf{u})$ is defined as

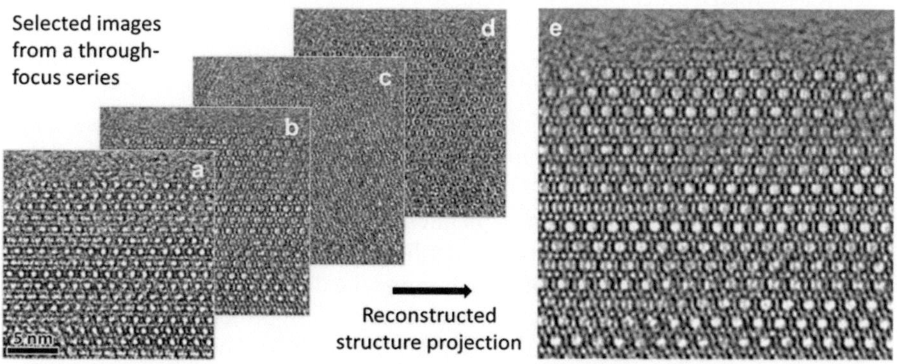

Fig. 5.2 (**a–d**) HRTEM images of zeolite ITQ-39 taken along [100] from the same crystal area but with different defocus values, (**a**) −678 Å, (**b**) −358 Å, (**c**) −38 Å, and (**d**) +282 Å. (**e**) A structure projection image reconstructed from 20 HRTEM images. The defocus determination and structure projection reconstruction were carried out using the program QFocus [63] (Reprinted by permission from Macmillan Publishers Ltd: Ref. [40], Copyright 2012)

$$F(\mathbf{u}) = \sum_{j=1}^{N} f_j(\mathbf{u}) \exp 2\pi i(\mathbf{u} \cdot \mathbf{r}_j) \tag{5.2}$$

where $f_j(\mathbf{u})$ is the atomic scattering factor of the jth atom in the unit cell and N is the total number of atoms. $F(\mathbf{u})$ is a complex and contains an amplitude part $|F(\mathbf{u})|$ and a phase part $\phi(\mathbf{u})$

$$F(\mathbf{u}) = F(\mathbf{u})\left|\exp\left[i\phi(\mathbf{u})\right]\right| \tag{5.3}$$

For thin crystals, the HRTEM image is related to the crystal projection in a simple way, and the crystallographic structure factors $F(\mathbf{u})$ can be obtained from the HRTEM image by

$$F(\mathbf{u}) = \frac{I(\mathbf{u})}{kT(\mathbf{u})} \left(\text{for } T(\mathbf{u}) \neq 0\right) \tag{5.4}$$

where $I(\mathbf{u})$ is the Fourier transform of the image and k is a constant. Equation (5.4) shows that the phases of $F(\mathbf{u})$ and $I(\mathbf{u})$ are the same if $T(\mathbf{u})>0$ and differ by 180° if $T(\mathbf{u})<0$ [64]. An HRTEM image can be interpreted directly in terms of the structure projection only if it is taken under the optimum defocus conditions (i.e., near the Scherzer defocus) (Fig. 5.2b). For HRTEM images taken under other conditions, different image processing approaches can be used for retrieve the structure projection.

One approach to reconstruct structure images is crystallographic image processing of HRTEM images [61, 62]. The defocus and astigmatism values can be determined from the Fourier transform of the HRTEM image, if there exists amorphous region on the crystal. The effects of the CTF on the HRTEM image can be compensated for by image processing, according to Eqs. (5.1) and (5.4). This image can be further improved by imposing the crystallographic symmetry. These procedures can be carried out by the crystallographic image processing software CRISP [61] and work for single HRTEM images [72]. Crystallographic image processing is especially helpful for HRTEM images of periodic objects with low signal to noise ratios, where the image quality can be improved by averaging the unit cells and imposing the symmetry.

For crystals where the periodicity is interrupted by faults and boundaries, unit cell averaging and symmetry cannot be applied. If there is little amorphous region in the HRTEM image, it is difficult to determine the defocus and astigmatism from a single image. Recently we have developed a structure projection reconstruction method from through-focus series of HRTEM images [63]. The defocus values are determined jointly from the through-focus series of HRTEM images with a known defocus step by evaluating the phase similarities of $F(\mathbf{u})$ over the entire u-values (pixels) in the Fourier transforms of the HRTEM images. $F(\mathbf{u})$ is calculated according to Eq. (5.4) using a trial defocus value for the first image in the series.

The defocus values for other images are calculated from the trial defocus value and the defocus step. The trial defocus value with the highest phase similarity will be the correct defocus. The twofold astigmatism in the HRTEM images can also be estimated. A projected electrostatic potential image is obtained from each HRTEM image in the series by the CTF correction based on Eqs. (5.1) and (5.4) using the following equation:

$$\varphi(\mathbf{r}) = k\sum_{\mathbf{u}} F(\mathbf{u})\exp\left[-2\pi i(\mathbf{u}\cdot\mathbf{r})\right] = \sum_{\mathbf{u}} \frac{I(\mathbf{u})}{T(\mathbf{u})}\exp\left[-2\pi i(\mathbf{u}\cdot\mathbf{r})\right] \quad (5.5)$$

The final corrected images were added together to reconstruct an image that is closer to the structure projection than the individual images (Fig. 5.2e). The structure projection reconstruction method has been implemented into the computer program QFocus [63]. The number of HRTEM images in a through-focus series is typically 10–20.

The through-focus structure projection reconstruction method using the program QFocus is more general and can be applied to both periodic and nonperiodic objects. This is especially useful on beam-sensitive and disordered materials, because the signal to noise ratio can be greatly enhanced by averaging reconstructed images taken at different defocus values. It is also beneficial for non-TEM experts and save time, because there is no need in finding the optimum defocus value when recording an HRTEM image, which also avoids unnecessary exposure of the sample to electrons. The chance to obtain a good structure image after image processing is greatly increased. The structure image can be used to build structural model and determine the crystallographic structure factors, which are important for structure determination. The later can be done using the crystallographic image processing program CRISP [61, 62, 64, 72].

5.2.3 STEM Imaging of Zeolites

The basic principle of scanning transmission electron microscopy (STEM) imaging is that the electron beam is focused onto the specimen, forming a small probe. The probe is scanned in a raster over the sample by coils above the specimen. The distance between each raster point defines the magnification. The resolution of a STEM image is determined by the size of the probe. Modern aberration-corrected microscopes have pushed the limit of the probe size into a new era, allowing resolution in the sub-angstrom range. As the electron beam passes through the sample, it interacts with the sample and the transmitted beam will be scattered. There are a number of different detectors available detecting electrons scattered at different angles. A bright-field detector will detect the intensity of the unscattered electrons and electrons scattered at low angles. An annular dark field (ADF) detector will detect the scattered electrons at high angles. If an ADF detector is designed to collect electrons at high scattering angles, it is called high-angle annular dark field (HAADF) detector. The electrons detected by the HAADF detector are generally inelastically scattered by the atomic nuclei. Hence the contrast formed by such events is related to the atomic number and therefore called Z contrast. The contrast of a STEM image is not altered by any contrast transfer function and is normally more easily interpreted as the projection of the specimen.

A second advantage of STEM imaging is that it allows collection of spectroscopic data from each of the raster points. This gives the possibility to construct a 2D map containing information about, for example, the composition. This is especially powerful in combination with atomic resolution imaging.

During the very recent years STEM imaging has been successfully applied to studies of zeolites [73–76]. STEM-HAADF images are especially suitable for studies of, e.g., zeolites loaded with atoms or clusters of atoms. Mayoral and coworkers have shown that clusters of silver ions can be imaged inside the pores of zeolite A; see Fig. 5.3 [73, 74].

5.2.4 3D Reconstruction of HRTEM and STEM Images

Because each HRTEM or STEM image is only a 2D projection of a 3D structure, it is necessary to combine images from different projections to reconstruct the 3D structure. The 3D reconstruction is performed in reciprocal space by determining the crystallographic structure factor F(*hkl*) from a series of projections. The 3D electrostatic potential distribution φ (*xyz*) inside a crystal can be calculated from the structure factors F(*hkl*) by inverse Fourier transformation:

$$\varphi(xyz) = \frac{\lambda}{\sigma\Omega}\sum_{hkl} \mathrm{F}(hkl)\exp\left[-2\pi i(hx + ky + lz)\right] \tag{5.6}$$

where σ is the interaction constant and Ω the unit cell volume.

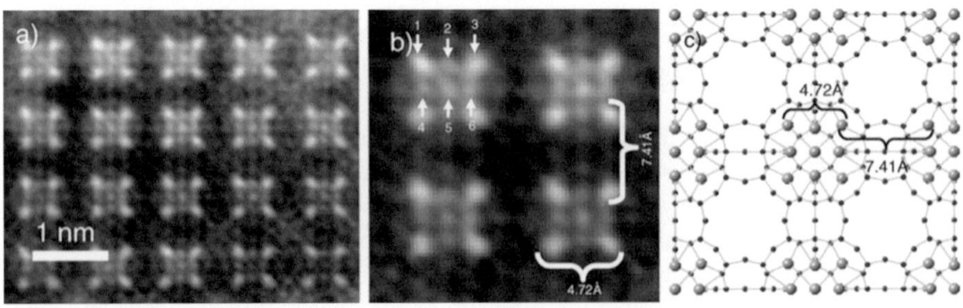

Fig. 5.3 STEM-HAADF images of zeolite A loaded with silver ions viewed along [100] (**a**) and (**b**). The structural model of zeolite A with silver ions inside (**c**) for comparison (Reprinted from Ref. [74], Copyright 2013, with permission from Elsevier)

A 3D electrostatic potential map can be obtained by Fourier summation of structure factors for all reflections using Eq. (5.6). If the resolution is high enough so that peaks in the reconstructed 3D electrostatic potential map are well resolved, the atomic positions can be directly located from the 3D map. The resolution that is needed to resolve the desired atoms should be higher than the interatomic distances, for example, 1.6 Å for Si-O and 3.1 Å for Si-Si distances. For zeolites, it is often sufficient to find the Si/Al positions in order to build a framework model, since oxygen atoms are always located between the Si and Al atoms. Light atoms are always difficult to be seen in the presence of heavy elements.

Theoretically, one might expect that it is necessary to collect all reflections within a desired resolution range in order to obtain the 3D potential map from which atomic positions are determined. However, it is very time consuming and practically impossible to collect all the projections. The number of projections that is needed for a 3D structure determination depends on the symmetry of the crystal and how the strong reflections are distributed. It is enough to determine the structure factors of symmetry-independent reflections; other reflections can be generated according to the symmetry [64]. The higher the symmetry, the fewer number of projections must be measured. Figure 5.4 illustrates the significance of strong and weak reflections, phases, and amplitudes on a structure solution. It can be seen that strong reflections are more important than weak reflections for structure solution, because strong reflections contribute more to the potential in a 3D potential map than weak reflections. Missing strong reflections may result in a wrong 3D map. In fact, weak reflections can be ignored during structure solution, but they are important for the structure refinement. Furthermore, structure factor phases are more important than amplitudes, because they determine the locations of the peak maxima and thus the atomic positions. Amplitudes, on the other hand, mainly affect the relative peak heights.

In conclusion, in order to obtain a 3D electrostatic potential map from which the most significant (heavy) atomic positions can be located, we need to collect all strong reflections within the asymmetric unit and be sure that the phases of the strongest reflections are correctly determined. A number of complex structures have been solved by 3D reconstruction from HRTEM images, including oxides, intermetallic phases, mesoporous silica, and zeolites. This method is rather demanding and often requires extensive experimental work by well-trained electron microscopists to take good HRTEM images. One example is the structure solution of the quasicrystal approximant υ-AlCrFe, where HRTEM images from 13 zone axes were used for the 3D reconstruction [77]. It is one of the most complex inorganic structures solved by electron crystallography, with 129 symmetry-independent atoms. A simpler and more feasible approach for structure determination of complex structures is to combine electron crystallography and powder X-ray diffraction. This is especially useful when the number of structure factor amplitudes and phases obtained from the HRTEM images is not enough or the quality is high enough to reach the structure solution. The structure factor phases obtained from the HRTEM images can be used as initial phases for structure determination by PXRD. The combination of electron crystallography and powder X-ray diffraction has been essential for the structure solution of three most complex zeolite structures [34–36].

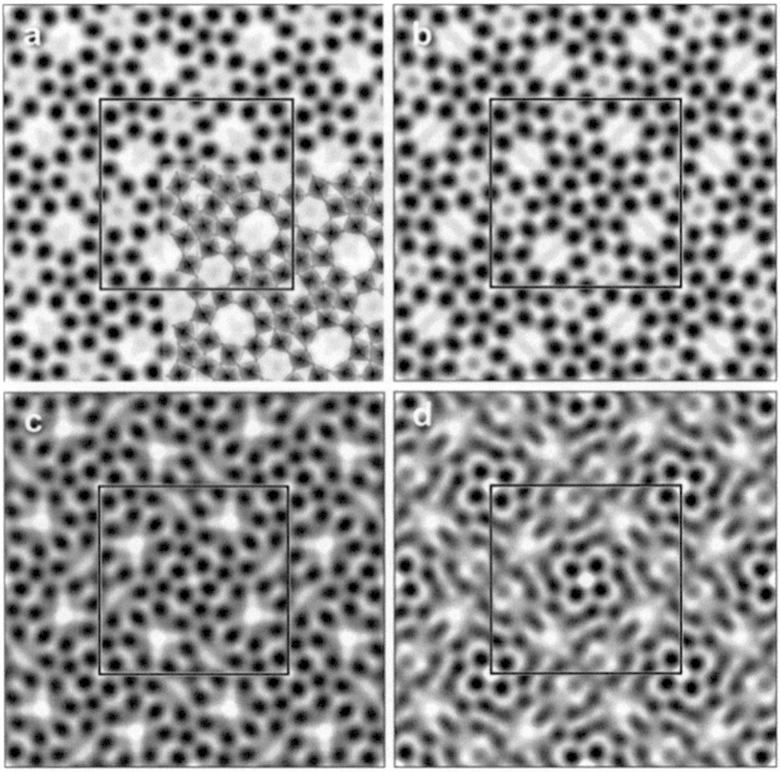

Fig. 5.4 Significance of strong and weak reflections, phases, and amplitudes for the structure solution, demonstrated here using the potential map of $K_2O{\bullet}7Nb_2O_5$ projected along the c-axis. (**a**) Projected electrostatic potential map reconstructed from an HRTEM image. *Insert*: the structural model. All Nb atoms are resolved. The positions of oxygen atoms are between the Nb atoms. (**b**) A map using only the strongest half of the reflections. The map is very similar to that in (**a**), showing that the weak reflections are less important for structure solution. (**c**) A map with all amplitudes set equal and phases unchanged from those used in (**a**). The Nb peaks are sharp and clearly resolved. This shows that as long as the phases are correct, the amplitudes can have large errors in structure solution. (**d**) A map with the phase of one strongest reflection shifted by 180° while all other amplitudes and phases are kept the same as in (**a**). The map now differs significantly from (**a**). The peaks are no longer sharp and some have shifted from those seen in (**a**) (Reproduced from Ref. [55] by permission of John Wiley & Sons Ltd)

5.3 Different Routes for Structure Determination by Electron Crystallography

We have described the different electron crystallographic methods for structure determination of zeolites. Because each new material has its own unique characteristics, it gives great challenges to the structure determination process. Different routes for structure solution may be needed for different zeolites. Some common obstacles are small crystal size, low beam stability causing degradation in the electron beam, high complexity of the crystal structure, and presence of elements interrupting the translational symmetry such as twinning, stacking disorder, or other kinds of disorder. Electron crystallography has the tools to overcome many of these obstacles.

The process of determining a crystal structure can be divided into two main steps. The first step is to find a plausible structural model, i.e., the structure solution. There are a number of existing methods available to obtain a structural model. Ab initio structure solution by well-established methods directly from diffracted intensities is just as feasible as model building, with just general information about the specimen as a base; see Fig. 5.5 for an overview of some possible routes for structure solution including electron crystallographic methods. This means that for structure solution it is important to have a broad knowledge about different available methods for structure solution as well as deep understanding of zeolite structures.

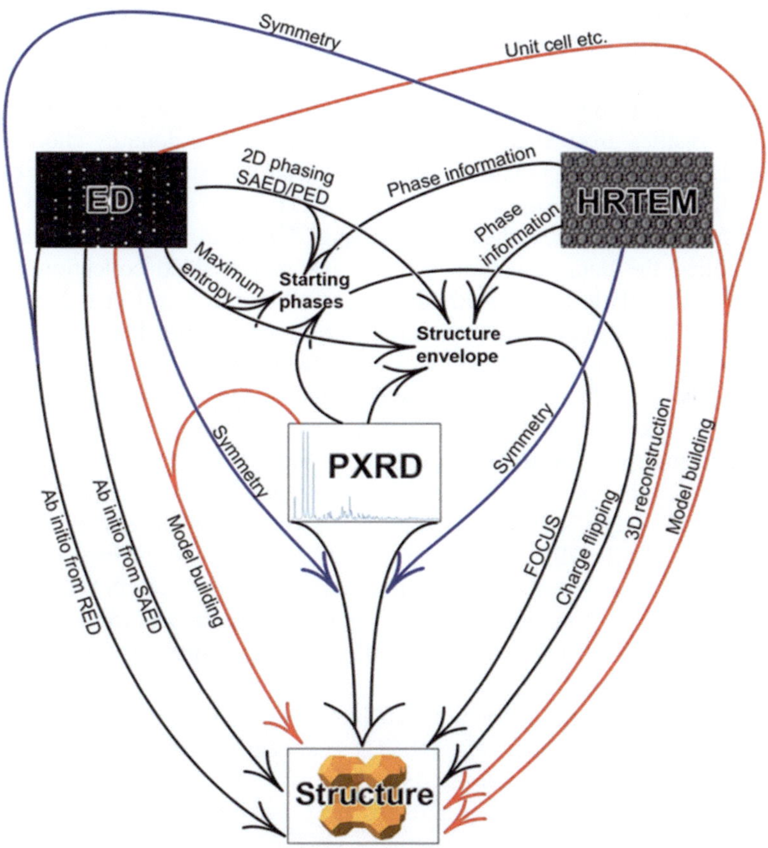

Fig. 5.5 A map illustrating some possible routes for structure solution from electron diffraction patterns, HRTEM images, and PXRD data. Routes marked by *black lines* are suitable for ordered structures and routes marked by *red lines* for disordered structures. *Blue routes* indicate that additional symmetry information may be needed

Many zeolite structures are built from the same building units, assembled in different ways. The second step of the structure determination process is to verify the structure. This process has to be done with great care and perfection. For ordered structures the structure verification is done mainly by a least-squares refinement between experimental and calculated diffracted intensities. These refinements should converge to reasonable levels according to conventions. In some cases a proper refinement is not doable, for example, for heavily disordered materials. In this case the structure verification involves obtaining as much information as possible from different sources and using them to confirm the proposed model.

5.3.1 Routes Based on Electron Diffraction

For materials with ordered structures electron diffraction is often the method of choice. Collecting electron diffraction data requires a lower electron dose compared to HRTEM imaging and is hence more suitable for beam-sensitive materials. Because of the dynamical effects, care has to be taken during data collection to achieve intensities as kinematical as possible. This can be done by choosing thin crystals. Dynamical effects can be further reduced by using either PED or a 3D electron diffraction technique such as RED or ADT.

Over the last half decade, the development of 3D electron diffraction techniques has simplified the data collection significantly and enabled collection of more complete data. With these new methods, it is now possible to collect and reconstruct the reciprocal lattice of crystalline materials. From the reciprocal lattice, the intensities of reflections can be extracted and used, for example, for ab initio structure solution. Structure solution from RED or ADT has shown to be very powerful. It has now evolved into a method for routine structure solution of materials that have ordered structures and are stable enough for collection of electron diffraction data. Ab initio structure determination can be made using different available methods developed for SXRD. Direct method is the most well-used method available in, e.g., SHELX [78] and SIR [79] software packages. Another way of phasing the electron diffraction intensities is to use the zeolite-specific program FOCUS [10], originally designed for X-ray diffraction, which has been modified to fit electron diffraction data [80]. It is also possible to use charge flipping of real-space methods to obtain a structural model.

Over the years electron diffraction data has also been used to complement structure solution by other methods, often PXRD. From electron diffraction, the unit cell can be easily determined since the data does not suffer from peak overlapping. Together with the unit cell, systematic absences as well as point group symmetry can be deduced. This information is crucial since it can be very difficult to obtain from PXRD data. Electron diffraction can also be used to pre-partition the intensities of the PXRD peaks.

Recently a new approach to use electron diffraction data has shown to be powerful. By applying the charge-flipping algorithm [81, 82] implemented in the program

Superflip [83] on intensities extracted from a 2D precession electron diffraction (PED) pattern, the projected potential of the crystal can be determined [84]. This means that the structure factor phases of reflections in one reciprocal lattice plane can be determined. These phases can be used either as starting phases for a structure solution by charge flipping or to produce an envelope function that can be used in, for example, the zeolite-specific software FOCUS [10]. An envelope function is a function defining the boundary between voids and matter inside the porous structure and can be used as a form of preknowledge in the structure solution process.

5.3.2 Routes Based on HRTEM Imaging

Taking good HRTEM images are more demanding than to collect electron diffraction data, but it is often more rewarding. HRTEM images have several benefits over diffraction techniques. Firstly the crystallographic structure factor phases can be extracted, which is of outermost importance during structure determination. Secondly HRTEM images give local information, whereas diffraction techniques always average over a whole crystal or even a large number of crystals. The local information is crucial when studying disordered materials. The occurrence of disorder in crystals will create diffuse scattering which in many cases obstruct structure solution ab initio from diffraction data. One common form of disorder in zeolite structures is the stacking of layers. This will generate diffuse streaks perpendicular to the layer in reciprocal space. Important information about, e.g., the stacking vector can be determined, by examining the sharp spots and diffuse streaks of the reciprocal lattice. But in order to obtain a structural model, it is often necessary to study the local structure atom by atom, which can only be done using images.

Structure solution from HRTEM images mainly follows two routes, either model building or 3D reconstruction. Model building utilizes HRTEM images along one or more projections and is sometimes combined with additional information from NMR, adsorption, electron diffraction, and other methods. From all the available information, a feasible model is built. Model building is a technique that needs knowledge and experience in the type of materials studied. It has shown to work well for structure solution of numerous zeolite structures over the years. A more general route is to use 3D reconstruction of HRTEM images acquired along several different zone axes. Each of the images shows the projected potential of the specimen. By merging information from several projections, a 3D reconstruction of the projected potential can be constructed; see Sect. 5.2.4. From the 3D potential, the atomic positions in the material can be found.

The structure factor phase information extracted from HRTEM images can also be used as prior knowledge for ab initio structure solution from PXRD data. The structure factor phases can then be used as starting phases for structure solution using charge flipping. This method has played a crucial role in the structure solution of several of the most complex zeolite structures. The phase information can be used as prior knowledge for ab initio structure solution from PXRD data.

The following sections will describe these different routes for structure determination using electron crystallography in more details. This will be done through examples taken from recently reported structure determinations.

5.4 Structure Determination by Electron Diffraction

5.4.1 Structure Determination of Zeolites by Combining Zone Axis ED Patterns

Structure determination can be made by combining ED patterns along different zone axes. Conventional selected-area electron diffraction (SAED) and precession electron diffraction (PED) patterns are often taken along a certain crystal direction (called zone axis) and the patterns appear symmetrical (Fig. 5.6a–c). ED patterns from several different zone axes can be collected from one or several crystals and

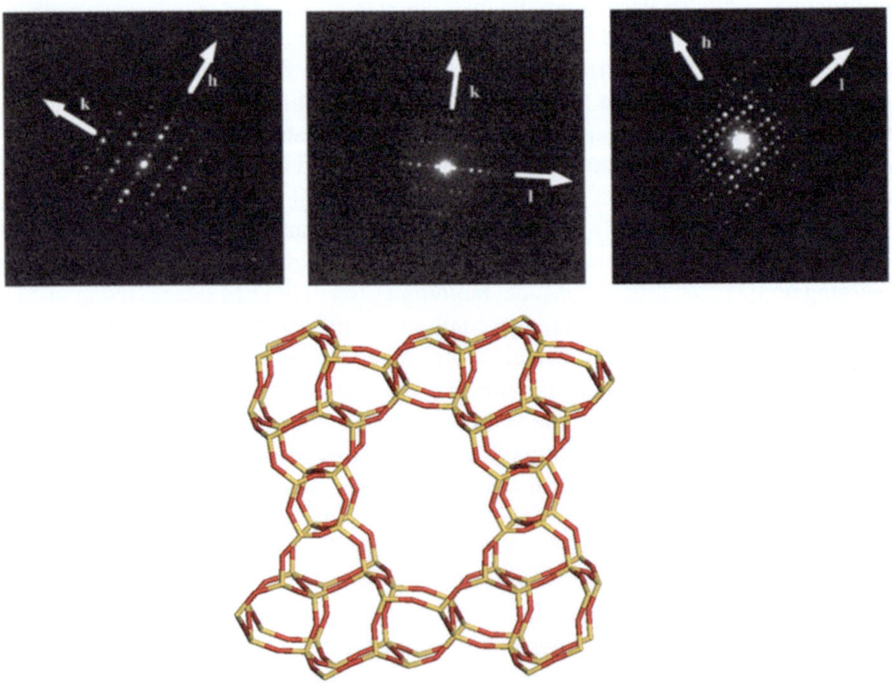

Fig. 5.6 Electron diffraction data from three main zone axes of SSZ-48 and the structural model obtained from electron diffraction by direct methods completed with the oxygen atoms that were missing in the structure solution (Reprinted with the permission from Ref. [86]. Copyright 1999 American Chemical Society)

then merged into a 3D dataset. Several zeolite structures such as MCM-22 [85], SSZ-48 [86], ITQ-40 [41], and ZSM-10 [87] have been solved by SAED.

SSZ-48 is a large-pore high-silica zeolite synthesized using N,N-diethyl-decahydroquinolinium as the organic structure-directing agent (OSDA). It was the first unknown zeolitc material solved by electron diffraction (Fig. 5.6). SAED patterns from 11 zone axes were collected from needlelike crystals of ca. 0.05 μm × 0.25 μm × 10 μm in size. Integrated ED intensities were extracted from SAED patterns and merged together to generate 326 symmetry-independent reflections ($d < 0.99$ Å). The unit cell parameters ($a = 11.19$ Å, $b = 4.99$ Å, $c = 13.65$ Å, and $\beta = 100.7°$) and space group ($P2_1$) were deduced from the 3D electron diffraction data. The ED intensities were used for structure solution of SSZ-48 by direct methods using atomic scattering factors for electrons implemented into SHELX. All seven silicon positions and five of the 14 oxygen positions were located from the structure solution. In addition 9 peaks in the pore of the structure were located which was believed to be carbon positions from the SDA. Oxygen atoms were positioned between the Si atoms in order to complete the silicate framework. The geometry of the framework, i.e., the Si-O distances and O-Si-O angles, was refined using distance least-squares refinement. The structure was then refined against the electron diffraction intensities with restrains on the Si-O and O-O bond lengths. The refinement converged with an agreement factor R_F of 0.3283. As a final stage the obtained structure was refined against high-resolution synchrotron powder X-ray diffraction data which converged with good agreement confirming the structural model obtained from SAED data [86].

Precession electron diffraction can reduce significantly dynamical effects and secondary scattering in electron diffraction [68–70]. Dorset et al. compared structure determination of various zeolites by PED with that by SAED and showed that PED data can be used for ab initio structure determination of zeolites [71]. The maximum entropy and likelihood phasing method has shown to be powerful in phasing both the SAED and PED data and solving the zeolite structures [71].

Structure determination using zone-axis SAED or PED patterns is not straightforward. The crystals have to be aligned manually along the individual zone axes. The new 3D electron diffraction techniques ADT and RED have made a breakthrough for the structure determination using ED data. The automated 3D ED data collection combined with the fast data processing procedures makes the structure determination from electron diffraction as feasible as that by single-crystal X-ray diffraction, but from crystals millions times smaller.

5.4.2 Structure Determination of Zeolites by Combining ED and PXRD

Electron diffraction has been combined with PXRD for structure determination of many zeolites. Structure solution from PXRD data can often be challenging, mainly due to the large degree of peak overlap for complex structures with large unit cells.

Single-crystal electron diffraction data can be collected from very small crystals, which give important additional information such as unit cell and symmetry. Different strategies have been applied to combine the ED and PXRD data. Below two notable examples will be described.

The structure determination of the germanosilicate ITQ-37 utilized electron diffraction for several purposes [38]. The high degree of reflection overlap (>94 %) made the structure solution by PXRD alone difficult. ITQ-37 was very beam sensitive, which made it difficult to obtain high-quality HRTEM images. Instead SAED patterns were taken along two different zone axes revealing four- and six-fold rotational symmetry (Fig. 5.7). This indicated a cubic unit cell which was later confirmed by PXRD, $a = 26.5126(3)$ Å. Systematic absences in the ED patterns narrowed down the possible space groups to the two related candidates $P4_132$ and $P4_332$. In the next step intensities from SAED patterns of four zone axes ($d > 3.3$ Å) were extracted and used for pre-repartitioning of the overlapping reflections in the PXRD pattern. ED patterns are very helpful for pre-partitioning of overlapping reflections because Bragg peaks do not overlap. For example, two reflections with similar d-spacings will overlap in the PXRD pattern, and the measured intensity is often split equally and assigned to the two reflections. With ED data the intensities of the two reflections can be distinguished and assigned according to the intensities in the ED pattern. Assigning correct intensities to reflections is often crucial for a successful structure solution by PXRD. After pre-partitioning of the PXRD peaks, the powder charge-flipping (pCF) algorithm [12] was applied using the computer program Superflip [83]. All ten symmetry-independent Si/Ge positions and 18 out of 19 oxygen positions could be located from the best electron density map (Fig. 5.7). The missing oxygen position could be identified and placed between two T-sites. Rietveld refinement was performed to confirm the final structural model. ITQ-37 is the first chiral zeolite with one single gyroidal channel system and pore openings defined by 30-ring windows. It had the lowest framework density (10.3 T-atoms/1000 Å^3) of all existing 4-coordinated crystalline oxide frameworks.

Another strategy is to retrieve rough structure factor phases directly from the ED data. The obtained structure factor phases can be used as the initial phases for the

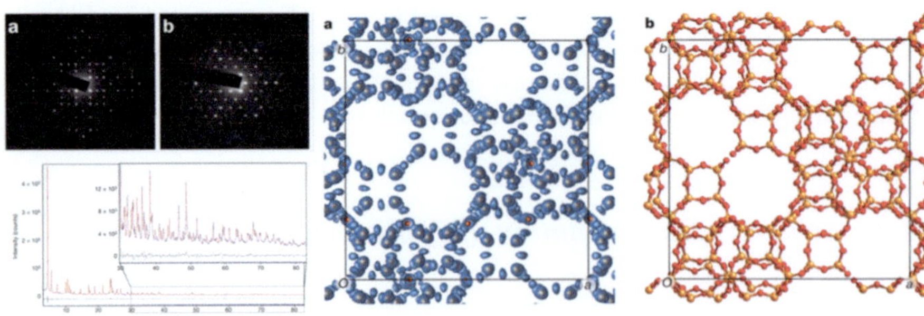

Fig. 5.7 (*Left*) The SAED and PXRD patterns from ITQ-37 and (*right*) the electron density maps derived by the powder charge-flipping (pCF) algorithm and the final structural model, viewed along *c*-axis (Reprinted by permission from Macmillan Publishers Ltd: Ref. [38], Copyright 2009)

structure solution using PXRD data. Such a strategy was used for the structure solution of the large-pore germanosilicate ITQ-26 [37]. The space group (I4/mmm) and unit cell parameters ($a=26.7769(8)$ Å, $c=13.2505(5)$ Å) of ITQ-26 were determined from a tilt series of ED patterns. Structure factor phases were deduced from the integrated ED intensities of the ($hk0$) and ($0kl$) reflections by maximum entropy and likelihood method [71] using the program MICE [88]. The structure factor phases obtained were used as constraints for subsequent phasing trials on the PXRD data using the zeolite-specific structure solution program FOCUS. The final model was refined by Rietveld refinement. The electrostatic potential maps obtained from electron diffraction matched with the final structural model. The structure of ITQ-26 has seven symmetry-independent T-sites. The framework contains 3D channel systems with straight 12-ring channels along [001] and two other 12-ring channels tilted with respect to the *c*-axis.

5.4.3 Structure Determination of Zeolites by 3D Electron Diffraction: ADT and RED

The methods described above in Sects. 5.4.1 and 5.4.2 involve combining data from several sources and are often rather complex. During the recent years, new methods have been developed for collection of 3D electron diffraction data, as represented by the automated diffraction tomography (ADT) and rotation electron diffraction (RED). Ab initio structure determination of zeolites has been shown to be successful from both ADT [42] and RED [43–48, 89] methods which have been very successful in solving structures of zeolites with very small crystal size. The structure solution methods applied on the ADT/RED data include direct methods (e.g., programs SHELX, SIR), charge flipping (e.g., programs Superflip, Jana) and simulated annealing (e.g., program SIR). The structural models can be refined by full-matrix least-squares refinement using the program SHELX. The kinematical approach has been applied, i.e., the ED intensities are proportional to the square of the structure factor amplitudes. Atomic scattering factors for electrons are used instead of those for X-rays. Due to the presence of dynamical scattering, the agreement factors for ADT/RED data are still not very good. Structures determined from RED data are however in good agreement with structures refined by Rietveld refinement. Here we describe the structure determination of one novel zeolite, ITQ-51, and the structure determination of two interlayer expanded zeolites solved from the RED data.

The structure of the new silicoaluminophosphate ITQ-51 was solved ab initio from RED data [43]. Two RED datasets were collected from submicrometer-sized crystals. Each of the datasets contains >1300 individual electron diffraction pattern and covered >130° tilt range with a tilt step of 0.1°. The data collection is controlled by the RED software and the entire data collection can be completed within 1–2 h. The ED patterns were processed using the software RED in order to reconstruct the reciprocal lattice of the material. From the reciprocal lattice, the diffracted intensities can be extracted and used for ab initio structure solution. From the two datasets collected of ITQ-51, the unit cell and possible space group could be determined. The unit cell matched well with the PXRD pattern (a=23.345(2) Å, b=16.513(2) Å, c=4.9814(5) Å, α=90°, β=90.620(5)°, γ=90°, $P2_1/n$). The scattered intensities extracted from each of the two RED datasets could be used for ab initio structure determination by direct methods using the software SHELX. All eight symmetry-independent framework T-atoms (Al, Si, P) and 16 oxygen atoms could be located by direct methods using the RED data. Two RED datasets from two different ITQ-51 crystals merged together in order to improve the completeness of the data. The Al and P positions could be assigned based on the difference in the Al-O and P-O bond distances. The structural model was further refined by Rietveld refinement against the PXRD data. The atomic positions obtained from the RED data (as-made ITQ-51 sample) deviated on average by 0.11 Å for Al/P and 0.13 Å for O from those obtained by the Rietveld refinement of the calcined ITQ-51 sample. This means that although the R-value after refinement is as high as 37 %, the atomic positions obtained from RED data show good correspondence with the positions obtained from PXRD data.

Recently a novel zeolite, PKU-16, with a straight 3D 11×11×12-ring channel system was synthesized using a simple OSDA. The structure was solved from RED data; see Fig. 5.8 [47]. PKU-16 is the first stable zeolite with large odd-number rings. It can be considered as a modified zeolite beta polymorph C by rotating half of single 4-rings by 90°. This relation reveals a possible route to finely tune pore sizes to form new zeolites with odd-number rings.

Structure determination of interlayer expanded zeolites is challenging because the materials often form as nano- and submicron-sized crystals too small to be studied by single-crystal X-ray diffraction. In addition, the materials often have poor crystallinity and contain stacking disorder, which lead to low resolution and peak broadening in the PXRD pattern. COE-3 and COE-4 are intralayer expanded zeolites with crystal structures built from the ferrierite layer with a CDO stacking (IEZ-CDO). COE-4 is the calcined form of the as-made COE-3 material. The structures of these two materials have been solved by RED [89].

In order to solve the structure of CEO-3, RED datasets were collected from three different crystals. From each of the datasets, the unit cell and three possible space groups could be determined (7.2 Å, 22.4 Å, 13.6 Å, 90°, 90°, 90°, $Cmc2_1$, $C2cm$, $Cmcm$). The structure of COE-3 could not be solved from any of the single datasets due to the high incompleteness of each RED data. The three datasets from different crystals were merged in order to obtain a more complete dataset. For COE-4, two datasets were collected from two different crystals, which contain 270 and 1239 ED frames and cover the tilt range of 108.71° and 119.66°, respectively. The reconstructed 3D reciprocal lattice is shown in Fig. 5.9. 2466 reflections (513 unique) were obtained from dataset 1, while 2752 reflections (477 unique) were obtained from dataset 2. The orientations of the two datasets are complementary to each other, as seen from the 3D reciprocal lattices reconstructed from the two datasets (Fig. 5.9a–d). The structure of COE-4 could be solved by direct methods from the intensities merged from these two datasets using the program SHELX (Fig. 5.9g).

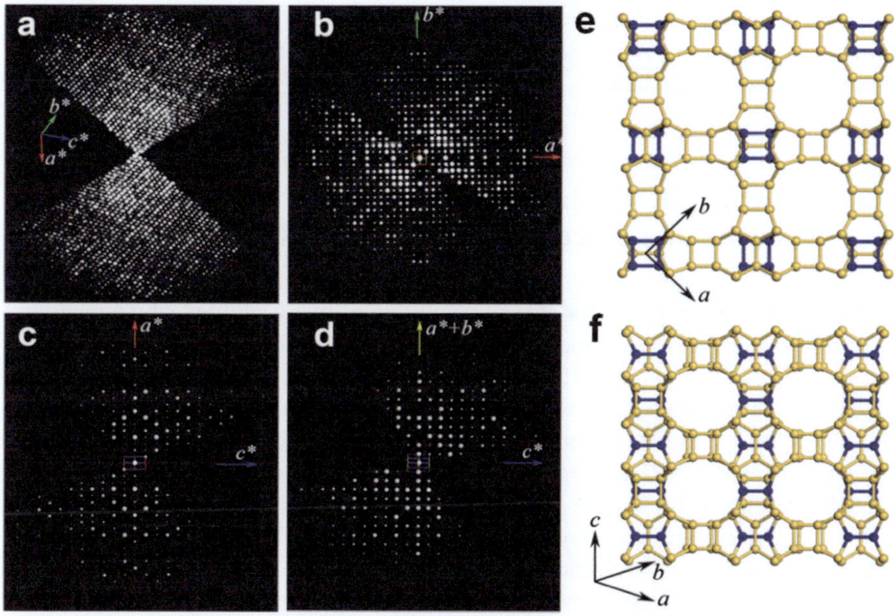

Fig. 5.8 Structure determination of PKU-16 from the RED data. (**a**) Reconstructed 3D reciprocal lattice from the RED data. (**b–d**) (*hk*0), (*h*0*l*), and (*hhl*) planes cut from the reconstructed reciprocal lattice. The reflection conditions can be obtained as *h*0*l*, $h+l=2n$; *h*00, $h=2n$; 00*l*, $l=2n$. Due to the dynamical effects, some reflections violate the reflection conditions. (**e–f**) The 3D structural model obtained from the RED data viewed (**e**) along the *c*-axis showing the 12-ring channels and (**f**) perpendicular to the *c*-axis showing the 11-ring channels. Only T-T connections are shown for clarity. The structure of PKU-16 is closely related to BEC. The differences are that half of the single 4-rings (highlighted in *blue*) are rotated by 90° (Reproduced from Ref. [47] by permission of John Wiley & Sons Ltd)

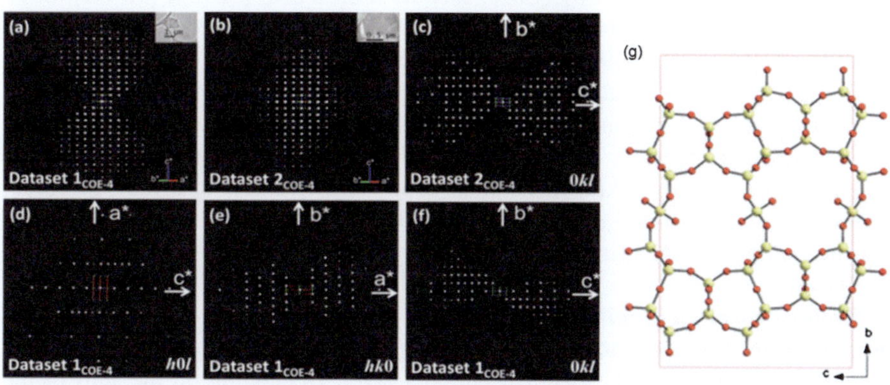

Fig. 5.9 (**a–b**) Reconstructed 3D reciprocal lattices of COE-4 (**a**) dataset $1_{COE\text{-}4}$ and (**b**) dataset $2_{COE\text{-}4}$ taken from two crystals. The crystal size and morphology are shown as an insert. (**c**) 2D slices (0*kl*) cut from the dataset $2_{COE\text{-}4}$. (**d–f**) Three 2D (*h*0*l*), (*hk*0), and (0*kl*) slices cut from the dataset $1_{COE\text{-}4}$. (**g**) The structure of the layered silicate COE-4 determined by RED (Reproduced from Ref. [89] by permission of The Royal Society of Chemistry)

The atomic positions of the framework atoms refined against RED data differ on average by 0.05–0.07 Å for Si atoms and 0.14–0.22 Å for O atoms compared to those from the PXRD data.

5.5 Structure Determination of Zeolites by High-Resolution Transmission Electron Microscopy (HRTEM)

HRTEM has been essential for structure solution of many zeolites which could not be solved by diffraction methods. Among them are crystals that are too small and/or the samples that contain impurity or disorders. The early efforts of structure solution of zeolites have been by combining HRTEM images with model building. During the past years, a number of complex zeolite structures have been solved by combining HRTEM with PXRD, where the structure factor phases obtained from HRTEM images have been the key for the structure solution. A more general approach of solving zeolite structures is by 3D reconstruction of HRTEM images taken from different zone axes, which has been used successfully to solve complex zeolite structures. The 3D reconstruction method is especially useful for solving disordered zeolites, which give poor PXRD patterns with broad peaks. We will describe these different approaches to solve zeolite structures by HRTEM below.

5.5.1 Structure Determination of Zeolites by Combining HRTEM and Model Building

Many zeolite structures were solved by model building based on HRTEM images taken from one or several projections, such as zeolite beta (*BEA) [90–92], SSZ-26/SSZ-33 (CON) [93, 94], SSZ-31 (*STO) [95, 96], ITQ-38 [39], MCM-22 (MWW) [97], Ti-YNU-1 [98], ETS-10 [99, 100], UTD-1 (DON) [101], and ITQ-15 (UTL) [102]. Model building has been especially useful for zeolites containing stacking disorders. Knowing the various types of stacking disorders in zeolites and the characteristic features in the corresponding diffraction patterns is therefore important for solving unknown zeolites containing stacking disorders. A recent review by us summarized the structure elucidation and analysis of stacking disorders in zeolites and open frameworks by electron crystallography and X-ray diffraction [103].

Zeolite beta has a heavily faulted structure which inhibited structure determination of the material for more than 20 years. The structure is formed by epitaxial stacking of a topologically identical building layer. Depending on the shift of the neighboring layers, different polytypes of zeolite beta can be constructed. Figure 5.10a–c presents three polytypes with different stacking sequences viewed in parallel to the layer, denoted as polytypes A, B, and C. The structural models were

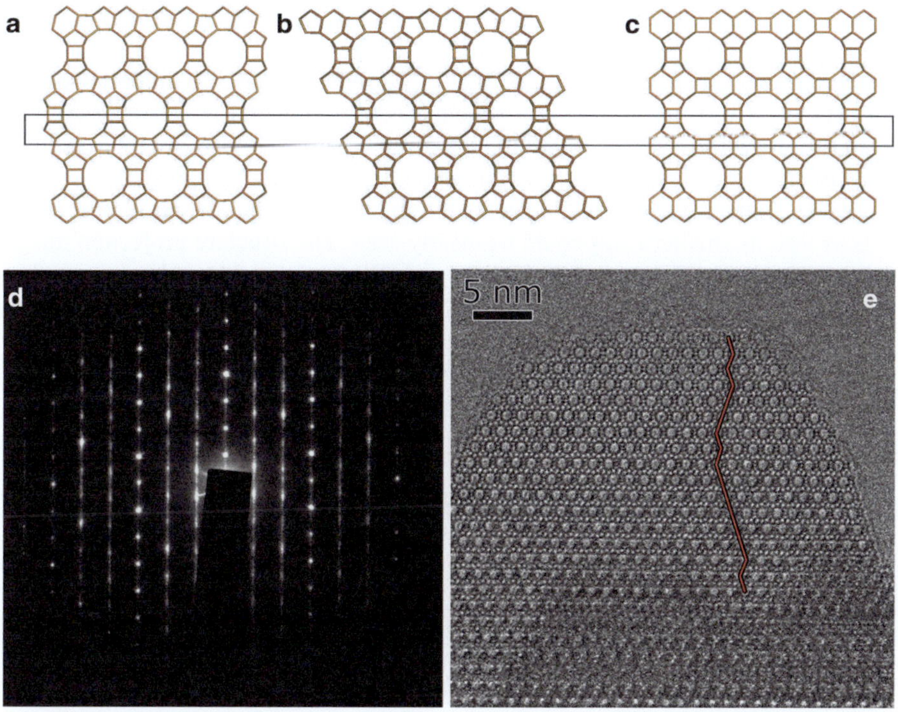

Fig. 5.10 (**a–c**) The structure of zeolite beta viewed along $[100]_A$ for polytypes (**a**) A, (**b**) B, and (**c**) C. (**d**) Selected-area electron diffraction pattern along $[100]_A$ showing both streaks and sharp spots. (**e**) HRTEM image along $[100]_A$ revealing the intergrowth between ABCA… and ABA… type stackings

determined by HRTEM, electron diffraction, and computer-assisted modeling [90–92]. 4-, 5-, 6-, and 12-rings as well as the pore stacking sequences could be identified from the HRTEM images, from which a structural model of the projection was derived. The pore stacking was identified to be two types ABAB… and ABCABC…, with the shift of one-third of the intralayer pore spacing. The unit cell and possible space groups were determined from a series of SAED patterns. Electron diffraction also showed diffuse streaks in parallel to the c^*-axis for reflections with $h \neq 3n$ or $k \neq 3n$ and sharp spots for reflections with $h = 3n$ and $k = 3n$, which agrees with the stacking sequences observed by HRTEM. Computer-assisted modeling was used to derive the 3D models of polytypes A and B, from the structure projection deduced from the HRTEM image. In addition, a hypothetical model of polytype C, with AA… stacking, was proposed. The real materials with the framework of polytype C (framework code BEC) were first reported as a germanate FOS-5 by our group [104] and later as a silicate [105] and silicogermanate [106].

HRTEM combined with model building was used for the structure solution of a new germanosilicate ITQ-38 ($P2/m$, $a = 13.02$ Å, $b = 12.70$ Å, $c = 21.25$ Å, $\beta = 96.87°$) with 10- and 12-ring channels along [010] and 10-ring channels along [100] and

[101] [39]. The unit cell of ITQ-38 was determined from electron diffraction. The unit cell showed to be related to the unit cell of the already known zeolite ITQ-22 [107]. The two structures share the *a*- and *b*- unit cell axes, with the third axis halved for ITQ-38. A structure projection image of ITQ-38 along [010] was reconstructed from a through-focus series of 20 HRTEM images using the software QFocus [63]. The structure projection reconstruction method is helpful to enhance the signal to noise level in the reconstructed image. In the structure projection image larger channels as well as smaller rings could be clearly seen. The structure projection image confirmed the similarity between ITQ-38 and ITQ-22. It could be clearly seen that ITQ-38 was built from the same layer as ITQ-22 with every second layer rotated by 180° around the c^*-axis. This will result in two related structures where the neighboring layers are related either by an inversion center as in ITQ-38 or by a glide plane as in ITQ-22. The structure of ITQ-38 could later be confirmed by structure projection image along [101] as well as Rietveld refinement against the PXRD data.

Recently we reported a new zeolite SSZ-52 (**SFW**) [21]. The structure was determined by combining PXPD, HRTEM, and model building. The framework structure consists of an 18-layer stacking sequence of hexagonally arranged $(Si,Al)_6O_6$ rings (6-rings). The structure has a 3D 8-ring channel system and is a member of the ABC-6 family of zeolites like SSZ-13 (CHA), but it has cavities that are twice as large. These large cavities contain pairs of the bulky organic cations. Because SSZ-52 was beam sensitive, low-dose conditions were applied for HRTEM imaging and a series of 20 HRTEM images were taken under the same conditions. A structure projection image of SSZ-52 along [100] was reconstructed from these images using the software QFocus [63]. The structure projection reconstruction method is helpful to enhance the signal to noise level. The HRTEM image taken from a defect-free region of an SSZ-52 crystal is clearly consistent with the refined framework model (Fig. 5.11b–c). The 8-ring channels can be clearly seen. HRTEM images from some crystals show the presence of some stacking disorder (Fig. 5.11d).

5.5.2 Structure Determination of Zeolites by Combining HRTEM and X-Ray Diffraction

High-resolution electron microscopy and powder X-ray diffraction are complementary methods. Complete and kinematical PXRD data can be obtained easily, while the acquisition of HRTEM images is very time consuming. The structure factor phase information obtained from HRTEM images can facilitate the ab initio structure solution processes by PXRD. Several most complex zeolite structures were solved by combining HRTEM and PXRD, such as TNU-9 (TNU) [34], IM-5 (IMF) [35], and SSZ-74 (–SVR) [36]. HRTEM was also combined with single-crystal X-ray diffraction to solve the structure of a disordered germanosilicate SU-78 [49].

IM-5 is also a high-silica zeolite with 2D 10-ring channels. The structure (*Cmcm*, a=14.299 Å, b=57.413 Å, c=20.143 Å) could not be solved from PXRD data

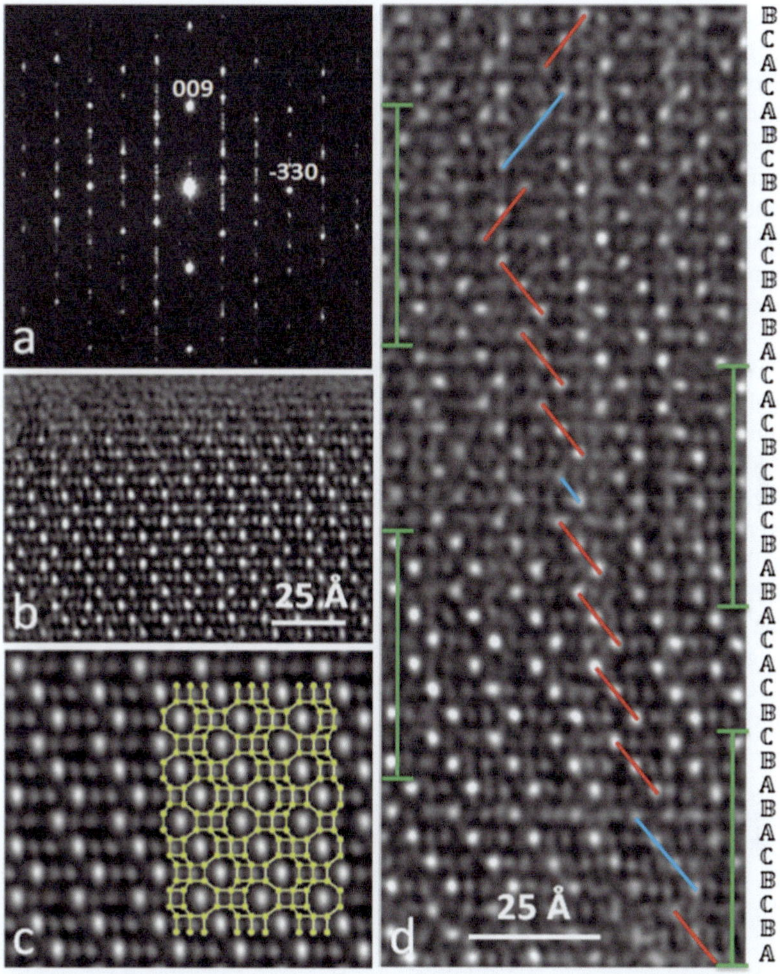

Fig. 5.11 SAED pattern (**a**) and HRTEM images (**b–d**) of the [100] projection of SSZ-52. (**b**) The projection of an ordered part of a crystal. (**c**) The averaged image (*p*1) from (**b**) with the structure overlaid. (**d**) The [100] projection taken from a disordered part of an SSZ-52 crystal with a schematic description of the stacking disorder overlaid. The stacking sequence of the *d*6*r* layers is given on the right. Each HRTEM image was averaged from 20 HRTEM images taken under identical conditions in order to enhance the signal-to-noise ratio (Reprinted with the permission from Ref. [21]. Copyright 2013 American Chemical Society)

alone using standard ab initio structure solution methods such as FOCUS [10] or powder charge flipping (pCF) [12]. Instead a combination of electron diffraction, HRTEM images, and PXRD was utilized. The unit cell parameters were determined from a tilt series of SAED patterns using the program Trice [108]. HRTEM images along the three main zone axes of IM-5 were acquired (Fig. 5.12a–c). The projection symmetry was determined and the structure factor amplitudes and phases were extracted from the HRTEM images using the program CRISP [61]. Ninety-five structure factors obtained from the three HRTEM images were used to reconstruct a 3D electrostatic potential map, from which an initial structural model was obtained. However, the model deviated to some extent from the final structure and could not be refined by PXRD. When the 95 structure factor phases calculated from the initial structural model were included as the starting phase sets in pCF [12] implemented in the program Superflip [83], a better structural model could be obtained from the electron density map (Fig. 5.12). The structural model could be finally refined against the PXRD data using the Rietveld method. IM-5 contains 24 symmetry-independent T-atoms and has an unusual 2D medium-pore channel system. It was shown later that the complex structure of IM-5 could also be solved by 3D reconstruction from HRTEM images alone [109].

Single-crystal X-ray diffraction (SXRD) is a powerful technique to solve crystal structures, but it can sometimes encounter difficulties when the structures are disordered. SU-78 is a germanosilicate with crystals large enough (10 μm × 10 μm × 30 μm) for single-crystal X-ray diffraction at a synchrotron facility [49]. The SXRD data however showed streaks for reflections with indices *hkl* ($h=3n\pm 1$) and sharp spots for the 0*kl* reflections. In order to determine the structure, a combination of ab initio structure solution from SXRD and HRTEM imaging was applied. A 2D charge-flipping algorithm was applied to the sharp spots of the 0*kl* reflections of the SXRD data to generate a projected electron density map along [100]. The streaks in the *h*0*l* plane correspond to the presence of stacking disorders, which was confirmed by HRTEM images along [010]. By combining the structure information along [100]

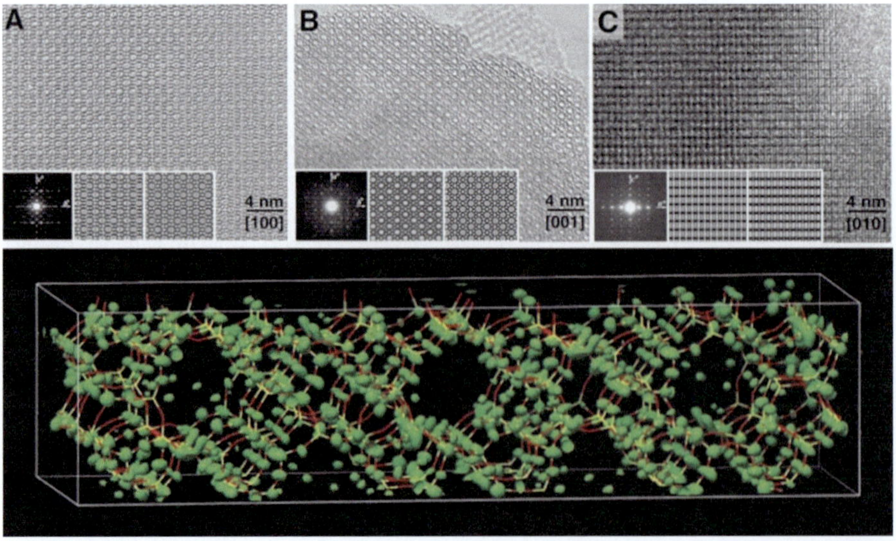

Fig. 5.12 (*Above*) HRTEM images of IM-5 taken along (**a**) [100], (**b**) [001], and (**c**) [010]. (*Below*) The electron density map (*in green*) obtained after pCF with the final structural model superimposed (From Baerlocher et al. [35]. Reprinted with permission from AAAS)

from SXRD and those along [010] from HRTEM, a structural model of SU-78 could be built (Fig. 5.13). SU-78 is an intergrowth of two polytypes SU-78A and SU-78B and contains interconnected 12-ring channels in three directions. The two polytypes are built from the same building layer, similar to those in the zeolite beta family. However, the layer stacking in SU-78 is different from those in zeolite beta polytypes A, B, and C. SU-78 turned out to be an intergrowth of two new polytypes of zeolite beta, which was previously proposed as polytypes D and E.

5.5.3 *Structure Determination of Zeolites by 3D Reconstruction of HRTEM Images*

A more general method to solve unknown zeolite structures is by 3D reconstruction of HRTEM images. Structure factor amplitudes and phases were extracted from HRTEM images along different projections and combined into a 3D electrostatic potential map, as described in Sect. 5.2.4. 3D reconstruction has been used to solve the structures of the high-silica zeolite beta polytype C (BEC) [110] and polytype B [111]. It has been recently used to solve the structure of an intergrown zeolite ITQ-39 [40].

Corma and coworkers synthesized a highly enriched form of zeolite beta polytype B[111]. SAED did not show diffuse streaks, as is usual for crystals of zeolite beta. Instead it showed twinning of the unit cell proposed for beta polytype B. The unit cell parameters ($a=17.97$ Å, $b=17.97$ Å, $c=14.82$ Å, $\beta=113.7°$) were determined from a tilt series of SAED patterns using the programs ELD [66] and Trice [108]. The space group ($C2/c$) was deduced from the systematic absences and by

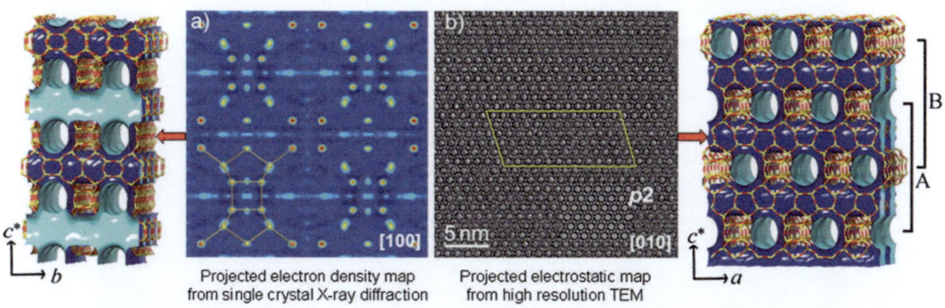

Fig. 5.13 Illustration of the structure of SU-78 solved by combining single-crystal X-ray diffraction and HRTEM. Electron density projected along [100] was obtained by 2D charge flipping from the X-ray diffraction data and resembles the projection of beta polytype C. The HRTEM image taken along [010] is similar to that of zeolite beta with stacking disorders. A 3D structural model was built by combining the two projections. The resulting structural model of SU-78 shows the BEC-type channels along [100] and the beta-type channel systems [010]. The cyan surfaces are toward the pores, while the blue surfaces are toward the framework. The parts with SU-78A and SU-78B are marked (Reprinted with the permission from Ref. [49]. Copyright 2012 American Chemical Society)

examining the plane group symmetry of the HRTEM image. The HRTEM image taken along [1–10] clearly shows 12-ring channels arranged in an ABCA… stacking sequence, characteristic for polytype B (Fig. 5.14). Structure factor amplitudes and phases of 39 symmetry-independent *hhl* reflections were extracted from the HRTEM image using the program CRISP [61]. Because the *h-hl* and *-hh-l* reflections in the [110] projection are symmetry related to the *hhl* and *-h-h-l* reflections in the [1–10] projection, their amplitudes and phases could be deduced from those of *hhl* and *-h-h-l* reflections. In such a way 152 structure factors could be generated, from which a 3D electrostatic potential map could be constructed. All the 9 independent Si atoms could be located from the 3D potential map (Fig. 5.14). This

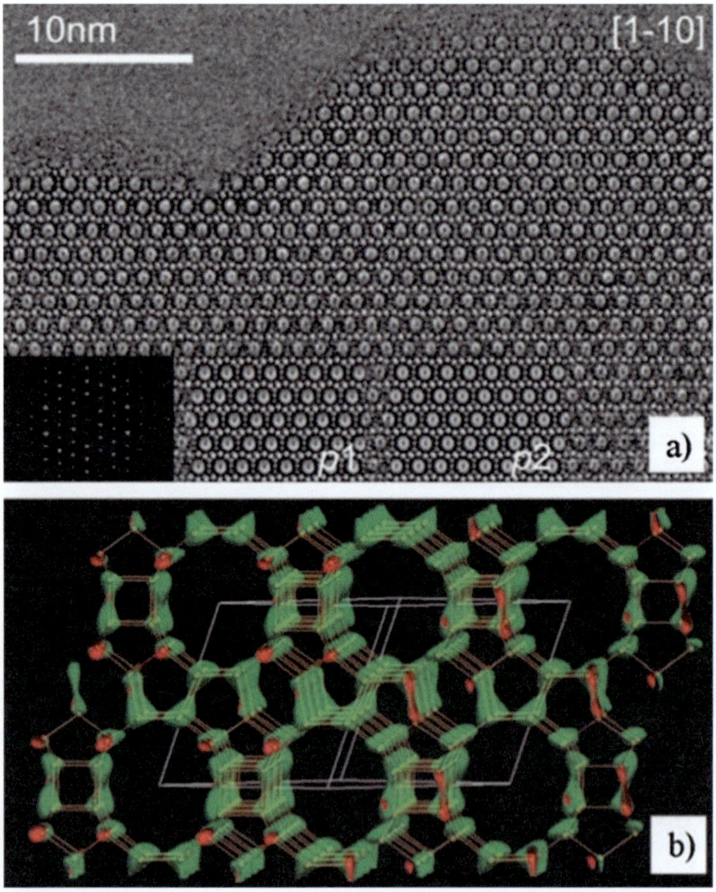

Fig. 5.14 (**a**) An HRTEM image of polytype B enriched beta taken along the [1–10] direction. (**b**) The 3D electrostatic potential map constructed from the [1–10] HRTEM image by utilizing the crystal symmetry (Reprinted with the permission from Ref. [111]. Copyright 2008 American Chemical Society)

example demonstrates that a 3D structure can be solved from one single projection, if the symmetry of the crystal is utilized.

The structure of the aluminosilicate ITQ-39 was solved by electron crystallography from crystal domains only a few unit cells in size [40]. ITQ-39 is an intergrowth of three different polytypes (called ITQ-39A, ITQ-39B, and ITQ-39C), built from the same layer but with different stacking sequences. ITQ-39 contains stacking faults and twinning with nano-sized domains, being one of the most complex zeolite ever solved. Because several types of disorders were present simultaneously in the ITQ-39 crystals, it was necessary to obtain complete 3D data from the same crystal in order to find the orientation and relations of the disorders and determine the unit cells. A 3D RED dataset with 880 SAED patterns was collected with 0.15° interval between the consecutive ED frames from one single crystal. Twining and stacking disorders could be identified from two perpendicular 2D cuts of the 3D reciprocal lattice reconstructed from the RED data; see Fig. 5.15a, b. The 3D RED data revealed that the different ITQ-39 polytypes share two common axes, *a* and *b*. HRTEM images were thus collected along these two common axes *a* and *b*, because they allow the observation of possible channels in a faulted sample. A through-focus series of 20 HRTEM images were acquired along each of the two axes and used to reconstruct the structure projection images (Fig. 5.15c, d). Pair-wise 12-ring channels and twinning can be observed from the image along [010] (Fig. 5.15c). 10-ring channels and stacking faults can be identified from the image along [100] (Fig. 5.15d). The ordered crystal domains in ITQ-39 are only a few unit cells in size. Structure factor amplitudes and phases were extracted from the Fourier transforms of the ordered regions of ITQ-39B. A 3D electrostatic potential map of ITQ-39B was reconstructed from the 53 strongest reflections (with amplitudes >8 % of the largest amplitude). All 28 unique Si atoms could be located from the 3D electrostatic potential map; see Fig. 5.15e. Oxygen atoms was added halfway in between nearest neighbor Si atoms in order to complete the silicate framework. The structures of two other polytypes ITQ-39A and ITQ-39C were constructed based on the stacking sequences found in Fig. 5.15d. The structures were geometrically refined using a distance least-squares refinement of the Si-O, O-O, and Si-Si distances. The final structural model could be confirmed from PXRD refinements from the disordered model. The structure of ITQ-39 contains straight pair-wise 12-ring channels along *b*, which are interconnected by three 10-ring channels. The pair-wise 12-ring channels are unique and have not been observed in any other zeolites.

5.6 Conclusions

In this chapter, we have presented different electron microscopic techniques for structure determination of unknown zeolites. We demonstrated a number of new electron crystallographic methods that have made structure determination of zeolites more feasible, even for non-TEM experts. These include the structure projection reconstruction from through-focus series of HRTEM images, automated

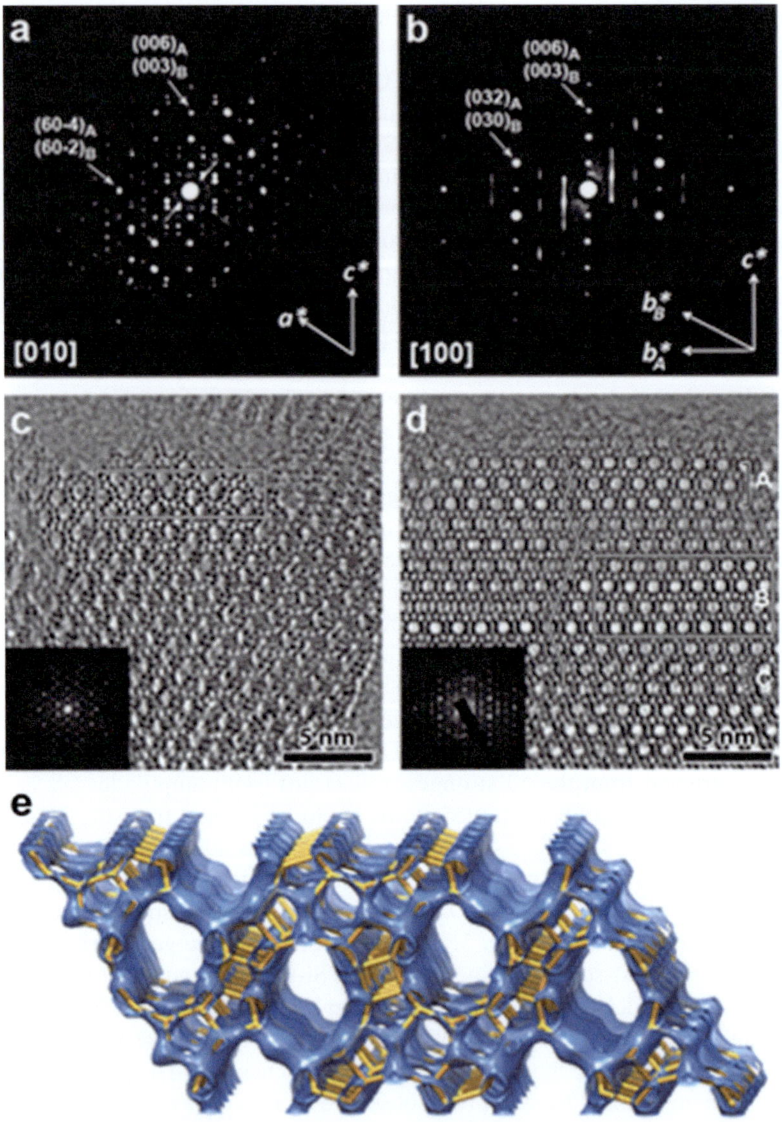

Fig. 5.15 (**a–b**) Two perpendicular cuts of the reconstructed 3D reciprocal lattice of ITQ-39 from the RED data from one single crystal. (**a**) The (*h*0*l*) slice and (**b**) the (0*kl*) slice. The (*h*0*l*) pattern can be indexed using two twin components A and B. The (0*kl*) pattern can be indexed using two different lattices, corresponding to ITQ-39A and ITQ-39B. (**c–d**) Structure projection images reconstructed from 20 HRTEM images along (**c**) [010] and (**d**) [100]. Twins can be identified from the orientation of the pair-wise 12-ring channels in (**c**). 10-ring channels are observed in (**d**) and the channel stacking is traced by a line. The different stacking leads to three polymorphs ITQ-39A, ITQ-39B, and ITQ-39C. The domains used for structure factor determination are outlined by rectangles. (**e**) The 3D electrostatic potential map of ITQ-39B reconstructed from the structure factor amplitudes and phases extracted from the marked nano-domains in (**c**) and (**d**). All 28 symmetry-independent Si atoms could be located directly from the 3D map. The refined structural model of ITQ-39 is superimposed (Reprinted by permission from Macmillan Publishers Ltd: Ref. [40], Copyright 2012)

diffraction tomography (ADT), and rotation electron diffraction (RED). The ADT and RED methods have shown to be very powerful and efficient in solving structures of unknown zeolites. HRTEM has unique advantages for studying disordered structures which are difficult to be studied by other techniques. The structure projection reconstruction from through-focus series of HRTEM images using the program QFocus makes it easier to obtain high-quality structural images. It also enhances the image contrast for especially beam-sensitive materials and materials containing defects. The crystallographic structure factor amplitudes and phases can be obtained from HRTEM images and used for reconstruction of the 3D electrostatic potential map, from which the 3D structure can be determined. The phase information from HRTEM images can also be used to assist the structure solution of complex zeolites from powder X-ray diffraction. Electron crystallography and PXRD are complementary for structural analysis of zeolites. PXRD patterns contain information about the bulk sample, while electron crystallography provides detailed information on individual crystalline particles down to the nanometer sizes. The development of ADT and RED methods makes the collection of 3D single-crystal diffraction data on a TEM almost as feasible and fast as that on a single-crystal X-ray diffractometer, but from crystals millions times smaller.

Acknowledgments This work was supported by the Swedish Research Council (VR), the Swedish Governmental Agency for Innovation Systems (VINNOVA), and the Knut and Alice Wallenberg Foundation through a grant for purchasing the TEM and the project grant 3DEM-NATUR.

References

1. Su J, Wang Y, Wang Z, Lin J (2009) PKU-9: an aluminogermanate with a new three-dimensional zeolite framework constructed from CGS layers and spiro-5 units. J Am Chem Soc 131(17):6080–6081
2. Tang L, Shi L, Bonneau C, Sun J, Yue H, Ojuva A, Lee B-L, Kritikos M, Bell RG, Bacsik Z, Mink J, Zou X (2008) A zeolite family with chiral and achiral structures built from the same building layer. Nat Mater 7(5):381–385
3. Xu Y, Li Y, Han Y, Song X, Yu J (2013) A gallogermanate zeolite with eleven-membered-ring channels. Angew Chem Int Ed 52(21):5501–5503
4. Han Y, Li Y, Yu J, Xu R (2011) A gallogermanate zeolite constructed exclusively by three-ring building units. Angew Chem Int Ed 50(13):3003–3005
5. Song X, Li Y, Gan L, Wang Z, Yu J, Xu R (2009) Heteroatom-stabilized chiral framework of aluminophosphate molecular sieves. Angew Chem Int Ed 48(2):314–317
6. Shao L, Li Y, Yu J, Xu R (2012) Divalent-metal-stabilized aluminophosphates exhibiting a new zeolite framework topology. Inorg Chem 51(1):225–229
7. Liu Z, Song X, Li J, Li Y, Yu J, Xu R (2012) $|(C_4NH_{12})_4|[M_4Al_{12}P_{16}O_{64}]$ (M = Co, Zn): new heteroatom-containing aluminophosphate molecular sieves with two intersecting 8-ring channels. Inorg Chem 51(3):1969–1974
8. Armstrong JA, Weller MT (2010) Beryllosilicate frameworks and zeolites. J Am Chem Soc 132(44):15679–15686
9. Baerlocher C, Weber T, McCusker LB, Palatinus L, Zones SI (2011) Unraveling the perplexing structure of the zeolite SSZ-57. Science 333(6046):1134–1137

10. Grosse-Kunstleve RW, McCusker LB, Baerlocher C (1997) Powder diffraction data and crystal chemical information combined in an automated structure determination procedure for zeolites. J Appl Crystallogr 30:985–995
11. Altomare A, Burla MC, Camalli M, Carrozzini B, Cascarano GL, Giacovazzo C, Guagliardi A, Moliterni AGG, Polidori G, Rizzi R (1999) EXPO: a program for full powder pattern decomposition and crystal structure solution. J Appl Crystallogr 32:339–340
12. Baerlocher C, McCusker LB, Palatinus L (2007) Charge flipping combined with histogram matching to solve complex crystal structures from powder diffraction data. Z Krist 222(2):47–53
13. Cantin A, Corma A, Leiva S, Rey F, Rius J, Valencia S (2005) Synthesis and structure of the bidimensional zeolite ITQ-32 with small and large pores. J Am Chem Soc 127(33):11560–11561
14. Corma A, Diaz-Cabanas MJ, Luis Jorda J, Martinez C, Moliner M (2006) High-throughput synthesis and catalytic properties of a molecular sieve with 18-and 10-member rings. Nature 443(7113):842–845
15. Corma A, Diaz-Cabanas MJ, Jorda JL, Rey F, Sastre G, Strohmaier KG (2008) A zeolitic structure (ITQ-34) with connected 9-and 10-ring channels obtained with phosphonium cations as structure directing agents. J Am Chem Soc 130(49):16482–16483
16. Jiang J, Jorda JL, Diaz-Cabanas MJ, Yu J, Corma A (2010) The synthesis of an extra-large-pore zeolite with double three-ring building units and a low framework density. Angew Chem Int Ed 49(29):4986–4988
17. Hernandez-Rodriguez M, Jorda JL, Rey F, Corma A (2012) Synthesis and structure determination of a new microporous zeolite with large cavities connected by small pores. J Am Chem Soc 134(32):13232–13235
18. Xie D, McCusker LB, Baerlocher C (2011) Structure of the borosilicate zeolite catalyst SSZ-82 solved using 2D-XPD charge flipping. J Am Chem Soc 133(50):20604–20610
19. Elomari S, Burton AW, Ong K, Pradhan AR, Chan IY (2007) Synthesis and structure solution of zeolite SSZ-65. Chem Mater 19(23):5485–5492
20. McCusker LB, Baerlocher C, Burton AW, Zones SI (2011) A re-examination of the structure of the germanosilicate zeolite SSZ-77. Solid State Sci 13(4):800–805
21. Xie D, McCusker LB, Baerlocher C, Zones SI, Wan W, Zou XD (2013) SSZ-52, a zeolite with an 18-layer aluminosilicate framework structure related to that of the DeNOx catalyst Cu-SSZ-13. J Am Chem Soc 135(28):10519–10524
22. Elomari S, Burton A, Medrud RC, Grosse-Kunstleve R (2009) The synthesis, characterization, and structure solution of SSZ-56: an extreme example of isomer specificity in the structure direction of zeolites. Microporous Mesoporous Mater 118(1–3):325–333
23. Lorgouiux Y, Dodin M, Paillaud J-L, Caullet P, Michelin L, Josien L, Ersen O, Bats N (2009) IM-16: a new microporous germanosilicate with a novel framework topology containing d4r and mtw composite building units. J Solid State Chem 182(3):622–629
24. Dodin M, Paillaud J-L, Lorgouilloux Y, Caullett P, Elkaim E, Bats N (2010) A zeolitic material with a three-dimensional pore system formed by straight 12-and 10-ring channels synthesized with an imidazolium derivative as structure-directing agent. J Am Chem Soc 132(30):10221–10223
25. McCusker LB, Baerlocher C, Wilson ST, Broach RW (2009) Synthesis and structural characterization of the aluminosilicate LZ-135, a zeolite related to ZSM-10. J Phys Chem C 113(22):9838–9844
26. Han Z, Picone AL, Slawin AMZ, Seymour VR, Ashbrook SE, Zhou W, Thompson SP, Parker JE, Wright PA (2010) Novel large-pore aluminophosphate molecular sieve STA-15 prepared using the tetrapropylammonium cation as a structure directing agent. Chem Mater 22(2):338–346
27. Dorset DL, Kennedy GJ (2005) Crystal structure of MCM-70: a microporous material with high framework density. J Phys Chem B 109(29):13891–13898

28. Xie D, McCusker LB, Baerlocher C, Gibson L, Burton AW, Hwang S-J (2009) Optimized synthesis and structural characterization of the borosilicate MCM-70. J Phys Chem C 113(22):9845–9850
29. Broach RW, Kirchner RM (2011) Structures of the K^+ and NH_4^+ forms of Linde J. Microporous Mesoporous Mater 143(2–3):398–400
30. Verheyen E, Joos L, Van Havenbergh K, Breynaert E, Kasian N, Gobechiya E, Houthoofd K, Martineau C, Hinterstein M, Taulelle F, Van Speybroeck V, Waroquier M, Bals S, Van Tendeloo G, Kirschhock CEA, Martens JA (2012) Design of zeolite by inverse sigma transformation. Nat Mater 11(12):1059–1064
31. Broach RW, Greenlay N, Jakubczak P, Knight LM, Miller SR, Mowat JPS, Stanczyk J, Lewis GJ (2014) New ABC-6 net molecular sieves ZnAPO-57 and ZnAPO-59: framework charge density-induced transition from two- to three-dimensional porosity. Microporous Mesoporous Mater 189:49–63
32. Zanardi S, Millini R, Frigerio F, Belloni A, Cruciani G, Bellussi G, Carati A, Rizzo C, Montanari E (2011) ERS-18: a new member of the NON-EUO-NES zeolite family. Microporous Mesoporous Mater 143(1):6–13
33. Inge AK, Fahlquist H, Willhammar T, Huang Y, McCusker LB, Zou XD (2013) Solving complex open-framework structures from X-ray powder diffraction by direct-space methods using composite building units. J Appl Crystallogr 46:1094–1104
34. Gramm F, Baerlocher C, McCusker LB, Warrender SJ, Wright PA, Han B, Hong SB, Liu Z, Ohsuna T, Terasaki O (2006) Complex zeolite structure solved by combining powder diffraction and electron microscopy. Nature 444(7115):79–81
35. Baerlocher C, Gramm F, Massueger L, McCusker LB, He Z, Hovmoeller S, Zou XD (2007) Structure of the polycrystalline zeolite catalyst IM-5 solved by enhanced charge flipping. Science 315(5815):1113–1116
36. Baerlocher C, Xie D, McCusker LB, Hwang S-J, Chan IY, Ong K, Burton AW, Zones SI (2008) Ordered silicon vacancies in the framework structure of the zeolite catalyst SSZ-74. Nat Mater 7(8):631–635
37. Dorset DL, Strohmaier KG, Kliewer CE, Corma A, Diaz-Cabanas MJ, Rey F, Gilmore CJ (2008) Crystal structure of ITQ-26, a 3D framework with extra-large pores. Chem Mater 20(16):5325–5331
38. Sun J, Bonneau C, Cantin A, Corma A, Diaz-Cabanas MJ, Moliner M, Zhang D, Li M, Zou XD (2009) The ITQ-37 mesoporous chiral zeolite. Nature 458(7242):1154–1157
39. Moliner M, Willhammar T, Wan W, Gonzalez J, Rey F, Jorda JL, Zou XD, Corma A (2012) Synthesis design and structure of a multipore zeolite with interconnected 12-and 10-MR channels. J Am Chem Soc 134(14):6473–6478
40. Willhammar T, Sun J, Wan W, Oleynikov P, Zhang D, Zou XD, Moliner M, Gonzalez J, Martinez C, Rey F, Corma A (2012) Structure and catalytic properties of the most complex intergrown zeolite ITQ-39 determined by electron crystallography. Nat Chem 4(3):188–194
41. Corma A, Diaz-Cabanas MJ, Jiang J, Afeworki M, Dorset DL, Soled SL, Strohmaier KG (2010) Extra-large pore zeolite (ITQ-40) with the lowest framework density containing double four- and double three-rings. Proc Natl Acad Sci U S A 107(32):13997–14002
42. Jiang J, Jorda JL, Yu J, Baumes LA, Mugnaioli E, Diaz-Cabanas MJ, Kolb U, Corma A (2011) Synthesis and structure determination of the hierarchical meso-microporous zeolite ITQ-43. Science 333(6046):1131–1134
43. Martinez-Franco R, Moliner M, Yun Y, Sun J, Wan W, Zou XD, Corma A (2013) Synthesis of an extra-large molecular sieve using proton sponges as organic structure-directing agents. Proc Natl Acad Sci U S A 110(10):3749–3754
44. Yun Y, Hernández M, Wan W, Zou XD, Jordá JL, Cantín A, Rey F, Corma A (2015) The first zeolite with a tri-directional extra-large 14-ring pore system derived using a phosphonium-based organic molecule. Chem Commun. doi:10.1039/C4CC10317C
45. Jiang J, Yun Y, Zou XD, Jordá JL, Corma A (2014) ITQ-54: a multi-dimensional extra-large pore zeolite with $20 \times 14 \times 12$-ring channels. Chem Sci 6:480–485

46. Liang J, Su J, Wang Y, Chen Y, Zou XD, Liao F, Lin J, Sun J (2014) A 3D 12-ring zeolite with ordered 4-ring vacancies occupied by $(H_2O)_2$ dimers. Chem Eur J 49:16097–16101
47. Hua W, Chen H, Yu Z-B, Zou XD, Lin J, Sun J (2014) A germanosilicate structure with 11x11x12-ring channels solved by electron crystallography. Angew Chem Int Ed 53(23):5868–5871
48. Willhammar T, Burton AW, Yun Y, Sun J, Afeworki M, Strohmaier KG, Vroman H, Zou XD (2014) EMM-23: a stable high-silica multidimensional zeolite with extra-large trilobe-shaped channels. J Am Chem Soc 136:13570–13573
49. Yu Z-B, Han Y, Zhao L, Huang S, Zheng Q-Y, Lin S, Cordova A, Zou XD, Sun J (2012) Intergrown new zeolite beta polymorphs with interconnected 12-ring channels solved by combining electron crystallography and single-crystal X-ray diffraction. Chem Mater 24(19):3701–3706
50. Pan M (1996) High resolution electron microscopy of zeolites. Micron 27(3–4):219–238
51. Diaz I, Mayoral A (2011) TEM studies of zeolites and ordered mesoporous materials. Micron 42(5):512–527
52. Anderson MW, Ohsuna T, Sakamoto Y, Liu Z, Carlsson A, Terasaki O (2004) Modern microscopy methods for the structural study of porous materials. Chem Commun 8:907–916
53. Sun J, Zou XD (2010) Structure determination of zeolites and ordered mesoporous materials by electron crystallography. Dalton Trans 39(36):8355–8362
54. Liu Z, Fujita N, Miyasaka K, Han L, Stevens SM, Suga M, Asahina S, Slater B, Xiao C, Sakamoto Y, Anderson MW, Ryoo R, Terasaki O (2013) A review of fine structures of nanoporous materials as evidenced by microscopic methods. Microscopy 62(1):109–146
55. Willhammar T, Yun Y, Zou XD (2014) Structural determination of ordered porous solids by electron crystallography. Adv Funct Mater 24(2):182–199
56. Vincent R, Midgley P (1994) Double conical beam-rocking system for measurement of integrated electron-diffraction intensities. Ultramicroscopy 53(3):271–282
57. Kolb U, Gorelik T, Kuebel C, Otten MT, Hubert D (2007) Towards automated diffraction tomography: Part I - data acquisition. Ultramicroscopy 107(6–7):507–513
58. Kolb U, Gorelik T, Otten MT (2008) Towards automated diffraction tomography. Part II – cell parameter determination. Ultramicroscopy 108(8):763–772
59. Zhang D, Oleynikov P, Hovmöller S, Zou XD (2010) Collecting 3D electron diffraction data by the rotation method. Z Krist 225(2–3):94–102
60. Wan W, Sun J, Su J, Hovmöller S, Zou XD (2013) Three-dimensional rotation electron diffraction: software RED for automated data collection and data processing. J Appl Crystallogr 46:1863–1873
61. Hovmöller S (1992) CRISP – crystallographic image-processing on a personal-computer. Ultramicroscopy 41(1–3):121–135
62. Zou XD, Sundberg M, Larine M, Hovmöller S (1996) Structure projection retrieval by image processing of HREM images taken under non-optimum defocus conditions. Ultramicroscopy 62(1–2):103–121
63. Wan W, Hovmoeller S, Zou XD (2012) Structure projection reconstruction from through-focus series of high-resolution transmission electron microscopy images. Ultramicroscopy 115:50–60
64. Zou XD, Hovmöller S, Oleynikov P (2011) Electron crystallography – electron microscopy and electron diffraction. Oxford University Press, Oxford
65. Vainshtein BK (1964) Structure analysis by electron diffraction. Pergamon Press, Oxford
66. Zou XD, Sukharev Y, Hovmöller S (1993) Eld – a computer-program system for extracting intensities from electron-diffraction patterns. Ultramicroscopy 49(1–4):147–158
67. Weirich TE, Zou XD, Ramlau R, Simon A, Cascarano GL, Giacovazzo C, Hovmöller S (2000) Structures of nanometre-size crystals determined from selected-area electron diffraction data. Acta Crystallogr Sect A 56:29–35

68. Zhang D, Gruner D, Oleynikov P, Wan W, Hovmöller S, Zou XD (2010) Precession electron diffraction using a digital sampling method. Ultramicroscopy 111(1):47–55
69. Oleynikov P, Hovmöller S, Zou XD (2007) Precession electron diffraction: observed and calculated intensities. Ultramicroscopy 107(6–7):523–533
70. Gemmi M, Zou XD, Hovmöller S, Migliori A, Vennstrom M, Andersson Y (2003) Structure of Ti2P solved by three-dimensional electron diffraction data collected with the precession technique and high-resolution electron microscopy. Acta Crystallogr Sect A 59:117–126
71. Dorset DL, Gilmore CJ, Jorda JL, Nicolopoulos S (2007) Direct electron crystallographic determination of zeolite zonal structures. Ultramicroscopy 107(6–7):462–473
72. Weirich TE, Ramlau R, Simon A, Hovmöller S, Zou XD (1996) A crystal structure determined with 0.02 angstrom accuracy by electron microscopy. Nature 382(6587):144–146
73. Mayoral A, Carey T, Anderson PA, Lubk A, Diaz I (2011) Atomic resolution analysis of silver ion-exchanged zeolite A. Angew Chem Int Ed 50(47):11230–11233
74. Mayoral A, Carey T, Anderson PA, Diaz I (2013) Atomic resolution analysis of porous solids: a detailed study of silver ion-exchanged zeolite A. Microporous Mesoporous Mater 166:117–122
75. Mayoral A, Coronas J, Casado C, Tellez C, Diaz I (2013) Atomic resolution analysis of microporous titanosilicate ETS-10 through aberration corrected STEM imaging. ChemCatChem 5(9):2595–2598
76. Ortalan V, Uzun A, Gates BC, Browning ND (2010) Direct imaging of single metal atoms and clusters in the pores of dealuminated HY zeolite. Nat Nanotechnol 5(7):506–510
77. Zou XD, Mo ZM, Hovmöller S, Li XZ, Kuo KH (2003) Three-dimensional reconstruction of the nu-AlCrFe phase by electron crystallography. Acta Crystallogr Sect A 59:526–539
78. Sheldrick GM (2008) A short history of SHELX. Acta Crystallogr Sect A 64:112–122
79. Burla MC, Caliandro R, Camalli M, Carrozzini B, Cascarano GL, Giacovazzo C, Mallamo M, Mazzone A, Polidori G, Spagna R (2012) SIR2011: a new package for crystal structure determination and refinement. J Appl Crystallogr 45:357–361
80. Smeets S, McCusker LB, Baerlocher C, Mugnaioli E, Kolb U (2013) Using FOCUS to solve zeolite structures from three-dimensional electron diffraction data. J Appl Crystallogr 46:1017–1023
81. Oszlanyi G, Suto A (2004) Ab initio structure solution by charge flipping. Acta Crystallogr Sect A 60:134–141
82. Oszlanyi G, Suto A (2005) Ab initio structure solution by charge flipping. II. Use of weak reflections. Acta Crystallogr Sect A 61:147–152
83. Palatinus L, Chapuis G (2007) SUPERFLIP – a computer program for the solution of crystal structures by charge flipping in arbitrary dimensions. J Appl Crystallogr 40:786–790
84. Xie D, Baerlocher C, McCusker LB (2008) Combining precession electron diffraction data with X-ray powder diffraction data to facilitate structure solution. J Appl Crystallogr 41:1115–1121
85. Nicolopoulos S, Gonzalezcalbet J, Valletregi M, Corma A, Corell C, Guil J, Perezpariente J (1995) Direct phasing in electron crystallography – Ab-initio determination of a new Mcm-22 zeolite structure. J Am Chem Soc 117(35):8947–8956
86. Wagner P, Terasaki O, Ritsch S, Nery JG, Zones SI, Davis ME, Hiraga K (1999) Electron diffraction structure solution of a nanocrystalline zeolite at atomic resolution. J Phys Chem B 103(39):8245–8250
87. Dorset DL (2006) The crystal structure of ZSM-10, a powder X-ray and electron diffraction study. Z Krist 221(4):260–265
88. Gilmore CJ, Bricogne G (1997) MICE computer program. In: Carter CW, Sweet RM (eds) Macromolecular crystallography, Pt B, vol 277. Academic, San Diego, pp 65–78
89. Guo P, Liu L, Yun Y, Su J, Wan W, Gies H, Zhang H, Xiao F-S, Zou XD (2014) Ab initio structure determination of interlayer expanded zeolites by single crystal rotation electron diffraction. Dalton Trans 43:10593–10601

90. Treacy M, Newsam J (1988) 2 new 3-dimensional 12-ring zeolite frameworks of which zeolite beta is a disordered intergrowth. Nature 332(6161):249–251
91. Newsam J, Treacy M, Koetsier W, Degruyter C (1988) Structural characterization of zeolite-beta. Proc R Soc Lond Ser Math Phys Eng Sci 420(1859):375–405
92. Higgins J, Lapierre R, Schlenker J, Rohrman A, Wood J, Kerr G, Rohrbaugh W (1988) The framework topology of zeolite-beta. Zeolites 8(6):446–452
93. Lobo R, Pan M, Chan I, Li H, Medrud R, Zones S, Crozier P, Davis M (1993) Ssz-26 and Ssz-33 – 2 molecular-sieves with intersecting 10-ring and 12-ring pores. Science 262(5139):1543–1546
94. Lobo R, Pan M, Chan I, Medrud R, Zones S, Crozier P, Davis M (1994) Physicochemical characterization of zeolites Ssz-26 and Ssz-33. J Phys Chem 98(46):12040–12052
95. Lobo RF, Tsapatsis M, Freyhardt CC, Chan I, Chen CY, Zones SI, Davis ME (1997) A model for the structure of the large-pore zeolite SSZ-31. J Am Chem Soc 119(16):3732–3744
96. van Koningsveld H, Lobo RF (2003) Disorder in zeolite SSZ-31 described on the basis of one-dimensional building units. J Phys Chem B 107(40):10983–10989
97. Leonowicz M, Lawton J, Lawton S, Rubin M (1994) Mcm-22 – a molecular-sieve with 2 independent multidimensional channel systems. Science 264(5167):1910–1913
98. Ruan JF, Wu P, Slater B, Terasaki O (2005) Structure elucidation of the highly active titanosilicate catalyst Ti-YNU-1. Angew Chem Int Ed 44(41):6719–6723
99. Anderson M, Terasaki O, Ohsuna T, Philippou A, Mackay S, Ferreira A, Rocha J, Lidin S (1994) Structure of the microporous titanosilicate Ets-10. Nature 367(6461):347–351
100. Anderson M, Terasaki O, Ohsuna T, Malley P, Philippou A, Mackay S, Ferreira A, Rocha J, Lidin S (1995) Microporous titanosilicate Ets-10 – a structural survey. Philos Mag B Phys Condens Matter Stat Mech Electron Opt Magn Prop 71(5):813–841
101. Lobo RF, Tsapatsis M, Freyhardt CC, Khodabandeh S, Wagner P, Chen CY, Balkus KJ, Zones SI, Davis ME (1997) Characterization of the extra-large-pore zeolite UTD-1. J Am Chem Soc 119(36):8474–8484
102. Corma A, Diaz-Cabanas MJ, Rey F, Nicolooulas S, Boulahya K (2004) ITQ-15: the first ultralarge pore zeolite with a bi-directional pore system formed by intersecting 14- and 12-ring channels, and its catalytic implications. Chem Commun 12:1356–1357
103. Willhammar T, Zou XD (2013) Stacking disorders in zeolites and open-frameworks – structure elucidation and analysis by electron crystallography and X-ray diffraction. Z Krist 228(1):11–27
104. Conradsson T, Dadachov MS, Zou XD (2000) Synthesis and structure of $(Me_3N)_6[Ge_{32}O_{64}]\cdot(H_2O)_{4.5}$, a thermally stable novel zeotype with 3D interconnected 12-ring channels. Microporous Mesoporous Mater 41(1–3):183–191
105. Liu Z, Ohsuna T, Terasaki O, Camblor MA, Diaz-Cabanas MJ, Hiraga K (2001) The first zeolite with three-dimensional intersecting straight-channel system of 12-membered rings. J Am Chem Soc 123(22):5370–5371
106. Corma A, Navarro MT, Rey F, Rius J, Valencia S (2001) Pure polymorph C of zeolite beta synthesized by using framework isomorphous substitution as a structure-directing mechanism. Angew Chem Int Ed 40(12):2277–2280
107. Corma A, Rey F, Valencia S, Jordá JL, Rius J (2003) A zeolite with interconnected 8-, 10- and 12-ring pores and its unique catalytic selectivity. Nat Mater 2(7):493–497
108. Zou XD, Hovmöller A, Hovmöller S (2004) TRICE – a program for reconstructing 3D reciprocal space and determining unit-cell parameters. Ultramicroscopy 98(2–4):187–193
109. Sun J, He Z, Hovmöller S, Zou XD, Gramm F, Baerlocher C, McCusker LB (2010) Structure determination of the zeolite IM-5 using electron crystallography. Z Krist 225(2–3):77–85
110. Ohsuna T, Liu Z, Terasaki O, Hiraga K, Camblor MA (2002) Framework determination of a polytype of zeolite beta by using electron crystallography. J Phys Chem B 106(22):5673–5678
111. Corma A, Moliner M, Cantin A, Diaz-Cabanas MJ, Lorda JL, Zhang D, Sun J, Jansson K, Hovmöller S, Zou XD (2008) Synthesis and structure of polymorph B of zeolite beta. Chem Mater 20(9):3218–3223

Chapter 6
Structure Analysis in Zeolite Research: From Framework Topologies to Functional Properties

Hermann Gies and Bernd Marler

Abstract The solution and the refinement of crystal structures of microporous materials are of prime interest for the synthesis and application community in porous solid research. The structure of a material reveals not only the geometry of the bonding network of the rigid framework but also the location of the non-framework constituents responsible for the porosity of the material. Analyzing the structure, therefore, unveils the synthesis protocol including nucleation and growth. The properties related to the porous net correspond to the local structure, i.e., the composition and the geometry of local configurations. It thus inspires ideas about possible properties and modifications to improve functionality.

The chapter is intended to highlight the contribution of crystallographic analysis of microporous materials and in particular porous silicates to the success of this class of materials. It does not survey the continuous growth of materials in the field but show how the insight gained from crystallographic studies acted as catalyst in the further development of the field.

Keywords Zeolite framework topologies • Structure determination • Electron crystallography • Reciprocal and direct space methods • Defect structures • Localizing heteroatoms • Spiro-5 units

6.1 Introduction

The determination and analysis of the crystal structures of microporous materials has an important impact on the functional development of these materials and also on their applications. The most valuable and obvious information that can be gained from diffraction experiments is the general crystal structure, i.e., the geometry of

H. Gies (✉) • B. Marler
Department of Geology, Mineralogy and Geophysics, Ruhr University Bochum, Bochum, Germany
e-mail: hermann.gies@rub.de

F.-S. Xiao, X. Meng (eds.), *Zeolites in Sustainable Chemistry*, Green Chemistry and Sustainable Technology,
DOI 10.1007/978-3-662-47395-5_6

the framework structure, the pore openings and volume, and the connectivity of pores, i.e., the dimensionality of pore system. In addition, specific structural properties can be investigated including the distribution of active centers in the framework, the location of extra-framework constituents, the degree of hydration, etc. This paper is focused on diffraction techniques for the elucidation of structural properties. Other methods like NMR spectroscopy, SEM, EXAFS, thermal analysis, porosity measurements, etc., certainly add valuable, complementary information which may be essential for a successful structure solution and, on the other hand, may help to obtain a most complete picture of the zeolite structure.

Nevertheless, X-ray diffraction is the most widely used technique for structure elucidation (Fig. 6.1). Over time, zeolite science tremendously benefited from the technical development in diffraction experimentation and data analysis with regard to solving and refining ever more complicated crystal structures. But zeolite materials also played a key role as trigger in these advances, in instrumentation and methodology, serving as challenging materials for case studies, notably in the field of structure solution and structure refinement from powder X-ray diffraction (P-XRD) data and electron crystallography (EC). Crystal structure analysis has contributed significant information to foster zeolite science. The ultimate need to visualize the materials atomic arrangement in order to rationalize synthesis parameters and physical and chemical properties of the microporous material has also fueled the development of structure solution techniques and methods. Zeolitic materials most often are obtained as crystalline powders. In the past 25 years, the advance in structure solution from X-ray powder diffraction data is strongly correlated with zeolite science. At any time, the most complicated structures solved from P-XRD data were microporous materials continuously shifting the degree of complexity higher and higher.

If single crystals of sufficient size and high structural order are available, the structure analysis can be performed in a straightforward process. Usually, however, microporous materials crystallize as fine crystalline powders precluding single-crystal work. Therefore, for almost all synthetic zeolites, powder data have to be used. Only recently, electron diffraction (ED) on nano-crystallites as an independent technique extended the scope and contributed substantially to structure solution. Powder diffraction can be used alone or in combination with other techniques like modeling, electron diffraction, and imaging. In particular, electron diffraction has

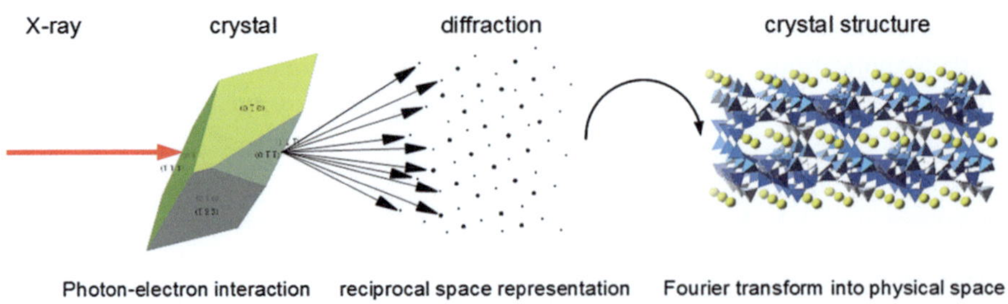

Fig. 6.1 Schematic representation of a structure analysis based on diffraction techniques

more recently been used in combination with P-XRD for structure solution. A subsequent Rietveld refinement of the diffraction data completes the structure analysis.

In general, all textbooks on zeolite science contain one chapter on the structures of zeolite frameworks to illustrate the close relationship between the geometric structure and property of this class of materials. The basic principles of diffraction techniques applied to zeolites can be found, for example, in a review of Morris and Wheatley [1]. Other relevant papers on the structure analysis of microporous materials are given as reference [2–4].

An extended discussion of the latest developments in strategies and concepts related to zeolite structure research is presented by Jihong Yu and Yi Li in a recent review [5]. An exhaustive evaluation of most recent highlights and their particular novelties is summarized there. This communication representing an expert's opinion on the state of the art in zeolite structure research is highly recommended for reading.

The Web-based *Atlas of Zeolite Framework Types*, the Database of Zeolite Structures, gives a complete overview on the structural information of all microporous materials with tetrahedral frameworks irrespective of composition [6]. It is regularly updated and contains a wealth of additional information related to the framework structures of zeolites. The acceptance of new materials in the Atlas requires the approval of the structure commission of the International Zeolite Association which is based on a careful evaluation of the structure analysis. The materials include defective and interrupted frameworks as well as structural disorder as stacking disorder of layers. An important example is the framework structure of zeolite beta. Mixed coordination or coordination higher than four in the framework structure is not included. The requirements for acceptance are accessible on the homepage (http://www.iza-structure.org/).

A recent example of the impact of structure analysis on the development of microporous materials is also the ever growing number of coordination polymers, e.g., metal-organic frameworks (MOFs). The results from structure analysis inspired chemists in the systematic exploration of this class of materials [7]. Today we know more than 100 structure variants; its number is still expanding dynamically with exciting new crystal structures. For potential applications, the knowledge of their structure and properties is very important.

This contribution, however, is focused on proper zeolites and mainly on silicate zeolites. Their importance in the field of adsorption, molecular sieving, ion exchange, and catalysis is outstanding, and the interaction of crystal structure and function is its key property. Selecting specific examples, the influence and importance of the knowledge of the crystal structure and, therefore, the impact of crystallography on the progress in zeolite structural science will be exemplified.

6.2 Structure Determinations

The structure determination of a new material is the fundamental process in the analysis of the properties of the investigated material. The knowledge of the geometry of the framework and the pore system together with the chemical composition is a prerequisite to envisage possible applications of the porous material. The crystal structure is also a protocol of the crystallization process and may give hints on an improved synthesis route.

6.2.1 *The Solution of Complex Zeolite Crystal Structures from Diffraction Data*

Since the advent of high-resolution powder diffractometers at synchrotron radiation sources, the refinement of crystal structures and later also the solution of crystal structures from powder X-ray diffraction data were pioneered by zeolite crystallographers. Various strategies were developed always involving most advanced instrumentation available at synchrotron light sources.

In order to benefit from a high-resolution powder diffractometer, materials of highest possible crystallinity are required facilitating the indexing of the pattern and data collection up to high Q-range in reciprocal space (Fig. 6.2).

Structure determination heavily relies on a careful analysis of the materials' metric and symmetry as well as on the precise extraction of intensities for as many

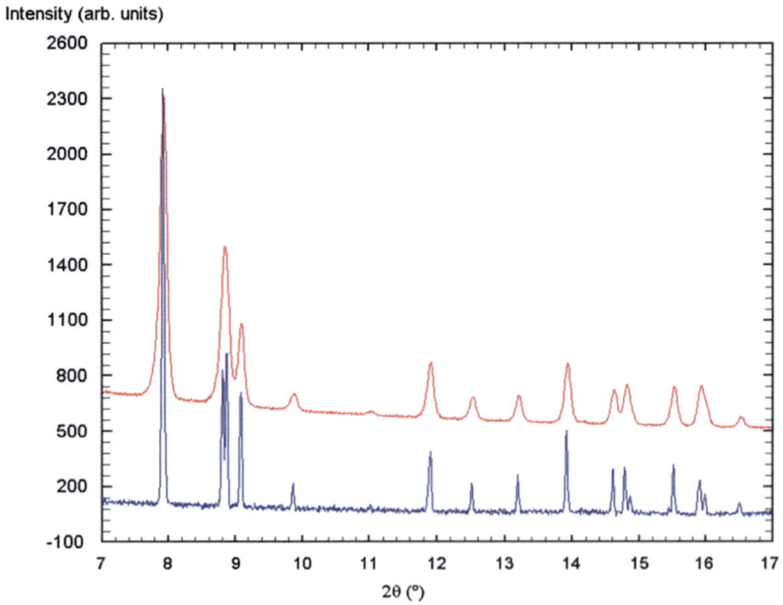

Fig. 6.2 Powder diffraction diagrams ($\lambda = 1.5406$ Å) of a highly crystalline silicalite sample recorded with a conventional diffractometer (*upper trace*) and at a synchrotron source (*lower trace*). The higher resolution of the synchrotron data facilitates, for example, the automatic indexation of the pattern, a precondition for structure solution

reflections as possible. The information on the material in diffraction experiments is obtained from reciprocal space and should be complemented by spectroscopic, thermal, or sorption experiments confirming the diffraction data analysis.

6.2.1.1 From P-XRD Data Using Direct Methods (Example: ITQ-22, Framework Code IWW)

The fundamental concept of direct methods is based on scatterers of equal kind and a diffraction data set showing atomic resolution. The requirements are known as Sheldrick's rule and demand diffraction data down to 1.2 Å resolution [8] which equals a diffraction angle of ca 80° 2θ for Cu Kα1 radiation ($\lambda = 1.5406$ Å). Considering the three-dimensional tetrahedral network, only the T centers are required to reconstruct the complete framework structure. This simplification softens Sheldrick's rule to ~2.4 Å resolution or ~40° 2θ resolved diffraction peaks for Cu radiation. However, the limitations of direct methods still require highly crystalline materials and well-resolved P-XRD diagrams where peak overlap is minimized.

Whereas most groups in zeolite crystallography used a complementary combination of methods like XRD, TEM, and NMR and modeling for structure solution, the group of J. Rius focused on diffraction information only and improved the power of the original direct methods introducing the modulus sum function [9] as implemented in structure solution program XLENS. Nowadays, direct methods are very efficient, and, given the right input information, the solution or at least the partial solution is obtained within seconds. Completing the partial model and refining the crystal structure from P-XRD data using the Rietveld technique, however, is a time-demanding and slow process.

As outstanding examples for the power of direct methods, the structure solution of ITQ-21 and ITQ-22 is presented [10, 11]. The structure solution of ITQ-21, a remarkable material, was straightforward, using highly crystalline material and a P-XRD data set from a high-resolution diffractometer. The structure refinement yielded a new, unusual central building block of the framework which can be describes as a [$4^6 6^{12}$] cage with an occluded 4-ring (Fig. 6.3). Modified building blocks of that type were discovered in a number of framework structures solved later, e.g., ITQ-26 [12], ITQ-13 [13], and ITQ-34 [14].

ITQ-22 is a much more complicated structure, the most complicated solved so far by direct methods alone. From the synchrotron-based P-XRD data set, the unit cell dimensions and symmetry were retrieved first, before the intensities of indexed reflections could be obtained for the direct method runs. From the electron density map of the best starting set, the 16 symmetrically independent T atoms of ITQ-22 were found, and, by interpolation, the oxygen atoms were added to complete the three-dimensional framework (Fig. 6.4). It is the first material with 12-ring, 10-ring, and 8-ring channels which intersect. The material shows important shape-selective properties in catalysis introduced through the channel hierarchy. In all these examples,

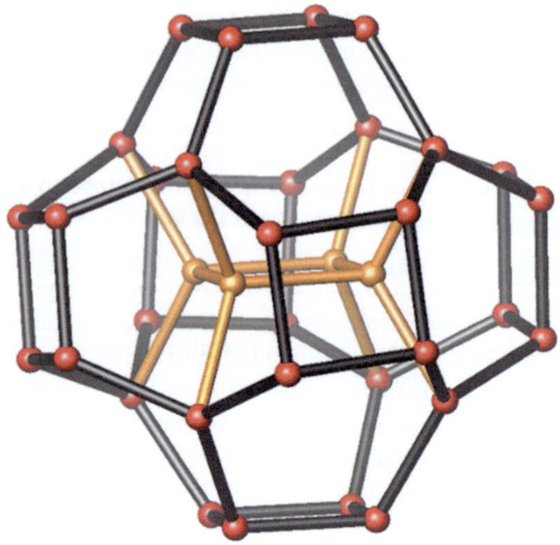

Fig. 6.3 The $[4^6 6^{12}]$ cage with an occluded 4-ring represents a very unusual building block of zeolite frameworks. *Spheres* indicate Si atoms; oxygen atoms are omitted for clarity

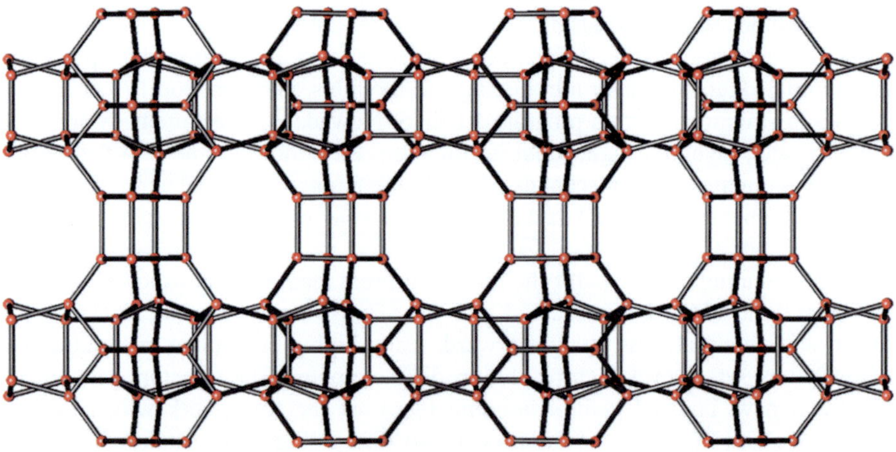

Fig. 6.4 Projection of the framework structure of zeolite ITQ-22 (framework type *IWW*) along [010] with [99] across and [001] down. The D4R can be seen connecting layer-like complex building units

structure analysis also showed that Ge segregates on specific T sites of the zeolite framework, namely, as part of the cube-like double-4-ring unit (D4R). The refinement of the site occupancies showed that 50 % of the T sites in this subunit are taken by Ge. However, because of symmetry reasons, no ordering of Ge in the D4R has been analyzed.

6.2.1.2 From Electron Diffraction: Recent Advances in Structure Analysis Concerning Zeolite Science in the Future—The Breakthrough of Electron Crystallography

A major breakthrough in structure determination is the introduction of the precession technique in electron crystallography. Kolb and coworkers showed, also using zeolite materials as examples, that reliable intensity data can be extracted automatically from electron diffraction experiments of nanocrystals and be processed as single-crystal intensity data using direct methods for structure solution [15, 16]. The most complicated crystal structures could be solved directly from TEM experiments using nanocrystalline powder materials without further involvement of complementary X-ray experiments. This advancement in electron crystallography instrumentation supersedes crystal structure solution from P-XRD data and opens new vista in structural science of nanocrystalline materials including zeolites.

An impressive example of a successful structure determination based on ED data is the structure analysis on zeolite ITQ-43 [17]. ADT data (automated diffraction tomography) of an as-made sample were collected at low temperature (~100 K). The data were sufficient for a reliable reflection intensity integration. The structure solution was performed by direct methods based on a fully kinematic approach. From the structure solution, 20 T (Si/Ge) and 24 O positions were identified. Most of the remaining O positions were later located by subsequent Fourier map refinements. ITQ-43 has a framework structure of very low density (11.4 T atoms per 1000 Å^3) but very complex structure. The pores can be described as cloverleaf-like channels formed by very wide 28-rings with large pore diameters of 21.9 Å × 19.6 Å and two types of 12-ring channels with apertures of 7.8 Å × 5.7 Å and 6.8 Å × 6.1 Å [17].

The principal advantages of electron crystallography are the reduction of crystal size to some 10 nm and the resolution of diffraction intensities in three-dimensional space. Using the nano-single crystals for data collection, the inherent problem of systematic peak overlap of the P-XRD experiment is overcome. An additional advantage is the much increased sensitivity of the electron-electron interaction in the diffraction process allowing the experiment with nanocrystals; however, because of the sensitivity, multiple diffraction and dynamic effects can obscure the data set. Since most zeolite materials are obtained as micro- to nanocrystalline powders, electron crystallography widens the range of materials including zeolites accessible for structure determination. However, the structure refinement and the geometrical optimization of the zeolite framework are still unsatisfactory using intensities from electron diffraction alone. Therefore, a combination of electron and powder X-ray diffraction is used to obtain best results.

6.2.1.3 Combining TEM and P-XRD in Reciprocal Space with Search Techniques in Direct Space: The Structure of TNU-9, Framework Code TUN

Another step of complexity constitutes the solution of the structure of zeolite TNU-9 [18]. With 24 symmetrically independent atoms on T sites, the unit cell volume is almost twice as big as in ITQ-22. The structure solution of TUN-9 predates the automated electron crystallography; however, it shows how crucial the information of electron diffraction was for the success. Despite best-quality data from the P-XRD experiment and using search techniques in direct space implemented in the FOCUS program [19], no solution was obtained in the beginning. From additional electron diffraction and imaging data, structural information of the material had been obtained analyzing projections along the channel axes. In addition, phases of reflections have been determined from these experiments. Adding this as starting information into the structure solution process using the full data set from P-XRD, the crystal structure was solved and refined using the Rietveld technique (Fig. 6.5). As can be learned from the original publication, the process of structure solution was a very time-consuming process, not only in setting up the physical collaboration between electron crystallography and powder diffraction but also in achieving the goal. In the successful final computer run, 16 days of CPU time was required just for structure solution. The successive Rietveld refinement still laid ahead. The proportion of time needed for reciprocal space structure solution, as, e.g., for ITQ-22, and time needed for direct space techniques (as, e.g., for TNU-9) is typical. Despite basic, but partial, structure information from diffraction analysis is provided, the search in direct space using the FOCUS approach is extremely costly. As perspective for the future, however, direct space modeling will become ever more competitive as computer power increases. With improved code and modern equipment, one might cut CPU time to much shorter periods.

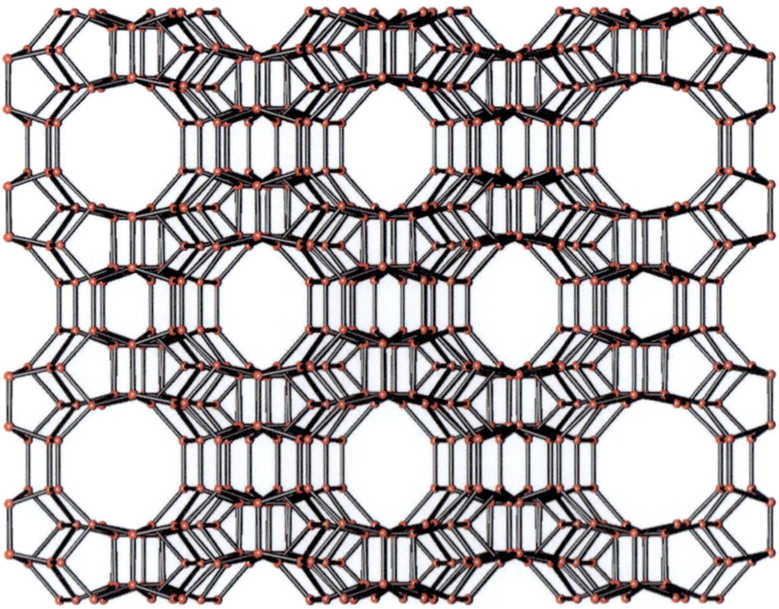

Fig. 6.5 Projection of the framework structure of zeolite TNU-9 (TUN) seen along [-101] with [101] across and [010] down

6.2.1.4 Extending the Approach of the Combination of Complementary Techniques Using Reciprocal and Direct Space: P-XRD, Electron Crystallography, Charge Flipping, and Structure Envelope

IM-5 was an enigma for more than 10 years before in 2007 its structure was solved [6]. Similar to the preceding material TNU-9, IM-5 has 24 symmetrically distinct T atoms on T sites in the unit cell. Different from the FOCUS approach, in this case, direct space becomes involved through the charge-flipping algorithm. In a recursive way, the electron density distribution in direct space as calculated from a first set of structure factors with phase values (obtained from direct methods) is repartitioned avoiding negative electron densities and back-transformed into structure factors with new phase values. These new structure factors are then used to calculate an improved electron density distribution. Structure factors in reciprocal space and electron density distribution in direct space are optimized in a cyclic process. This process is very efficient and fast recycling is possible.

In addition, partial information about the crystal structure was obtained from electron diffraction experiments leading to phase information for ca. 95 reflections. Feeding this information into the charge-flipping structure solution process using intensity information from the extracted P-XRD data, the crystal structure was retrieved. The final electron density map obtained from the charge-flipping calculations precisely represents the tetrahedral network of the microporous IM-5 material (Fig. 6.6). The material has an unusual 2D 10-ring (10R) channel system which is

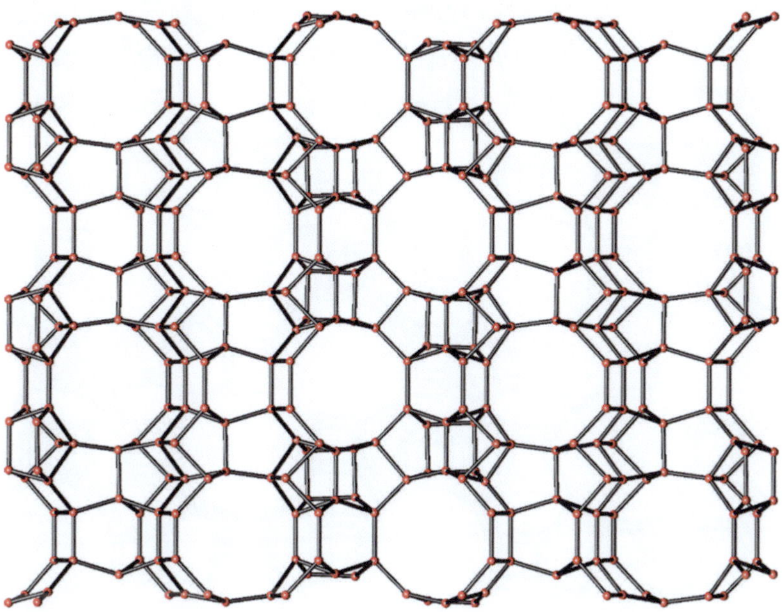

Fig. 6.6 Projection of the framework of zeolite IM-5 (*IMF*) seen along [100] with [010] across and [001] down

connected through a larger cavity. Thus, the material has potential catalytic activity and shape selectivity which should reflect the structural peculiarities.

ITQ-37 (framework type -ITV) is a chiral zeolite crystallizing in the cubic enantiomeric space group P $4_1 32$ and is the first chiral zeolite with one type of gyroidal channel [20]. In order to solve the structure having 10 T atoms in symmetrically inequivalent position, reciprocal space and direct space methods were applied to diffraction data from P-XRD and ED experiments. It is an exceptional material with 30-ring pores and possesses the lowest density of all framework oxides. Only structure analysis explained to full extent the unique structural properties which can be rationalized considering the synthesis conditions. In order to stabilize this framework of low density, the structure contains a high proportion of D4R units (directed by the fluoride anion and by Ge atoms at T sites) which form a rigid and rather dense backbone of the porous framework (Fig. 6.7).

As can be seen in ITQ-40 (framework type -IRY), another example of a framework with large 16- and 15-rings and compact 3-ring units together with D4R units containing Ge atoms and fluoride anions, similar structural features were obtained

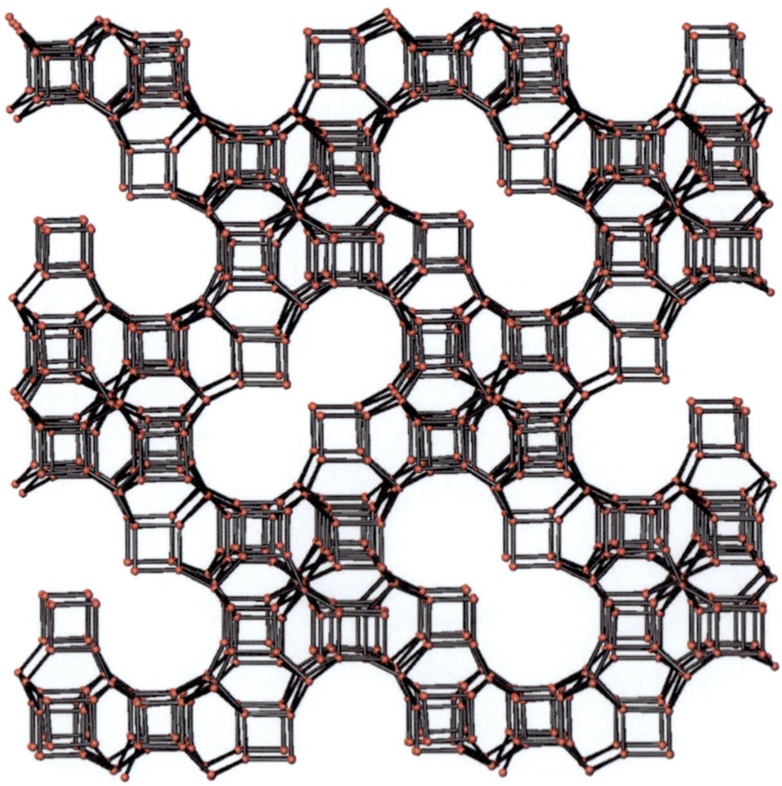

Fig. 6.7 Projection of the framework of zeolite ITQ-37 (-ITV) seen along [100] with [010] across and [001] down

by similar reaction conditions [21]. The crystallization of the particular framework type (e.g., -ITV or -IRY) is mainly determined by the use of a specific SDA. Feedback from structure analysis stimulates the synthesis efforts toward new zeolites with unexpected framework structures. However, the complexity of these structures can only be mastered using structure solution techniques at the most advanced level.

A particular feature of zeolite frameworks is structural disorder. TEM experiments show that ITQ-39 material is an intergrown framework having three ordered end-member polytypes. Using electron crystallography, the structure of polymorph C has been determined allowing for the modeling of the remaining polymorphs A and B. The 12R by 10R pairwise channels intersect to give a three-dimensional pore system making the materials a good candidate for petrol refining [22]. The example shows convincingly that only the high spatial resolution of electron diffraction made structure solution possible. The information obtained from the diffraction pattern of an undisturbed array of a few unit cells only revealed the structure and explained the stacking disorder obvious in the diffraction pattern (Fig. 6.8).

As precaution should be added, the advancement made by the introduction of charge flipping in the structure solution process is the efficiency with which direct space information is extracted and positively used and fed into the refinement of the structure factor phases. Still as a precondition for a successful structure solution, materials of highest quality are required for the diffraction experiment, thus limiting structure analysis.

6.2.1.5 Zeolite Materials of Limited Crystallinity

All examples of structure solution and refinement described so far required materials of perfect crystallinity. However, many materials exist which are of limited crystal quality. Still, their crystal structures are of interest for reasons of synthesis and

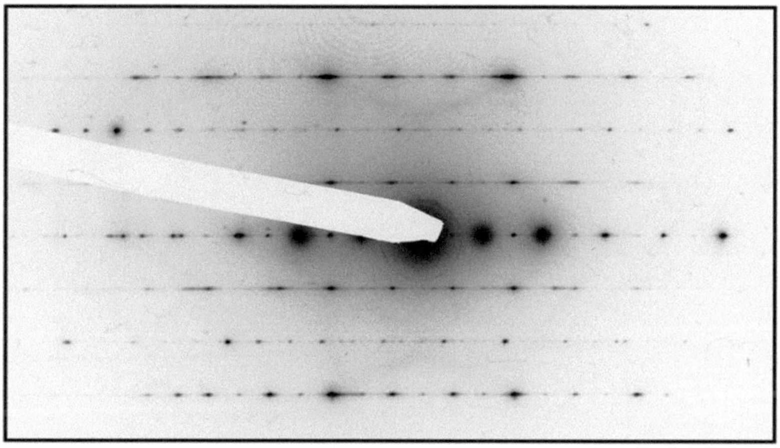

Fig. 6.8 Electron diffraction pattern of a disordered zeolite. The 3D data of such a pattern allow for a detailed analysis of the type of disorder

application-oriented research strategies. As examples, two studies are mentioned describing zeolite crystal structure determinations from diffraction data of limited resolution. In these examples, standard direct methods failed, and computer-assisted model building was involved as important tool. The RUT framework type belongs to the clathrasil family and could never be obtained as well-crystallized material. Considering the $[SiO_4]$ unit as the invariant and fundamental building block of the structure and connecting those units to a three-dimensional four-connected silicate framework, the requirements of "atomic resolutions" are shifted from ca. 1.2 Å to a larger value of ca. 2.4 Å. Introducing the new limits into the structure solution software, the RUT crystal structure was solved (Fig. 6.9) [23]. From the distribution of T centers, which showed up in the electron density map, the silicate framework could be built. Finally, the structure was refined from P-XRD data using the Rietveld method.

Whereas the crystal structure of RUB-10 has been solved and refined from P-XRD data, COK-14 [24] and SSZ-31 [25] posed a much more difficult problem. The structure of COK-14 was solved by model building based on chemical knowledge and simulation of diffraction patterns.

Another example is the disassembly of proper zeolites and its reassembly to a more or less ordered three-dimensional framework. The disassembly (delamination) of UTL-type zeolite into layered subunits and the reassembly of these subunits through topotactic condensation are described by Roth et al. [182] and Verheyen et al. [24]. Since the structure of the layers was known from the precursor material, the structure of the condensation product could be derived and geometrically optimized. Since the scattering experiment contains all structural information of the material even when poorly crystallized, the refinement of the structure based on the

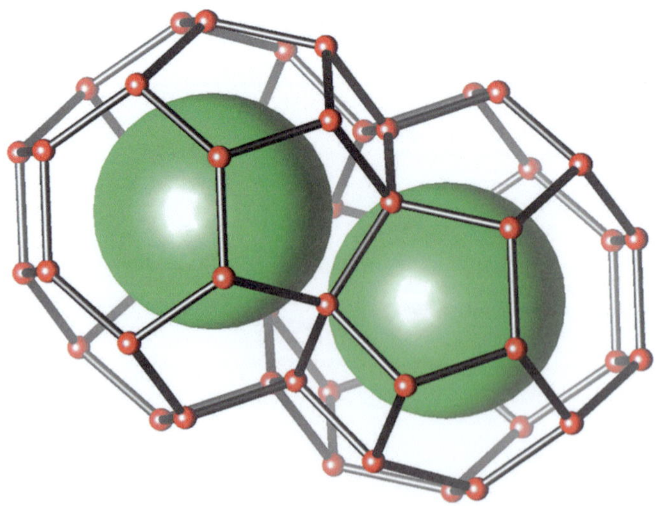

Fig. 6.9 Structure plot of the characteristic double cage of RUB-10 (RUT). *Large green spheres* represent the occluded TMA cations acting as structure directors

experimental model is a valuable procedure to confirm the proposed structure and to gain detailed information on structural details.

SSZ-31 was obtained as a crystalline sample showing sharp and broad diffraction peaks. From an analysis of the anisotropic peak broadening, it was obvious that the structural disorder is an inherent feature of the material. Based on additional TEM experiments and after indexing of the diffraction pattern of the material, a projection of the structure was obtained which showed close similarity to a projection of the structure of zeolite ß as well as zeolite ZSM-12. This led to structure models which were derived from an ordered polytype of zeolite ZSM-48 applying a σ-transformation [26] to its framework structure increasing the pore size from 10-ring to 12-ring dimension. The structure model agreed very well with other physicochemical characterization results; however, because of the structural disorder of the material, no refinement of the structure against the experimental diffraction data was possible.

Wherever limited crystallinity is an inherent property of the material or electron beam-sensitive materials degrade in the TEM, direct space ab initio methods come into play. With information on the space group symmetry, chemical composition, information on building blocks, or structural subunits from IR or NMR spectroscopy, structure models can be generated using, e.g., Monte Carlo techniques and tested against the P-XRD diagram. There are commercial programs such as TOPAS distributed by Bruker and Endeavour distributed by Crystal Impact and public domain programs such as FOX [27], ZEFSAII [28], or FullProf [29] which address inorganic materials in particular and are well suited to solve zeolite structures.

6.2.2 *Disordered Materials*

6.2.2.1 Zeolite ß and the Search for a Chiral Zeolite Framework

A most interesting case of the impact of structure analysis on applications is the structure solution of zeolite ß ("beta" for short). In 1967, Mobil Corporation filed the patent [30]. However, because of its structural disorder, the P-XRD diagram of beta looked as if the crystallinity of the material was poor and contained an amorphous fraction. This led to the neglect of further characterization and testing studies, although adsorption data indicated high porosity similar to zeolite Y. Only in 1989, structure models were published in literature by Treacy et al. [31] and Higgins et al. [32] (Fig. 6.10). Immediately after, zeolite ß became one of the most interesting and commercially important microporous materials for catalytic applications.

In the course of structure determination, TEM experiments, diffraction, and imaging played a crucial role. Despite its poor powder diagram, TEM images of the zeolite revealed its crystalline nature; however, there was disorder in the stacking sequence of layer-like building units ("beta-sheets").

Assuming stacking disorder, Treacy et al. succeeded in solving the structure by model building and by comparing the simulated powder patterns and TEM images

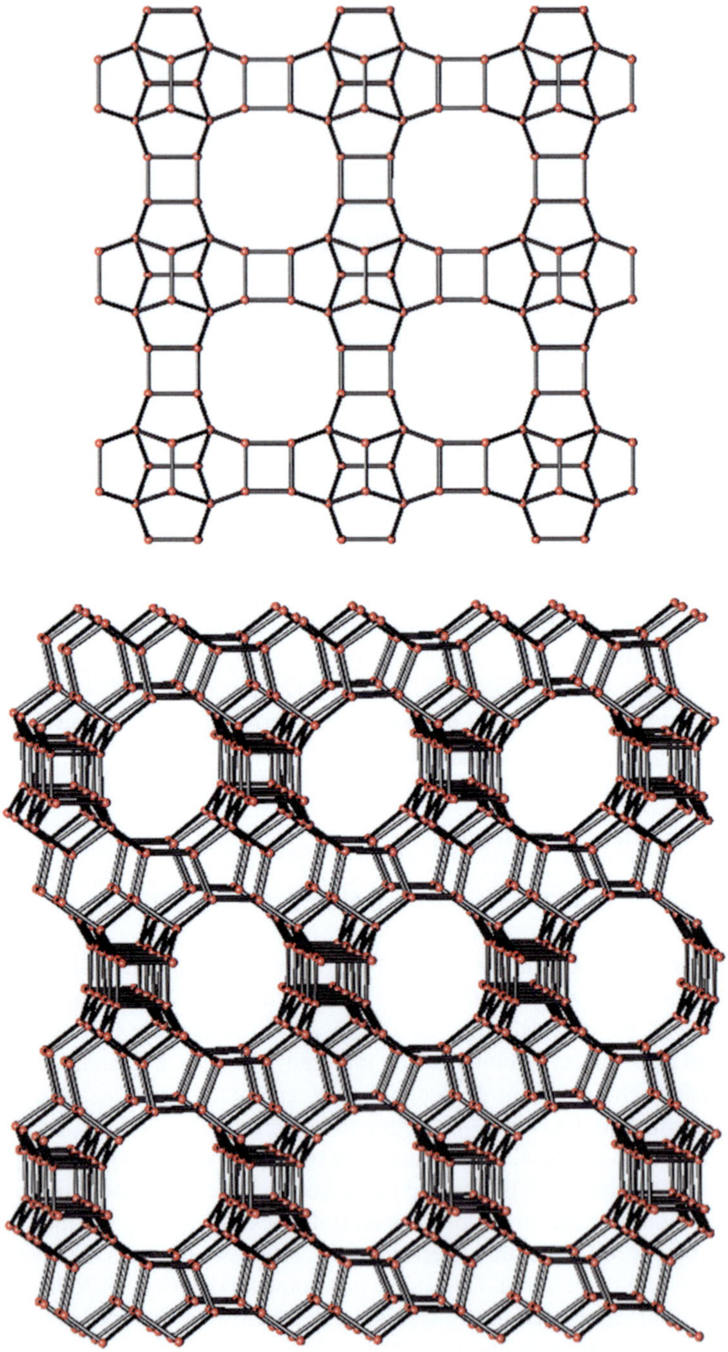

Fig. 6.10 *Top*, projection of the beta-sheet on (001); *bottom*, the framework of zeolite ß, polymorph A, seen along [100] with [010] across and [001] down

of the structure models with the experimental one. This led to the structural characterization of the material and revealed its unperturbed three-dimensional 12-ring pore system. Using the beta-sheet as building blocks, Higgins et al. showed in the same year that there are ordered simplest end-members, one of which has a chiral framework structure, polymorph A. Its space group symmetry would be $P4_122$ or $P4_322$. Not only the revelation of the crystal structure of beta immediately led to applications and commercialization but also the search for the pure chiral ordered end-member materials started. Based on the SDA synthesis approach, for more than a decade, synthesis chemists tried to achieve the goal without success. The improvement of synthesis conditions stimulated by the knowledge of the crystal structure and the very successful SDA-based synthesis protocol led to the isolation of polymorph B [33, 34] and polymorph C [35, 36]. An enrichment of polymorph A, the chiral variety, in the synthesis of zeolite ß was reported only recently [37]; however, there is no evidence of the influence of the chiral framework on the catalytic properties of the material.

In contrast to the exhaustive synthesis attempts chasing for the chiral zeolite framework, application-oriented research transformed the disordered beta material from a laboratory curiosity into a valuable industrial catalyst very quickly, after the crystal structure was known. It became immediately clear that zeolite beta has a three-dimensional, interconnected, open pore system even as disordered material. Again, the visualization of the framework structure inspired the zeolite science community in all disciplines leading to year-long extensive studies of the many beta materials which have been synthesized meanwhile. The latest progress in the field is the template-free synthesis of beta materials which has less structural defects and lower Si/Al ratio, challenging the SDA concept and turning beta into an even more interesting material for industrial applications [38].

Other chiral frameworks have been synthesized until now [20, 39]; however, there is no positive impact on catalytic reactions resulting from the chiral nature of the framework structure. The product spectrum of catalyzed reactions including chiral molecules shows no specific preference for one enantiomer, at least for those systems tested so far.

6.2.2.2 Data Mining: Generating a Starting Structure Model on the Basis of a Closely Related Material of Known Structure (Examples: Interlayer Expanded Zeolites, New Materials with Reactive Centers)

Very recently, structure elucidation of interlayer expanded zeolites was reported. The synthesis method of interlayer expansion uses hydrous layer silicates as precursor materials to built three-dimensional framework silicates interconnecting the layers with silylation reagents. In this way, expanded channel structures of variable diameter are created and functionalized T centers are introduced, allowing for the fine-tuning of structural and chemical properties of the expanded product. Although the silylation procedure of IEZ exercises little control on the general crystallinity of the product, structure analysis proved that the IEZ materials are locally well ordered with uniform pores and periodic framework structure [40, 41]. For IEZ materials also, the future will show whether or not the information on structural details will help to develop useful microporous materials for applications in sorption, separation, or catalysis.

Interlayer expanded zeolites were introduced as post-synthesis derivatives of layered precursors. The materials are homogeneous; however, their crystallinity is low since the bridging linker unit introduced flexibility into the framework. An example is the material IEZ-RRO (Fig. 6.11). Knowing the crystal structure of the precursor and also the molecular structure of the linker, a structure model can be derived and used as starting model for the Rietveld analysis of the P-XRD data. Because there is no strict periodicity in the framework of the IEZ material, reflections in the powder diagram are broadened. In addition, stacking disorder might be present in the sample, further complicating the analysis. Still, it is worthwhile to carry out the structure refinement in order to obtain the most detailed information about the structure from experimental analysis [41]. This is the best precondition for simulation studies using structure modeling.

Interlayer expanded zeolites (IEZ) in fact comprise a family of materials which are obtained through interconnecting layered silicate precursors with reactive linkers through chemical bonds. The directed chemical bond of the linker unit with the

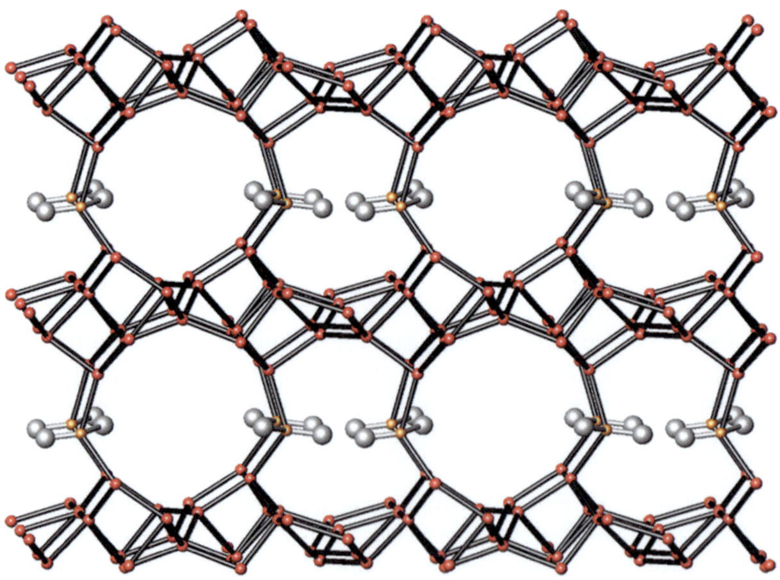

Fig. 6.11 Projection of the framework of interlayer expanded zeolite (IEZ) of RRO type along [001] with [100] across and [010] down. The *orange sphere* is the center (Si) of the linker which was introduced by silylation. *Gray spheres* represent the ligands which can be of different types (–OH, $–CH_3$, –Cl, etc.)

surface silanol of neighboring layers makes these materials zeolite-like with the linker units as carrier of a functional group.

In the simplest example, a [SiO_4] tetrahedron connects the layers providing two reactive silanol groups for chemical manipulations. The first attempt to interlayer expansion goes back to Kuroda and coworkers who modified hydrous layer silicates such as RUB-18, kanemite, and kenyaite by silylation to noncrystalline porous framework materials [42, 43] and literature cited therein. Since the diffraction diagrams showed no sharp diffraction peaks, i.e., no translational periodicity, no structural studies were carried out on these materials. Later, the group of Tatsumi for the first time used the term "interlayer expansion" and showed that silylation of the hydrous silicate layer PSL-1 yielded a microporous solids with uniform porosity of the product [44]. Prior to coining the term interlayer expansion, Ruan et al. [40] discovered in zeolite Ti-YNU-1 the insertion of one [SiO_4] tetrahedron connecting neighboring MWW layers to build an expanded, crystalline, still thermally stable microporous framework. The study was extended by this group and others to various layer silicates, and expanded, reasonably periodic structures made from mww, fer, and heu layers were synthesized [40, 41, 45, 46]. The X-ray structure analysis from bulk powder samples confirmed the TEM studies showing the well-ordered active center at the linker site [41, 47]. In explorative studies, it could be shown that other, functionalized linker groups can be inserted and that by increasing the linker dimensions, the pore size could be expanded [48–50]. The latest Rietveld analysis of an IEZ material showed that using Fe^{3+} in aqueous solution as linker species, a microporous, metal-expanded, thermally stable silicate framework could be formed exposing the reactive metal linker site to the pore space [51]. The direction of an active component on a specific framework site is demonstrated with the concept of interlayer expanded zeolites and proven based on various structure analyses. This should open a new field of exploration in custom-made zeolite catalysts with tailored functionality.

6.2.2.3 PDF Analyses of Highly Disordered Materials

In X-ray diffraction diagrams, there is information beyond the Bragg peaks, which are the signature of the periodic crystal structure. The Bragg intensities are used in conventional structure analysis to solve and refine the crystal structure after subtracting the pattern background. However, the X-ray scattering process, in general, also involves intensities from disordered, glassy, or amorphous material giving rise to intensity contributions in the diffraction pattern off the Bragg peaks [52]. The same holds for neutron and electron scattering with the specific analytical power of the respective technique.

An example in zeolite structural science is the delamination of layered precursors [53, 54] and their three-dimensional arrangement as house of cards. The central question is about the structure of the layer slab after the chemical treatment: is it maintained and extended or is it partially condensed through folding or perhaps is it without any structural order? Since there are almost no Bragg peaks left in the

P-XRD diagram, only the analysis of the pair distribution function can reveal the structure of the layer slab. In a feasibility study, delaminated MCM-22(P) was investigated [55]. The analysis of the high-energy, high-resolution X-ray data set ($Q = 20$ Å^{-1}) confirmed the structure of the layer known from the framework slab, but also showed that the layer is slightly bent, despite its thickness of ~25 Å.

6.3 Distribution of Elements in the Framework

So far, the geometry of the zeolite framework was the focus of the discussion. Since the scattering process is proportional to the number of electrons of the scattering center, the atom, the intensity of a diffraction peak also contains precise information on the chemical composition of the crystal. However, the Bragg peak is the result of periodic translational order in the crystal, and, therefore, its intensity reflects the spatial average of the composition of every unit cell. In addition, data collection with conventional diffractometers averages dynamic disorder. This leads to severe limitations in the analysis of local structure and composition.

In the current literature, there are a number of interesting examples where structure analysis has contributed to the location of active centers, in particular in cases where compositional ordering was achieved or the scattering contribution of the active site had a major impact.

6.3.1 Acidic Sites (Al, B)

6.3.1.1 Loewenstein's Avoidance Rule

Since diffraction studies by nature have difficulties to discern Al and Si scatterers on the T sites of the silicate framework, the advent of high-resolution solid-state magic-angle-spinning NMR spectroscopy (HR SS MAS NMR) as element-specific experiment probing local order promised fresh insight into the atomic ordering of framework silicates. Loewenstein's empirical avoidance rule stated that there are no direct linkages between [AlO_4] tetrahedra [56]. The neighboring T atoms should be another tetravalent cation such as Si or higher valent such as P. This explains the minimum Si/Al ratio of 1 in zeolite frameworks which leads to strictly ordered silicate frameworks with alternating Si and Al occupancies. Loewenstein's rule also takes influence on properties typical for zeolites, limiting the ion exchange capacity and the number of acid sites. For catalytic applications, high-silica zeolites are most important. Again, the distribution of Al in the framework becomes of particular importance. Until the 1980s of the last century, structure analyses used geometrical features such as the different d(T–O) of Si and Al to analyze the Al distribution over T sites if at all the ordering was studied.

Early ^{29}Si NMR experiments showed that Si in siloxanes has nonoverlapping chemical shift ranges which are typical for their connectivity [57]. The findings were shown to also apply for silicate anions in solution and in solids [58, 59]. In addition, the quantitative nature of the experiments allowed for the analysis of Al distribution in the silicate framework of zeolite NaA leading to the conclusion that there are anti-Loewenstein centers [60, 61]. The NMR results were approved by electron and neutron diffraction experiments [62, 63]. FAU-type materials with higher Si/Al ratio (1.18–67), however, showed no deviation from Loewenstein's rule, and the five signals in the ^{29}Si spectrum represent all possible Si environments from 0 Al to 4 Al on the next nearest T site [64, 65]. This is not only in agreement with Loewenstein's rule but also fulfills Dempsey's rule, whereupon the number of Al neighbors of Si sites should be minimized [66]. In particular, high-silica zeolites should not show clustering of Al near one Si center. Based on these new findings, the NMR spectra of LTA-type zeolite with Si/Al ratio of one were reinterpreted later and consolidated with complete Si/Al order in agreement with Loewenstein's rule [67, 68]. In the following years, Loewenstein's rule was the basis for the interpretation of the distribution of Al or other trivalent substituents on T framework sites. Theoretical studies complemented the experimental findings [69]. Based on Loewenstein's avoidance rule, an ordering of Al on T sites has been deduced for several zeolite frameworks which is in agreement with the intensity ratios of NMR signals in ^{29}Si spectra [70, 71]. These results confirm the complementary nature of diffraction analysis probing long-range periodicity and spectroscopic analysis probing short-range local order. The combination and the quantitative interpretation fully explain the atomic structure of many zeolite frameworks including the reactive center, here, as a representative example, Al order on T sites.

6.3.1.2 Heteroatom Localization and the Problem of Reactive Centers in the Porous Frameworks: Where Is Al?

Besides the porosity with its pore size and dimension, the distribution of electrical charge in the framework is of major importance for applications. Charge is introduced by substitution of Si on T sites by cations of lower valence, mainly by Al. The charge is compensated by extra-framework mineral cations which can be exchanged, also by protons introducing acidity. Cation sites and also the sites of acidity are dominated by the local geometry of the framework which exerts, therefore, considerable influence on the particular properties, ion exchange, and acid catalysis.

In the early period of zeolite science and applications, the discovery of new materials and the geometrical analysis of framework topologies of the silicate zeolites were in the focus of interest. Pore openings and pore geometries were related to the functional properties in ion exchange, sorption, and catalysis. Si/Al order on T sites in the silicate framework was not discussed mainly because of the poor scattering contrast of Al and Si in X-ray diffraction experiments. This led to averaged structure models where T site occupancies were simulated by Si/Al partitioning. Therefore, the space group symmetries of these structures are typically too high, further hiding the local structure of silicate zeolite frameworks. Examples are faujasite-, chabazite-, or gmelinite-type zeolites with Si/Al ~ 2. In structure analyses, highest space group symmetry is assumed, and T positions are simulated by Si/Al ratios obtained from chemical analysis. Only in such cases where Loewenstein's rule required an ordered distribution for the T atoms (Si/Al = 1), distinct sites for Si and Al were considered in the description of the silicate host (Fig. 6.12).

The importance of the knowledge of the Al distribution in zeolite frameworks and of the control over it is the reason that even today new studies are published and controversial discussions about new findings take place. Loewenstein's rule is generally accepted now; still there are discussions on Dempsey's rule [72, 73]. The examples show the impact of structure analysis on the atomic scale on the understanding of properties of zeolites, whether ion exchange or catalysis. This also extends to the synthesis of particular materials with tailored properties, i.e., the design of materials. In the next paragraph, the even more subtle structural description of the siting of active centers will be discussed.

With the advent of high-silica zeolites, tantamount to the introduction of organic structure directors and the invention of ZSM-5, the demand for a structural under-

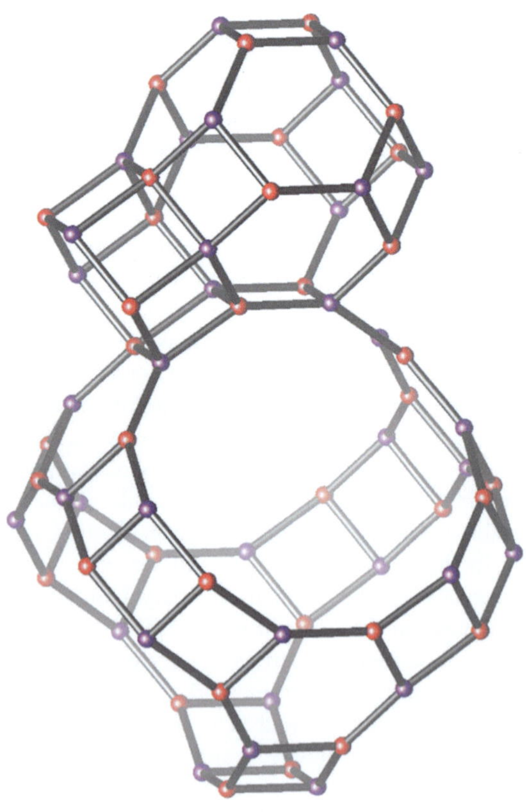

Fig. 6.12 Section of the faujasite structure (*FAU*) with Si/Al = 1 showing a complete ordering of Si (*red*) and Al (*violet*) on distinct T sites

standing of the distribution of Al on T sites became more and more urgent. In addition to structure analysis from X-ray diffraction data, neutron and electron diffraction methods became heavily involved in structural characterization, and, in addition, spectroscopic methods for the analysis of the short-range order added important contributions to the understanding. Because of the limited availability of neutron beam time and the problem of beam sensitivity in TEM imaging experiments, solid-state NMR became the experimental technique of choice for the study of local order. In addition, structure simulation using numerical methods of increasing complexity was involved in the discussion of order/disorder of active sites in zeolite host frameworks. Often, analytical techniques probing long-range and short-range order in the host framework were combined because of the precarious nature of the single results and of the complementary nature of the information acquired.

The complexity of the problem is exemplified using MFI-type materials. Shortly after the structure of the materials was solved, a gradient in Al concentration across the single crystal was established [74, 75]. Obviously, large crystals of ZSM-5 zeolite showed a significantly increased Al content near the surfaces of the crystal. Since the catalytic activity is correlated with the Al content, the surface near part of the crystal should be more active than the center. In a number of structure analyses, spectroscopic experiments, and numerical simulations, an attempt was made to localize Al on specific T sites. The detailed X-ray diffraction analysis of ZSM-5 [76] explicitly investigated structural features related to the particular properties of the zeolite. Because of the low Al content of the sample studied and because of the given framework type, the authors concluded that even if all Al would occupy one particular T site of the framework, still only a partial occupancy of that T site is realized. Therefore, because of the small scattering contrast of the Si-Al pair, it is very unlikely that Al could be identified. The claim of Al ordering in ZSM-5 material was intensively discussed over many years based on the results of many different analytic methods.

A hint for Al ordered on T site was achieved by ^{27}Al MQ-MAS NMR experiments [77]. For the first time, nonequivalent Al sites were detected in NMR experiments indicating the distribution of Al on specific T sites. However, computational studies showed that simple correlations of the isotropic chemical shifts with geometrical features of the local order of the structure are not unambiguous [78, 79]. Also, using ion-exchanged materials and introducing the strongly scattering cation Cs, X-ray diffraction studies were carried out again in order to assign the ordering pattern observed in spectroscopic analyses to distinct T site available from structure refinement [80]. Different to previous studies, the most likely T sites for Al are T12, T7, and T10 using the numbering of Olson et al. [76], to which those oxygen atoms are connected with the shortest Cs–O distances (Fig. 6.13). The findings on the site-specific distribution of Al on T sites of the silicate framework were transferred then into synthesis recipes with the aim to direct Al on distinct T sites [81, 82]. In catalytic experiments, the impact of distinct Al distribution pattern on the performance and product distribution was exemplified [83]. Similarly, studies on SCR catalysis (selected catalytic reduction) using MFI-type materials with distinct Al distribution and, therefore, specific cation sites were published [84]. More recent investigations

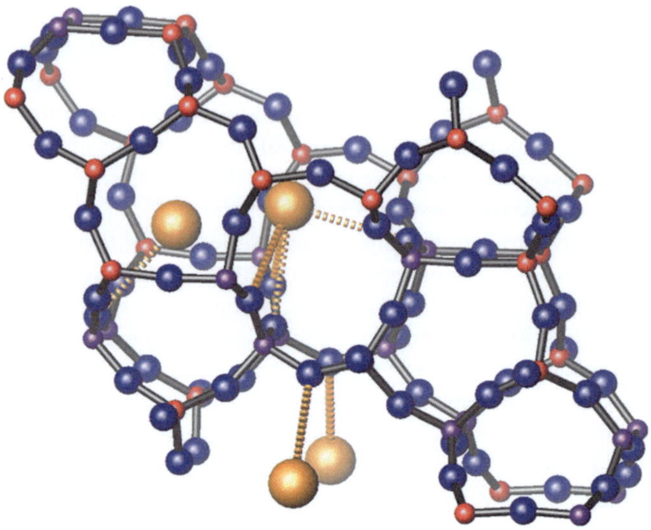

Fig. 6.13 Section of the structure of Cs-ZSM-5. The T sites which are coordinated by oxygen atoms with short contacts to Cs cations (*orange*) are indicated in *violet*. These T sites are likely to be partially occupied by Al^{3+}. T sites which are exclusively occupied by Si are represented by *red spheres*. Oxygen atoms are displayed in *blue*

using NMR and computational techniques [85] again confirmed that there is ordering of Al on T sites. There is evidence that 10 out of 24 T sites of the monoclinic unit cell of ZSM-5 contain Al, and hence, 14 are occupied by Si only [83, 86].

Contrary to the diffraction study of Al-containing MFI-type material, Rietveld analysis of synchrotron-based diffraction data of B- and Ga-containing MFI-type materials shows close to random distributions of the trivalent T atom on the 12 T sites of the silicate framework [87].

Al already indicated, the identification of Al on specific sites in the zeolite framework establishes the reactive center for acid catalysis. As trivalent cation, Al introduces negative charge into the tetrahedral oxide framework which is localized on the oxygen atoms of the coordination polyhedron. The charge-compensating cation, therefore, is proximately placed interacting with Coulomb forces. Both species are reactive centers, and their reactivity and selectivity depend strongly on the local geometric structure. The active sites might be diluted in the porous framework, preventing molecules from multiple reactions on their way through the pore system. In order to judge, simulate, and predict the specific activity of the reactive center, either the acid center in conventional acid catalysis or the metal center for, e.g., in SCR catalysis of NOx, the knowledge of the site is a prerequisite, and the controlled synthesis of appropriate materials is based on this knowledge. Because of the impact of this challenging problem on the performance of the catalyst, the controlled placing of active metal centers in microporous frameworks was called a major challenge for the future [88, 89].

6.3.1.3 Where Is B

The substitution of Si on T sites by B results in weaker acidity of the framework on the one hand, and the amount of B introduced into the framework is much more limited compared to Al on the other hand. There is no equivalent to the low Si/Al ratio materials for B; in high-silica materials, the amount of B in the framework relates to the number of organic cations in the pore system [90]. The evidence for B on specific T sites is as difficult as for Al with the exception of some framework structure types with B ordered on specific T sites. Single-crystal structure analysis from synchrotron-based diffraction data of defect-free MFI-type crystals showed almost random distribution of B on T sites [87] even if there is a slight preference for T1. In contrast, in the clathrasil RUB-10 having a borosilicate framework, B, occupies one T site exclusively and, therefore clearly, shows up in structure analysis [23]. In this way, it is optimally positioned to compensate the charge of the organic SDA in the cages confirming the correlations of SDA and B content in the material. Obvious ordering of boron on T sites is also observed in SSZ-82, a 12-ring by 10-ring zeolite [91].

6.3.1.4 Redox Sites: Fe and Ti in MFI-Type Frameworks

The role of Fe in catalysis is not restricted to the isomorphously substituted species on T sites but more importantly to extra-framework iron. There is one specific structural study from synchrotron-based single-crystal experiments on the iron location at T sites [92]. The interpretation of the electron density map including the charge-compensating non-framework Na cation indicates that Fe segregates on the T sites 9 and 10. This result is in contrast to an earlier energy-minimization calculation where an almost statistical distribution of Fe on T sites is proposed [93] which is in agreement with experimental EXAFS spectra. The energy differences relevant for the location of iron by EXAFS spectroscopy, however, are very small. Therefore, no site preference can be attributed. The hard experimental evidence given by the diffraction experiment sheds new light on the structural organization in this complicated framework and suggests preferential site occupation.

Fe on extra-framework sites and in its catalytically active form is present as small oxide cluster or as ionic species [94, 95] which is not accessible for diffraction analysis. For this purpose, various spectroscopic techniques such as EXAFS, UV, IR, etc., were applied to study the structure of the iron oxide nanocluster.

Ti-substituted MFI-type materials are the most important zeolitic materials used for oxidation catalysis. Ti concentration at T sites is in the ~1 % range, i.e., approximately 1 Ti atom is contained in one unit cell. Neutron powder diffraction experiments on Ti containing MFI-type materials, which were either isotopically enriched in Ti-46 or in Ti-48, showed the segregation of Ti on specific T sites. These experiments used the contrasting positive (Ti-46) and negative scattering length (Ti-48) of the respective isotope. The T sites which contain Ti are T3, T8, and T10, whereas all other sites are essentially free of Ti [96]. In a different neutron powder diffraction analysis [97], the authors suggested that T3, T7, T8, T10 or T6, T7, and T11, respectively, are preferentially occupied by Ti, with T10 showing only weak evidence [98]. However, the authors did not use samples isotopically enriched in Ti isotopes and, thus, were not able to differentiate contributions due to structural defects from those of Ti occupancy at specific T sites. Attempts to refine the Ti segregation from X-ray diffraction data failed [99].

6.3.1.5 Defect Structures

Despite the high crystallinity of zeolitic materials in X-ray diffraction experiments, ^{29}Si NMR spectra of all-silica zeolite frameworks show rather poor resolution. This is due to defects in the silicate network leading to local distortions which strongly disturb the NMR signals. The interrupted framework or the T-site vacancies change the material properties significantly, e.g., rendering a hydrophobic all-silica framework into a hydrophilic material. Calcination might heal certain defects if neighboring T sites are of the -O_3T-OH type, and the thermal treatment leads to the condensation of the OH groups. If there are missing T atoms, calcination will not improve the NMR spectrum. All these defective sites are introduced during synthesis and are generally distributed over particular framework sites. Neutron diffraction experiments of silicalite material showed that in the MFI-type material, the T sites T7, T10, and T11 are involved in clustered defects leading to hydroxyl nests [100]. Ordered defects as systematic part of the microporous framework have been reported for RUB-22, a derivative framework structure of the RUT-type framework. In the as-made material, one T site is missing systematically in comparison to a RUT-type zeolite, and the T atom (B or Si) is replaced by hydrogen atoms forming OH groups [101] (Fig. 6.14). A similar material has also been described by Plevert and Tatsumi [102].

SSZ-74 is another zeolite framework with ordered vacancies in its framework structure [103]. The material with a three-dimensional 10-ring channel system are promising for applications in catalysis. Considering that the ordered vacancy might be filled in with functional T atoms, new possibilities are open.

6.3.1.6 Conclusions

The detailed structural analysis of elements on specific sites is still a challenging analytical problem. Using techniques probing long-range periodic order, i.e., diffraction experiments based on X-rays, neutrons, or electrons, will always remain difficult and specific for the respective technique. Depending on the scattering cross section, elements contribute differently to the diffraction diagram, and in some instances such as Si and Al for X-ray and electron diffraction experiments, there is

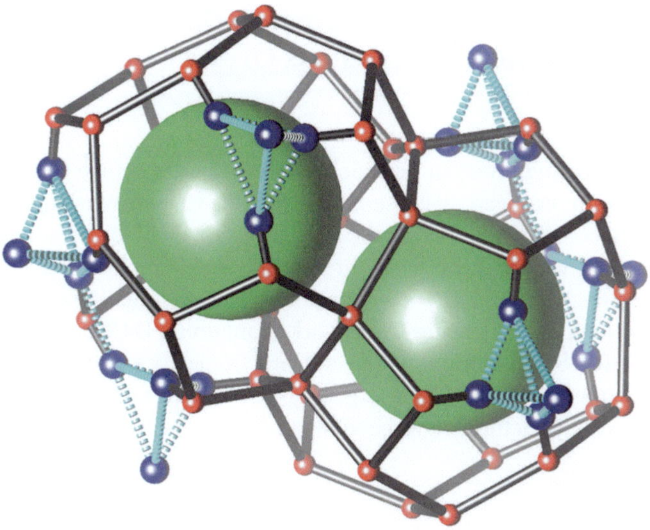

Fig. 6.14 The structure of RUB-22. Specific T sites of the (hypothetical) non-interrupted framework are missing and replaced by hydrogen atoms forming hydroxyl nests. The hydrogen bonds between them are displayed as *dotted blue lines*. *Large green spheres* represent the occluded TMA cations

almost no difference between the nuclei. In such cases, the complementary nature of spectroscopic experiments probing short-range local order is very helpful. In particular, the combination of element-sensitive NMR experiments with diffraction experiments has proved to be very useful. Wherever the analysis was successful, computational analysis can complete the understanding of the material in shedding light on reaction mechanism and structural dynamics.

Still, there are many problems concerning the local structure of the reactive centers in porous framework which remain unsolved. Structure analysis will definitely play a key role in the experimental register, together and complementary to spectroscopic methods and numerical simulation studies. One way to include diffraction analysis in the characterization of active centers is the resonant scattering experiment. Such experiments use X-rays with a wavelength near the absorption edge of the active element. The reflection intensities of the scattered light are dominated in particular by the contribution of the stimulated T element. Elastic resonant scattering experiments have been demonstrated to be useful for the analysis of extra-framework cations [104], but also for the distribution and structure of particular active elements in nanoclusters, e.g., HgSe, inside the pore system of zeolites, such as FAU- and LTL-type frameworks [105].

6.4 Specific Building Units: Structure Direction with Framework Constituents Ordered on Distinct T Sites

The structure-directing effect of organic molecules in the course of seed formation and crystal growth has been discussed widely [106, 107]. However, there are other non-framework constituents such as the fluoride anion and Ge as framework T atom which play a decisive role in structure direction. Besides the function as mineralizer, the fluoride anion preferentially occupies double-4-ring units (D4R), thus directing the crystallization process toward frameworks containing D4R units (Fig. 6.15). Ge as substituent for Si on T sites of the silicate framework also exerts substantial structure-directing influence. Using organic SDA together with fluoride and/or Ge, various framework structures have been obtained.

6.4.1 The Clathrasil Octadecasil and the F Synthesis of Zeolites

The fluoride route in zeolite synthesis had been introduced in order to use other mineralization media than the basic hydroxyl anion. The first example of a zeolite obtained via the fluoride route was silicalite-1, the all-silica analogue of zeolite ZSM-5. It was known that fluoride binds to silicon and might also stabilize coordination states of Si higher than four; however, the role of fluoride during zeolite synthesis was not recognized. The crystal structure analysis of the clathrasil octadecasil for the first time revealed that fluoride occupies the double-4-ring unit, thus stabilizing the AST framework type which is of rather low framework density (Fig. 6.16) [108]. The structure refinement from single-crystal data immediately

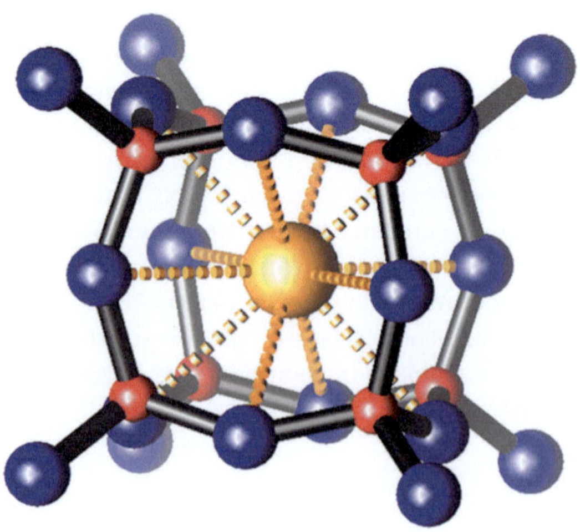

Fig. 6.15 The double-4-ring unit (D4R) containing a fluoride anion (*orange sphere*)

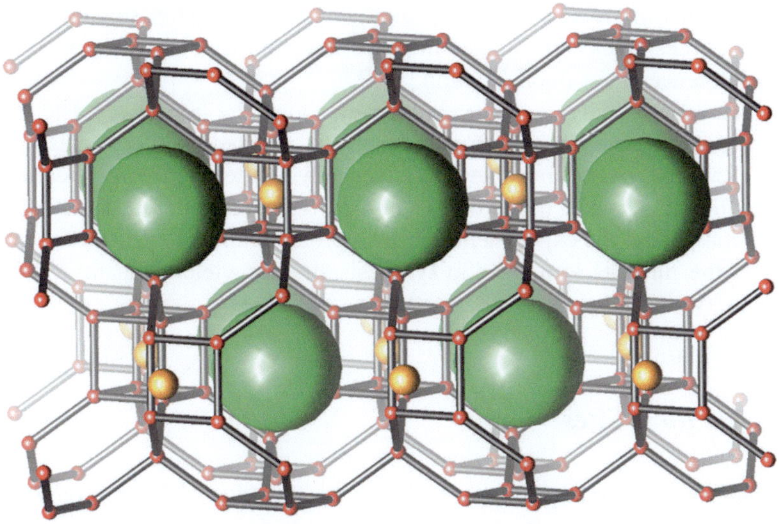

Fig. 6.16 *AST* framework showing the fluoride anion (*orange sphere*) occupying the D4R and the TMA cation (*green*) occupying the larger cage

also explained the unusual chemical shift values in the ^{19}F MAS NMR spectra at ~ -40 ppm as fluoride entity in the double-4-ring confinement (D4R confinement). Later, in NMR experiments monitoring the synthesis of zeolite ITQ-7, it has been shown that the signal characteristic for the fluoride anion in the D4R confinement shows up long before the crystalline material could be detected in P-XRD experiments [109].

Based on this insight, in particular, Avelino Corma and his group of the Instituto de Tecnología Química, ITQ, very successfully used fluoride as mineralizer and structure director for the synthesis of many new silicate zeolite framework types all containing the D4R unit as characteristic structural feature. These are the framework structure types containing D4R units and the corresponding type materials which have been obtained by the ITQ synthesis group where the fluoride route was developed: ISV (ITQ-7) [110], ITW (ITQ-12) [12], ITH (ITQ-13) [13], BEC (beta polymorph C or ITQ-17) [36], IWW (ITQ-22) [11], IWR (ITQ-24) [11], IWS (ITQ-26) [12], IWV (ITQ-27) [111], ITR (ITQ-34) [14], -ITV (ITQ-37) [112], and IRR (ITQ-44) [113]. In addition, the all-silica LTA (ITQ-29) [114] was also obtained using fluoride as one of the structure directors. The long list of different and new framework structure types was not exclusively obtained by the fluoride SDA but in combination with other molecular SDA, most often organic amines, and by the addition of Ge as substituent for Si on T positions. However, for the successful synthesis, fluoride was essential, and the prime intention of the synthesis chemists was to stabilize D4R units as they were first detected in octadecasil with AST framework type. Interestingly, before the fluoride anion was established experimentally in the D4R unit, it was speculated from theoretical calculations that the sodium cation should fit inside the D4R and thus stabilize the structure. After some controversy, new simulation studies consolidated the experimental result and also found the stabilizing influence of the fluoride anion [115].

The octadecasil example very conclusively shows the impact of basic research on the progress of science. The structure analysis of octadecasil crystallizing in the AST framework type became a key experiment in zeolite science after realizing the structure-directing role of the fluoride anion during synthesis. The crystal structure analysis constituted the experimental growth protocol and allowed for the generalization of the findings to a specific synthesis strategy yielding many new zeolite framework types.

6.4.2 Ge on T Sites

Germanium was reported first as substituent on T sites in FAU-type zeolite in the early 1970s [116] and later as germanate zeolite [117]. In the late 1990s, the structure-directing effect of structural germanate subunits which fit into the silicate framework was recognized [118]. Shortly after this report, the synthesis of ITQ-7 was drastically improved by adding Ge which was supposed to facilitate the formation of D4R units [119]. Now, these silico-germanates showed a high enough thermal and chemical stability essential for ion exchange and catalytic applications. As a consequence, systematic experiments were carried out to synthesize new zeolite framework structures containing D4R units with the aid of other, organic SDA. The first example was the D4R-containing polymorph C of the beta family of framework structures [36, 120]. The zeolite groups of the ITQ de Valencia systematically exploited the findings of Ge as structure director and discovered a series of new zeolite framework structures. The detailed structural analysis of ITQ-7 [121] and AST-type zeolite [122] showed that Ge preferentially occupies sites of the D4R unit. The analysis of the solid solution series of AST-type zeolite which can be obtained as all-silica and all-germanium framework structure revealed the stepwise substitution pattern of Ge for Si following first an avoidance rule similar to Loewenstein's rule for aluminosilicates, before an all-germanium framework structure is formed. The local order is particularly obvious from ^{19}F SS-NMR experiments which show the coexistence of definite ordering schemes in the framework avoiding random distribution of Ge on T sites. The Ge order on T sites and its electron density is significantly different from those of the main constituents on framework T sites, Si, and Al, also alleviating the task of structure determination [36]. As shown in several examples, conventional direct methods yielded the complete zeolite framework from powder diffraction data [10, 123–125]. Structure refinement then confirmed Ge ordering on T sites, e.g., the preferential occupation of D4R T sites.

The theoretical and structural explanation for the unique property of Ge as structure director and codirector concurrently with organic SDA has been ascribed to the ability to accommodate O–T–O angles in a wider range than other T atoms, in particular Si, to smaller values down to close to 90°. This reflects the geometry in the preferred six-fold coordination of Ge in oxides in its thermodynamically stable form [121, 126].

6.4.2.1 Secondary Building Units as Framework Constituents in Germanates

Germanium on T sites of porous framework materials has been classified by O'Keefe [118]. Random substitution in silicates resulted in typical silicate zeolites with no specific structural or chemical features. Substitution on specific T site of structural subunits like the D4R leads to the new framework structure types which were discussed above. The third category are those porous frameworks made predominantly by Ge or by Ge only which include other, new subunits as secondary building units which are geometrically not compatible with silicate frameworks [39, 127–131] (Fig. 6.17).

Allowing for an extension of the four-fold coordination to five- or six-fold coordination for [GeOx] primary building units extends the number of subunits leading to complex oxide framework structures which are modularly constructed from few basic units in varying ratio and number [132]. In particular, enormous ring sizes and pore volumes have been stabilized leading into the mesopore dimension still maintaining an ordered, crystalline structure of the pore wall [20, 130, 133]. Unfortunately, the chemical and thermal stability of these materials is lower than that of mesoporous silicas comparable in pore size and pore volume.

From the available literature data, there is no other T atom with such decisive influence on the framework structure during synthesis. As already discussed, Al is heavily disordered with a tendency of preference for particular T sites B or Ga show hardly any segregation.

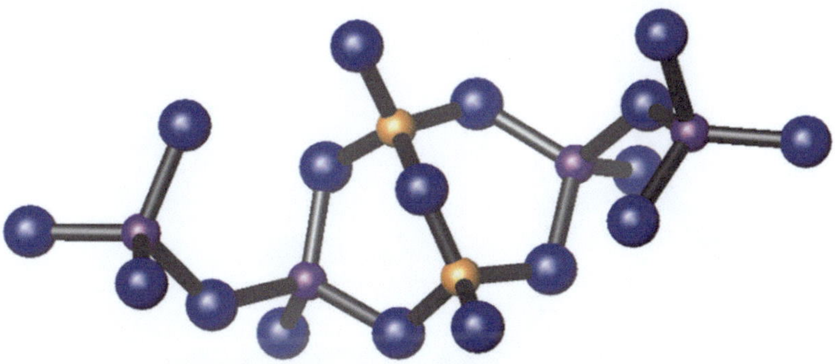

Fig. 6.17 Example of a secondary building unit which is not compatible with silicate frameworks but requires a high Ge content (*violet* Ge, *orange* B, *blue* O)

6.4.3 Three Rings and Spiro-5 Units

Li, Be, and Zn are additional T atoms which have a structure-directing influence, replacing Si on T sites. The solution of the crystal structure of the beryllium containing mineral lovdarite showed that the spiro-5 unit with two 3-rings connected with one spiro-Si atom at the center was a so far unknown feature of these silicate materials (Fig. 6.18, left). Be is part of the spiro-5 unit [134, 135] which is the characteristic structural unit of this group of materials. Since three rings are correlated with low framework density and, therefore, large pore volume of the framework structure, much attention had been paid for the synthesis of new, 3-ring-containing materials [136]. Nabesite is another beryllosilicate mineral, and OSB-1, OSB-2, and LSJ-10 are synthetic materials which all contain spiro-5 units [137–139]. A related material to lovdarite is the zinc containing mineral gaultite [140] and the isostructural synthetic materials VPI-7 [141] and VPI-7* [142] with Zn as part of the spiro-5 unit, however, in a different arrangement of the layer-like building blocks. Here the chemical cross relationship of the elements Be and Zn shows up and enables the synthesis of similar framework structures with the catalytically interesting and less poisonous element Zn. The spiro-5 unit is also a structural feature in other zinc silicate materials such as RUB-17 [143] and VPI-9 [144].

The same spiro-5 unit is also formed in the presence of Li; however, the central position of the spiro-5 unit is now taken by Li as part of the tetrahedral silicate framework (Fig. 6.18, right). Representative materials are RUB-23 [145], RUB-29 [146], and RUB-12 [147]. Apart from these examples of T atoms in spiro-5 units in silicates, there are in all 23 type materials listed in the zeolite framework data base [148] with spiro-5 units in aluminosilicate, germanate, phosphate, arsenate, etc., microporous frameworks.

Single 3-rings as typical features of silicate framework structures are rare; however, they cannot be related to specific substitution patterns or single elements. They are found in silicates, germanosilicates, aluminosilicates, etc., and are made of Si as

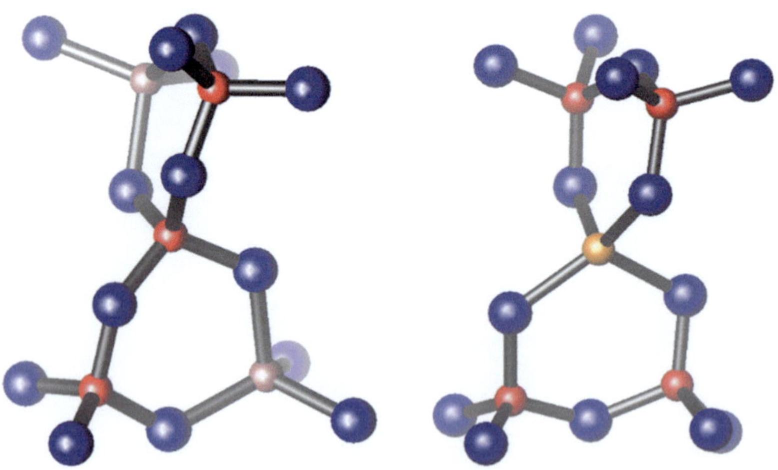

Fig. 6.18 The spiro-5 units containing Be or Zn (*violet spheres*) or Li (*orange sphere*). Si atoms are displayed as *red* and O as *blue spheres*

T atoms exclusively. Why such a strained structural environment is stable cannot be related to specific compositional features of the material. Where possible, substitution of Si in 3-rings by other T atoms releases strain, thus stabilizing the 3-ring or double-3-ring units.

6.4.4 Zeolite Frameworks with Mixed Pore Sizes

Zeolites possessing framework structures with mixed pore sizes show an improved catalytic performance and a broader applicability since the different pore sizes allow for a variety of diffusion pass ways. The prominent example is framework structure-type MWW with a 10-ring and 12-ring channel system which interconnects. An evaluation of the catalytic performance shows that the properties are similar to those of 10-ring zeolite ZSM-5 and 12-ring zeolite ß [149]. Other examples are the framework structure types of CON with a three-dimensional pore system of 12-ring and 10-ring channels (12R × 12R × 10R) [150], UTL with 14R × 12R [151] (see Fig. 6.19), MSE with again 12R × 10R × 10R [152], ITQ-39 with 12R × 10R

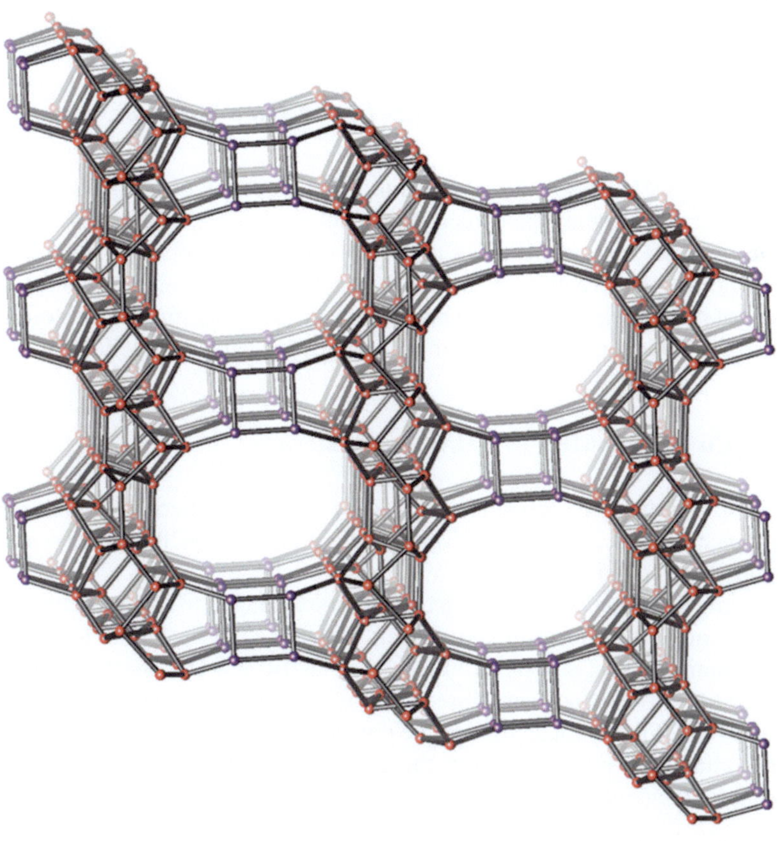

Fig. 6.19 Projection of the UTL-type framework structure along [010] with [100] across and [001] down. 12-ring channels are running parallel [010] and 14-ring channels along [001]

[22, 153], and ITQ-52 with 10R × 8R [154]. Only after the crystal structures were solved, a rational understanding of their properties in catalysis and separation was obvious. Since the differences in absolute pore dimensions are rather small and quite different from those which are classified as hierarchical pore sizes, the geometry-related properties are more hybrid type reflecting a fine-tuning in size and shape selectivity.

From the synthesis point of view, there is no clear concept which links the mixed pore size to structure directing by organic SDA. It seems more that building blocks of higher packing density are linked to three-dimensional frameworks giving rise to the particular channel systems. However, the questions whether these building blocks exist in solution and whether "Lego-" type construction of zeolite frameworks from building blocks occurs during crystal growth are discussed controversially and have not been proven convincingly by experiments. In the future, high-energy X-ray scattering experiments and their analysis using the PDF technique might help to understand the formation of these framework structures from atomistic principles. The feasibility of such analyses has been shown in a delamination experiment of the MWW layered precursor. The PDF analysis of the scattering diagram of the layered structure slab has been refined and proved the integrity and stability of the subunit in its isolated state [55].

6.4.5 *Positions of Active Cations in the Pore Volume*

Structure analysis has also paid attention to non-framework constituents such as the structure directors or the charge-compensating cations. In ion exchange, adsorption, and catalytic processes, the cations are of major importance as carriers of the functionality active in the chemical reactions. Recently, SCR catalysis and latent heat storage were in the focus of interest, two processes, where the nature of the non-framework constituents and their distribution in the pore system are strongly related to the applications.

6.4.5.1 Cu or Fe in MFI- or CHA-Type Materials

Cu or Fe ion-exchanged high-silica MFI- and CHA-type materials showed superior performance and stability in selective catalytic reduction (SCR) and also in oxidation catalysis [94, 155–158].

The analysis of the local structure of the ion-exchanged MFI-type zeolites, however, relies on probes for local and element-specific analysis and no longer on diffraction studies. EXAFS and XANES as well as IR and UV spectroscopies were employed to study the local structure of nano-oxide particles, their behavior during reaction cycles, their embedding into the pore system, and interaction with the reactant. The citation given above covers these aspects of the non-framework species in relation with their function.

CHA-type zeolites are much less studied. However, the Cu-exchanged materials play an important role in Diesel-exhaust SCR, where the Cu cations and their location in the zeolite framework are important for the excellent performance [159–163]. Again, diffraction has only supportive character in the localization of the Cu ions, although evidence from diffraction analysis indicates that three different Cu sites exist. Because of the low concentration of the metal cation and the high symmetry of the framework structure, only the statistical distribution of the cation over the different extra-framework sites can be determined.

6.4.5.2 Water Adsorption at Specific Sites (Example: Li-LSX), Use of Neutron Diffraction Data

The storage of latent heat in zeolite materials requires cations with high hydration energies, pore space for the deposition of water, high framework charge for a large number of cations, and high stability against hydrothermal decay. In order to understand the process of hydration, low-silica X zeolite in its Li form, Li-LSX, has been studied extensively with neutron and X-ray diffraction analyses and numerical simulations (see Fig. 6.20). Since the cation sites of dehydrated zeolites are different from those of the hydrated form, and since the process of hydration proceeds in steps depending on the energy gain by the hydration process, a structure analysis of starting and final state of hydration plus intermediate steps of hydration is required.

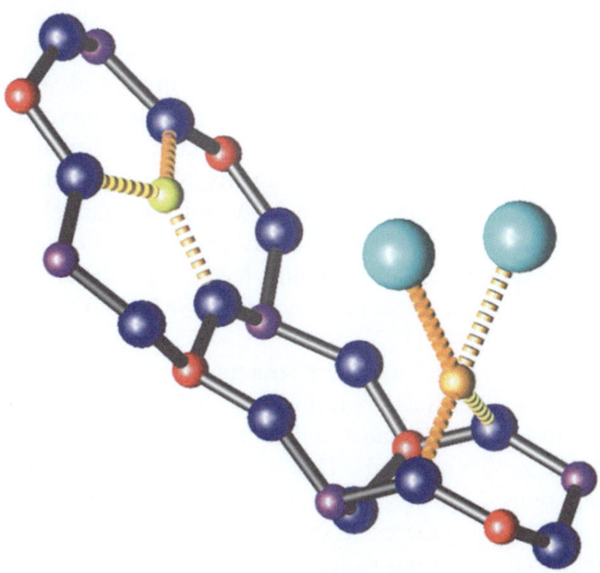

Fig. 6.20 Section of the Li-LSX structure at a low degree of hydration. Note the different coordination spheres of the Li atoms (*orange*): One Li is three coordinated by framework oxygen atoms (*dark blue*) only, while the other is four coordinated by oxygen atoms and water molecules (*light blue*)

Surprisingly, there are only a few studies on the dehydrated faujasite-type materials, even less on hydrated and partially hydrated phases [164–169]. The structure analysis of the specific hydration behavior of Li-LSX reflects the calorimetric study showing distinct steps in the heat of adsorption [170–172]. The combination of the different techniques nicely complements each other showing the atomistic structural and macroscopic thermodynamic evolution of the material upon hydration.

6.5 Conclusions

The history of structure analysis of microporous materials and particularly silicas, aluminosilicates, and aluminophosphates shows impressively that crystal chemistry and the search for new synthesis routes have strongly benefited from the insight into the details of the framework structure. Once structure analysis revealed the synthesis mechanism, the modularity of structures, the ordering scheme of SDA, etc., the door for exploration of the features was opened.

For example, the visualization of the crystal structure of ZSM-5 (framework type MFI) in combination with data from nitrogen adsorption, chemical analyses, NMR spectroscopy, etc., substantiated the 10-ring pore dimension and the three-dimensional diffusion pathway of the pores in the material, even if twinning or intergrowth occurs. Based on the knowledge of its crystal structure, MFI-type materials became the most researched and best studied zeolites in the following years. Successful and important applications in the field of catalysis followed and emphasized the importance and the impact of structure analysis. Structure analysis of a synthesized tetrapropylammonium silicalite, TPA-silicalite, very convincingly underlined the concept of structure directing by organic templates, the structure-directing agents, SDA. The intimate correspondence of the molecular geometry of the SDA and the channel geometry in the silicate framework of ZSM-5 [173] certainly inseminated chemists working in the synthesis of zeolites and subsequently lead to the discovery of numerous zeolite materials, the most important ones are ZSM-12 [174], ZSM-22 [175], ZSM-23 [176], and ZSM-48 [177].

On the other hand, zeolite framework structures played the role of guinea pigs for the development of diffraction techniques and structure solution theory of microcrystalline materials. The most complicated structures solved from powder diffraction data are zeolite materials. This is also true for the most complex structures which were refined from X-ray powder pattern. It is no surprise that electron crystallography and its latest success in solving new structures from nano-crystallites is strongly related to microporous materials. The near future will bring the distribution of the technique to more and more laboratories and work through materials structures which have not been able to solve so far. Most of these materials still derive from OSDA-driven synthesis which has been exploited by a number of laboratories extensively. There must exist more, unsolved structural problems which now are ready to be worked up. The method bears also hope for poorly crystalline materials such as the topotactically condensed phases or the interlayer expanded phases. Since the electron diffraction experiment requires only nano-sized crystals which scatter coherently, the chances of solving these structures improve significantly. Therefore, technically speaking, the future is in electron crystallography for structure solution combined with X-ray powder diffraction for the refinement of the framework geometry.

As far as new strategies for the synthesis of new microporous materials are concerned, there are interesting concepts beyond the structure-directing effect of organic molecules which have not yet been exploited exhaustively. The use of supramolecular assemblies of organic SDA was introduced by Corma [114] and successfully demonstrated for the synthesis of all-silica LTA. Later, few follow-up reports were published in the open literature, demonstrating the feasibility of the approach [178–180]. A breakthrough of supramolecular assemblies as organic SDA for the synthesis of new materials has not been reported. The use of framework slabs as modular units for the synthesis of new framework structures was successfully shown for the cubic octamer or D4R, stabilized with fluoride anions or Ge substituents for Si. There are many more subunits analyzed as dominant species in silicate solutions which show similar potential in stabilizing new framework structures. Purposeful experiments should be directed toward the idea in order to prove or disapprove the approach. A more recent structural subunit for three-dimensional framework structures, the layered precursor, has shown remarkable variability, and several new structure types have been obtained [181]. The technique of topotactic condensation should be extended to other already known hydrous layer silicates, and, in addition, new hydrous silicate precursors should expand the potential. Using different inorganic linker groups, the layered precursors can also be obtained as thermally stable three-dimensional framework structures [41, 47]. In this field, the exploration of product space has just begun, and a wide range of new microporous framework structures are expected.

For all new synthetic crystalline products, the structure analysis is the final protocol of the successful experiment. Conclusions from the analysis of the framework geometry and the host-guest interaction feed back in new experiments developing the synthesis strategy and refining the synthesis experimental protocol.

It is expected that, with respect to structure solution, electron crystallography will take over the role of powder X-ray crystallography in the future. In this sense, the review also highlights the achievements of P-XRD and sets the starting point for the new electron crystallography era in zeolite sciences. Extrapolating from the past, TEM equipment with precession instrumentation will substitute for the synchrotron experiment in the future; however, the technique still needs experts until automation will substitute them.

References

1. Morris RE, Wheatley PS (2007) Diffraction techniques. In: Introduction to zeolite molecular sieves. Stud Surf Sci Catal (Introduction to zeolite science and practice) 168:375–402
2. Xu R (2007) Chemistry of zeolites and related porous materials. Wiley (Asia), Singapore, p 679
3. Wright PA (2008) Microporous framework solids. Royal Society of Chemistry, Cambridge, p 429
4. Čejka J, Corma A, Zones S (2010) Zeolites and catalysis: synthesis, reactions and applications. Wiley-VCH, Weinheim, p 1056
5. Li Y, Yu J (2014) New stories of zeolite structures: their descriptions, determinations, predictions, and evaluations. Chem Rev 114:7268–7316
6. Baerlocher C, McCusker LB, Olson DH (2007) Atlas of zeolite framework Types, 6th rev. edn. Elsevier, Amsterdam
7. Horiuchi Y, Ura H, Kamegawa T, Mori K, Yamashita H (2011) Low-temperature synthesis of highly hydrophilic Ti-containing mesoporous silica thin films on polymer substrates by photocatalytic removal of structure-directing agents. J Mater Chem 21:23–241
8. Sheldrick GM (1990) Phase annealing in SHELX-90: direct methods for larger structures. Acta Crystallogr A: Found Crystallogr 46:467–473
9. Rius J, Frontera C (2007) Application of the constrained S-FFT direct-phasing method to powder diffraction data. XIII. J Appl Crystallogr 40:1035–1038
10. Corma A, Diaz-Cabanas M, Martinez-Triguero J, Rey F, Rius J (2002) A large-cavity zeolite with wide pore windows and potential as an oil refining catalyst. Nature 418:514–517
11. Corma A, Rey F, Valencia S, Jordá JL, Rius J (2003) A zeolite with interconnected 8-, 10- and 12-ring pores and its unique catalytic selectivity. Nat Mater 2:493–497
12. Dorset DL, Strohmaier KG, Kliewer CE, Corma A, Díaz-Cabañas MJ, Rey F, Gilmore CJ (2008) Crystal structure of ITQ-26, a 3D framework with extra-large pores. Chem Mater 20:5325–5331
13. Corma A, Puche M, Rey F, Sankar G, Teat SJ (2003) A zeolite structure (ITQ-13) with three sets of medium-pore crossing channels formed by 9- and 10-rings. Angew Chem Int Ed 42:1156–1159
14. Corma A, Diaz-Cabanas MJ, Jorda JL, Rey F, Sastre G, Strohmaier KG (2008) A zeolitic structure (ITQ-34) with connected 9- and 10-ring channels obtained with phosphonium cations as structure directing agents. J Am Chem Soc 130:16482–16483
15. Mugnaioli E, Kolb U (2013) Applications of automated diffraction tomography (ADT) on nanocrystalline porous material. Microporous Mesoporous Mater 166:93–101
16. Lu Q, Zhang Z, Dong C, Zhu X (2010) Catalytic upgrading of biomass fast pyrolysis vapors with nano metal oxides: an analytical Py-GC/MS study. Energies 3:1805–1820
17. Jiang J, Jorda JL, Yu J, Baumes LA, Mugnaioli E, Diaz-Cabanas MJ, Kolb U, Corma A (2011) Synthesis and structure determination of the hierarchical meso-microporous zeolite ITQ-43. Science 333:1131–1134
18. Gramm F, Baerlocher C, McCusker LB, Warrender SJ, Wright PA, Han B, Hong SB, Liu Z, Ohsuna T, Terasaki O (2006) Complex zeolite structure solved by combining powder diffraction and electron microscopy. Nature 444:79–81
19. Grosse-Kunstleve RW, McCusker LB, Baerlocher C (1997) Powder diffraction data and crystal chemical information combined in an automated structure determination procedure for zeolites. J Appl Crystallogr 30:985–995
20. Sun J, Bonneau C, Cantín A, Corma A, Díaz-Cabañas MJ, Moliner M, Zhang D, Li M, Zou X (2009) The ITQ-37 mesoporous chiral zeolite. Nature 458:1154–1157
21. Corma A, Diaz-Cabanas MJ, Jiang J, Afeworki M, Dorset DL, Soled SL, Strohmaier KG (2010) Extra-large pore zeolite (ITQ-40) with the lowest framework density containing double four- and double three-rings. Proc Natl Acad Sci U S A 107(32):13997–14002

22. Willhammar T, Sun J, Wan W, Oleynikov P, Zhang D, Zou X, Moliner M, Gonzalez J, Martínez C, Rey F, Corma A (2012) Structure and catalytic properties of the most complex intergrown zeolite ITQ-39 determined by electron crystallography. Nat Chem 4:188–194
23. Gies H, Rius J (1995) Ab-initio structure determination of zeolite RUB-10 from low resolution X-ray powder diffraction data. Z Kristallogr 210:475–480
24. Verheyen E, Joos L, van Havenbergh K, Breynaert E, Kasian N, Gobechiya E, Houthoofd K, Martineau C, Hinterstein M, Taulelle F, van Speybroeck V, Waroquier M, Bals S, van Tendeloo G, Kirschhock CEA, Martens JA (2012) Design of zeolite by inverse sigma transformation. Nat Mater 11:1059–1064
25. Lobo RF, Tsapatsis M, Freyhardt CC, Chan I, Chen C, Zones SI, Davis ME (1997) A model for the structure of the large-pore zeolite SSZ-31. J Am Chem Soc 119:3732–3744
26. Shoemaker D, Robson HE, Broussar L, Ogilvie RA, Manor PC (1974) Zeolite RHO and sigma transformation. Am Mineral 59:1127–1128
27. Černý R, Favre-Nicolin V (2007) Direct space methods of structure determination from powder diffraction: principles, guidelines and perspectives. Z Kristallogr 222:105–113
28. Falcioni M, Deem MW (2000) Library design in combinatorial chemistry by Monte Carlo methods. Phys Rev E 61:5948–5952
29. Rodriguez-Carvajal J. Fullprof, crystallographic tools for Rietveld, profile matching and integrated intensity refinements of X-ray and/or neutron data. http://www.ill.eu/sites/fullprof/
30. Wadlinger RL et al (1967) Catalytic composition of a crystalline zeolite. US Patent 3308069
31. Newsam JM, Treacy MM, Koetsier WT, Degruyter CB (1988) Structural characterization of zeolite beta. Proc R Soc Lond A Math 420:375–405
32. Higgins JB, Lapierre RB, Schlenker JL, Rohrman AC, Wood JD, Kerr GT, Rohrbaugh WJ (1988) The framework topology of zeolite beta. Zeolite 8:446–452
33. Corma A, Moliner M, Cantin A, Diaz-Cabanas MJ, Lorda JL, Zhang D, Sun J, Jansson K, Hovmoller S, Zou X (2008) Synthesis and structure of polymorph B of zeolite beta. Chem Mater 20:3218–3223
34. Kadgaonkar MD, Kasture MW, Bhange DS, Joshi PN, Ramaswamy V, Kumar R (2007) NCL-7, A novel all silica analog of polymorph B rich member of BEA family: synthesis and characterization. Microporous Mesoporous Mater 101:108–114
35. Chen Y, Yang Z, Guo C, Ni C, Li H, Ren Z, Lang J (2011) Using alcohols as alkylation reagents for 4-cyanopyridinium and N, N′-dialkyl-4,4′-bipyridinium and their one-dimensional iodoplumbates. Cryst Eng Commun 13:243–250
36. Corma A, Navarro MT, Rey F, Rius J, Valencia S (2001) Pure polymorph C of zeolite beta synthesized by using framework isomorphous substitution as a structure-directing mechanism. Angew Chem Int Ed 40:2337–2340
37. Tong M-Q, Yan W-F, Yu J-H, Xu R-R (2013) Crystallization process of polymorph A enriched beta-zeolite. Chem J Chin Univ 34:494–498
38. Xie B, Song J, Ren L, Ji Y, Li J, Xiao F (2008) Organotemplate-free and fast route for synthesizing Beta zeolite. Chem Mater 20:4533–4535
39. Tang L, Shi L, Bonneau C, Sun J, Yue H, Ojuva A, Lee B, Kritikos M, Bell RG, Bacsik Z, Mink J, Zou X (2008) A zeolite family with chiral and achiral structures built from the same building layer. Nat Mater 7:381–385
40. Ruan JF, Wu P, Slater B, Terasaki O (2005) Structure elucidation of the highly active titanosilicate catalyst Ti-YNU-1. Angew Chem Int Ed 44:6719–6723
41. Gies H, Mueller U, Yilmaz B, Tatsumi T, Xie B, Xiao F, Bao X, Zhang W, de Vos D (2011) Interlayer expansion of the layered zeolite precursor RUB-39: a universal method to synthesize functionalized microporous silicates. Chem Mater 23:2545–2554
42. Mochizuki D, Kuroda K (2006) Design of silicate nanostructures by interlayer alkoxysilylation of layered silicates (magadiite and kenyaite) and subsequent hydrolysis of alkoxy groups. New J Chem 30:277–284
43. Mochizuki D, Shimojima A, Imagawa T, Kuroda K (2005) Molecular manipulation of two- and three-dimensional silica nanostructures by alkoxysilylation of a layered silicate octosilicate and subsequent hydrolysis of alkoxy groups. J Am Chem Soc 127:7183–7191

44. Inagaki S, Yokoi T, Kubota Y, Tatsumi T (2007) Unique adsorption properties of organic-inorganic hybrid zeolite IEZ-1 with dimethylsilylene moieties. Chem Commun 48:5188–5190
45. Wu P, Ruan J, Wang L, Wu L, Wang Y, Liu Y, Fan W, He M, Terasaki O, Tatsumi T (2008) Methodology for synthesizing crystalline metallosilicates with expanded pore windows through molecular alkoxysilylation of zeolitic lamellar precursors. J Am Chem Soc 130:8178–8187
46. Ruan J, Wu P, Slater B, Zhao Z, Wu L, Terasaki O (2009) Structural characterization of inter-layer expanded zeolite prepared from ferrierite lamellar precursor. Chem Mater 21:2904–2911
47. Gies H, Müller U, Yilmaz B, Feyen M, Tatsumi T, Imai H, Zhang H, Xie B, Xiao F, Bao X, Zhang W, de Baerdemaeker T, de Vos D (2012) Interlayer expansion of the hydrous layer silicate RUB-36 to a functionalized, microporous framework silicate: crystal structure analysis and physical and chemical characterization. Chem Mater 24:1536–1545
48. Xu H, Fu L, Jiang J, He M, Wu P (2014) Preparation of hierarchical MWW-type titanosilicate by interlayer silylation with dimeric silane. Microporous Mesoporous Mater 189:41–48
49. Yilmaz B, Mueller U, Feyen M, Zhang H, Xiao F, de Baerdemaeker T, Tijsebaert B, Jacobs P, de Vos D, Zhang W, Bao X, Imai H, Tatsumih T, Gies H (2012) New zeolite Al-COE-4: reaching highly shape-selective catalytic performance through interlayer expansion. Chem Commun 48:11549–11551
50. de Baerdemaeker T, Vandebroeck W, Gies H, Yilmaz B, Müller U, Feyen M, de Vos D (2014) Shape-selective organic–inorganic zeolitic catalysts prepared via interlayer expansion. Catal Today 235:169–175
51. de Baerdemaeker T, Gies H, Yilmaz B, Müller U, Feyen M, Xiao F, Zhang W, Yokoi T, Bao X, Vos D, Dirk E (2014) A new class of solid Lewis acid catalysts based on interlayer expansion of layered silicates of the RUB-36 type with heteroatoms. J Mater Chem A 2:9709–9717
52. Billinge SJL, Kanatzidis MG (2004) Beyond crystallography: the study of disorder, nano-crystallinity and crystallographically challenged materials with pair distribution functions. Chem Commun 7:749–760
53. Corma A, Fornes V, Pergher SB, Maesen TLM, Buglass JG (1998) Delaminated zeolite precursors as selective acidic catalysts. Nature 396:353–356
54. Roth WJ (2005) MCM-22 zeolite family and the delaminated zeolite MCM-56 obtained in one-step synthesis. In: Cejka J, Zilkova N, Nachtigall P (eds) Molecular sieves: from basic research to industrial applications, Pts A and B, vol 158. Elsevier Science BV, Amsterdam, pp 19–26
55. Narkhede VV, Gies H (2009) Crystal structure of MCM-22 (MWW) and its delaminated zeolite ITQ-2 from high-resolution powder X-ray diffraction data: an analysis using rietveld technique and atomic pair distribution function. Chem Mater 21:4339–4346
56. Loewenstein W (1954) The distribution of aluminum in the tetrahedra of silicates and aluminates. Am Mineral 39:92–96
57. Engelhardt G, Jancke H, Mägi M, Pehk T, LIPPMAA E (1971) Über die ^{1}H-, ^{13}C- und ^{29}Si-NMR chemischen Verschiebungen einiger linearer, verzweigter und cyclischer Methylsiloxan-Verbindungen. J Organomet Chem 28:293–300
58. Marsmann HC (1974) SI-29 NMR-studies on aqueous silicate solutions in water. Z Naturforsch B 29:495–499
59. Lippmaa E, Maegi M, Samoson A, Engelhardt G, Grimmer AR (1980) Structural studies of silicates by solid-state high-resolution silicon-29 NMR. J Am Chem Soc 102:4889–4893
60. Engelhardt G, Zeigan D, Lippmaa E, Mägi M (1980) Zur Frage der Verteilung der Silicium- und Aluminiumatome im Kristallgitter von Zeolithen. Ein alternativer Strukturvorschlag für Zeolithe vom Typ NaA. Z Anorg Allg Chem 468:35–38
61. Klinowski J, Thomas JM, Fyfe CA, Hartman JS (1981) Applications of magic-angle-spinning silicon-29 nuclear magnetic resonance: evidence for two different kinds of silicon-aluminum ordering in zeolitic structures. J Phys Chem 85:2590–2594

62. Lodge EA, Bursill LA, Thomas JM (1980) Have the structure and space-group of A-type zeolite been correctly assigned? J Chem Soc Chem Commun 18:875–876
63. Bursill LA, Lodge EA, Thomas JM, Cheetham AK (1981) New light on the crystal structure of zeolite A. J Phys Chem 85:2409–2421
64. Engelhardt G, Lohse U, Lippmaa E, Tarmak M, Mägi M (1981) ^{29}Si-NMR-Untersuchungen zur Verteilung der Silicium- und Aluminumatome im Alumosilicatgitter von Zeolithen mit Faujasit-Struktur. Z Anorg Allg Chem 478:239–240
65. Engelhardt G, Lohse U, Lippmaa E, Tarmak M, Mägi M (1981) ^{29}Si-NMR-Untersuchungen zur Verteilung der Silicium-und Aluminiumatome im Alumosilicatgitter von Zeolithen mit Faujasit-Struktur. Z Anorg Allg Chem 482:49–64
66. Dempsey E (1974) Acid strength and aluminum site reactivity of Y zeolites. J Catal 33:497–499
67. Klinowski J, Ramdas S, Thomas JM, Fyfe CA, Hartman JS (1982) A re-examination of Si, Al ordering in zeolites NaX and NaY. J Chem Soc Faraday Trans 2 78:1025–1050
68. Bennett JM, Blackwell CS, Cox DE (1983) High-resolution silicon-29 nuclear magnetic resonance and neutron powder diffraction study of Na-A zeolite. Loewenstein's rule vindicated. J Phys Chem 87:3783–3790
69. Tossell JA (1993) A theoretical study of the molecular basis of the Al avoidance rule and the spectral characteristics of the Al-O-Al linkages. Am Mineral 78:911–920
70. Takaishi T, Kato M, Itabashi K (1995) Determination of the ordered distribution of aluminum atoms in a zeolitic framework. Part II. Zeolites 15:21–32
71. Takaishi T (1998) Ordered distribution of Al atoms in the framework of analcimes. Faraday Trans 94:1507–1518
72. Schroder KP, Sauer JC (1993) Preferred stability of the Al-O-Si-O-Al linkages in high-silica zeolite catalysts – theoretical predictions contrary to Dempsey's rule. J Phys Chem 97:6579–6581
73. Ruiz-Salvador AR, Grau-Crespo R, Gray AE, Lewis DW (2013) Aluminium distribution in ZSM-5 revisited: The role of Al–Al interactions. J Solid State Chem 198:330–336
74. von Ballmoos R, Meier WM (1981) Zoned aluminium distribution in synthetic zeolite ZSM-5. Nature 289:782–783
75. Derouane E (1981) Concerning the aluminum distribution gradient in ZSM-5 zeolites. J Catal 71:447–448
76. Olson DH, Kokotailo GT, Lawton SL, Meier WM (1981) Crystal structure and structure-related properties of ZSM-5. J Phys Chem 85:2238–2243
77. Sarv P, Fernandez C, Amoureux JP, Keskinen K (1996) Distribution of tetrahedral aluminium sites in ZSM-5 type zeolites: An Al-27 (Multiquantum) magic angle spinning NMR study. J Phys Chem 100:19223–19226
78. Kucera J, Nachtigall P (2005) A simple correlation between average T-O-T angles and Al-27 NMR chemical shifts does not hold in high-silica zeolites. Microporous Mesoporous Mater 85:279–283
79. Kucera J, Nachtigall P (2005) Al-27 NMR chemical shifts do not correlate with average T-O-T angles: theoretical study of MCM-58 zeolite. In: Cejka J, Zilkova N, Nachtigall P (eds) Molecular sieves: from basic research to industrial applications, Pts A and B, vol 158. Elsevier Science BV, Amsterdam, pp 917–924
80. Olson DH, Khosrovani N, Peters AW, Toby BH (2000) Crystal structure of dehydrated CsZSM-5 (5.8A1): evidence for nonrandom aluminum distribution. J Phys Chem B 104:4844–4848
81. Gábová V, Dědeček J, Čejka J (2003) Control of Al distribution in ZSM-5 by conditions of zeolite synthesis. Chem Commun 10:1196–1197
82. Dedecek J, Balgová V, Pashkova V, Klein P, Wichterlová B (2012) Synthesis of ZSM-5 zeolites with defined distribution of Al atoms in the framework and multinuclear MAS NMR analysis of the control of Al distribution. Chem Mater 24:3231–3239
83. Dědeček J, Sobalík Z, Wichterlová B (2012) Siting and distribution of framework aluminium atoms in silicon-rich zeolites and impact on catalysis. Catal Rev 54:135–223

84. Dědeček J, Čapek L, Sazama P, Sobalík Z, Wichterlová B (2011) Control of metal ion species in zeolites by distribution of aluminium in the framework: from structural analysis to performance under real conditions of SCR-NOx and NO, N2O decomposition. Appl Catal A Gen 391:244–253
85. Dedecek J, Sklenak S, Li C, Wichterlova B, Gabova V, Brus J, Sierka M, Sauer J (2008) Aluminum siting in the framework of silicon rich zeolites. A ZSM-5 study. In: Gedeon A, Massiani P, Baboneau F (eds) Zeolites and related materials: trends, targets and challenges. Proceedings of the 4th international FEZA conference, vol 174, pp 781–786
86. Sklenak S, Dedecek J, Li C, Wichterlová B, Gábová V, Sierka M, Sauer J (2007) Aluminum siting in silicon-rich zeolite frameworks: a combined high-resolution (27)Al NMR spectroscopy and quantum mechanics / molecular mechanics study of ZSM-5. Angew Chem Int Ed 46:7286–7289
87. Palin L, Lamberti C, Kvick A, Testa F, Aiello R, Milanesio M, Viterbo D (2003) Single-crystal synchrotron radiation X-ray diffraction study of B and Ga silicalites compared to a purely siliceous MFI: a discussion of the heteroatom distribution. J Phys Chem B 107:4034–4042
88. Davis ME (2002) Ordered porous materials for emerging applications. Nature 417:813–821
89. Davis ME (2014) Zeolites from a materials chemistry perspective. Chem Mater 26:239–245
90. Millini R, Perego G, Bellussi G (1999) Synthesis and characterization of boron-containing molecular sieves. Top Catal 9:13–34
91. Xie D, McCusker LB, Baerlocher C (2011) Structure of the borosilicate zeolite catalyst SSZ-82 solved using 2D-XPD charge flipping. J Am Chem Soc 133:20604–20610
92. Milanesio M, Lamberti C, Aiello R, Testa F, Piana M, Viterbo D (2000) Iron location in Fe-silicalites by synchrotron radiation single crystal X-ray diffraction. J Phys Chem B 104:9951–9953
93. Lewis DW, Catlow CRA, Sankar G, Carr SW (1995) Structure of iron-substituted ZSM-5. J Phys Chem 99:2377–2383
94. Joyner RW, Stockenhuber M (1999) Preparation, characterization, and performance of Fe-ZSM-5 catalysts. J Phys Chem B 103:5963–5976
95. Schwidder M, Kumar M, Klementiev K, Pohl M, Bruckner A, Gruenert W (2005) Selective reduction of NO with Fe-ZSM-5 catalysts of low Fe content I. Relations between active site structure and catalytic performance. J Catal 231:314–330
96. Henry PF, Weller MT, Wilson CC (2001) Structural investigation of TS-1: determination of the true nonrandom titanium framework substitution and silicon vacancy distribution from powder neutron diffraction studies using isotopes. J Phys Chem B 105:7452–7458
97. Hijar CA, Jacubinas RM, Eckert J, Henson NJ, HAY PJ, Ott KC (2000) The siting of Ti in TS-1 is non-random: powder neutron diffraction studies and theoretical calculations of TS-1 and FeS-1. J Phys Chem B 104:12157–12164
98. Lamberti C, Bordiga S, Zecchina A, Artioli G, Marra G, Spanò G (2001) Ti location in the MFI framework of Ti – silicalite-1: a neutron powder diffraction study. J Am Chem Soc 123:2204–2212
99. Lamberti C, Bordiga S, Zecchina A, Carati A, Fitch AN, Artioli G, Petrini G, Salvalaggio M, Marra GL (1999) Structural characterization of Ti-silicalite-1: A synchrotron radiation X-ray powder diffraction study. J Catal 183:222–231
100. Bordiga S, Roggero I, Ugliengo P, Zecchina A, Bolis V, Artioli G, Buzzoni R, Marra G, Rivetti F, Spanò G, Lamberti C (2000) Characterisation of defective silicalites. J Chem Soc Dalton Trans 21:3921–3929
101. Latzel S, Marler B, Oberhagemann U, Osterhoff C, Gies H (2007) Synthesis and general characterisation of RUB-22: a new microporous silicate possessing an interrupted framework structure. In: Xu R, Gao Z, Chen J, Yan W (eds) From zeolites to MOF materials – the 40th anniversary of the international zeolite conference, proceedings of the 15th international zeolite conference, vol 170. Elsevier Science BV, Amsterdam, pp 199–205
102. Plévert J, Tatsumi T (2012) Silicon vacancy ordering in pure silica TMA-Nu-1: opening the pores of a clathrasil. Abstract P-026 of the international symposium on zeolites and MicroPorous crystals, Hiroshima, Japan

103. Baerlocher C, Xie D, McCusker LB, Hwang S, Chan IY, Ong K, Burton AW, Zones SI (2008) Ordered silicon vacancies in the framework structure of the zeolite catalyst SSZ-74. Nat Mater 7:631–635
104. Wilkinson AP, Cheetham AK, Tang SC, Reppart WJ (1992) Cation distribution in zeolite zinc/sodium-Y by resonant X-ray diffraction. J Chem Soc Chem Commun 20:1485–1486
105. Abeykoon A, Castro-Colin M, Anokhina EV, Iliev MN, Donner W, Brunelli M, Jacobson AJ, Moss SC (2008) X-ray scattering studies of HgSe nanoclusters in zeolite. Metall Mater Trans A 39:3179–3183
106. Gies H, Marler B (1992) The structure controlling role of organic templates for the synthesis of porosils in the system SiO_2/Template/H_2O. Zeolites 12:42–49
107. Davis ME, Lobo RF (1992) Zeolite and molecular sieve synthesis. Chem Mater 4:756–768
108. Caullet P, Guth JL, Hazm J, Lamblin JM, Gies H (1991) Synthesis, characterisation and crystal-structure of the new clathrasil phase octadecasil. Eur J Solid State Inorg Chem 28:345–361
109. Song JQ, Marler B, Gies H (2005) Synthesis of ITQ-7 with a new template molecule and its crystal structure analysis in the as synthesized form. C R Chim 8:341–352
110. Villaescusa L, Barrett PA, Camblor MA (1999) ITQ-7: a new pure silica polymorph with a three-dimensional system of large pore channels. Angew Chem Int Ed 38:1997–2000
111. Dorset DL, Kennedy GJ, Strohmaier KG, Diaz-Cabanas MJ, Rey F, Corma A (2006) P-derived organic cations as structure-directing agents: synthesis of a high-silica zeolite (ITQ-27) with a two-dimensional 12-ring channel system. J Am Chem Soc 128:8862–8867
112. Chang X, He L, Liang H, Liu X, Yan Z (2010) Screening of optimum condition for combined modification of ultra-stable Y zeolites using multi-hydroxyl carboxylic acid and phosphate. Catal Today 158:198–204
113. Parvulescu AN, Hausoul PJC, Bruijnincx PCA, Gebbink RJMK, Weckhuysen BM (2010) Synthesis of octyl-ethers of biomass-based glycols through two competitive catalytic routes: telomerization and etherification. Catal Today 158:130–138
114. Corma A, Rey F, Rius J, Sabater MJ, Valencia S (2004) Supramolecular self-assembled molecules as organic directing agent for synthesis of zeolites. Nature 431:287–290
115. George AR, Catlow CR (1997) A computational study of the role of F^- ions in the octadecasil structure. Zeolites 18:67–70
116. Lerot L, Poncelet G, Fripiat JJ (1974) Synthesis, structural characterization and stability of a germanic (Al-Ge) near faujasite zeolite. Mater Res Bull 9:979–987
117. Sturua GI, Belokoneva EL, Simonov MA, Belov NV (1978) Crystal structure of Ge-zeolite, $K(H_3O)_3[Ge_7O_{16}]KH_3[Ge_7O_{16}]\cdot 3H_2O$. Dokl Akad Nauk SSSR+ 242 5:1078–1081
118. O'Keeffe M, Yaghi OM (1999) Germanate zeolites: contrasting the behavior of germanate and silicate. Chem Eur J 5:2796–2801
119. Corma A, Díaz-Cabañas MJ, Domine ME, Rey F (2000) Ultra fast and efficient synthesis of Ti-ITQ-7 and positive catalytic implications. Chem Commun 18:1725–1726
120. Corma A, Navarro MT, Rey F, Valencia S (2001) Synthesis of pure polymorph C of Beta zeolite in a fluoride-free system. Chem Commun 16:1486–1487
121. Blasco T, Corma A, Díaz-Cabañas MJ, Rey F, Vidal-Moya JA, Zicovich-Wilson CM (2002) Preferential location of Ge in the double four-membered ring units of ITQ-7 zeolite. J Phys Chem B 106(10):2634–2642
122. Wang Y, Song J, Gies H (2003) The substitution of germanium for silicon in AST-type zeolite. Solid State Sci 5:1421–1433
123. Castaneda R, Corma A, Fornes V, Rey F, Rius J (2003) Synthesis of a new zeolite structure ITQ-24, with intersecting 10-and 12-membered ring pores. J Am Chem Soc 125:7820–7821
124. Blasco T, Corma A, Diaz-Cabanas MJ, Rey F, Rius J, Sastre G, Vidal-Moya JA (2004) Synthesis, characterization, and framework heteroatom localization in ITQ-21. J Am Chem Soc 126:13414–13423
125. Cantin A, Corma A, Leiva S, Rey F, Rius J, Valencia S (2005) Synthesis and structure of the bidimensional zeolite ITQ-32 with small and large pores. J Am Chem Soc 127:11560–11561

126. Sastre G, Vidal-Moya JA, Blasco T, Rius J, Jordá JL, Navarro MT, Rey F, Corma A (2002) Preferential location of Ge atoms in polymorph C of beta zeolite (ITQ-17) and their structure-directing effect: a computational, XRD, and NMR spectroscopic study. Angew Chem Int Ed 41:4722–4726
127. Bu X, Feng P, Gier TE, Zhao D, Stucky GD (1998) Hydrothermal synthesis and structural characterization of zeolite-like structures based on gallium and aluminum germanates. J Am Chem Soc 120:13389–13397
128. Gier TE, Bu X, Feng P, Stucky GD (1998) Synthesis and organization of zeolite-like materials with three-dimensional helical pores. Nature 395:154–157
129. Li H, Yaghi OM (1998) Transformation of germanium dioxide to microporous germanate 4-connected nets. J Am Chem Soc 120:10569–10570
130. Zou X, Conradsson T, Klingstedt M, Dadachov MS, O'Keeffe M (2005) A mesoporous germanium oxide with crystalline pore walls and its chiral derivative. Nature 437:716–719
131. Peskov MV, Zou X (2011) Germanates built from Ge 10 (O, OH) 27 – 28 and Ge 7 (O, OH, F) 19 secondary building units: from systematic study of reported compounds to rational design of novel structures. J Phys Chem C 115:7729–7739
132. Christensen KE, Shi L, Conradsson T, Ren T, Dadachov MS, Zou X (2006) Design of open-framework germanates by combining different building units. J Am Chem Soc 128:14238–14239
133. Han Y, Zhang D, Chng LL, Sun J, Zhao L, Zou X, Ying JY (2009) A tri-continuous mesoporous material with a silica pore wall following a hexagonal minimal surface. Nat Chem 1:123–127
134. Merlino S (1981) Lovdarite: three membered rings in a framework silicate. Acta Crystallogr. (Suppl)(A37):C189
135. Merlino S (1990) Lovdarite, $K_4Na_{12}(Be_8Si_{28}O_{72})\cdot 18H2O$, a zeolite-like mineral-structural features and OD character. Eur J Mineral 2:809–817
136. Brunner GO, Meier WM (1989) Framework density distribution of zeolite-type tetrahedral nets. Nature 337:146–147
137. Petersen OV, Giester G, Brandstatter F, Niedermayr G (2002) Nabesite, $Na_2BeSi_4O_{10}\cdot 4H2O$, a new mineral species from the Ilimaussaq alkaline complex, South Greenland. Can Mineral 40:173–181
138. Cheetham AK, Fjellvag H, Gier TE, Kongshaug KO, Lillerud KP, Stucky GD (2001) Very open framework materials: From concept to reality. Stud Surf Sci Catal (From zeolites to porous MOF-materials)
139. Armstrong JA, Weller MT (2010) Beryllosilicate frameworks and zeolites. J Am Chem Soc 132:15679–15686
140. Ercit TS, Vanvelthuizen J (1994) Gaultite, a new zeolite-like mineral species from Mont-Saint-Hilaire, Quebec, and its crystal structure. Can Mineral 32:855–863
141. Röhrig C, Gies H, Marler B (1994) Rietveld refinement of the crystal structure of the synthetic porous zincosilicate VPI-7. Zeolites 14:498–503
142. Röhrig C, Dierdorf I, Gies H (1995) X-ray powder diffraction and NMR-spectroscopic investigations on a porous zincosilicate related to the zeolite VPI-7 (VSV). J Phys Chem Solids 56:1369–1376
143. Röhrig C, Gies H (1995) A new zincosilicate zeolite with nine-ring channels. Angew Chem Int Ed Engl 34:63–65
144. McCusker LB, Grosse-Kunstleve RW, Baerlocher C, Yoshikawa M, Davis ME (1996) Synthesis optimization and structure analysis of the zincosilicate molecular sieve VPI-9. Microporous Mater 6:295–309
145. Park SH, Daniels P, Gies H (2000) RUB-23: a new microporous lithosilicate containing spiro-5 building units. Microporous Mesoporous Mater 37:129–143
146. Park SH, Parise JB, Gies H, Liu HM, Grey GP, Toby BH (2000) A new porous lithosilicate with a high ionic conductivity and ion-exchange capacity. J Am Chem Soc 122:11023–11024

147. Marler B, Grünewald-Lüke A, Gies H (2104) The average structure of RUB-12: a new lithosilicate zeolite. In: Book of abstracts, 22nd annual Conference of the German crystallographic society, Berlin, pp 118–119
148. Baerlocher C, McCusker LB, Database of zeolite structures. http://www.iza-structure.org/databases/
149. Corma A, Gonzàlez-Alfaro V, Orchillès AV (1995) Catalytic cracking of alkanes on MCM-22 zeolite: comparison with ZSM-5 and beta zeolite and its possibility as an FCC cracking additive. Appl Catal A Gen 129:203–215
150. Corma A (1997) Cracking behavior of zeolites with connected 12- and 10-member ring channels: the Influence of pore structure on product distribution. J Catal 167:438–446
151. Paillaud J, Harbuzaru B, Patarin J, Bats N (2004) Extra-large-pore zeolites with two-dimensional channels formed by 14 and 12 rings. Science 304:990–992
152. Inagaki S, Takechi K, Kubota Y (2010) Selective formation of propylene by hexane cracking over MCM-68 zeolite catalyst. Chem Commun 46:2662
153. Moliner M, Corma A (2012) Direct synthesis of a titanosilicate molecular sieve containing large and medium pores in its structure. Microporous Mesoporous Mater 164:44–48
154. Simancas R, Jordá JL, Rey F, Corma A, Cantín A, Peral I, Popescu C (2014) A new microporous zeolitic silicoborate (ITQ-52) with interconnected small and medium pores. J Am Chem Soc 136:3342–3345
155. Kumar MS, Schwidder M, Gruenert W, Brückner A (2004) On the nature of different iron sites and their catalytic role in Fe-ZSM-5 DeNOx catalysts: new insights by a combined EPR and UV/VIS spectroscopic approach. J Catal 227:384–397
156. Gruenert W, HAYES NW, Joyner RW, Shpiro ES, Siddiqui MR, Baeva GN (1994) Structure, chemistry and activity of Cu-ZSM-5 catalysts for the selective reduction of NOX in the presence of oxygen. J Phys Chem 98:10832–10846
157. Woertink JS, Smeets PJ, Groothaert MH, Vance MA, Sels BF, Schoonheydt RA, Solomon EI (2009) A [Cu2O]2+ core in Cu-ZSM-5, the active site in the oxidation of methane to methanol. Proc Natl Acad Sci U S A 106:18908–18913
158. Bordiga S, Buzzoni R, Geobaldo F, Lamberti C, Giamello E, Zecchina A, LeofantiI G, Petrini G, Tozzola G, Vlaic G (1996) Structure and reactivity of framework and extraframework iron in Fe-silicalite as investigated by spectroscopic and physicochemical methods. J Catal 158:486–501
159. Bates SA, Verma AA, Paolucci C, Parekh AA, Anggara T, Yezerets A, Schneider WF, Miller JT, Delgass WN, Ribeiro FH (2014) Identification of the active Cu site in standard selective catalytic reduction with ammonia on Cu-SSZ-13. J Catal 312:87–97
160. Giordanino F, Vennestrøm PNR, Lundegaard LF, Stappen FN, Mossin S, Beato P, Bordiga S, Lamberti C (2013) Characterization of Cu-exchanged SSZ-13: a comparative FTIR, UV-Vis, and EPR study with Cu-ZSM-5 and Cu-β with similar Si/Al and Cu/Al ratios. Dalton Trans 42:12741–12761
161. McEwen J, Anggara T, Schneider WF, Kispersky VF, Miller JT, Delgass WN, Ribeiro FH (2012) Integrated operando X-ray absorption and DFT characterization of Cu–SSZ-13 exchange sites during the selective catalytic reduction of NOx with NH3. Catal Today 184:129–144
162. Fickel DW, Fedeyko JM, Lobo RF (2010) Copper coordination in Cu-SSZ-13 and Cu-SSZ-16 investigated by variable-temperature XRD. J Phys Chem C 114:1633–1640
163. Xie L, Liu F, Ren L, Shi X, Xiao F, He H (2014) Excellent performance of one-pot synthesized Cu-SSZ-13 catalyst for the selective catalytic reduction of NOx with NH3. Environ Sci Technol 48:566–572
164. Plévert J, Di Renzo F, Fajula F, Chiari G (1997) Structure of dehydrated zeolite Li–LSX by neutron diffraction: evidence for a low-temperature orthorhombic faujasite. J Phys Chem B 101:10340–10346
165. Feuerstein M, Lobo RF (1998) Characterization of Li cations in zeolite LiX by solid-state NMR spectroscopy and neutron diffraction. Chem Mater 10:2197–2204

166. Pichon C, Méthivier A, Simonot-Grange M, Baerlocher C (1999) Location of water and xylene molecules adsorbed on prehydrated Zeolite BaX. A low-temperature neutron powder diffraction study. J Phys Chem B 103:10197–10203
167. Beta IA, Bohlig H, Hunger B (2004) Structure of adsorption complexes of water in zeolites of different types studied by infrared spectroscopy and inelastic neutron scattering. Presented at the 81st international Bunsen discussion meeting on? Interfacial water in chemistry and biology?, Velen, Germany, 2003. Phys Chem Chem Phys 6:1975–1981
168. Wozniak A, Marler B, Angermund K, Gies H (2008) Water and cation distribution in fully and partially hydrated Li-LSX zeolite. Chem Mater 20:5968–5976
169. Kim H, Ko S, Lim W (2012) Single-crystal structures of Li^+ -exchanged zeolite X (FAU, Si/Al = 1.09) from aqueous solution depends on ion-exchange temperatures at 293 and 333 K. Bull Kor Chem Soc 33:3303–3310
170. Moise JC, Bellat JP, Methivier A (2001) Adsorption of water vapor on X and Y zeolites exchanged with barium. Microporous Mesoporous Mater 43:91–101
171. Boddenberg B, Rakhmatkariev GU, Hufnagel S, Salimov Z (2002) A calorimetric and statistical mechanics study of water adsorption in zeolite NaY. Phys Chem Chem Phys 4:4172–4180
172. Boddenberg B, Rakhmatkariev GU, Wozniak A, Hufnagel S (2004) A calorimetric and statistical mechanics study of ammonia adsorption in zeolite NaY. Phys Chem Chem Phys 6:2494–2501
173. Price GD, Pluth JJ, Smith JV, Bennett JM, Patton RL (1982) Crystal structure of tetrapropylammonium fluoride-containing precursor to fluoride silicalite. J Am Chem Soc 104:5971–5977
174. Lapierre RB, Rohrman AC, Schlenker JL, Wood JD, Rubin MK, Rohrbaugh WJ (1985) The framework topology of zeolite ZSM-12 – a high silica zeolite. Zeolites 5:346–348
175. Kokotailo GT, Schlenker JL, Dwyer FG, Valyocsik EW (1985) The framework topology of ZSM-22 – a high silica zeolite. Zeolites 5:349–351
176. Rohrman AC, Lapierre RB, Schlenker JL, Wood JD, Valyocsik EW, Rubin MK, Higgins JB, Rohrbaugh WJ (1985) The framework topology of ZSM-23 – a high silica zeolite. Zeolites 5:352–354
177. Schlenker JL, Rohrbaugh WJ, Chu P, Valyocsik EW, Kokotailo GT (1985) The framework topology of ZSM-48 – a high silica zeolite. Zeolites 5:355–358
178. Gomez-Hortigueela L, Lopez-Arbeloa F, Cora F, Perez-Pariente J (2008) Supramolecular chemistry in the structure direction of microporous materials from aromatic structure-directing agents. J Am Chem Soc 130:13274–13284
179. Gomez-Hortiguela L, Cora F, Catlow CR, Perez-Pariente J (2006) Computational study of a chiral supramolecular arrangement of organic structure directing molecules for the AFI structure. Phys Chem Chem Phys 8:486–493
180. Gomez-Hortiguela L, Cora F, Perez-Pariente J (2008) Supramolecular assemblies of fluoroaromatic organic molecules as structure directing agents of microporous materials: different effects of fluorine. Microporous Mesoporous Mater 109:494–504
181. Marler B, Gies H (2012) Hydrous layer silicates as precursors for zeolites obtained through topotactic condensation: a review. Eur J Mineral 24:405–428
182. Roth WJ, Shvets OV, Shamzhy M, Chlubna P, Kubu M, Nachtigall P, Cejka, J (2011) Postsynthesis transformation of three-dimensional framework into a lamellar zeolite with modifiable architecture. J Am Chem Soc 133(16): 6130–6133

Chapter 7
Solid-State NMR Studies of Zeolites

Shenhui Li and Feng Deng

Abstract Due to having unique pore structures and intrinsic acid–base properties, zeolites have been widely used in ion exchange, adsorption, and catalysis in chemical and petrochemical industry. Solid-state NMR is a well-established tool for the structural characterization of zeolites. The detailed information about the zeolite framework can be obtained from multinuclear and multidimensional ^{1}H, ^{29}Si, ^{27}Al, and ^{17}O MAS NMR spectroscopy. The structure and communication of cages and channels in zeolites can be extracted by ^{129}Xe NMR spectroscopy. The acidic properties of zeolites can be well characterized through solid-state NMR probe molecule techniques. In addition, two-dimensional (2D) ^{1}H–^{1}H and ^{27}Al–^{27}Al double-quantum (DQ) MAS NMR experiments are utilized to probe the spatial proximities and the synergy effect between different acid sites in zeolites. Moreover, in situ solid-state NMR is able to explore the mechanism of zeolite-catalyzed reactions by monitoring the evolution of the reactants, intermediates, and products. The crystallization mechanism for synthesis of zeolites can be elucidated by detecting the variation of framework ^{29}Si, ^{27}Al, and ^{31}P NMR signals at different crystallization stages.

Keywords Zeolite • Solid-state NMR • Acidity characterization • Heterogeneous catalysis • Crystallization mechanism

7.1 Introduction

Zeolites, which are built of the corner-sharing SiO_4 and AlO_4 tetrahedral units, represent the most important class of known porous materials which are widely used as catalysts, sorbents, and ion exchangers. The wide application of zeolites in chemical and petrochemical industry could be mainly ascribed to their unique pore topology and tunable acid–base property. Incorporation of heteroatoms, such

S. Li • F. Deng (✉)
State Key Laboratory of Magnetic Resonance and Atomic and Molecular Physics, Wuhan Center for Magnetic Resonance, Wuhan Institute of Physics and Mathematics, The Chinese Academy of Sciences, Wuhan 430071, China
e-mail: dengf@wipm.ac.cn

F.-S. Xiao, X. Meng (eds.), *Zeolites in Sustainable Chemistry*,
Green Chemistry and Sustainable Technology,
DOI 10.1007/978-3-662-47395-5_7

as Al into siliceous zeolites, will introduce a charge imbalance in the framework which is balanced by a proton, thus generating bridging hydroxyl groups (SiOHAl, Brønsted acid sites). Lewis acid sites can be formed by either the dealumination of H-form zeolites or introduction of metal species into zeolites, in which extra-framework Al (EFAL) and metal species act as the Lewis acid sites [1, 2]. The catalytic performances and adsorption behavior of zeolites can be improved by post-modification of as-synthesized zeolites. Thermal or hydrothermal treatment of H-type zeolites could largely improve not only their stability but also their catalytic activity [3, 4]. Bifunctional zeolite catalysts (such as Zn/ZSM-5 and Mo/ZSM-5) exhibit significantly higher catalytic performances for alkane activation and conversion compared to conventional zeolite catalysts [5–7]. Both the catalytic activity and selectivity of zeolites in various heterogeneously catalyzed processes are strongly dependent on their acidic property as well as their pore structure [8–10]. All these fundamental issues could be comprehensively addressed through the characterization of zeolites, investigation of acidic properties of zeolites, and elucidation of catalytic reactions via various advanced spectroscopic techniques.

With the development of both solid-state NMR methodologies and high-field spectrometers, solid-state NMR is emerging as an important tool for the determination of structures and dynamic behaviors of multifunctional materials [11–15]. In the past decades, various *state-of-the-art* solid-state NMR techniques facilitate the investigation of the framework structure of zeolites and elucidation of catalytic reaction mechanisms to clarify the structure–property relationship in zeolites [16–20]. Here, the applications of solid-state NMR techniques to characterization of the framework structures and acidic properties of zeolites and to elucidation of crystallization process and catalytic reaction mechanisms of zeolites were briefly introduced.

7.2 Structural Characterization of Zeolites

7.2.1 ^{1}H MAS NMR

^{1}H MAS NMR can provide direct information about various hydroxyl groups in zeolites [21–24]. The chemical shift ranges for the Brønsted acidic proton, nonacidic SiOH, and AlOH groups in various zeolites were comprehensively summarized by Hunger et al. [18] (see Table 7.1). In addition, their concentrations could be directly obtained by measuring the integrated areas of the corresponding ^{1}H signals. In the ^{1}H MAS NMR spectrum of parent HY zeolite (Fig. 7.1a), two signals at 5.0 and 4.3 ppm are assigned to bridging SiOHAl groups (Brønsted acid sites) in the sodalite and the supercage of HY zeolite, respectively. The peak at 2.2 ppm is due to nonacidic SiOH groups. For dealuminated HY (Fig. 7.1b), the resonances at 2.8 and 1.0 ppm are associated with two different types of AlOH hydroxyl groups.

Table 7.1 Assignments of ^{1}H NMR chemical shift for various hydroxyl groups in dehydrated zeolites [18]

δ_{1H}/ppm	Hydroxyl group	Assignment
1.2–2.2	SiOH	Silanol groups on the external surfaces or lattice defect sites
0.6–3.6	AlOH	Non-framework Al hydroxyl groups in the pore channels
3.6–4.3	SiOHAl	Bridging hydroxyl groups in large channels
4.6–5.2	SiOHAl	Bridging hydroxyl group in small channels

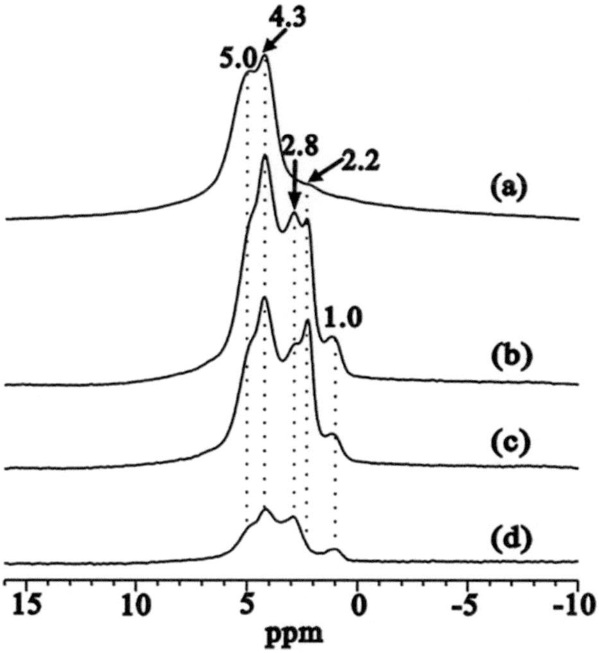

Fig. 7.1 ^{1}H NMR spectrum of (**a**) HY, dealuminated HY without ^{27}Al irradiation (**b**) and with ^{27}Al irradiation (**c**), and difference spectra (**d**) of (b) and (c) (Reprinted with permission from Ref. [26]. Copyright 2007 American Chemical Society)

Under on-resonance ^{27}Al irradiation in a TRAPDOR experiment [25], apart from the signals at 5.0 and 4.3 ppm, the signals at 2.8 and 1.0 ppm are largely reduced as well (see Fig. 7.1c, d), indicating that these hydroxyl groups are in close proximity to aluminum and are due to extra-framework AlOH groups [26]. ^{1}H MAS NMR has been extensively used to characterize various hydroxyl groups in various types of zeolites such as H–mordenite, H–ZSM-5, H–beta, etc. [27, 28].

7.2.2 ^{27}Al MAS NMR

Zeolite framework consists of SiO_4 and AlO_4 tetrahedra linked by oxygen atoms. These tetrahedra can link together by their corners to form cages or channels, which are consisting of different types of zeolite topology. Solid-state ^{27}Al and ^{29}Si MAS NMR have been widely used to characterize the local environments in zeolites [29–33]. The coordination environments of aluminum species in zeolite could be clearly distinguished from ^{27}Al MAS NMR spectroscopy. Figure 7.2 shows the ^{27}Al MAS NMR spectra of hydrated H–Y zeolites with calcination treatments at various designated temperatures. In the ^{27}Al MAS NMR spectrum of parent hydrated H–Y zeolite as shown in Fig. 7.2a, only a single peak at 60 ppm arising from four-coordinated framework Al species can be observed [34]. Whereas in calcined H–Y zeolite as shown in Fig. 7.2b, d, additional resonances at 30 and −2 ppm are assigned to five- and six-coordinated EFAL species, respectively. The signals of five-coordinated and six-coordinated EFAL species gradually grow up at the expense of the signal of four-coordinated framework Al with the increase of calcination temperature, indicating a progressive increase in the extent of dealumination [34].

For non-hydrated zeolites, spin echo is usually applied to collect the NMR signals from various Al species [35–37]. A series of steam-treated dealuminated H–Y

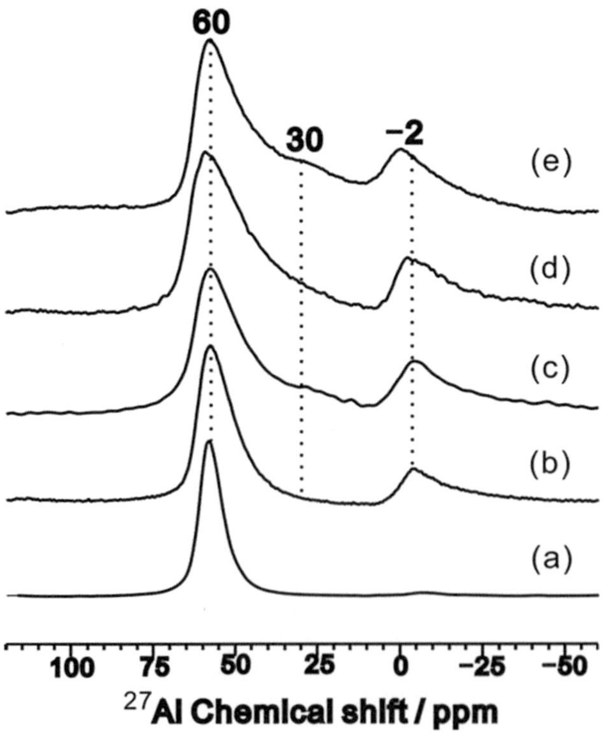

Fig. 7.2 ^{27}Al MAS NMR spectra of parent H–Y (**a**) and dealuminated H–Y zeolites upon calcination treatment at 400 (**b**), 500 (**c**), 600 (**d**), and 700 °C (**e**) (Reprinted with permission from Ref. [34]. Copyright 2008 American Chemical Society)

zeolites in anhydrous state were thoroughly investigated by the spin-echo method [36]. (see Fig. 7.3) Signal 1 with a QCC (quadrupole coupling constant) value of 14.4–15.0 MHz is assigned to the tetrahedrally coordinated framework Al of the bridging hydroxyl (SiOHAl) groups, whereas the signal 2 with a QCC value of 6.4–9.1 MHz is due to the tetrahedrally coordinated framework Al species which is compensated by residual sodium cation. It is revealed that the relative signal intensity of framework Al species on the bridging OH groups decreases dramatically with increasing severity of dealumination [36].

7.2.3 ^{27}Al MQ MAS NMR

Multiple-quantum magic-angle spinning (MQ MAS) is a solid-state NMR method for obtaining a high-resolution spectrum of half-integer spin quadrupolar nuclei [38, 39]. It is performed as a two-dimensional NMR experiment, establishing the correlations between isotropic (high resolution) dimension and anisotropic central

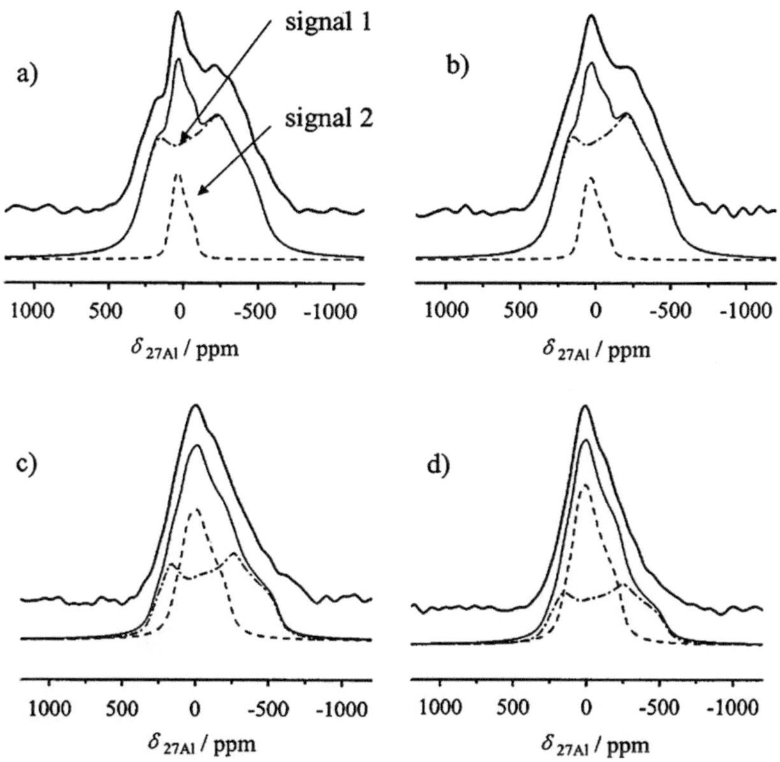

Fig. 7.3 ^{27}Al spin-echo NMR spectra of anhydrated H-Y zeolites' (**a**) parent sample, (**b**) deH–Y/7.4, (**c**) deH–Y/31.1, and (**d**) deH–Y/81.5. Experimental spectra (*top*) are compared with simulated spectra (*bottom*) (Reprinted with permission from Ref. [36] Copyright 2004 American Chemical Society)

transition MAS dimension. 2D ^{27}Al MQ MAS NMR can provide a more resolved pattern compared with 1D ^{27}Al MAS NMR after averaging out the second-order quadrupole interaction. ^{27}Al MQ MAS NMR was employed to investigate the variation of the chemical environment of Al species in zeolites upon calcination treatment [40]. Figure 7.4 shows the ^{27}Al MQ MAS spectra of H–Y zeolites calcined at different temperatures. Four types of aluminum species can be clearly distinguished, which can be assigned to four-coordinated framework Al, six-coordinated EFAL, five-coordinated EFAL, and four-coordinated EFAL, respectively [40]. As shown in Fig. 7.4, it is notable that the five-coordinated EFAL and four-coordinated EFAL are present at high calcination temperature [40]. The isotropic chemical shift δ_{iso} as well as the second-order quadrupolar interaction parameter P_Q of each aluminum species can be calculated according to the position of signals in the 2D MQ MAS spectra. Combined with ^{27}Al MAS NMR, ^{27}Al MQ MAS NMR has been extensively used to investigate the dealumination process, the water hydration process, and the aluminum states in various zeolites [41–43].

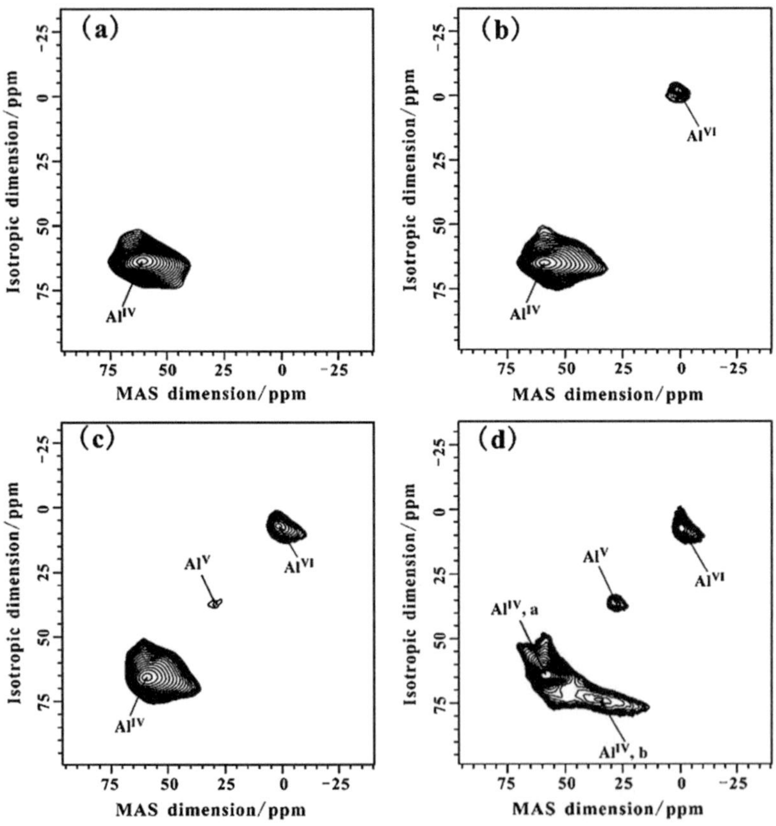

Fig. 7.4 ^{27}Al MQ MAS NMR spectra of parent HY zeolite (**a**) and dealuminated H–Y zeolites upon calcination treatment at 500 (**b**), 600 (**c**), and 700 °C (**d**) (Reproduced from Ref. [40] by permission of John Wiley & Sons Ltd)

7.2.4 ^{29}Si MAS NMR

^{29}Si NMR chemical shifts normally vary from different chemical environments in the zeolite framework. Thus, the local framework structure can be reflected from the ^{29}Si NMR chemical shifts [44–48]. Figure 7.5 shows the ^{29}Si MAS NMR spectrum of H–Y zeolite upon calcination treatments at different temperatures [34]. As shown in Fig. 7.5, four different peaks at −89, −95, −101, and −106 ppm arising from Si[3Al, 1Si], Si[2Al, 2Si], Si[1A1,3Si], and Si[0Al,4Si] could be clearly resolved. As suggested by Lippmaa et al. [44], the framework Si/Al ratio could be quantitatively determined after careful deconvolution of these four different peaks. The corresponding framework Si/Al ratios were determined to be 2.8, 3.0, 3.5, 4.6, and 5.3 for parent HY, CAL-400, CAL-500, CAL-600, and CAL-700 zeolites, respectively [34]. In addition, ^{29}Si MAS NMR is well suited for observing the chemically distinct and the crystallographic nonequivalent silicon sites in a topologically ordered framework of zeolite and the mapping of various sites to specific tetrahedral locations in the unit cell [48, 49]. Figure 7.6 shows the ^{29}Si NMR spectra of nitrogen substituted HY zeolites. Using ^{29}Si MAS NMR spectra in conjunction with quantum chemical calculation, the ^{29}Si resonances at −89, −80, and −68 ppm have been ascribed to $SiNH_2Al$, Al–OH–Si–NH_2–Al, and $Si(NH_2Al)_2$, respectively. The

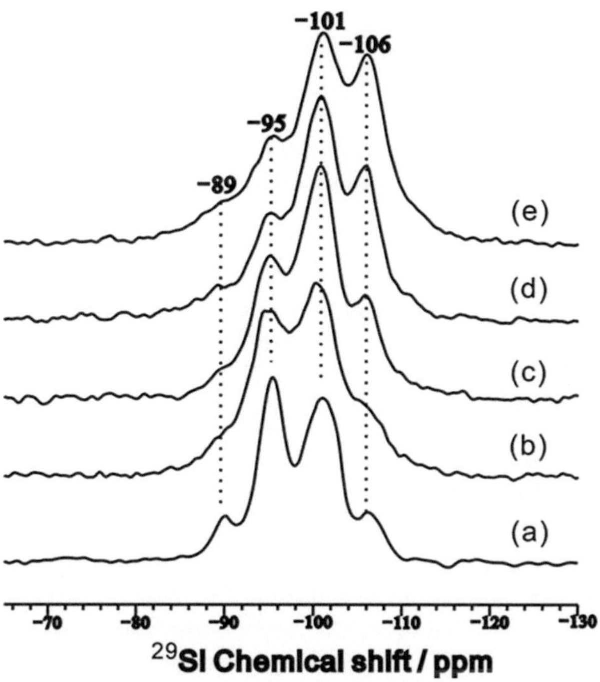

Fig. 7.5 ^{29}Si MAS NMR spectra of parent H–Y (**a**) and dealuminated H–Y zeolites upon calcination treatment at 400 (**b**), 500 (**c**), 600 (**d**), and 700 °C (**e**) (Reprinted with permission from Ref. [34]. Copyright 2008 American Chemical Society)

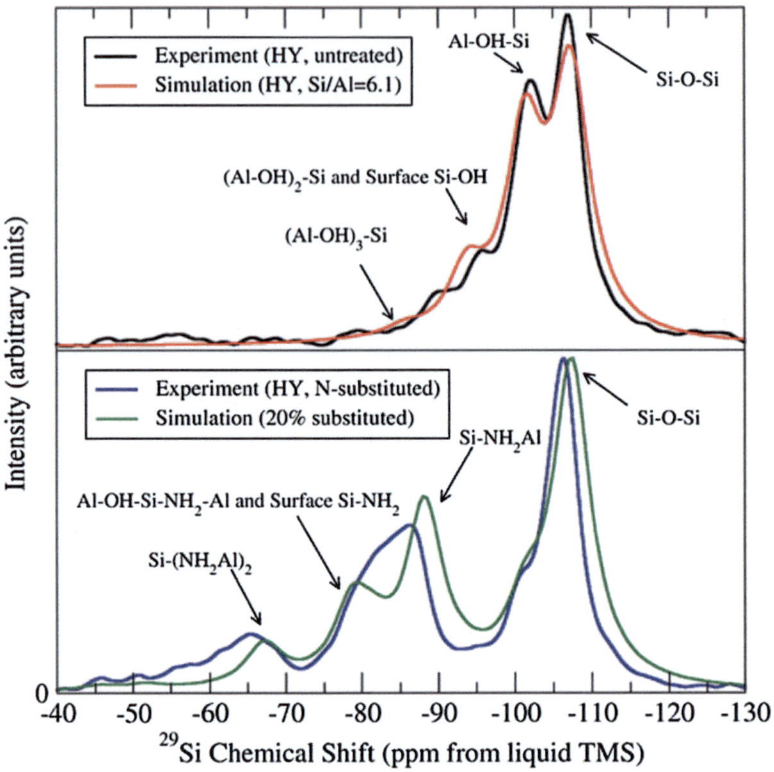

Fig. 7.6 ^{29}Si NMR spectra of HY zeolite with 0 % (*top*) and 20 % (*below*) of oxygen substituted for NH groups (Reprinted with permission from Ref. [50] Copyright 2008 American Chemical Society)

framework structure of newly designed N-doped zeolites could be elucidated accordingly [50]. For the pure siliceous zeolite, the ^{29}Si NMR chemical shift strongly depends on chemical environment of different silicon sites, which could be further verified from DFT theoretical calculations [51].

7.2.5 ^{29}Si DQ MAS NMR

The intramolecular distance constraints are of great importance to solve the 3D topology structure of zeolites. To establish the ^{29}Si–^{29}Si correlations and probe the long-range Si–Si distances in the zeolite frameworks, Levitt and coworkers introduced a new ^{29}Si solid-state MAS NMR experiment by incorporating the homonuclear dipolar recoupling sequence $SR26_4^{11}$ [52, 53]. In the 2D ^{29}Si DQ MAS NMR spectrum of a zeolite sigma-2 with natural abundance as shown in Fig. 7.7a, the spatial proximity among different ^{29}Si sites has been demonstrated. In addition, long-range ^{29}Si–^{29}Si distances of different Si pairs can be elucidated from the ^{29}Si DQ

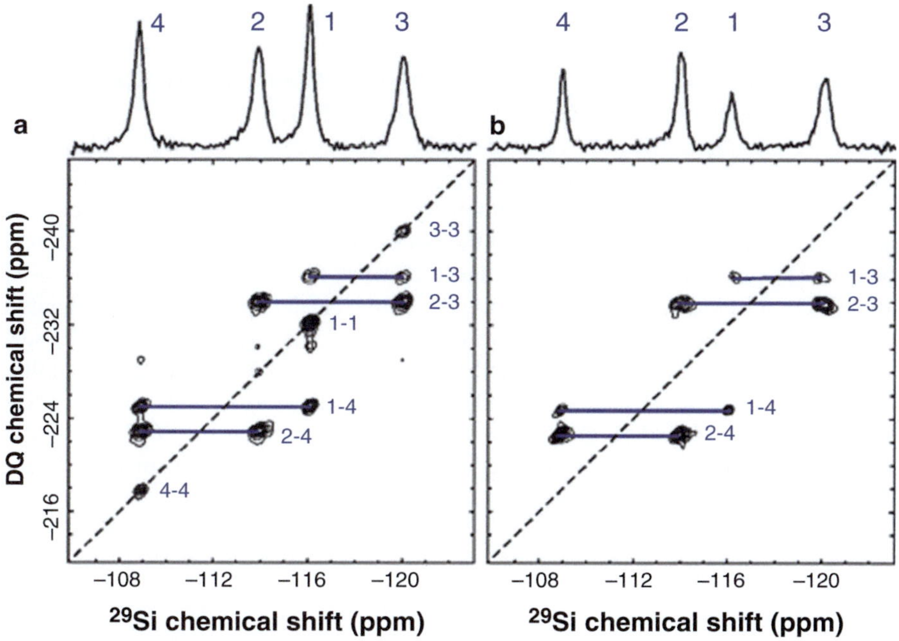

Fig. 7.7 ^{29}Si DQ correlation spectra of sigma-2 obtained with (**a**) the $SR26_4^{11}$ dipolar recoupling sequence and (**b**) the J-coupling-based refocused INADEQUATE experiment (Reprinted with permission from Ref. [53]. Copyright 2005 American Chemical Society)

built-up curves as a function of homonuclear recoupling time [53]. Moreover, as manifested in Fig. 7.7b, the two-bond ^{29}Si–^{29}Si connectivity in sigma-2 could be obtained through 2D J-coupling-based refocused INADEQUATE experiment [54]. Both the spatial proximity and bond connection information can be successively achieved separately from the ^{29}Si–^{29}Si correlations through dipolar and J-coupling interactions [53]. By incorporation with the unit cell parameters and space group from the powder XRD, the 3D topology structure of zeolites can be accurately determined based on the long-range and short-range ^{29}Si–^{29}Si distance constraints which were obtained from the 2D ^{29}Si DQ MAS NMR experiments. It was also demonstrated that the structure of two purely siliceous zeolites could be solved by the 2D ^{29}Si DQ MAS NMR technique [53]. This method opens up a new window for high-resolution structure determination of zeolites by solid-state NMR spectroscopy.

7.2.6 ^{17}O MAS NMR

^{17}O MAS NMR has been used to observe the zeolite framework oxygen signals since the 1980s [55, 56]. The framework oxygen sites (Si–O–Si) could be directly detected from ^{17}O MAS and MQ MAS NMR techniques. However, the ^{17}O signal of the Brønsted acid ^{17}O site (Al–OH–Si) cannot be clearly distinguished due to its relatively low concentration. To detect the ^{17}O site in HY, H–ZSM-5, and H–mordenite zeolites, several double-resonance solid-state NMR techniques including ^{1}H–^{17}O TRAPDOR [25], ^{17}O–^{1}H REDOR [57], ^{1}H–^{17}O REDOR, and ^{1}H–^{17}O HETCOR experiments were utilized by Grey and coworkers [58–61].

Figure 7.8 shows the $^{17}O\{^{1}H\}$ REDOR NMR spectra of ^{17}O-enriched HY zeolite. The intensity of the shoulder at −24 ppm decreases significantly in the double-resonance experiment, whereas the intensity of the broad peak at 21 ppm remains almost unchanged [58]. According to the double-resonance experiment, the broad peak at 21 ppm could be assigned to the framework oxygen atoms in the Si–O–Al and Si–O–Si linkages, whereas the shoulder signal at −24 ppm could be associated to the oxygen site in Brønsted acid site (Si–OH–Al) [58]. ^{1}H–^{17}O CP–REDOR NMR experiment was further applied to determine the O–H distance in zeolite HY, which is around 0.98–1.01 Å [60]. The Brønsted acid sites in other zeolites such as H–ZSM-5 and H–mordenite [61] were also studied by ^{17}O MAS NMR spectroscopy, which provides insights into the active center in acid-mediated zeolite catalysts.

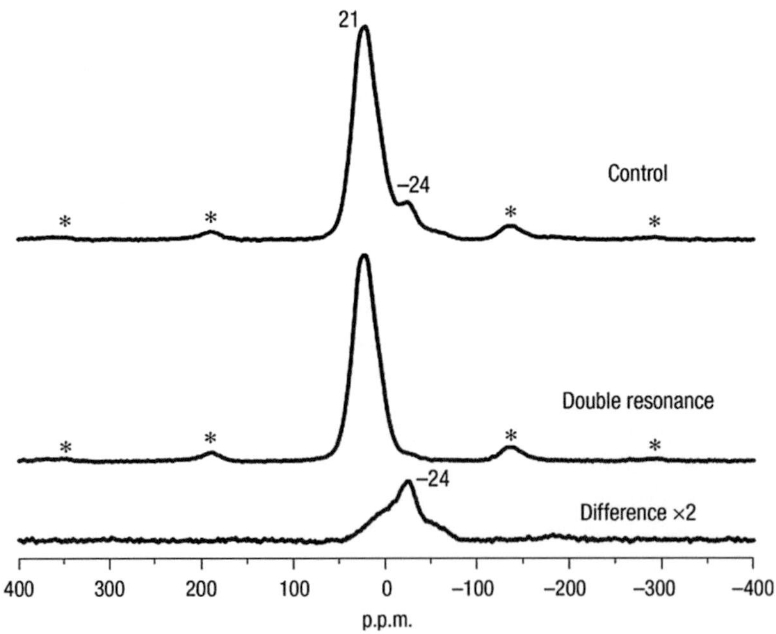

Fig. 7.8 ^{17}O–^{1}H REDOR NMR spectra of ^{17}O-enriched HY zeolite. The difference spectrum is obtained by subtracting the double-resonance spectrum from the control spectrum (Reprinted by permission from Macmillan Publishers Ltd: Ref. [58], copyright 2005)

7.2.7 ^{129}Xe NMR

The structure of cages and channels of zeolites play key roles in their adsorption and desorption behavior and their catalytic activity. ^{129}Xe NMR chemical shifts are very sensitive to the pore geometry, which can serve as a probe for studying the microstructure of zeolites [62–64]. For xenon adsorbed in a porous material, the observed ^{129}Xe chemical shift is a weighted average of different types of interactions on NMR time scale. ^{129}Xe NMR has been used to characterize the pore structure and adsorption properties in various zeolites such as MCM-22, MAS-7, ZSM-5, etc. [65–68, 6].

By using ^{129}Xe NMR, Chen et al. [65] found that Xe atoms were preferentially adsorbed in the supercages of zeolite MCM-22 at low Xe pressure, while Xe atoms could penetrate into the two-dimensional sinusoidal channels at high Xe pressure. As Xe adsorption pressure is greater than 6 atm, two different kinds of xenon exchange could be observed. Exchange of xenon at different adsorption sites in the same supercage occurs at lower temperature ranges of 170–122 K (see Fig. 7.9). Exchange between xenon in the supercage and the sinusoidal channel would take place through gaseous xenon at higher temperatures (280–350 K) as shown in Fig. 7.9. The ^{129}Xe NMR results provided some new insights into the adsorption properties of the MCM-22 zeolite [65]. Since the chemical shift of the xenon is very sensitive to the types of atoms with which it collides, ^{129}Xe NMR is especially successful in investigating charge, location, and state of cations in the metal-exchanged zeolites [6, 67].

7.3 Acidity Characterization of Zeolites

7.3.1 Acidity Type and Strength

The acidity of zeolites dictates their catalytic performance during various hydrocarbon reactions [69]. Thus, a comprehensive investigation of the acidic properties is crucial for the design, modification, and application of zeolites. The most important features during the acidity characterization lie in the precise determination of the acid type (Brønsted vs Lewis), strength, concentration, and distribution (or location) of acid sites.

The utilization of probe molecules was widely used to investigate the acidity of zeolites. Pioneered by Lunsford et al. [70], trimethylphosphine (TMP) was adopted as a probe molecule to characterize the acidity of H-form faujasite-type zeolites based on the observed ^{31}P NMR chemical shifts. Thereafter, the technique has been widely used for acidity characterization of other zeolites [28, 71, 72]. As a TMP molecule adsorbed onto a Brønsted acid site, the formation of a $TMPH^+$ ionic

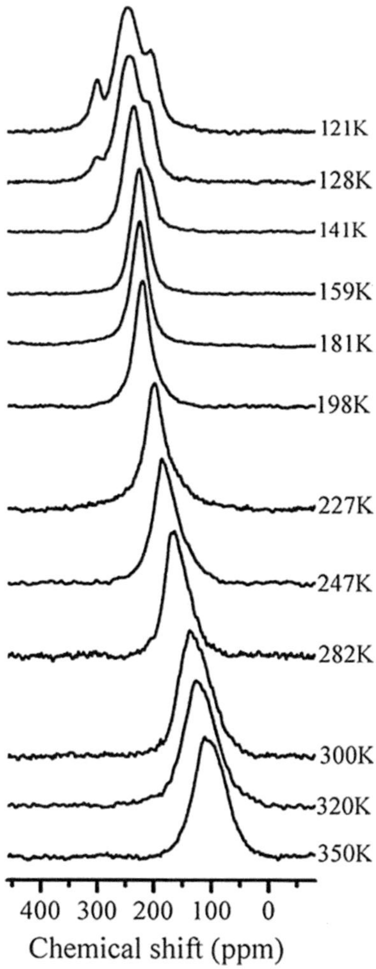

Fig. 7.9 Temperature-dependent ^{129}Xe NMR spectra of Xe adsorbed in the H–MCM-22 zeolite with a Xe pressure of 6.0 atm (Reprinted with permission from Ref. [65]. Copyright 2001 American Chemical Society)

complex would give rise to a ^{31}P chemical shifts ranging from ca. −1 to −6 ppm, whereas TMP molecules bound to Lewis acid sites normally result in ^{31}P chemical shifts in the range of −30 to −70 ppm [70]. Therefore, Brønsted and Lewis acid sites in zeolites can be clearly distinguished, and their concentrations can be quantitatively determined. Recent DFT calculation results indicate that the ^{31}P NMR chemical shifts of adsorbed TMP molecules can be used as a scale to quantitatively measure the Lewis acid strength, i.e., the Lewis acid strength linearly increases with the increase of the ^{31}P NMR chemical shift [73].

In order to accurately determine the Brønsted acid strength of zeolites, several spectroscopic and analytical methods such as IR spectroscopy, temperature-programmed desorption (TPD), and microcalorimetry have been employed. However, in the conventional NH_3–TPD and pyridine IR experiments, ammonia and pyridine probe molecules are normally too basic to distinguish the differences of

Brønsted acid strengths of zeolites. It is very promising to characterize the acid strength of zeolites by utilizing solid-state MAS NMR spectroscopy of various probe molecules. ^{13}C MAS NMR of adsorbed 2-^{13}C-acetone [74–76], ^{31}P MAS NMR of adsorbed trimethylphosphine oxide (TMPO) [77, 78], and ^{1}H MAS NMR of adsorbed pyridine-d_5 [79, 27, 80] were used to characterize the acidity of various zeolites.

Besides the experimental observations, theoretical DFT calculations were recently employed to establish the correlation between the ^{1}H, ^{13}C, and ^{31}P NMR chemical shifts of probe molecules and the acid strength of zeolites [81–88]. Proton affinity (PA) or deprotonation energy (DPE) can serve as an indicator for the intrinsic acid strengths of Brønsted acid sites. A smaller PA or DPE corresponds to a stronger Brønsted acid. In the DFT calculations, Brønsted acid sites with varying acid strengths were built by a series of 8 T zeolite models with varying terminal Si–H bond lengths. As such, proton affinities of Brønsted acid sites varying from 246.7 to 310.8 kcal/mol could be generated from these selected models, covering from weak, medium strong, strong, to super acid [83]. Furthermore, the ^{1}H, ^{13}C, and ^{31}P NMR chemical shifts could be calculated after optimizing the complex structures of the probe molecules adsorbed on the Brønsted acid sites, and thus, the correlations between the Brønsted acid strengths and NMR chemical shifts were established [83, 84, 82, 81].

For pyridine-d_5 probe molecule, it was found that the ^{1}H chemical shift of pyridinium ions decreases linearly with the decrease of PA or the increase of Brønsted acid strength, indicating that the ^{1}H chemical shift of adsorbed pyridine-d_5 can be used as a scale for quantitatively measuring the Brønsted acid strength [83]. Moreover, by taking a PA value of 250 kcal/mol as the threshold for superacidity, an observed ^{1}H chemical shift of 12.6 ppm could be inferred as the corresponding chemical shift threshold for pyridine-d_5 probe molecule (see Fig. 7.10), which is in well agreement with the experimental values. As pyridine-d_5 is adsorbed on a strong Brønsted acid site, a complete transfer of the Brønsted acidic proton to the pyridine molecule prevails; the pronounced shielding effect arising from the pyridine ring thus results in variation in the observed δ_{1H} toward a lower value [83]. The linear correlation between δ_{1H} and acidic strength based on the ^{1}H-pyridine-d_5 NMR approach affords quantitative assessment of Brønsted acid strength in solid acid catalysts.

For 2-^{13}C-acetone probe molecule, it was found that three adsorption conformations (hydrogen bonded, proton shared, and ion pair) exist, corresponding to different extents of proton transfer from the Brønsted acid site to the adsorbed acetone [84]. A correlation of three-broken lines was obtained for the ^{13}C chemical shift of acetone versus DPE values [84]. As shown in Fig. 7.11, trifolded correlation between δ_{13C} and DPE may be described as followed: Region I (DPE $<$ 270 kcal/mol) is associated with the formation of ion-pair adsorption complexes, corresponding to the presence of strong Brønsted acid sites. Region II (270 $\leq$ DPE $\leq$ 276 kcal/mol) may be related to proton-shared adsorption scheme with medium acidic strength, whereas Region III (DPE $>$ 276 kcal/mol) belongs to hydrogen-bonded adsorption scheme with relatively weak acidity. A ^{13}C chemical shift threshold of

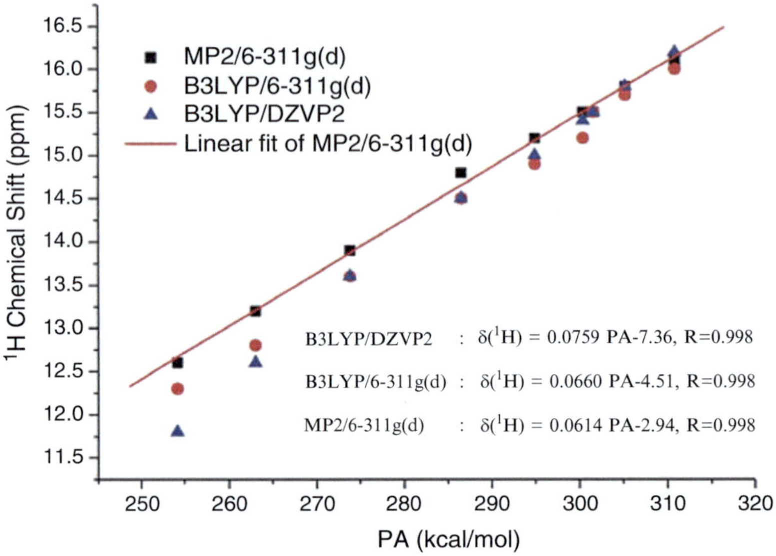

Fig. 7.10 Correlations of ^{1}H NMR chemical shift with proton affinity (PA) of Brønsted acid sites in solid acids while in the presence of adsorbed pyridine-d_5 (Reprinted with permission from Ref. [83]. Copyright 2007 American Chemical Society)

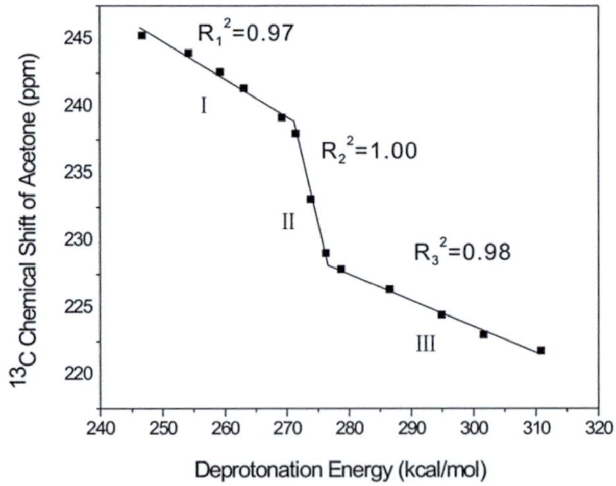

Fig. 7.11 Correlations of ^{13}C chemical shift of adsorbed 2-^{13}C-acetone with deprotonation energy (DPE) of Brønsted acidity in solid acid catalysts (Reprinted with permission from Ref. [84]. Copyright 2010 American Chemical Society)

ca. 245 ppm was obtained for superacidity in this case [84]. The correlation can be used as a scale for quantitatively measuring the Brønsted acid strength of various solid acid catalysts.

Similarly, the ^{31}P chemical shift of TMPO probe molecule also exhibits a linear correlation with Brønsted acid strength (see Fig. 7.12) [81]. In this case, the ^{31}P chemical shift increases linearly with the decrease of PA or the increase of Brønsted acid strength, a larger ^{31}P chemical shift corresponds a strong acid strength, and a ^{31}P chemical shift of 86 ppm was obtained as the threshold for superacidity [81]. It is noteworthy that the acidic scales have been extensively utilized to measure the acid strength of other solid acids [89–91], including heteropolyacids [92–94].

A recent study theoretically investigated the influence of acid strength on solid-acid-catalyzed protonation of olefins. This revealed that with increasing acid strength, the extent of energy reduction generally follows the order of the ionic character of these species. Ionic species are much more sensitive to acid strength than covalent species, regardless of intermediates or transition states. Therefore, an increase in acid strength can result in the preferential formation of ion-pair intermediates with greater stability than π complex and alkoxy species intermediates [95]. The rule that ionic species are more sensitive than covalent species to acid strength can be used to interpret and predict the effects of acid strength on catalytic activity for other solid-acid-catalyzed reactions [96]. Comprehensive studies have been carried out on the product selectivity and reaction pathway during alkane activation (hydrogen exchange, dehydrogenation, and cracking) and alkene hydrogenation reactions over zeolite catalysts with varied acid characteristics [96, 97]. It is well known that catalysts with greater acidic strength are more favorable for enhanced alkane activation during propane cracking, but hydrogen exchange reactions are not sensitive to the acidic strength of the catalyst, which is also well explained by the ionic character of the transition states [97].

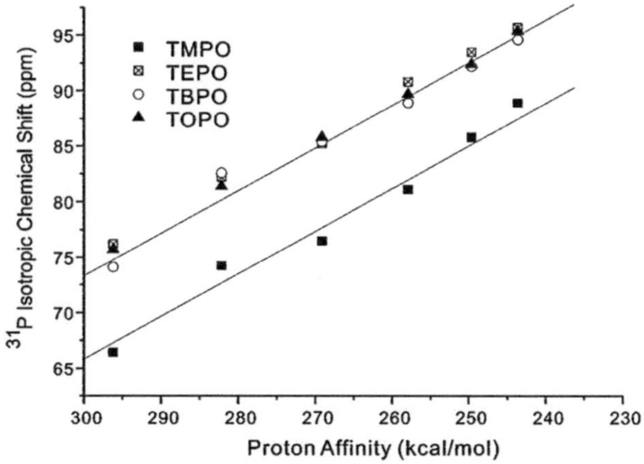

Fig. 7.12 Correlations of calculated ^{31}P chemical shift of adsorbed R_3POH^+ complexes and proton affinity (PA) (Reprinted with permission from Ref.[82]. Copyright 2008 American Chemical Society)

7.3.2 Acid Site Location

The characterization of acid site distributions is of great importance to understand the detailed mechanism of catalytic reactions occurring on different sites. In order to distinguish the acid sites located in the internal voids and on the external surfaces of zeolites, a solid-state ^{31}P NMR technique was proposed by using trialkylphosphine probe molecules with different sizes, such as trimethylphosphine oxide (TMPO) and tributylphosphine oxide (TBPO) [98]. The size of TMPO (kinetic diameter ca. 0.55 nm) is smaller than the typical pore aperture of the ten-membered ring (ca. 0.60 nm) of zeolite ZSM-5. The small size of TMPO enables it to diffuse into the intracrystalline channels and pores of the zeolite. Both the internal and external acid sites are accessible to TMPO, whereas the size of TBPO (ca. 0.82 nm) is too large to penetrate into the channels and can only detect acid sites located on the external surface of the zeolite. Therefore, the concentration of the internal acid sites can be obtained from the difference between those determined from TMPO and TMBO. With the aid of DFT calculation, it was demonstrated that the ^{31}P chemical shifts of various trialkylphosphine probe molecules (triethylphosphine oxide, TEPO; tributylphosphine oxide, TBPO; and trioctylphosphine oxide, TOPO) could be utilized as scales for quantitatively measuring the Brønsted acid strength, and a ^{31}P chemical shift of 92–94 ppm corresponded to the chemical shift threshold for superacidity (also see Fig. 7.12) [82]. Similarly, Zhang et al.[99] adopted perfluorotributylamine as a probe molecule and employed ^{1}H MAS NMR to distinguish the internal and external acidic sites in zeolites. By using DFT calculations and solid-state NMR of adsorbed TMPO, Zheng and coworkers [100] found that the accessible Brønsted acidic protons reside in both the supercages (at the Al8–OH–Si8 and Al1–OH–Si2 sites) and the external surface pockets (at the Al8–OH–Si8 site) rather than in the sinusoidal channels (at the Al5–OH–Si7 site) of zeolite H–MCM-22, with the Al1–OH–Si2 site having the strongest acid strength. This finding may partially explain the special selectivity of acid-catalyzed reactions occurring inside the channels of the zeolite.

7.3.3 Spatial Proximity Among Different Acid Sites

In order to better understand the possible synergy effect between Brønsted and Lewis acids, advanced 2D solid-state NMR methods, such as ^{1}H–^{1}H [26], ^{27}Al–^{27}Al [40], and ^{31}P–^{31}P [101] double-quantum (DQ) magic-angle spinning (MAS) NMR spectroscopy methods, have been developed and applied to investigate the spatial proximity among different acid sites in various zeolites.

2D ^{1}H DQ MAS NMR experiment is a robust technique for probing proton–proton proximities in various functional materials [102]. It was first employed by Li et al. to explore the spatial proximities among various acid sites in dealuminated HY zeolites [26]. As shown in Fig. 7.13, several types of correlation peaks can be clearly discerned

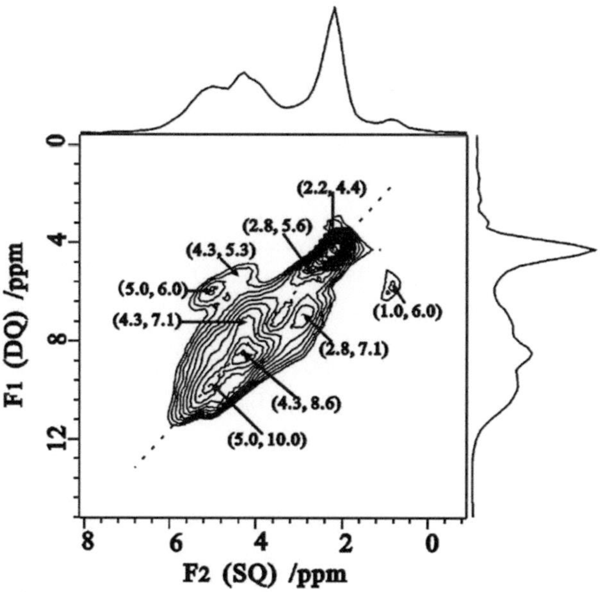

Fig. 7.13 ^{1}H DQ MAS NMR spectra of dealuminated HY (Reprinted with permission from Ref. [26]. Copyright 2007 American Chemical Society)

in the ^{1}H DQ MAS spectrum of dealuminated HY. The autocorrelation peaks at (4.3, 8.6) ppm and (5.0, 10.0) ppm suggest the spatial proximity of Brønsted acid sites in the supercage and sodalite cage, respectively. The autocorrelation peak at (2.2, 4.4) ppm results from the formation of silanol groups during the dealumination process. Another autocorrelation peak at (2.8, 5.6) ppm is due to EFAL species containing more than one hydroxyl group such as $Al(OH)_3$ and $Al(OH)_2^+$. Additionally, the off-diagonal peak pair at (1.0, 6.0) and (5.0, 6.0) ppm corresponds to the correlation between the extra-framework AlOH group and the bridging hydroxyl group in the sodalite cage, suggesting the spatial proximity between Brønsted and Lewis acid sites. The appearance of another off-diagonal peak pair at (2.8, 7.1) and (4.3, 7.1) ppm confirms the spatial proximity between the Lewis and Brønsted acid sites in the supercage [26].

Meanwhile, ^{1}H–^{1}H DQ MAS NMR experiments were performed with varying DQ recoupling time to determine the average ^{1}H–^{1}H distance between various hydroxyl groups [34]. Accordingly, the ^{1}H–^{1}H distance between a Brønsted acidic proton and an extra-framework AlOH species in both the supercage and the sodalite cage was determined to be 4.3 Å, whereas the average distance between two nearby Brønsted acidic protons in the supercages was measured to be 5.0 Å. By using the similar techniques, a comprehensive study was carried out to probe the spatial proximities between different acid sites in H–Y zeolites modified with different dealumination treatments (including calcination, steaming, and acid leaching) [26]. It was found that the Brønsted/Lewis acid synergy effect was always present in the samples prepared by thermal and hydrothermal treatments but absent in the samples prepared by acid leaching treatment [34]. Furthermore, extra-framework aluminum

(EFAL) species in hydrated dealuminated HY zeolite were also investigated by ^{1}H–^{1}H DQ MAS NMR experiments [103]. The ^{1}H–^{1}H DQ MAS NMR spectroscopy is also employed to investigate the spatial proximity of acid sites in highly siliceous zeolites, such as H–ZSM-5, H–mordenite, and H–MCM-22 [104, 105]. All these findings would provide insights into the roles of Lewis acid and its synergy with the Brønsted acid in zeolite-mediated hydrocarbon reactions.

Apart from ^{1}H–^{1}H DQ MAS NMR, ^{27}Al–^{27}Al DQ MAS NMR can also be employed to investigate the Brønsted/Lewis acid synergy in dealuminated zeolites as well [40]. By utilizing a sensitivity-enhanced 2D ^{27}Al DQ MAS NMR technique [106], the spatial proximities among various Al species in dealuminated HY zeolites were demonstrated.

Figure 7.14 shows 2D ^{27}Al DQ MAS NMR spectra of parent HY and calcined HY zeolites. For the parent HY as shown in Fig. 7.14a, the autocorrelation peak at

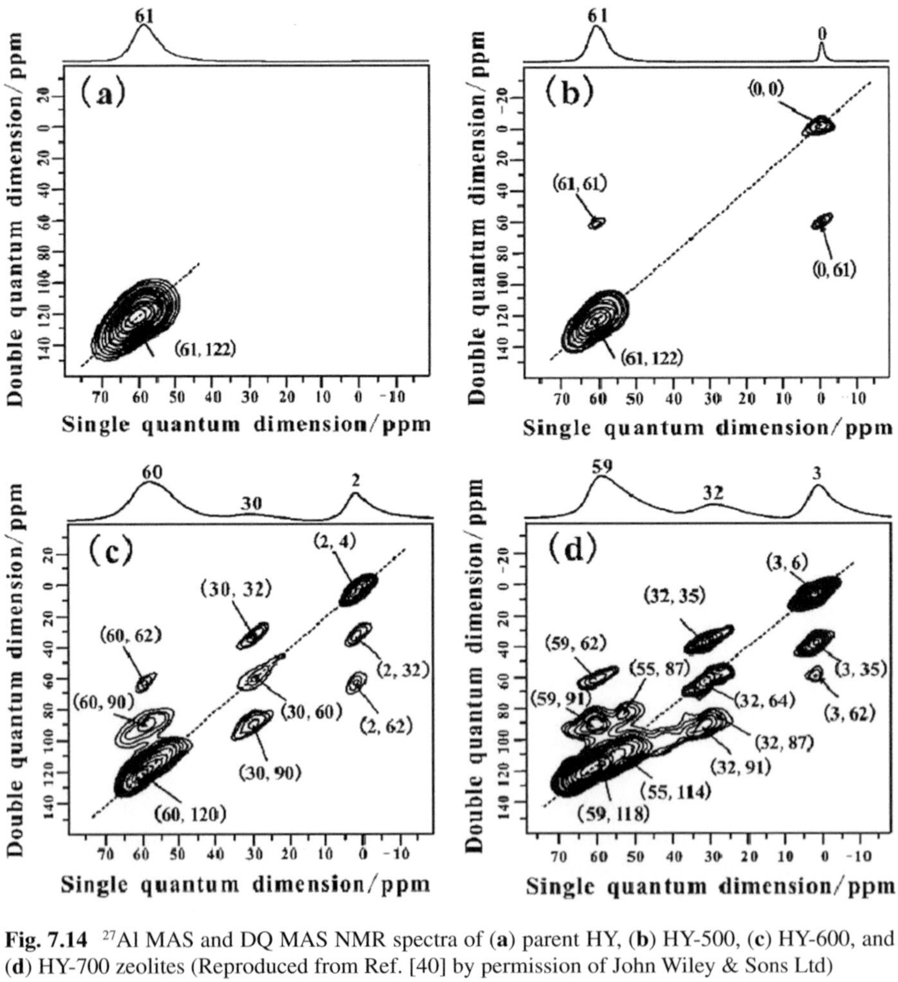

Fig. 7.14 ^{27}Al MAS and DQ MAS NMR spectra of (**a**) parent HY, (**b**) HY-500, (**c**) HY-600, and (**d**) HY-700 zeolites (Reproduced from Ref. [40] by permission of John Wiley & Sons Ltd)

(61, 122) ppm in the ^{27}Al–^{27}Al DQ MAS NMR spectrum indicates that four-coordinated framework Al (FAL) species are in close proximity to one another. For the HY-500 zeolite, the cross peak pair at (61, 61) and (0, 61) ppm results from the spatial proximity between the four-coordinated FAL and the six-coordinated EFAL (see Fig. 7.14b), implying the existence of Brønsted/Lewis acid synergy in the dealuminated HY zeolite [40]. For the HY-600 zeolite, it is revealed that three kinds of aluminum species including four-coordinated FAL, five-coordinated EFAL and six-coordinated EFAL are in close proximity one another (Fig. 7.14c). In the 2D ^{27}Al DQ MAS NMR spectrum of HY-700 (Fig. 7.14d), the correlation peak pair at (55, 87) and (32, 87) ppm is ascribed to the spatial proximity between four-coordinated EFAL species and five-coordinated EFAL species. On the basis of the experimental observations, the detailed spatial correlations among various aluminum species in hydrated HY zeolites after dealumination treatment are clearly identified [40]. With the assistance of DFT theoretical calculation, a new dealumination mechanism was proposed, and three types of EFAL species, $Al(OH)_3$, $Al(OH)_2^+$, and $Al(OH)^{2+}$, in close proximity to framework aluminum, were identified in dealuminated HY zeolites [40]. The spatial proximities of Brønsted and Lewis acid sites in the highly siliceous zeolites, such as H–MOR, H–ZSM-5, and MCM-22 zeolites, were also investigated by 2D ^{27}Al DQ MAS experiments [104, 105]. It was found that the Brønsted/Lewis acid synergy was present in these highly siliceous zeolites as well.

To measure the Brønsted acid densities in zeolites, Peng et al. [101] proposed a new method by using diphenyldiphosphines, $Ph_2P(CH_2)nPPh_2$ molecules having two basic sites. 2D ^{31}P DQ MAS NMR experiments were employed to probe ^{31}P–^{31}P internuclear distances and distinguish the non-protonated, single-protonated, or double-protonated diphenyldiphosphines binding sites in the zeolite. By using these techniques, the density and distribution of Brønsted acid sites in HY and H–ZMS-5 zeolites with different Si/Al ratios were comprehensively investigated [107].

7.4 Elucidation of Reaction Mechanisms in Heterogeneous Catalysis

Understanding the reaction mechanisms involved in heterogeneously catalyzed reactions is helpful to refine zeolite catalysts and design new ones. In order to elucidate the catalytic reaction mechanisms, the nature of reactant, intermediate, and product formed on the working catalysts during catalyzed reaction process needs to be clearly identified. The reaction pathway in heterogeneous catalysis involves in the several important stages: adsorption of the reactant molecules onto active sites, activation of the adsorption molecules, formation of possible reaction intermediates and products, and desorption of the product molecules from the catalyst [16]. The adsorption states of reactant, reaction intermediates, and final product formed on zeolite catalysts during reaction process can be identified from in situ MAS NMR spectroscopy [17, 108, 109]. Meanwhile, the thermodynamic behavior and kinetic property, strongly subject to the catalyzed performance, could be theoretically described on the basis of activation barriers and reaction energies derived from the reaction pathway calculations [110–112]. A combined in situ MAS NMR and theoretical DFT calculation study could provide deep insight into the structure–activity relationship of zeolites in heterogeneously catalyzed reactions [6, 113].

Wang et al. [114] studied the formation and decomposition of N, N, N-trimethylanilinium cations in the methylation of aniline on acidic zeolite catalysts through in situ ^{13}C MAS NMR spectroscopy under continuous-flow and stop-flow conditions. Figure 7.15 shows the formation of N, N, N-trimethylanilinium cations during methylation of aniline by methanol on H–Y zeolite performed under continuous-flow conditions. As shown in Fig. 7.15a, the signal at 50 ppm was due

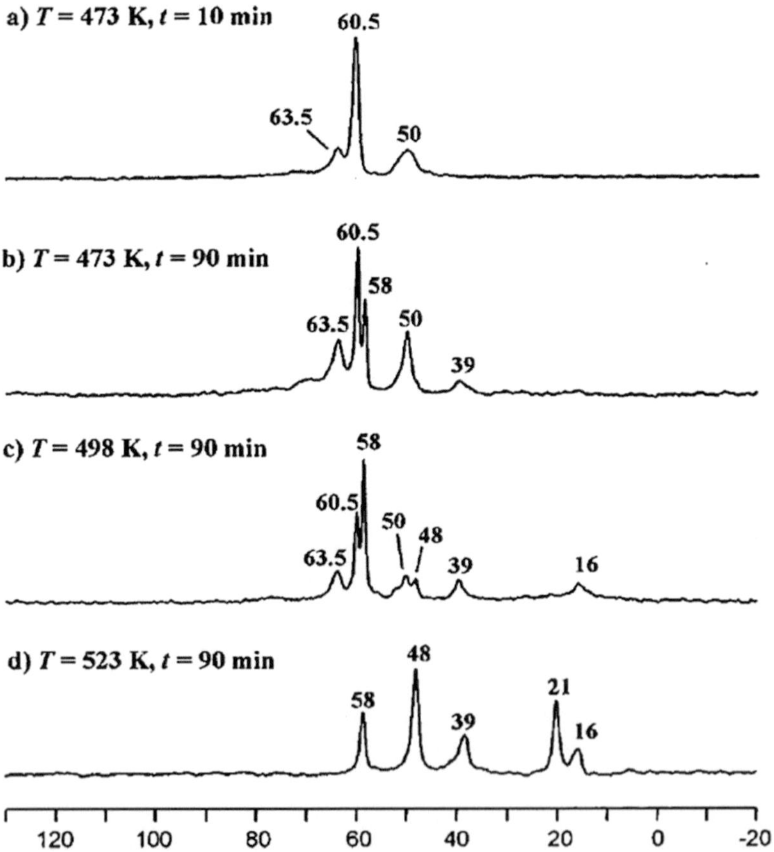

Fig. 7.15 In situ ^{13}C MAS NMR spectra recorded during aniline methylation on zeolite H–Y under continuous-flow conditions at reaction temperatures of 473 K for 10 min (**a**), 473 K for 90 min (**b**), 498 K for 90 min (**c**), and 523 K for 90 min (**d**) (Reprinted with permission from Ref. [114]. Copyright 2002 American Chemical Society)

to methanol adsorbed on Brønsted acid site, while the peaks at 63.5 and 60.5 ppm were ascribed to the formation of dimethyl ether (DME). After 90 min reaction at 473 K (Fig. 7.15b), a new signal appeared at 39 ppm, corresponding to the formation of N-methylanilinium cations ($[PhNH_2CH_3]^+$). After further increasing the reaction temperature, the intensity of the signal at 58 ppm due to N,N,N-trimethylanilinium cations increased, while signal intensities of methanol and DME decreased. During the aniline methylation reaction at temperatures of 498–523 K, other signals at 48 and 16 ppm were assigned to *N*, *N*-dimethylanilinium cations ($[PhNH(CH_3)_2]^+$) and ring-alkylated reaction products, respectively [114]. The role of N,N,N- trimethylanilinium cations in the methylation of aniline process was further investigated by stopped-flow ^{13}C MAS NMR techniques. As N,N,N-trimethylanilinium cations were formed on zeolite H–Y, the reactant flow was stopped and the progressive reaction of these cations on the catalyst was investigated by stopped-flow MAS NMR spectroscopy at different reaction temperatures.

On the basis of the results of in situ ^{13}C NMR investigations, a mechanism of aniline methylation on acidic zeolite H–Y was proposed (see Fig. 7.16) [114]. In the first step of the reaction, methanol is converted to DME and surface methoxy groups. Then, alkylation starts as a consequence of the formation of *N*-methylanilinium, *N*,*N*-dimethylanilinium, and *N*,*N*,*N*-trimethylanilinium cations on H–Y zeolite. The products of the N-alkylation of aniline, *N*-methylaniline, and *N*,*N*-dimethylaniline are further formed via deprotonation of the *N*-methylanilinium and *N*,*N*-*dimethylanilinium* cations [114].

Recently, in situ ^{13}C MAS NMR was also employed to monitor the carbonylation of methane with CO over a Zn-modified ZSM-5 zeolite catalyst [115]. Figure 7.17 shows the ^{13}C CP/MAS NMR spectra of products formed from carbonylation reaction of ^{13}C-labeled methane with CO on ZnZSM-5 catalyst heated at different temperatures for 1 h [115].

At 523 K, the signal at −7 ppm was assigned to unreacted methane, and the intense peak at 173 ppm was due to formate species (Fig. 7.17a). In addition, two

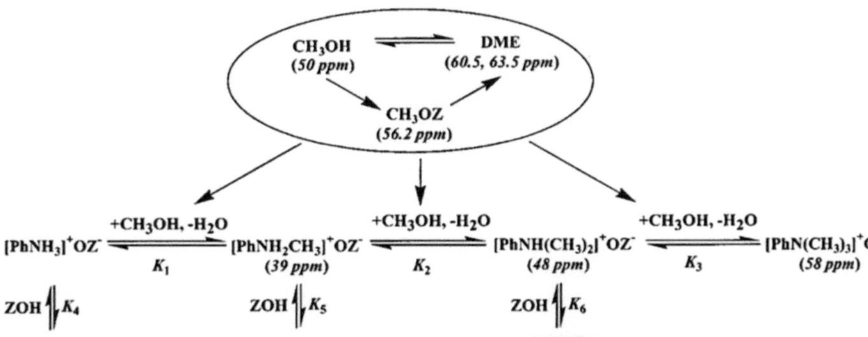

Fig. 7.16 Aniline methylation mechanism on zeolite H–Y (Reprinted with permission from Ref. [114]. Copyright 2002 American Chemical Society)

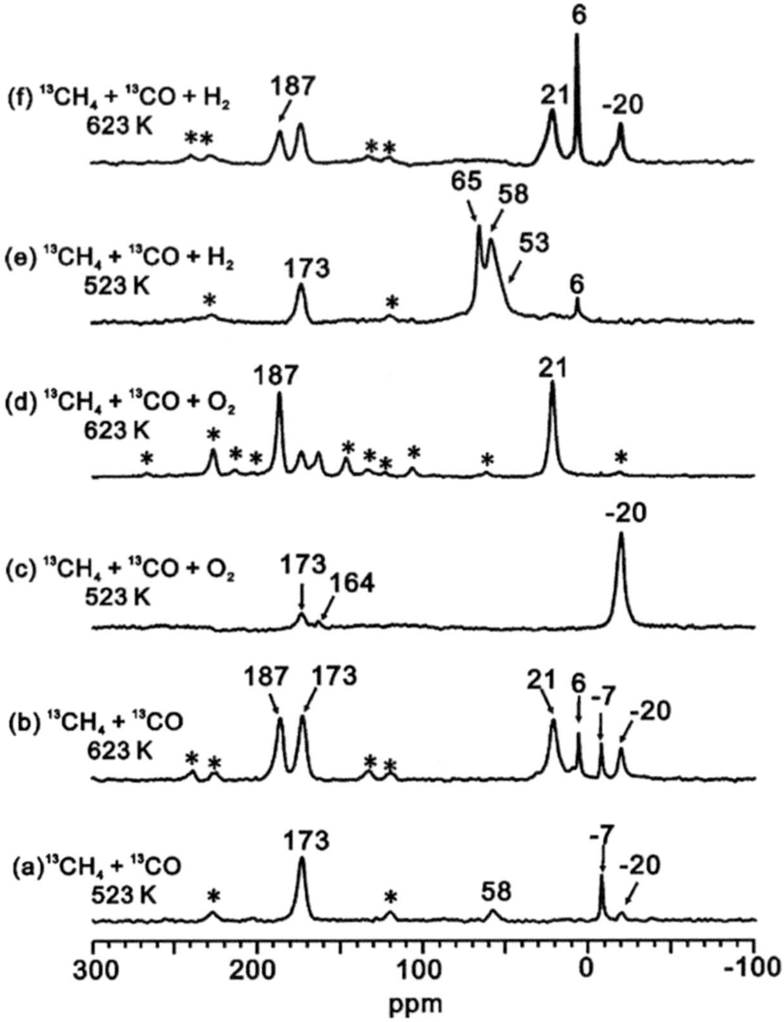

Fig. 7.17 ^{13}C CP/MAS NMR spectra of products formed from co-adsorption of CH_4 and CO on ZnZSM-5 catalyst heated for 1 h. (**a**, **b**) $^{13}CH_4$ and ^{13}CO at 523 K and 623 K. (**c**, **d**) $^{13}CH_4$, ^{13}CO, and O_2 at 523 K and 623 K. (**e**, **f**) $^{13}CH_4$, ^{13}CO, and H_2 at 523 K and 623 K (Reproduced from Ref. [115] by permission of John Wiley & Sons Ltd)

other weak signals were observable at 58 and −20 ppm, which could be assigned to surface methoxy species and zinc methyl species, respectively [115]. The appearance of methyl (21 ppm) and carbonyl groups (187 ppm) of acetic acid indicated the occurrence of the carbonylation reaction on the ZnZSM-5 catalyst at 623 K (see Fig. 7.17b). Interestingly, when a trace of molecular oxygen was added, the amount of zinc methyl species (−20 ppm) was significantly increased at 523 K while the methoxy species (58 ppm) disappeared (Fig. 7.17c). Meanwhile, a small amount of formate (173 ppm) and carbonate species (164 ppm) was observable as well. Further increasing the reaction temperature to 623 K resulted in selective formation of acetic acid (see Fig. 7.17d). In contrast, when a trace of H_2 was adsorbed, the signals of the methoxy species and its derivatives dimethyl ether (65 ppm) and methanol (53 ppm) were remarkably increased at 523 K (Fig. 7.17e). Acetic acid was formed along with the zinc methyl species and ethane (Fig. 7.17f). It was demonstrated that acetic acid could be generated directly under a mild reaction condition (573–623 K) through two parallel reaction pathways (see Fig. 7.18) [115]. In one reaction pathway as manifested in Fig. 7.18, methane is activated into methoxy intermediates, which can further interact with CO to generate acetic acid. In another reaction pathway, methane is activated into zinc methyl intermediates that can be consequently transformed into the methyl group of acetic acid with CO_2 through a typical organometallic reaction [115].

In combination with other spectroscopic tools such as GC–MS and UV–vis, in situ MAS NMR spectroscopy has been extensively used in the elucidation of the reaction mechanisms in various zeolite-mediated heterogeneously catalyzed reaction processes such as the activation of alkane C–H bonds by Brønsted acid sites over Ga-modified beta zeolite [116], co-aromatization of methane with propane on Mo-containing zeolite H–BEA [117], propane aromatization on Zn-modified BEA

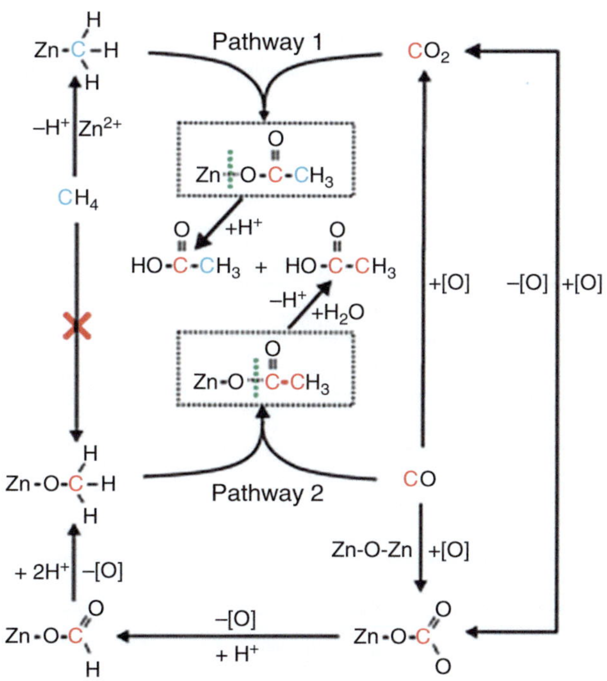

Fig. 7.18 Proposed reaction pathways for the formation of acetic acid from methane and carbon monoxide on ZnZSM-5 catalyst (Reproduced from Ref. [115] by permission of John Wiley & Sons Ltd)

zeolite [118], methanol-to-hydrocarbon conversion over SAPO-41 [119], methanol-to-olefin conversion on silicoaluminophosphate catalyst [120], and Beckmann rearrangement of cyclohexanone oxime on MFI-type zeolites [121]. *In-situ* solid-state NMR spectroscopy has also been employed to monitor the reaction intermediates such as alkoxy species [16] and carbenium ion intermediates [108, 122, 113] such as alkyl-substituted cyclopentenyl cations [123], methylindanyl carbenium ions [124], *gem*-dimethylbenzenium cations [125], and heptamethylbenzenium cation [126] in a variety of zeolite-catalyzed reactions.

In addition to the detection of reactants, intermediates, and products, the adsorption and desorption behavior and reaction kinetics in the reaction process can also be studied by in situ solid-state NMR spectroscopy. Xu et al. [127] proposed an approach of in situ continuous-flow laser-hyperpolarized (HP) ^{129}Xe MAS NMR in combination with ^{13}C MAS NMR to simultaneously obtain the adsorption states and the reaction kinetics on working zeolites. ^{13}C-labeled methanol and laser-hyperpolarized ^{129}Xe were introduced into CHA zeolite. During the catalytic reaction process, the reaction intermediates and products which were detected by in situ ^{13}C CP/MAS NMR experiments are helpful for the elucidation of reaction mechanisms. Meanwhile, the adsorption and reaction kinetics could be investigated by in situ HP ^{129}Xe MAS NMR. Figure 7.19 shows in situ HP ^{129}Xe MAS NMR and ^{13}C CP/MAS NMR spectra of methanol reaction in CHA nanocages at various temperatures.

In the ^{13}C CP/MAS NMR spectra as shown in Fig. 7.19a, only methanol signal with a chemical shift at 50.5 ppm was observed at temperature below 120 °C. By

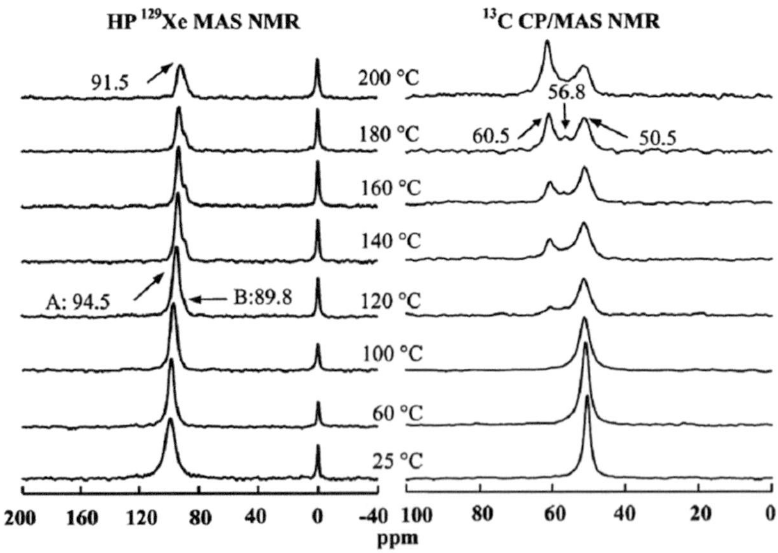

Fig. 7.19 In situ hyperpolarized (HP) ^{129}Xe MAS NMR and ^{13}C CP/MAS NMR spectra of methanol reaction in CHA nanocages at various temperatures (Reprinted with permission from Ref. [127] Copyright 2009 American Chemical Society)

increasing the temperature beyond 120 °C, methanol was converted into dimethyl ether (DME, 60.5 ppm) via methoxy species (56.8 ppm). In the HP ^{129}Xe NMR spectra as shown in Fig. 7.19b, the main peak A at 94.5 ppm was due to Xe coadsorbed with methanol in the CHA cages. The peak B at 89.8 ppm was ascribed to xenon in the methanol reaction cage, in which methanol molecule was converted to DME on the Brønsted acid site. By analysis of the ^{129}Xe MAS NMR spectra as a function of time, the evolution of methanol conversion in the CHA nanocages at different temperatures was identified. It was revealed that empty cage signal at 84 ppm decreased in intensity, while the signals of methanol adsorbed cage (92 ppm) and methanol reaction cage (88 ppm) increased with the time. By measuring the reaction kinetics at three different reaction temperatures, the activation energy of active site in methanol reaction could be calculated to be ca. 57 kJ/mol according to the Arrhenius equation [127]. This technique might be promising to investigate other reaction kinetics inside restricted zeolite nanocages.

7.5 Crystallization Mechanism of Zeolites

Solid-state NMR has been employed to characterize the synthesis processes of silicoaluminum, aluminophosphate, and silicoaluminophosphate molecular sieves [128, 129]. The detailed structural information on the evolution of intermediate phases is of great importance for better understanding the crystallization mechanisms of molecular sieves.

Figure 7.20 shows in situ ^{29}Si and ^{27}Al MAS NMR spectra of silicoaluminate zeolite A synthesis from gel at 65 °C as a function of crystalline periods [130]. ^{27}Al and ^{29}Si NMR were used to monitor the growth of the crystalline phase. As shown in Fig. 7.20a, a rapid increase in crystalline is evidenced from decreasing line width of the peak at −89 ppm assigned to Si(3Al) site as a function of the crystalline time. Meanwhile, the peak intensity increase of the $Al(OSi)_4$ at ~60 ppm at the expense of $[Al(OH)_4]^-$ species (Fig. 7.20b) indicates that $[Al(OH)_4]^-$ species is gradually transformed into $Al(OSi)_4$ and zeolite A framework is formed during the crystalline process [130]. Therefore, an ordering of the local environment of silicon tetrahedra and formation of aluminum tetrahedral $Al(OSi)_4$ can be clearly evidenced from the in situ ^{29}Si and ^{27}Al NMR spectroscopy [130].

$AlPO_4$-n is usually prepared by hydrothermal synthesis, and the crystalline molecular sieves are produced from the gel phases. Huang et al. [131] investigated the hydrothermal synthesis mechanism of $AlPO_4$-18 molecular sieves using multinuclear solid-state NMR spectroscopy. As shown in Fig. 7.21, the evolution of the local environments for various P and Al sites during the crystallization process could be manifested from ^{31}P and ^{27}Al MAS NMR. The states and the coordination of ^{31}P and ^{27}Al species can be tentatively assigned according to the NMR chemical shifts and line width. The assignments can be further confirmed by two-dimensional ^{27}Al MQ MAS and $^{27}Al/^{31}P$ CP-HETCOR experiments. Several representative intermediate phases could be clearly identified according to ^{27}Al and ^{31}P assignments as

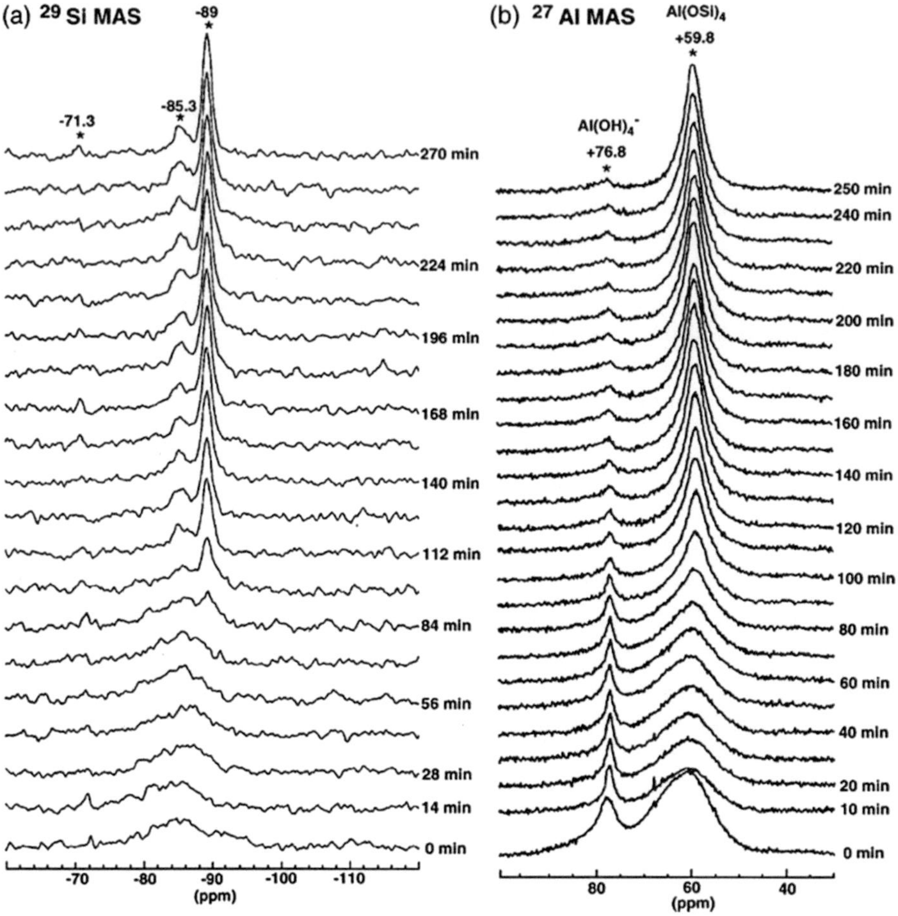

Fig. 7.20 (**a**) ^{29}Si and (**b**) ^{27}Al MAS NMR spectra of zeolite A synthesis from gel at 65 °C (Reprinted with permission from Ref. [130]. Copyright 1996 American Chemical Society)

a function of crystallization time. It could be concluded that the formation of $AlPO_4$-18 undergoes several distinct stages [131]. Firstly, the ordered aluminophosphate phase and a phosphate material were converted into an amorphous AlPO phase under hydrothermal conditions. Subsequently, crystalline microporous $AlPO_4$-5 was transformed from the amorphous phase after continuous heating. Finally, a further increase in heating time resulted in the formation of $AlPO_4$-18 [131].

The synthesis of zeolites usually requires the presence of solvents and organic templates. Very recently, Xiao et al. [132] developed a novel strategy for sustainable synthesis of zeolites without addition of both organotemplates and solvents. Zeolite beta has been successfully synthesized and thoroughly characterized by XRD patterns, SEM images, UV–Raman spectra, and ^{29}Si and ^{27}Al MAS NMR spectra.

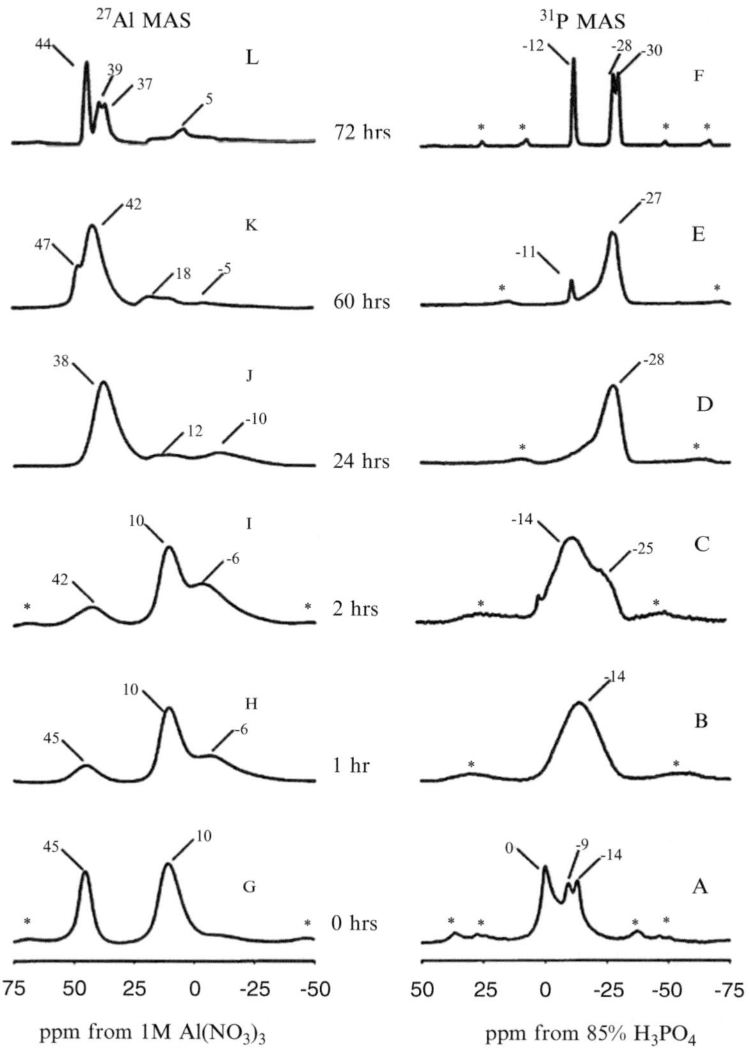

Fig. 7.21 ^{31}P and ^{27}Al MAS spectra of the selected gel samples as a function of crystallization process (Reprinted with permission from Ref. [131]. Copyright 2003 American Chemical Society)

Figure 7.22 shows ^{29}Si NMR spectra of synthesized beta zeolites as a function of crystallization time [132]. Before crystallization, the Si(2Si) signal is dominant. After crystallization for 2 days, Si(3Si) signal intensity increases at the expense of that of Si(2Si). Here, Si(2Si) includes Si(2Si, 2Al), Si(2Si,1Al,1OH), and Si(2Si, 2OH) structural units, while Si(3Si) includes Si(3Si, 1Al) and Si(3Si, 1OH) structural units. The observed phenomenon might be related to the rearrangement of zeolite building units. Meanwhile, a broad signal in the range of 50–70 ppm is observed in the ^{27}Al NMR spectrum, mainly attributed to Al(3Si) and Al(4Si) spe-

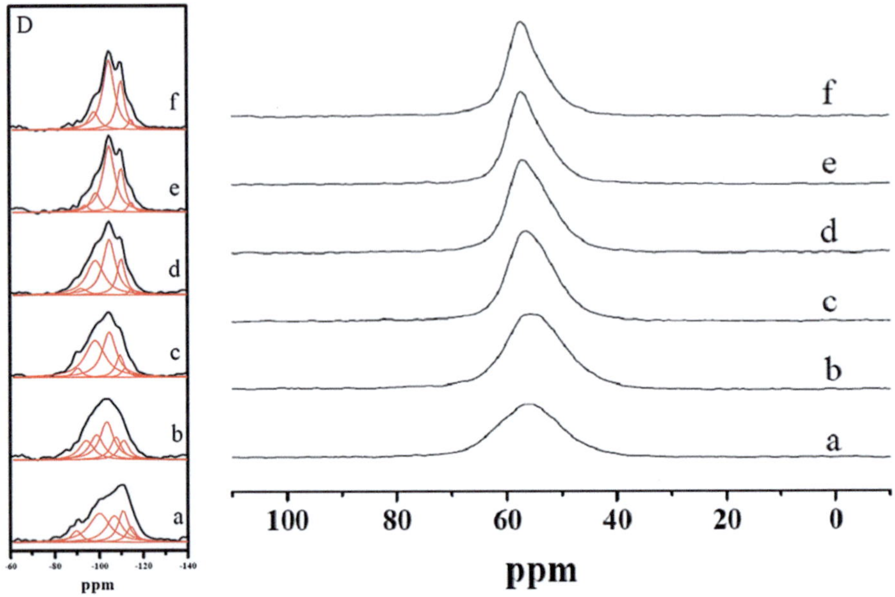

Fig. 7.22 ^{29}Si (*left*) and ^{27}Al (*right*) MAS NMR spectra of S–beta samples crystallized at (**a**) 0, (**b**) 2, (**c**) 4, (**d**) 5, (**e**) 7, and (**f**) 9 days, respectively (Reprinted with permission from Ref. [132]. Copyright 2014 American Chemical Society)

cies. It is noted that the concentration of Si(2Si) species is increasing with the crystallization time from 2 to 4 days while decreasing with the crystallization time from 4 to 9 days. The enrichment of Si(2Si) species should be assigned to the hydrolysis of silica species due to the presence of a little amount of water in the solid mixture, while the significant reduction of Si(2Si) species at longer crystallization time could be due to the crystallization of zeolites by condensation of Si(2Si) species containing hydroxyl groups. After crystallization for 9 days, the ^{29}Si NMR spectrum exhibits major peaks at −104.7 and −110.5 ppm associated with Si(3Si) and Si(4Si) species. Meanwhile, a major peak at 57 ppm assigned to tetrahedral coordination of framework Al is predominant in the ^{27}Al MAS NMR spectrum [132]. These results are well consistent with those of beta zeolite synthesized under hydrothermal conditions. Using similar methods, ^{31}P, ^{27}Al, and ^{29}Si MAS NMR spectroscopy was also used to characterize solvent-free as-synthesized S–SAPO-34 and calcined S–SAPO-34 zeolites [133].

Solid-state NMR spectroscopy has also been used in the investigation of crystallization mechanism of other microporous and mesoporous aluminophosphate or silicoaluminophosphate zeolites [134–136]. Huang et al. [137] directly monitored the involvement of ^{17}O-enriched water vapor in the formation of molecular sieve process of $AlPO_4$-11 by dry-gel conversion methods through ^{17}O solid-state NMR spectroscopy. Fluorine ion is a well-known mineralizer in the synthesis of aluminophosphate molecular sieves [138]. By using ^1H–^1H DQ MAS NMR spectroscopy, Shen et al. [139] proposed a model for surfactant aggregation and phase transformation during the crystalline process of mesostructured aluminophosphate materials on the basis of the spatial proximities among surfactant polar head groups, cetyl chains, and nonpolar tail CH_3 groups. Additionally, the crystallization of SAPO-34 molecular sieve under hydrothermal condition and dry-gel conversion condition was comprehensively examined by multinuclear ^{31}P, ^{27}Al, and ^{29}Si solid-state NMR techniques [134].

7.6 Conclusion and Outlook

Solid-state NMR spectroscopy has been demonstrated to be a powerful tool for the characterization of topology structures, crystallization processes, acidic properties, and catalytic reaction mechanisms of zeolites. In order to obtain deeper insights into the structure–property relationship of zeolites, it is still desirable to further develop solid-state NMR techniques. Enhancement of sensitivity and resolution in solid-state NMR spectroscopy would be achieved from various methods such as wide-band excitation [140], inverse detection [141] for the low-gamma nuclei [142], homo- and heteronuclear dipole recoupling and decoupling techniques via symmetry-based [143–145] and optimized-control [146] scheme under magic-angle spinning. Recently, utilization of ultrahigh-field NMR spectrometers has made the observation of a wide range of low-γ quadrupolar nuclei with low sensitivity feasible [147–149]. Even though progress to identify the low-γ active centers over zeolite catalysts has been gained by using high magnetic field solid-state NMR spectrometer [150], there are still lots of challenges and opportunities to design advanced solid-state NMR methods to enhance both sensitivity and resolution of the NMR signal of low-γ nuclei in real zeolite catalysts, in which the loading of active metal species is extremely low (usually less than 5 %). Nowadays, dynamic nuclear polarization (DNP) [151, 152] is favorably considered to be very promising techniques to remarkably enhance NMR signals in materials and biological systems. As external paramagnetic centers must be introduced, DNP technique has been only applied to study the surface property of mesoporous materials [153–157]. It is still a great challenge to study the heterogeneously catalytic reactions occurring in zeolites by using DNP technique since the externally introduced paramagnetic centers might cover the reaction active sites. The existence of various interactions in solids sometimes hinders the resolution and sensitivity of solid-state NMR spectroscopy. In this case, multiple-dimensional NMR spectroscopy is usually applied to gain enough sensitivity and resolution. However, the time-consuming drawback of the multiple-dimensional NMR techniques usually limits their wide application in zeolites. Thus, it is desirable to develop fast acquisition methods [158–161] in the multiple-dimensional NMR spectroscopy to remarkably reduce the acquisition time of NMR experiments, especially for the homonuclear and heteronuclear correlation experiments involving quadrupolar nuclei.

References

1. Coster D, Blumenfeld AL, Fripiat JJ (1994) Lewis acid sites and surface aluminum in aluminas and zeolites: a high resolution NMR study. J Phys Chem 98:6201–6211
2. Kiricsi I, Flego C, Pazzuconi G, Parker WO, Millini R, Perego C, Bellussi G (1994) Progress toward understanding zeolite-beta acidity: an IR and ^{27}Al NMR spectroscopic study. J Phys Chem 98:4627–4634
3. DeCanio SJ, Sohn JR, Fritz PO, Lunsford JH (1986) Acid catalysis by dealuminated zeolite-Y: I. Methanol dehydration and cumene dealkylation. J Catal 101:132–141
4. Beyerlein RA, McVicker GB, Yacullo LN, Ziemiak JJ (1988) The influence of framework and nonframework aluminum on the acidity of high-silica, proton-exchanged FAU-framework zeolites. J Phys Chem 92:1967–1970
5. Biscardi JA, Meitzner GD, Iglesia E (1998) Structure and density of active Zn species in Zn/H-ZSM-5 propane aromatization catalysts. J Catal 179:192–202
6. Xu J, Zheng A, Wang X, Qi G, Su J, Du J, Gan Z, Wu J, Wang W, Deng F (2012) Room temperature activation of methane over Zn modified H-ZSM-5 zeolites: insight from solid-state NMR and theoretical calculations. Chem Sci 3:2932–2940
7. Wu J-F, Yu S-M, Wang WD, Fan Y-X, Bai S, Zhang C-W, Gao Q, Huang J, Wang W (2013) Mechanistic insight into the formation of acetic acid from the direct conversion of methane and carbon dioxide on zinc-modified H–ZSM-5 zeolite. J Am Chem Soc 135:13567–13573
8. Breysse M, Afanasiev P, Geantet C, Vrinat M (2003) Overview of support effects in hydrotreating catalysts. Catal Today 86:5–16
9. Katada N, Igi H, Kim JH, Niwa M (1997) Determination of the acidic properties of zeolite by theoretical analysis of temperature-programmed desorption of ammonia based on adsorption equilibrium. J Phys Chem B 101:5969–5977
10. Wang Q, Cui ZM, Cao CY, Song WG (2011) 0.3 angstrom makes the difference: dramatic changes in methanol-to-olefin activities between H-ZSM-12 and H-ZSM-22 zeolites. J Phys Chem C 115:24987–24992
11. Brown SP (2009) Recent advances in solid-state MAS NMR methodology for probing structure and dynamics in polymeric and supramolecular systems. Macromol Rapid Commun 30:688–716
12. Morris K, Serpell L (2010) From natural to designer self-assembling biopolymers, the structural characterisation of fibrous proteins & peptides using fibre diffraction. Chem Soc Rev 39:3445–3453
13. Fu DW, Cai HL, Li SH, Ye Q, Zhou L, Zhang W, Zhang Y, Deng F, Xiong RG (2013) 4-Methoxyanilinium perrhenate 18-crown-6: a new ferroelectric with order originating in swinglike motion slowing down. Phys Rev Lett 110:257601
14. Zhang Y, Zhang W, Li SH, Ye Q, Cai HL, Deng F, Xiong RG, Huang SPD (2012) Ferroelectricity induced by ordering of twisting motion in a molecular rotor. J Am Chem Soc 134:11044–11049
15. Sun ZH, Luo JH, Zhang SQ, Ji CM, Zhou L, Li SH, Deng F, Hong MC (2013) Solid-state reversible quadratic nonlinear optical molecular switch with an exceptionally large contrast. Adv Mater 25:4159–4163
16. Wang W, Hunger M (2008) Reactivity of surface alkoxy species on acidic zeolite catalysts. Acc Chem Res 41:895–904
17. Zhang WP, Xu ST, Han XW, Bao XH (2012) In situ solid-state NMR for heterogeneous catalysis: a joint experimental and theoretical approach. Chem Soc Rev 41:192–210
18. Jiang YJ, Huang J, Dai WL, Hunger M (2011) Solid-state nuclear magnetic resonance investigations of the nature, property, and activity of acid sites on solid catalysts. Solid State Nucl Magn Reson 39:116–141
19. Zhang L, Ren YH, Yue B, He HY (2012) Recent development in in situ NMR study on heterogeneous catalysis: mechanisms of light alkane functionalisation. Chem Commun 48:2370–2384

20. Li S, Deng F (2013) Recent advances of solid-state NMR studies on zeolites. Annu Rep NMR Spectrosc 78:1–54
21. Freude D, Hunger M, Pfeifer H, Schwieger W (1986) ^{1}H MAS NMR studies on the acidity of zeolites. Chem Phys Lett 128:62–66
22. Freude D, Klinowski J (1988) Solid state ^{1}H NMR studies of the structure of the active site in zeolite H-ZSM-5. J Chem Soc Chem Commun 1411–1413
23. Hunger M, Freude D, Frohlich T, Pfeifer H, Schwieger W (1987) ^{1}H MAS NMR studies of ZSM-5 type zeolites. Zeolites 7:108–110
24. Hunger M, Freude D, Pfeifer H (1991) Magic angle spinning nuclear magnetic resonance studies of water molecules adsorbed on Bronsted acid and Lewis acid sites in zeolites and amorphous silica aluminas. J Chem Soc Faraday Trans 87:657–662
25. Grey CP, Vega AJ (1995) Determination of the quadrupole coupling constant of the invisible aluminum spins in zeolite HY with ^{1}H/^{27}Al TRAPDOR NMR. J Am Chem Soc 117:8232–8242
26. Li SH, Zheng AM, Su YC, Zhang HL, Chen L, Yang J, Ye CH, Deng F (2007) Bronsted/Lewis acid synergy in dealuminated HY zeolite: a combined solid-state NMR and theoretical calculation study. J Am Chem Soc 129:11161–11171
27. Hunger M (1997) Bronsted acid sites in zeolites characterized by multinuclear solid-state NMR spectroscopy. Catal Rev Sci Eng 39:345–393
28. Hu W, Luo Q, Su YC, Chen L, Yue Y, Ye CH, Deng F (2006) Acid sites in mesoporous Al-SBA-15 material as revealed by solid-state NMR spectroscopy. Microporous Mesoporous Mater 92:22–30
29. Derouane EG, Nagy JB, Gabelica Z, Blom N (1982) An ^{27}Al NMR investigation of ZSM-5 type zeolite aluminosilicate precursor species. Zeolites 2:299–302
30. Engelhardt G, Fahlke B, Magi M, Lippmaa E (1983) High resolution solid state ^{29}Si and ^{27}Al NMR of aluminosilicate intermediates in zeolite-A synthesis. Zeolites 3:292–294
31. Klinowski J, Thomas JM, Fyfe CA, Gobbi GC, Hartman JS (1983) A highly siliceous structural analog of zeolite Y: high resolution solid state ^{29}Si and ^{27}Al NMR studies. Inorg Chem 22:63–66
32. Liu W, Xu Y, Wong ST, Wang L, Qiu J, Yang N (1997) Methane dehydrogenation and aromatization in the absence of oxygen on Mo/HZSM-5: a study on the interaction between Mo species and HZSM-5 by using ^{27}Al and ^{29}Si MAS NMR. J Mol Catal A Chem 120:257–265
33. Zhang WP, Bao XH, Guo XW, Wang XS (1999) A high-resolution solid-state NMR study on nano-structured H-ZSM-5 zeolite. Catal Lett 60:89–94
34. Li SH, Huang SJ, Shen WL, Zhang HL, Fang HJ, Zheng AM, Liu SB, Deng F (2008) Probing the spatial proximities among acid sites in dealuminated H-Y zeolite by solid-state NMR spectroscopy. J Phys Chem C 112:14486–14494
35. Seiler M, Wang W, Hunger M (2001) Local structure of framework aluminum in zeolite H-ZSM-5 during conversion of methanol investigated by in situ NMR spectroscopy. J Phys Chem B 105:8143–8148
36. Jiao J, Altwasser S, Wang W, Weitkamp J, Hunger M (2004) State of aluminum in dealuminated, nonhydrated zeolites Y investigated by multinuclear solid-state NMR spectroscopy. J Phys Chem B 108:14305–14310
37. Jiao J, Kanellopoulos J, Wang W, Ray SS, Foerster H, Freude D, Hunger M (2005) Characterization of framework and extra-framework aluminum species in non-hydrated zeolites Y by ^{27}Al spin-echo, high-speed MAS, and MQMAS NMR spectroscopy at $B_0 = 9.4$ to 17.6 T. Phys Chem Chem Phys 7:3221–3226
38. Medek A, Harwood JS, Frydman L (1995) Multiple-quantum magic-angle spinning NMR: a new method for the study of quadrupolar nuclei in solids. J Am Chem Soc 117:12779–12787
39. Amoureux JP, Fernandez C, Steuernagel S (1996) Z filtering in MQ-MAS NMR. J Magn Reson Ser A 123:116–118

40. Yu ZW, Zheng AM, Wang QA, Chen L, Xu J, Amoureux JP, Deng F (2010) Insights into the dealumination of zeolite HY revealed by sensitivity-enhanced ^{27}Al DQ-MAS NMR spectroscopy at high field. Angew Chem Int Ed 49:8657–8661
41. Sklenak S, Dedecek J, Li C, Wichterlova B, Gabova V, Sierka M, Sauer J (2007) Aluminum siting in silicon-rich zeolite frameworks: a combined high-resolution ^{27}Al NMR spectroscopy and quantum mechanics/molecular mechanics study of ZSM-5. Angew Chem Int Ed 46:7286–7289
42. van Bokhoven JA, Koningsberger DC, Kunkeler P, van Bekkum H, Kentgens APM (2000) Stepwise dealumination of zeolite beta at specific T-sites observed with ^{27}Al MAS and ^{27}Al MQ MAS NMR. J Am Chem Soc 122:12842–12847
43. Yan ZM, Ding MA, Zhuang JQ, Liu XC, Liu XM, Han XW, Bao XH, Chang FX, Xu L, Liu ZM (2003) On the acid-dealumination of USY zeolite: a solid state NMR investigation. J Mol Catal A Chem 194:153–167
44. Lippmaa E, Magi M, Samoson A, Tarmak M, Engelhardt G (1981) Investigation of the structure of zeolites by solid state high resolution Si-29 NMR spectroscopy. J Am Chem Soc 103:4992–4996
45. Engelhardt G, Lohse U, Samoson A, Magi M, Tarmak M, Lippmaa E (1982) High resolution ^{29}Si NMR of dealuminated and ultrastable Y zeolites. Zeolites 2:59–62
46. Melchior MT, Vaughan DEW, Jarman RH, Jacobson AJ (1982) The characterization of Si-Al ordering in A-type zeolite (ZK4) by ^{29}Si NMR. Nature 298:455–456
47. Fyfe CA, Feng Y, Grondey H, Kokotailo GT, Gies H (1991) One dimensional and 2-dimensional high resolution solid state NMR studies of zeolite lattice structures. Chem Rev 91:1525–1543
48. Ramdas S, Klinowski J (1984) A simple correlation between isotropic ^{29}Si NMR chemical shifts and T-O-T angles in zeolite frameworks. Nature 308:521–523
49. Thomas JM, Klinowski J, Ramdas S, Hunter BK, Tennakoon DTB (1983) The evaluation of non-equivalent tetrahedral sites from ^{29}Si NMR chemical shifts in zeolites and related aluminosilicates. Chem Phys Lett 102:158–162
50. Hammond KD, Dogan F, Tompsett GA, Agarwal V, Conner WC Jr, Grey CP, Auerbach SM (2008) Spectroscopic signatures of nitrogen-substituted zeolites. J Am Chem Soc 130:14912–14913
51. Bussemer B, Schroder KP, Sauer J (1997) Ab initio predictions of zeolite structures and ^{29}Si NMR chemical shifts. Solid State Nucl Magn Reson 9:155–164
52. Brouwer DH, Darton RJ, Morris RE, Levitt MH (2005) A solid-state NMR method for solution of zeolite crystal structures. J Am Chem Soc 127:10365–10370
53. Brouwer DH, Kristiansen PE, Fyfe CA, Levitt MH (2005) Symmetry-based ^{29}Si dipolar recoupling magic angle spinning NMR spectroscopy: a new method for investigating three-dimensional structures of zeolite frameworks. J Am Chem Soc 127:542–543
54. Lesage A, Bardet M, Emsley L (1999) Through-bond carbon-carbon connectivities in disordered solids by NMR. J Am Chem Soc 121:10987–10993
55. Janes N, Oldfield E (1986) ^{17}O NMR study of bonding in silicates: the orbital controversy. J Am Chem Soc 108:5743–5753
56. Timken HKC, Turner GL, Gilson JP, Welsh LB, Oldfield E (1986) Solid state ^{17}O nuclear magnetic resonance spectroscopic studies of zeolites and related systems. J Am Chem Soc 108:7231–7235
57. Gullion T, Schaefer J (1989) Rotational echo double resonance NMR. J Magn Reson 81:196–200
58. Peng LM, Liu Y, Kim NJ, Readman JE, Grey CP (2005) Detection of Bronsted acid sites in zeolite HY with high-field O-17-MAS-NMR techniques. Nat Mater 4:216–219
59. Peng L, Huo H, Gan Z, Grey CP (2008) ^{17}O MQMAS NMR studies of zeolite HY. Microporous Mesoporous Mater 109:156–162
60. Peng L, Huo H, Liu Y, Grey CP (2007) ^{17}O magic angle spinning NMR studies of Bronsted acid sites in zeolites HY and HZSM-5. J Am Chem Soc 129:335–346

61. Huo H, Peng L, Gan Z, Grey CP (2012) Solid-state MAS NMR studies of Bronsted acid sites in zeolite H-mordenite. J Am Chem Soc 134:9708–9720
62. Demarquay J, Fraissard J (1987) ^{129}Xe NMR of xenon adsorbed on zeolites: relationship between the chemical shift and the void space. Chem Phys Lett 136:314–318
63. Ito T, Fraissard J (1982) ^{129}Xe NMR study of xenon adsorbed on zeolites. J Chem Phys 76:5225–5229
64. Fraissard J, Ito T (1988) 129Xe NMR-study of adsorbed xenon: a new method for studying zeolites and metal-zeolites. Zeolites 8:350–361
65. Chen F, Deng F, Cheng MJ, Yue Y, Ye CH, Bao XH (2001) Preferential occupation of xenon in zeolite MCM-22 as revealed by ^{129}Xe NMR spectroscopy. J Phys Chem B 105:9426–9432
66. Chen F, Zhang MJ, Han Y, Xiao FS, Yue Y, Ye CH, Deng F (2004) Characterization of microporosity in ordered mesoporous material MAS-7 by ^{129}Xe NMR spectroscopy. J Phys Chem B 108:3728–3734
67. Zeng D, Yang J, Wang J, Xu J, Yang Y, Ye C, Deng F (2007) Solid-state NMR studies of methanol-to-aromatics reaction over silver exchanged H-ZSM-5 zeolite. Microporous Mesoporous Mater 98:214–219
68. Huang SJ, Huh S, Lo PS, Liu SH, Lin VSY, Liu SB (2005) Hyperpolarized ^{129}Xe NMR investigation of multifunctional organic/inorganic hybrid mesoporous silica materials. Phys Chem Chem Phys 7:3080–3087
69. Lercher JA, Grundling C, EderMirth G (1996) Infrared studies of the surface acidity of oxides and zeolites using adsorbed probe molecules. Catal Today 27:353–376
70. Lunsford JH, Rothwell WP, Shen W (1985) Acid sites in zeolite Y: a solid state NMR and infrared study using trimethylphosphine as a probe molecule. J Am Chem Soc 107:1540–1547
71. Lunsford JH (1997) Characterization of acidity in zeolites and related oxides using trimethylphosphine as a probe. Top Catal 4:91–98
72. Luo Q, Deng F, Yuan ZY, Yang J, Zhang MJ, Yue Y, Ye CH (2003) Using trimethylphosphine as a probe molecule to study the acid states in Al-MCM-41 materials by solid-state NMR spectroscopy. J Phys Chem B 107:2435–2442
73. Chu Y, Yu Z, Zheng A, Fang H, Zhang H, Huang S-J, Liu S-B, Deng F (2011) Acidic strengths of Brønsted and Lewis acid sites in solid acids scaled by ^{31}P NMR chemical shifts of adsorbed trimethylphosphine. J Phys Chem C 115:7660–7667
74. Biaglow AI, Gorte RJ, Kokotailo GT, White D (1994) A probe of Bronsted site acidity in zeolites: ^{13}C chemical shift of acetone. J Catal 148:779–786
75. Xu T, Munson EJ, Haw JF (1994) Toward a systematic chemistry of organic reactions in zeolites: in situ NMR studies of ketones. J Am Chem Soc 116:1962–1972
76. Yang J, Janik MJ, Ma D, Zheng A, Zhang M, Neurock M, Davis RJ, Ye C, Deng F (2005) Location, acid strength, and mobility of the acidic protons in Keggin 12-H3PW12O40: a combined solid-state NMR spectroscopy and DFT quantum chemical calculation study. J Am Chem Soc 127:18274–18280
77. Rakiewicz EF, Peters AW, Wormsbecher F, Sutovich KJ, Mueller KT (1998) Characterization of acid sites in zeolitic and other inorganic systems using solid-state ^{31}P NMR of the probe molecule trimethylphosphine oxide. J Phys Chem B 102:2890–2896
78. Chen W-H, Ko H-H, Sakthivel A, Huang S-J, Liu S-H, Lo A-Y, Tsai T-C, Liu S-B (2006) A solid-state NMR, FT-IR and TPD study on acid properties of sulfated and metal-promoted zirconia: influence of promoter and sulfation treatment. Catal Today 116:111–120
79. Hunger M (1996) Multinuclear solid-state NMR studies of acidic and non-acidic hydroxyl protons in zeolites. Solid State Nucl Magn Reson 6:1–29
80. Xu M, Arnold A, Buchholz A, Wang W, Hunger M (2002) Low-temperature modification of mesoporous MCM-41 material with sublimated aluminum chloride in vacuum. J Phys Chem B 106:12140–12143

81. Zheng A, Zhang H, Lu X, Liu S-B, Deng F (2008) Theoretical predictions of ^{31}P NMR chemical shift threshold of trimethylphosphine oxide absorbed on solid acid catalysts. J Phys Chem B 112:4496–4505
82. Zheng A, Huang S-J, Chen W-H, Wu P-H, Zhang H, Lee H-K, L-Cd M, Deng F, Liu S-B (2008) ^{31}P chemical shift of adsorbed trialkylphosphine oxides for acidity characterization of solid acids catalysts. J Phys Chem A 112:7349–7356
83. Zheng A, Zhang H, Chen L, Yue Y, Ye C, Deng F (2007) Relationship between ^{1}H chemical shifts of deuterated pyridinium ions and Brønsted acid strength of solid acids. J Phys Chem B 111:3085–3089
84. Fang H, Zheng A, Chu Y, Deng F (2010) ^{13}C chemical shift of adsorbed acetone for measuring the acid strength of solid acids: a theoretical calculation study. J Phys Chem C 114:12711–12718
85. Zheng A, Deng F, Liu S-B (2014) Acidity characterization of solid acid catalysts by solid-state ^{31}P NMR of adsorbed phosphorus-containing probe molecules. Annu Rep NMR Spectrosc 81:47–108
86. Zheng AM, Huang SJ, Wang Q, Zhang HL, Deng F, Liu SB (2013) Progress in development and application of solid-state NMR for solid acid catalysis. Chin J Catal 34:436–491
87. Zheng AM, Liu SB, Deng F (2013) Acidity characterization of heterogeneous catalysts by solid-state NMR spectroscopy using probe molecules. Solid State Nucl Magn Reson 55–56:12–27
88. Zheng A, Huang S-J, Liu S-B, Deng F (2011) Acid properties of solid acid catalysts characterized by solid-state ^{31}P NMR of adsorbed phosphorous probe molecules. Phys Chem Chem Phys 13:14889–14901
89. Zhang XM, Zhao YP, Xu ST, Yang Y, Liu J, Wei YX, Yang QH (2014) Polystyrene sulphonic acid resins with enhanced acid strength via macromolecular self-assembly within confined nanospace. Nat Commun 5:3170
90. de Clippel F, Dusselier M, Van Rompaey R, Vanelderen P, Dijkmans J, Makshina E, Giebeler L, Oswald S, Baron GV, Denayer JFM, Pescarmona PP, Jacobs PA, Sels BF (2012) Fast and selective sugar conversion to alkyl lactate and lactic acid with bifunctional carbon–silica catalysts. J Am Chem Soc 134:10089–10101
91. Huang J, van Vegten N, Jiang Y, Hunger M, Baiker A (2010) Increasing the Brønsted acidity of flame-derived silica/alumina up to zeolitic strength. Angew Chem Int Ed 49:7776–7781
92. Feng ND, Zheng AM, Huang SJ, Zhang HL, Yu NY, Yang CY, Liu SB, Deng F (2010) Combined solid-state NMR and theoretical calculation studies of Bronsted acid properties in anhydrous 12-molybdophosphoric acid. J Phys Chem C 114:15464–15472
93. Huang SJ, Yang CY, Zheng AM, Feng ND, Yu NY, Wu PH, Chang YC, Lin YC, Deng F, Liu SB (2011) New insights into Keggin-type 12-tungstophosphoric acid from ^{31}P MAS NMR analysis of absorbed trimethylphosphine oxide and DFT calculations. Chem Asian J 6:137–148
94. Filek U, Bressel A, Sulikowski B, Hunger M (2008) Structural stability and Brønsted acidity of thermally treated $AlPW_{12}O_{40}$ in comparison with $H_3PW_{12}O_{40}$. J Phys Chem C 112:19470–19476
95. Fang H, Zheng A, Li S, Xu J, Chen L, Deng F (2010) New insights into the effects of acid strength on the solid acid-catalyzed reaction: theoretical calculation study of olefinic hydrocarbon protonation reaction. J Phys Chem C 114:10254–10264
96. Chu YY, Han B, Fang HJ, Zheng AM, Deng F (2012) Influence of acid strength on the reactivity of alkane activation on solid acid catalysts: a theoretical calculation study. Microporous Mesoporous Mater 151:241–249
97. Zheng AM, Liu SB, Deng F (2009) Chemoselectivity during propene hydrogenation reaction over H-ZSM-5 zeolite: insights from theoretical calculations. Microporous Mesoporous Mater 121:158–165
98. Zhao Q, Chen WH, Huang SJ, Wu YC, Lee HK, Liu SB (2002) Discernment and quantification of internal and external acid sites on zeolites. J Phys Chem B 106:4462–4469

99. Zhang WP, Ma D, Liu XC, Liu XM, Bao XH (1999) Perfluorotributylamine as a probe molecule for distinguishing internal and external acidic sites in zeolites by high-resolution ^{1}H MAS NMR spectroscopy. Chem Commun 1091–1092
100. Zheng AM, Chen L, Yang J, Zhang MJ, Su YC, Yue Y, Ye CH, Deng F (2005) Combined DFT theoretical calculation and solid-state NMR studies of Al substitution and acid sites in zeolite MCM-22. J Phys Chem B 109:24273–24279
101. Peng LM, Chupas PJ, Grey CP (2004) Measuring Bronsted acid densites in zeolite HY with diphosphine molecules and solid state NMR spectroscopy. J Am Chem Soc 126:12254–12255
102. Brown SP (2012) Applications of high-resolution ^{1}H solid-state NMR. Solid State Nucl Magn Reson 41:1–27
103. Li SH, Zheng AM, Su YC, Fang HJ, Shen WL, Yu ZW, Chen L, Deng F (2010) Extra-framework aluminium species in hydrated faujasite zeolite as investigated by two-dimensional solid-state NMR spectroscopy and theoretical calculations. Phys Chem Chem Phys 12:3895–3903
104. Yu ZW, Li SH, Wang Q, Zheng AM, Jun X, Chen L, Deng F (2011) Bronsted/Lewis acid synergy in H-ZSM-5 and H-MOR zeolites studied by ^{1}H and ^{27}Al DQ-MAS solid-state NMR spectroscopy. J Phys Chem C 115:22320–22327
105. Yu ZW, Wang Q, Chen L, Deng F (2012) Bronsted/Lewis acid sites synergy in H-MCM-22 zeolite studied by ^{1}H and ^{27}Al DQ-MAS NMR spectroscopy. Chin J Catal 33:129–139
106. Wang Q, Hu B, Lafon O, Trebosc J, Deng F, Amoureux JP (2009) Double-quantum homonuclear NMR correlation spectroscopy of quadrupolar nuclei subjected to magic-angle spinning and high magnetic field. J Magn Reson 200:251–260
107. Peng LM, Grey CP (2008) Diphosphine probe molecules and solid-state NMR investigations of proximity between acidic sites in zeolite HY. Microporous Mesoporous Mater 116:277–283
108. Haw JF, Song WG, Marcus DM, Nicholas JB (2003) The mechanism of methanol to hydrocarbon catalysis. Acc Chem Res 36:317–326
109. Hunger M, Weitkamp J (2001) In situ IR, NMR, EPR, and UV/Vis spectroscopy: tools for new insight into the mechanisms of heterogeneous catalysis. Angew Chem Int Ed 40:2954–2971
110. Svelle S, Tuma C, Rozanska X, Kerber T, Sauer J (2009) Quantum chemical modeling of zeolite-catalyzed methylation reactions: toward chemical accuracy for barriers. J Am Chem Soc 131:816–825
111. Chu YY, Han B, Zheng AM, Deng F (2012) Influence of acid strength and confinement effect on the ethylene dimerization reaction over solid acid catalysts: a theoretical calculation study. J Phys Chem C 116:12687–12695
112. Zheng AM, Deng F, Liu SB (2011) Regioselectivity of carbonium ion transition states in zeolites. Catal Today 164:40–45
113. Xu ST, Zheng AM, Wei YX, Chen JR, Li JZ, Chu YY, Zhang MZ, Wang QY, Zhou Y, Wang JB, Deng F, Liu ZM (2013) Direct observation of cyclic carbenium ions and their role in the catalytic cycle of the methanol-to-olefin reaction over chabazite zeolites. Angew Chem Int Ed 52:11564–11568
114. Wang W, Seiler M, Ivanova II, Sternberg U, Weitkamp J, Hunger M (2002) Formation and decomposition of N, N, N-trimethylanilinium cations on zeolite H−Y investigated by in situ stopped-flow MAS NMR spectroscopy. J Am Chem Soc 124:7548–7554
115. Wang X, Qi G, Xu J, Li B, Wang C, Deng F (2012) NMR-spectroscopic evidence of intermediate-dependent pathways for acetic acid formation from methane and carbon monoxide over a ZnZSM-5 zeolite catalyst. Angew Chem Int Ed 51:3850–3853
116. Gabrienko AA, Arzumanov SS, Toktarev AV, Freude D, Haase J, Stepanov AG (2011) H/D exchange and activation of C-1-n-C-4 alkanes on Ga-modified zeolite BEA studied with ^{1}H magic angle spinning nuclear magnetic resonance in situ. J Phys Chem C 115:13877–13886

117. Luzgin MV, Toktarev AV, Parmon VN, Stepanov AG (2013) Coaromatization of methane with propane on Mo-containing zeolite H-BEA: a solid-state NMR and GC-MS study. J Phys Chem C 117:22867–22873
118. Gabrienko AA, Arzumanov SS, Freude D, Stepanov AG (2010) Propane aromatization on Zn-modified zeolite BEA studied by solid-state NMR in situ. J Phys Chem C 114:12681–12688
119. Wang X, Dai WL, Wu GJ, Li LD, Guan NJ, Hunger M (2014) Verifying the dominant catalytic cycle of the methanol-to-hydrocarbon conversion over SAPO-41. Catal Sci Technol 4:688–696
120. Dai WL, Wang X, Wu GJ, Guan NJ, Hunger M, Li LD (2011) Methanol-to-Olefin conversion on silicoaluminophosphate catalysts: effect of Bronsted acid sites and framework structures. ACS Catal 1:292–299
121. Marthala VRR, Rabl S, Huang J, Rezai SAS, Thomas B, Hunger M (2008) In situ solid-state NMR investigations of the vapor-phase Beckmann rearrangement of ^{15}N-cyclohexanone oxime on MFI-type zeolites and mesoporous SBA-15 materials in the absence and presence of the additive ^{13}C-methanol. J Catal 257:134–141
122. Li JZ, Wei YX, Chen JR, Tian P, Su X, Xu ST, Qi Y, Wang QY, Zhou Y, He YL, Liu ZM (2012) Observation of heptamethylbenzenium cation over SAPO-Type molecular sieve DNL-6 under real MTO conversion conditions. J Am Chem Soc 134:836–839
123. Haw JF, Richardson BR, Oshiro IS, Lazo ND, Speed JA (1989) Reactions of propene on zeolite HY catalyst studied by in situ variable temperature solid-state nuclear magnetic resonance spectroscopy. J Am Chem Soc 111:2052–2058
124. Xu T, Haw JF (1994) NMR observation of indanyl carbenium ion intermediates in the reactions of hydrocarbons on acidic zeolites. J Am Chem Soc 116:10188–10195
125. Xu T, Barich DH, Goguen PW, Song W, Wang Z, Nicholas JB, Haw JF (1998) Synthesis of a benzenium ion in a zeolite with use of a catalytic flow reactor. J Am Chem Soc 120:4025–4026
126. Li J, Wei Y, Chen J, Tian P, Su X, Xu S, Qi Y, Wang Q, Zhou Y, He Y, Liu Z (2012) Observation of heptamethylbenzenium cation over SAPO-type molecular sieve DNL-6 under real MTO conversion conditions. J Am Chem Soc 134:836–839
127. Xu S, Zhang W, Liu X, Han X, Bao X (2009) Enhanced in situ continuous-flow MAS NMR for reaction kinetics in the nanocages. J Am Chem Soc 131:13722–13727
128. de Moor P, Beelen TPM, van Santen RA (1999) In situ observation of nucleation and crystal growth in zeolite synthesis. A small-angle X-ray scattering investigation on Si-TPA-MFI. J Phys Chem B 103:1639–1650
129. Verhoef MJ, Kooyman PJ, van der Waal JC, Rigutto MS, Peters JA, van Bekkum H (2001) Partial transformation of MCM-41 material into zeolites: formation of nanosized MFI type crystallites. Chem Mater 13:683–687
130. Alfredsson V, Anderson MW (1996) Structure of MCM-48 revealed by transmission electron microscopy. Chem Mater 8:1141–1146
131. Huang YN, Demko BA, Kirby CW (2003) Investigation of the evolution of intermediate phases of AIPO(4)-18 molecular sieve synthesis. Chem Mater 15:2437–2444
132. Wu Q, Wang X, Qi G, Guo Q, Pan S, Meng X, Xu J, Deng F, Fan F, Feng Z, Li C, Maurer S, Müller U, Xiao F-S (2014) Sustainable synthesis of zeolites without addition of both organotemplates and solvents. J Am Chem Soc 136:4019–4025
133. Jin Y, Sun Q, Qi G, Yang C, Xu J, Chen F, Meng X, Deng F, Xiao F-S (2013) Solvent-free synthesis of silicoaluminophosphate zeolites. Angew Chem Int Ed 52:9172–9175
134. Zhang L, Bates J, Chen DH, Nie HY, Huang YN (2011) Investigations of formation of molecular sieve SAPO-34. J Phys Chem C 115:22309–22319
135. Greer H, Wheatley PS, Ashbrook SE, Morris RE, Zhou W (2009) Early stage reversed crystal growth of zeolite A and its phase transformation to sodalite. J Am Chem Soc 131:17986–17992

136. Van Grieken R, Sotelo JL, Menendez JM, Melero JA (2000) Anomalous crystallization mechanism in the synthesis of nanocrystalline ZSM-5. Microporous Mesoporous Mater 39:135–147
137. Chen BH, Huang YN (2006) ^{17}O solid-state NMR spectroscopic studies of the involvement of water vapor in molecular sieve formation by dry-gel conversion. J Am Chem Soc 128:6437–6446
138. Xu J, Chen L, Zeng DL, Yang J, Zhang MJ, Ye CH, Deng F (2007) Crystallization of AlPO4-5 aluminophosphate molecular sieve prepared in fluoride medium: a multinuclear solid-state NMR study. J Phys Chem B 111:7105–7113
139. Shen WL, Li SH, Xu J, Zhang HL, Hu W, Zhou D, Zhang JA, Yu JH, Xu WJ, Xu Y, Deng F (2010) A novel phase transformation phenomenon in mesostructured aluminophosphate. J Phys Chem C 114:7076–7084
140. Tosner Z, Vosegaard T, Kehlet C, Khaneja N, Glaser SJ, Nielsen NC (2009) Optimal control in NMR spectroscopy: numerical implementation in SIMPSON. J Magn Reson 197:120–134
141. Zhou DH, Shah G, Cormos M, Mullen C, Sandoz D, Rienstra CM (2007) Proton-detected solid-state NMR spectroscopy of fully protonated proteins at 40 kHz magic-angle spinning. J Am Chem Soc 129:11791–11801
142. Smith ME (2001) Recent progress in solid-state NMR of low-gamma nuclei. Annu Rep NMR Spectrosc 43:121–175
143. Brinkmann A, Levitt MH (2001) Symmetry principles in the nuclear magnetic resonance of spinning solids: heteronuclear recoupling by generalized Hartmann-Hahn sequences. J Chem Phys 115:357–384
144. Carravetta M, Eden M, Zhao X, Brinkmann A, Levitt MH (2000) Symmetry principles for the design of radiofrequency pulse sequences in the nuclear magnetic resonance of rotating solids. Chem Phys Lett 321:205–215
145. Hohwy M, Jakobsen HJ, Eden M, Levitt MH, Nielsen NC (1998) Broadband dipolar recoupling in the nuclear magnetic resonance of rotating solids: a compensated C7 pulse sequence. J Chem Phys 108:2686–2694
146. Lesage A, Sakellariou D, Hediger S, Elena B, Charmont P, Steuernagel S, Emsley L (2003) Experimental aspects of proton NMR spectroscopy in solids using phase-modulated homonuclear dipolar decoupling. J Magn Reson 163:105–113
147. Xu J, Zhu P, Gan Z, Sahar N, Tecklenburg M, Morris MD, Kohn DH, Ramamoorthy A (2010) Natural-abundance ^{43}Ca solid-state NMR spectroscopy of bone. J Am Chem Soc 132:11504–11509
148. Wu G, Gan Z, Kwan ICM, Fettinger JC, Davis JT (2011) High-resolution 39 K NMR spectroscopy of bio-organic solids. J Am Chem Soc 133:19570–19573
149. Laurencin D, Wong A, Hanna JV, Dupree R, Smith ME (2008) A high-resolution ^{43}Ca solid-state NMR study of the calcium sites of hydroxyapatite. J Am Chem Soc 130:2412–2413
150. Zheng H, Ma D, Bao X, Hu JZ, Kwak JH, Wang Y, Peden CHF (2008) Direct observation of the active center for methane dehydroaromatization using an ultrahigh field ^{95}Mo NMR spectroscopy. J Am Chem Soc 130:3722–3723
151. Maly T, Debelouchina GT, Bajaj VS, Hu K-N, Joo C-G, Mak-Jurkauskas ML, Sirigiri JR, van der Wel PCA, Herzfeld J, Temkin RJ, Griffin RG (2008) Dynamic nuclear polarization at high magnetic fields. J Chem Phys 128:052211
152. Bajaj VS, Farrar CT, Hornstein MK, Mastovsky I, Vieregg J, Bryant J, Elena B, Kreischer KE, Temkin RJ, Griffin RG (2003) Dynamic nuclear polarization at 9T using a novel 250 GHz gyrotron microwave source. J Magn Reson 160:85–90
153. Lelli M, Gajan D, Lesage A, Caporini MA, Vitzthum V, Mieville P, Heroguel F, Rascon F, Roussey A, Thieuleux C, Boualleg M, Veyre L, Bodenhausen G, Coperet C, Emsley L (2011) Fast characterization of functionalized silica materials by 29Si surface-enhanced NMR spectroscopy using dynamic nuclear polarization. J Am Chem Soc 133:2104–2107

154. Rossini AJ, Zagdoun A, Lelli M, Gajan D, Rascon F, Rosay M, Maas WE, Coperet C, Lesage A, Emsley L (2012) One hundred fold overall sensitivity enhancements for ^{29}Si NMR spectroscopy of surfaces by dynamic nuclear polarization with CPMG acquisition. Chem Sci 3:108–115
155. Zagdoun A, Casano G, Ouari O, Lapadula G, Rossini AJ, Lelli M, Baffert M, Gajan D, Veyre L, Maas WE, Rosay M, Weber RT, Thieuleux C, Coperet C, Lesage A, Tordo P, Emsley L (2012) A slowly relaxing rigid biradical for efficient dynamic nuclear polarization surface-enhanced NMR spectroscopy: expeditious characterization of functional group manipulation in hybrid materials. J Am Chem Soc 134:2284–2291
156. Rossini AJ, Zagdoun A, Lelli M, Canivet J, Aguado S, Ouari O, Tordo P, Rosay M, Maas WE, Coperet C, Farrusseng D, Emsley L, Lesage A (2012) Dynamic nuclear polarization enhanced solid-state NMR spectroscopy of functionalized metal-organic frameworks. Angew Chem Int Ed 51:123–127
157. Lafon O, Rosay M, Aussenac F, Lu X, Trebosc J, Cristini O, Kinowski C, Touati N, Vezin H, Amoureux J-P (2011) Beyond the silica surface by direct ^{29}Si dynamic nuclear polarization. Angew Chem Int Ed 50:8367–8370
158. Freeman R, Kupce E (2003) New methods for fast multidimensional NMR. J Biomol NMR 27:101–113
159. Frydman L, Scherf T, Lupulescu A (2002) The acquisition of multidimensional NMR spectra within a single scan. Proc Natl Acad Sci U S A 99:15858–15862
160. Jiang B, Jiang XW, Xiao N, Zhang X, Jiang L, Mao XA, Liu ML (2010) Gridding and fast Fourier transformation on non-uniformly sparse sampled multidimensional NMR data. J Magn Reson 204:165–168
161. Jiang B, Luo F, Ding Y, Sun P, Zhang X, Jiang L, Li C, X-a M, Yang D, Tang C, Liu M (2013) NASR: an effective approach for simultaneous noise and artifact suppression in NMR spectroscopy. Anal Chem 85:2523–2528

Part III
Emerging Applications of Zeolites for Sustainable Chemistry

Chapter 8
Zeolites in Fluid Catalytic Cracking (FCC)

Vasileios Komvokis, Lynne Xin Lin Tan, Melissa Clough, Shuyang Shaun Pan, and Bilge Yilmaz

Abstract Looking back at the history of industrial chemistry, it can undoubtedly be declared that the impact of the introduction of zeolites into refinery operations has been revolutionary as it resulted in a significant increase in gasoline yield from catalytic cracking, leading to more efficient utilization of the petroleum feedstocks. Based on this successful entry into the industrial scene, zeolites quickly found many applications in petroleum processing in subsequent years; however, catalytic cracking still by far is the largest volume of zeolite catalyst consumer worldwide. Today, fluid catalytic cracking (FCC) catalysts account for more than 95 % of the global zeolite catalyst consumption (primarily USY zeolite). Fluid catalytic cracking units (FCCUs) around the world currently process around 15 million barrels per day and constitute the main conversion process in a typical fuels refinery. Primary output from FCC units is gasoline, and due to their pivotal role in fluid catalytic cracking, it can be stated that zeolite catalysts have been fueling our industrialized society by providing majority of the transportation fuels used, thus having direct impact on the global economy. Furthermore, in order to enhance the light olefin yields from FCCUs as a result of the increasing demand for building blocks for the chemical industry such as propylene and butylene, the FCC catalyst compositions are being modified by tuning their formulations and introducing different types of zeolite structures as additives. Using these zeolite-based catalyst formulations, many new FCC units worldwide also provide the critical starting points for (petro)chemical value chains. Hence, for the global supply of both fuels and chemicals, zeolite-based FCC catalysts are of paramount strategic and commercial importance.

V. Komvokis
BASF Refining Catalysts, SK8 6QG Cheadle, United Kingdom

L.X.L. Tan
BASF Refining Catalysts, 7 Temasek Blvd, #35-01, Suntec-1 038987, Singapore

M. Clough
BASF Refining Catalysts, 1111 Bagby St., Suite 2600, Houston, TX 77002, USA

S.S. Pan • B. Yilmaz (✉)
BASF Refining Catalysts, 25 Middlesex-Essex Turnpike, Iselin, NJ 08830, USA
e-mail: bilge.yilmaz@basf.com

F.-S. Xiao, X. Meng (eds.), *Zeolites in Sustainable Chemistry*,
Green Chemistry and Sustainable Technology,
DOI 10.1007/978-3-662-47395-5_8

Keywords Fluid catalytic cracking • FCC • Zeolite catalysis • FCC catalyst • Petroleum refining • Refining catalysts • Catalyst deactivation • FCC testing • Catalyst manufacturing

8.1 History of Fluid Catalytic Cracking (FCC)

Since the 1800s, thermal cracking had been in practice to break down crude oil into functional components. The first commercial use can be traced back to the days of Samuel M. Kier, a salt well entrepreneur, who started refining crude into lamp oil that can burn with lesser smoke and odor [1]. By 1854, he and his partner, John T. Kirkpatrick, pioneered the first oil refinery in Pittsburgh, Pennsylvania. Following their success, the market became saturated with refiners and by the end of 1860s, 58 refineries were in operation in Pittsburgh alone.

In the early 1900s, engineering advancement in combustion engines gave rise to a demand for gasoline fuel. Several patents of thermal cracking techniques emerged with the first commercial success in such a process by Dr. William M. Burton of Standard Oil Company of Indiana [2]. In the "Burton cracking process" (US Patent 1049667), gasoline production can be significantly increased by heating crude oil in excess of 750 °F (ca. 400 °C) in a still under pressure of around 90 psig [3]. This was the primary method of gasoline production before catalytic cracking was introduced to the refining industry in the 1940s.

Almer M. McAfee, a chemist in Gulf Refining Company, first introduced the commercial application of anhydrous aluminum chloride to catalytically crack heavy petroleum oil in 1915 [4]. Plagued by the high cost of the catalyst and the amount of corrosion waste it generated, the process was not widely accepted. It was later in the 1920s when a French engineer named Eugene J. Houdry discovered the use of Fuller's earth, a naturally occurring aluminosilicate (Al_2SiO_6) clay, that catalytic cracking achieved its full potential [5]. His development included a key economical step to regenerate the catalyst by burning off the accumulated carbon, a challenge faced by McAfee [6, 7]. In 1937, the "Houdry process" was first commercialized in a 15,000 bbl/d Sun Oil unit. The performance was an immediate success with almost half of the product as high octane gasoline [6, 7].

Engineering advances had since improved the original fixed-bed Houdry process units to a more economical fluidized-bed operation. The Thermofor catalytic cracking (TCC) unit was introduced in Socony-Vacuum's Paulsboro refinery in 1941 to enable the catalyst to be moved between the reactor and regeneration zones [8]. But the high cost of erecting a TCC unit and the limited feed it can take drove the need for a more efficient unit. In 1942, after a successful pilot plant test, the first commercial fluid catalytic cracker unit, PCLA No. 1 (Model 1), was built in Standard Oil of New Jersey's Baton Rouge, LA refinery [9]. This unit allows catalyst to flow through the reactor and regenerator, and catalyst and cracked products exit from the top to the cyclones for separation. Soon after, two other Model 1 FCC units were completed in the same year: in Standard Oil of New Jersey's Bayway, NJ refinery and in Humble Oil and Refinery Co. affiliate at Baytown, TX. These units mark the successful commercialization of the fluid catalytic cracking process.

8.2 Introduction to Cracking Catalysts

Concurrent to the development of catalytic cracking, catalyst technology had also evolved since the discovery of activated clay. To crack the organic C-C bond in crude oil, acid catalysts are required. McAfee first used aluminum chloride to crack heavy petroleum oil but was challenged by the cost of manufacturing and disposing the corrosion waste it generated as by-product. Houdry's introduction of Fuller's earth shed new light to the use of naturally occurring aluminosilicate clay as an active acid catalyst.

Fuller's earth comprises of montmorillonite, kaolinite, and attapulgite [10] and can be widely mined in areas with deposits of volcanic ash of Cretaceous age and younger [11]. Starting with the available Fuller's earth, Houdry processed it with acid to remove impurities and other undesired components such as iron to leave a residual structure comprising of mainly silica and alumina (US Patent 2551580A). This acid-activated catalyst significantly increased the cracking activity and improved overall product yield from the crude. This invention had set the foundation for future catalyst development.

In 1942, to support the "powdered catalyst" operation in PCLA No. 1, a synthetic fluid cracking catalyst was also developed [9]. The catalyst material was grinded into a fine powder to enable the fluidization of catalyst from the reactor to the regeneration section. The catalyst contained 13 % alumina. This development pioneered the fluid catalytic cracking technology, which has since given rise to more than 430 FCC units worldwide.

Advancement of fluid catalytic cracking catalysts continued, and in 1960s a major breakthrough came about when synthetic aluminosilicate zeolites were incorporated to further enhance catalytic activity and selectivity for outstanding cracking properties. Incorporation of zeolites in FCC catalysts results in strong Brønsted acid sites as well as easily accessible Lewis acid sites, which are key for hydrogen transfer reactions [12]. In 1962, Mobil Oil introduced a new catalyst containing Zeolite Y, which was prepared by adding a small amount of zeolite into the matrix of silica–alumina catalyst. The new catalyst outperformed all existing catalysts in the market and resulted in a significant increase in gasoline make. This marked the first commercial zeolite-based FCC catalysts and truly revolutionized the industry.

First examples of synthetic zeolites had actually come to the industrial scene earlier (such as those discovered by Union Carbide in the 1940s); however, these materials could not be converted to acid catalysts that would be thermally stable under harsh FCC operating conditions [13]. Zeolite Type X was later produced in the 1950s, having a faujasite framework (FAU), which has a three-dimensional aluminosilicate skeleton with a larger pore opening. It had a Si-to-Al ratio of 1.2 but was synthesized in Na form and had to be ion exchanged in order to achieve acid catalytic activity. Soon after, Zeolite Type Y (i.e., Zeolite Y) was produced also with the faujasite framework but this time with a Si-to-Al ratio of 2.5–3. The higher ratio has found to be more stable against acid treatment and under harsh hydrothermal conditions, paving the way for the new generation of FCC catalysts. Further advancement was made by treating Zeolite Y with steam calcination or chemical treatment to expel aluminum (i.e., dealumination) in a controlled manner and subsequently healing it with silicon, simultaneously creating a secondary mesoporous structure within the catalyst. This gave rise to a higher Si-to-Al ratio with a smaller unit cell size to create an ultra-stable zeolite, termed Zeolite USY. This type of zeolite has extremely high cracking activity and set the basis of FCC catalysts today.

Modern FCC catalysts are in the form of fine spherical particles with a bulk density of 0.80–0.96 g/cm^3 with average diameter of 60–100 μm [14]. As the heart of the FCC unit, the physical and chemical properties of the catalyst can greatly influence the operation of the unit. A good FCC catalyst is characterized by (1) high activity, (2) highly accessible catalytic sites for optimal cracking, (3) good resistance to attrition, (4) low production of coke and dry gas, and (5) high structural integrity even at high temperature and steaming condition. For the refiner it is vital to select the catalyst with the right properties to meet the various requirements of the refinery, and FCC catalyst formulations are typically tailored by the catalyst supplier for the specific needs of the FCC unit to meet the overall performance objectives.

8.3 Building Blocks of FCC Catalysts

A typical FCC catalyst consists of multiple components each to influence the activity, selectivity, structural integrity, and hydrothermal stability. The four major components are zeolite, matrix, filler, and binder. These components can vary in type and quantity and the catalyst can be made by incorporating these components in a single particle by carefully combining them to meet the functions required. A more advanced route for FCC catalyst preparation is the in situ process, where zeolite and active matrix are synthesized directly within a kaolin microsphere. The in situ method does not require additional fillers or binder and is therefore considered more effective.

8.3.1 Zeolite

Zeolite is the primary source of activity in today's FCC catalysts. Its concentration can range from 15 to 50 wt% in a catalyst. The zeolite most commonly used for cracking is Zeolite Type Y with the faujasite framework structure as described above. It comprises of tetrahedra with Si and Al joined by oxygen bridges. The tetrahedra form truncated octahedra (sodalite) that is joined in hexagonal prisms to form the faujasite (FAU) framework. It is a molecular sieve with a distinctive lattice structure that allows only a small range of hydrocarbon molecules to enter the lattice. The inner cage opening is typically about 7.4 Å. The larger oil molecule will be pre-cracked by matrix or outer zeolite crystal surface, and within the zeolitic pore openings, majority of the upgrading to naphtha and liquefied petroleum gas (LPG) occurs.

Zeolite Y (faujasite) is comprised of crystalline aluminosilicate frameworks in which tetrahedrally arranged silicon and aluminum atoms are joined by oxygen bridges. The insertion of Al in a zeolitic structure results in a net negative charge, which must be neutralized by a positive charge, i.e., cations. If the cation is H^+, it represents the strongest form of the zeolitic acid sites. The activity in the zeolite is dependent on the acidic sites provided by the alumina tetrahedra. This negative charge is neutralized by a positive charge (typically Na^+ in current commercial processes) during the zeolite synthesis. Zeolite at this stage is not hydrothermally stable and will be further treated to exchange the Na^+ with $NH4^+$, which upon heat treatment leaves behind the H^+ cation necessary to create the Lewis and Brønsted acidic sites. The Brønsted site can be further exchanged with rare earth cations such as lanthanum or cerium to further improve the stability of the zeolite structure [15].

8.3.2 Matrix

In general, the matrix component in an FCC catalyst refers to all inherent components in a catalyst particle except the zeolite. Matrix is typically in the form of active matrix which is related to the porous silica–alumina which were used in the cracking catalyst formulations before the development of zeolite-based FCC catalysts. The matrix provides the mesoporosity (30–500 Å) and macroporosity (above 500 Å) necessary for larger hydrocarbon molecules to access the catalyst. The broken down molecules can then penetrate into the zeolite pore/channel system for further cracking. This greatly reduces the bottoms yield in feedstock of higher boiling points.

Active matrix is rather stable under hydrothermal conditions. However, in the presence of feed with high contaminant metals such as nickel and vanadium, the active matrix may act as a catalyst support dispersing the metals, effectively promoting their dehydrogenation activity and resulting in the formation of excessive hydrogen and coke, which can be undesirable in an FCCU. Irregular pore size on the matrix surface may also result in poor stripping leading to increased accumulated coke in regenerator section. To optimize operation, a balanced zeolite-to-matrix ratio (i.e., Z/M) is necessary to satisfy the specific selectivity requirements of the FCC unit.

Matrix provides mesoporosity and macroporosity for the diffusion of reactants and products, as well as particle size and shape for fluidization. Two types of matrix components are used. Cracking matrix are acidic in nature and can crack bottoms to light cycle oil (LCO), which is critical for maximizing the concentration of product molecules in the diesel range from the FCC operation. It is less selective than zeolite cracking and can lead to higher coke and gas formation. Functional matrix components can provide passivation of contaminant metals in FCC operation with heavier residual feedstocks.

8.3.3 Filler and Binder

Both the filler and the binder provide the mechanical strength necessary for the FCC catalyst to fluidize in the unit. Filler usually comes in the form of clay such as kaolin to provide the microspheroidal form required for the FCC catalyst to properly fluidize in the unit. The filler also acts as a thermal heat storage medium to enable heat transfer from the reactor to the regenerator and prevent catalyst damage due to the severe hydrothermal conditions in the unit. Filler typically has no activity which helps to moderate the overall catalyst activity to prevent over-cracking which can result in excessive coke and dry gas formation.

The binder is the "glue" that holds the different components of the FCC catalyst together. A good binder is necessary in the incorporated synthesis route to ensure a sufficiently stable catalyst to prevent deterioration of the catalyst particles under the hydrothermal conditions during the operation of an FCC unit. It also plays a critical role in the mechanical integrity of the catalysts and thus attrition resistance. A commonly used binder is gel-based silica sol. Other types of binders such as alumina- or clay-based types are also utilized.

Catalysts synthesized by different manufacturers will vary in kind and amount of the components mentioned above. Conventional incorporated preparation method consists of zeolite and matrix separately produced and spray dried with a binder to make up the catalyst particle. Fillers like kaolin may be added to provide the particle size. This method results in a catalyst with components embedded in the microsphere. The in situ synthesis process enables the zeolite and active matrix to be synthesized within a preformed kaolin microsphere. As the zeolite crystals directly grow on the pores, they are typically more accessible for hydrocarbon molecules, which is a phenomenon that significantly enhances the cracking activity. These two manufacturing routes are explained in greater detail in the following section.

8.4 FCC Catalyst Manufacturing

As described above, the main source of activity in a typical FCC catalyst comes from the zeolite, which can be introduced into the catalyst particle by two methods. The first approach is generally referred to as the incorporated method. In this technique, zeolite is first crystallized and then incorporated into catalyst microspheres in a separate step with the use of binder and filler. The second method is referred to as the in situ method. Microspheres are first formed and the zeolite is crystallized in the microspheres themselves to provide microspheres containing both zeolite and other components [16].

8.4.1 *Incorporated Catalyst Technology*

Zeolite Y is produced by a reaction of silica and alumina under caustic conditions. The typical sources of silica and alumina are sodium silicate and sodium aluminate [17]. As-produced zeolite Y contains a high level of sodium, which is detrimental to the zeolite stability. Thus, ion exchanging of zeolite with rare earth and ammonium cations is needed to lower the Na level. To further enhance the zeolite stability, controlled steam calcination is used to replace some of the aluminum ions in the framework with silicon and to yield ultra-stable zeolite Y (USY).

Zeolite USY, matrix, filler, and binder are then mixed to form a slurry. The slurry is fed to a spray dryer. The spray-drying process results in the formation of the final incorporated microspheres. A schematic of the incorporated catalyst production process is shown in Fig. 8.1.

8.4.2 *In Situ Catalyst Technology*

In contrast to the incorporated approach, the in situ approach produces zeolite and matrix in one single step. The core of the in situ technique lays in the chemistry and structure of kaolin. The composition of raw kaolin contains silica and alumina, whose reactivity is strongly impacted by the calcination conditions of kaolin. The reactive silica and alumina are the key nutrients for the zeolite growth.

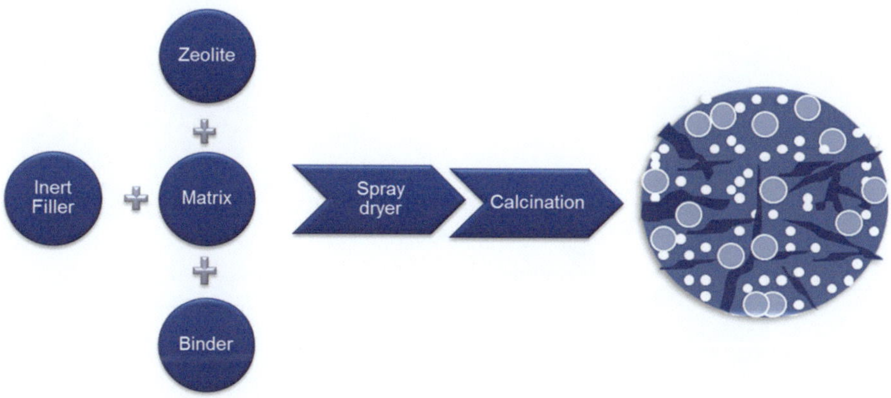

Fig. 8.1 Incorporated FCC catalyst production process

In situ catalyst production is composed of two major processes: (1) microsphere formation and (2) in situ crystallization of zeolite and subsequent treatments for its stabilization and catalytic functionality (Fig. 8.2). In the first process, an aqueous slurry containing raw kaolin (hydrous kaolin), powdered calcined kaolin (clay-based intermediate), non-clay functional matrix materials, and temporary binder is charged into a spray dryer to form microspheres. The microspheres then undergo further calcination to obtain the desirable reactivity and structure. In the second process, the calcined microspheres are reacted with a caustic sodium silicate solution to leach the reactive silica and alumina in the microspheres and convert them into the sodium form of zeolite Y. The as-crystallized microspheres are then ion exchanged with rare earth and ammonium ions and further calcined to enhance the zeolite stability.

In situ techniques achieve a desirable and unique combination of high zeolite content (levels >50 % can be reached) associated with high activity, enhanced selectivity, and thermal stability. Such a combination can be attributed to a number of features. First of all, the macropore walls in the catalyst particle are coated with the in situ grown zeolite (Fig. 8.3), and the zeolite is free of binder coating, in contrast to the case of incorporated catalyst. In fact, zeolite also plays the role of a binder in the in situ technology. Thus, the coated zeolite is readily accessible to the hydrocarbon feed, resulting in coke-selective cracking of the feedstock. Secondly, the catalyst exhibits high macroporosity. Large open pores are clearly visible in the electron microscopy studies of catalysts produced by the in situ route (Fig. 8.3). During the microsphere formation, spray-drying creates macropores in the microspheres. Additionally, the structural matrix in the calcined kaolin further enhances mesoporosity and macroporosity, resulting in a desirable porosity that reduces diffusional limitations during the cracking process, which leads to high conversion of the hydrocarbon feedstock.

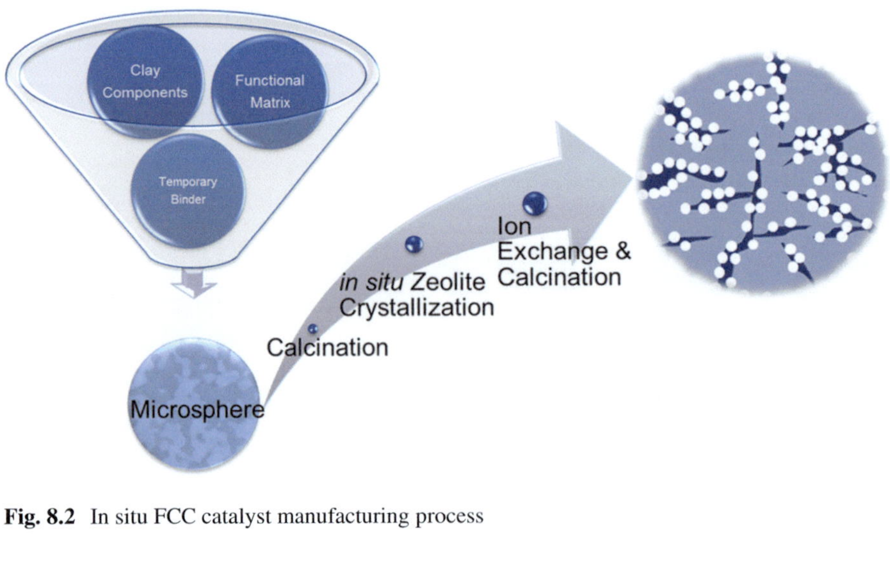

Fig. 8.2 In situ FCC catalyst manufacturing process

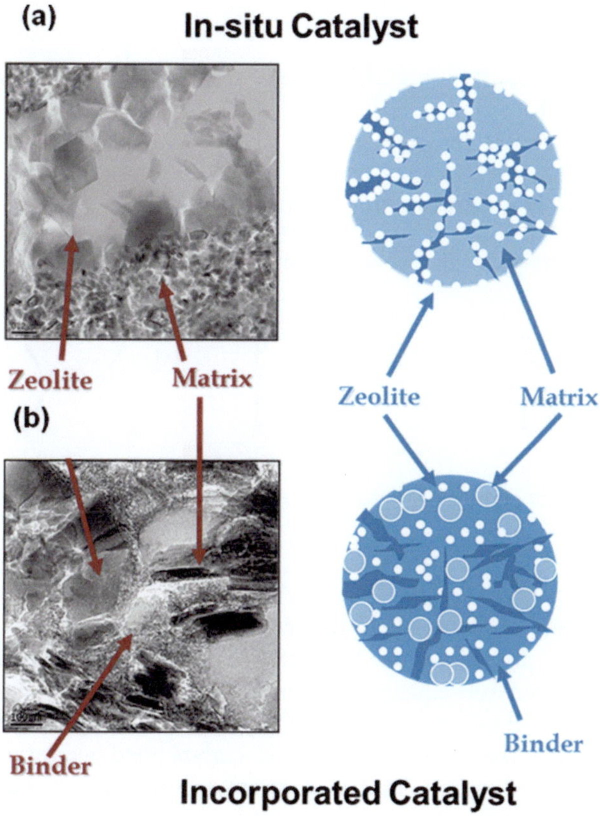

Fig. 8.3 Transmission electron microscopy (TEM) images (**a**) and schematic representation (**b**) showing catalysts from in situ and incorporated manufacturing processes

8.5 Catalytic Cracking Mechanism in the Zeolite

Cracking in an FCC unit involves the breaking of carbon–carbon bonds to make smaller fragments. Most cracking in an FCC unit is catalytic; however, thermal cracking does occur and the two routes have very different mechanisms. Both involve the formation of a reactive species, which initiates and propagates all cracking reactions. With thermal cracking, this involves first the formation of a radical. With catalytic cracking, this involves first the formation of a carbocation via interaction with the zeolite-based catalyst. Both give very different product slates. Since thermal cracking always occurs to some extent in an FCC unit, it is also briefly discussed here.

Thermal cracking is heavily dependent on temperature and occurs when temperatures are between 800 and 1200 °F (425 and 650 °C). The mechanism involves first the formation of free radicals as a result of the breaking of a carbon–carbon bond evenly, i.e., both carbons retain one electron from the broken bond (Fig. 8.4).

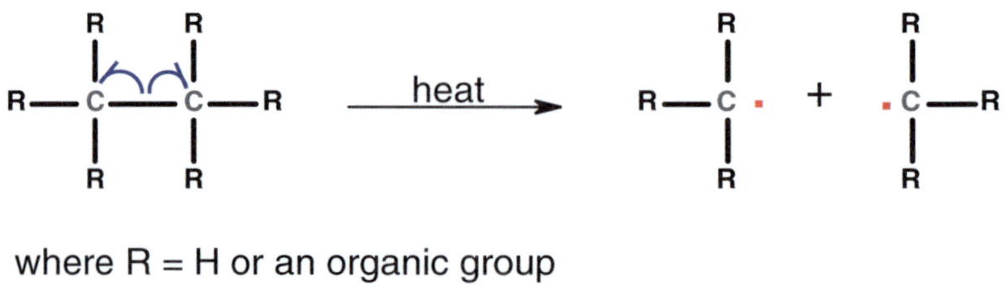

Fig. 8.4 Formation of radicals via thermal cracking

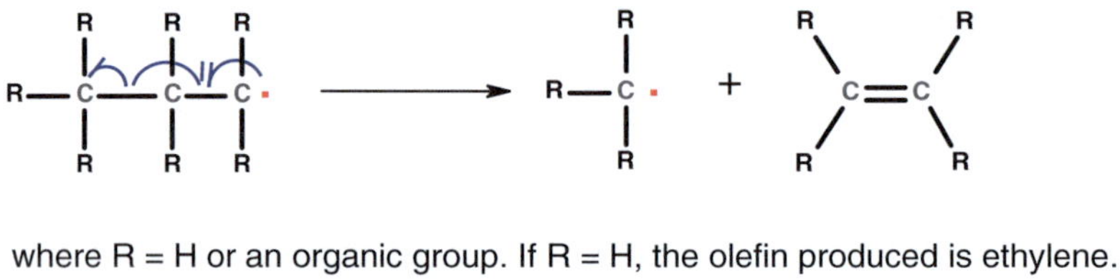

Fig. 8.5 Mechanism for beta-scission

The free radicals are then able to further react with other molecules or radicals and do so in a short amount of time because free radicals are very reactive and very short-lived. The main subsequent reactions include alpha-scission, beta-scission, and polymerization. Beta-scission is more thermodynamically favored than alpha-scission and involves the breaking of a carbon–carbon bond two carbons away from the radical containing carbon. The result is ethylene and a new radical, with the latter continuing the chain of reactions (Fig. 8.5).

Alpha-scission, resulting from the breaking of a carbon–carbon bond one carbon away from the radical, produces a methyl radical, H_3C, which can abstract a hydrogen atom from another molecule to produce methane and a new radical. Again, the new radical continues the chain of reactions.

Catalytic cracking, contrarily, involves interaction of the feed molecules with the zeolite-containing catalyst, which is already at high temperature, first resulting in the vaporization of feed molecules. The next step in catalytic cracking is the formation of a carbocation, or a carbon with a positive charge, and can be in the form of a carbenium or a carbonium ion. As carbonium ions are not stable, most cracking mechanisms involve carbenium ion reactions. A carbenium ion is formed (a) when an olefin, coming in with the feed or as a result of thermal cracking, interacts with a Brønsted acid site on the catalyst, in which the acid site donates a proton (H^+) to the olefin, and (b) when a paraffin interacts with a Lewis acid site on the catalyst, in which the acid site removes a hydride ion (H^-). Both of these mechanisms result in the formation of a positive charge on the carbon atom (Fig. 8.6).

Not all carbocations are created equally. The stability of a carbocation is dependent on the substituents of the positive charge containing carbon, namely, the more carbon atoms or organic groups (versus simply hydrogen atoms) on the positively charged carbon, the more stable the carbocation is. This is summarized in the diagram

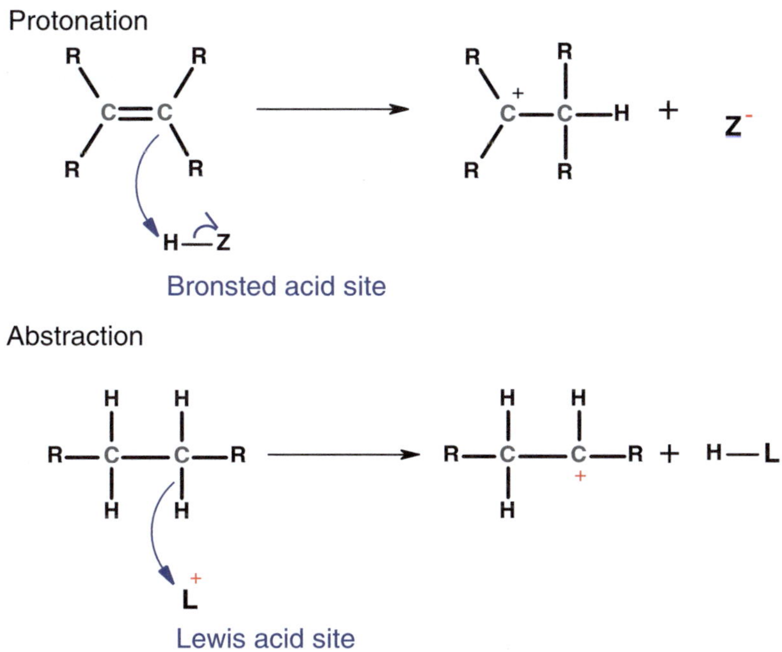

Fig. 8.6 Carbocation formation via two mechanisms: protonation and abstraction

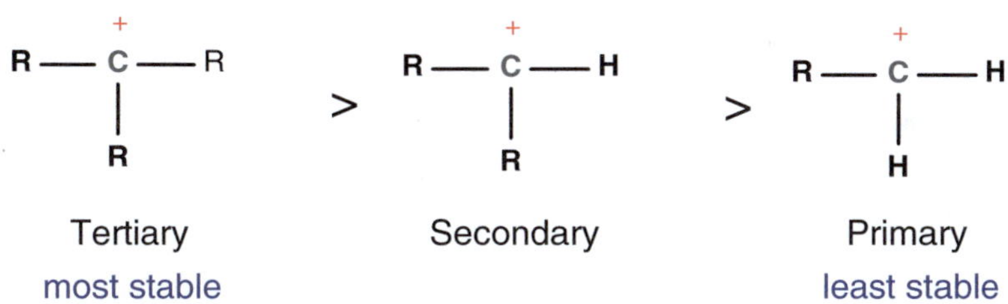

Fig. 8.7 Summary of the stability of carbocations

below (Fig. 8.7), where a tertiary (with no hydrogen atoms attached to the positively charged carbon) is the most stable carbocation species.

The more unstable carbocations are also capable of rearranging to form more stable secondary or tertiary carbocations. The rearrangement of less stable carbocations is a reason that branched products occur readily in an FCC unit (Fig. 8.8).

Following carbocation formation, a number of reactions occur next. The main reactions that occur include breaking of a carbon–carbon bond, hydrogen transfer, and isomerization.

Breaking a carbon–carbon bond is the heart of cracking. Like in thermal cracking, this can be in the form of alpha-scission or beta-scission. For the same reason, namely, alpha-scission being thermodynamically unfavorable, beta-scission accounts for the

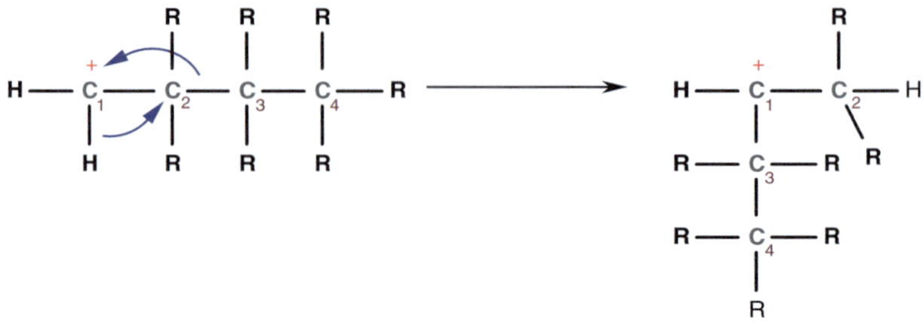

Fig. 8.8 Rearrangement of a primary carbocation to a secondary carbocation

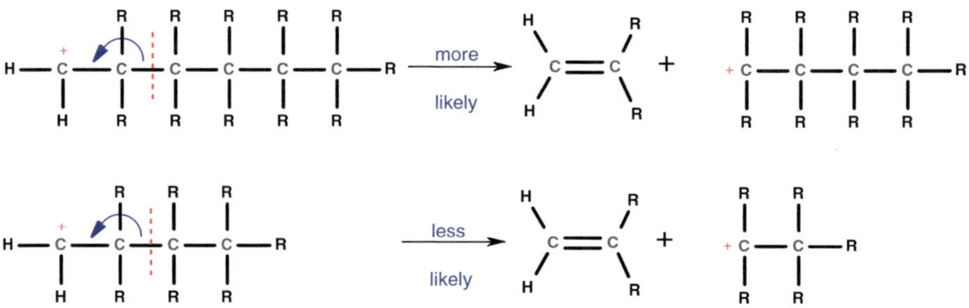

Fig. 8.9 Probability of beta-scission reactions based on chain length

bulk of the cracking reactions. Like carbocation stability ranking, beta-scission reactivity also trends with the size of the molecule involved. Longer chain hydrocarbons (up to maximum C16) tend to undergo beta-scission more readily, whereas a short chain hydrocarbon undergoing beta-scission would be less preferred, culminating in the shortest chain being unable to undergo beta-scission because of the lack of a beta-carbon. This trend is summarized in the graphic below (Fig. 8.9).

Also, the inherent stability of the products dictates the probability of the reaction occurring. In the example below, the first reaction produces an olefin and a tertiary carbocation, while the second reaction produces an olefin and a primary carbocation. Since tertiary carbocations are more stable than primary carbocations, the second reaction will proceed more readily (Fig. 8.10).

Beta-scission results in the formation of an olefin and a new carbocation, which continues the cracking reactions. The cracking mechanism is complete when the carbocation is desorbed from the surface of the catalyst, and hydrogen transfer from the carbocation to the zeolite forms a paraffin. As shown in the figure below, the Brønsted acid site is neutralized and can again participate in carbenium forming reactions (Figs. 8.6 and 8.11).

In addition to the breaking of carbon–carbon bonds, isomerization also occurs in an FCC unit. As described above, carbocations tend to rearrange to form the most stable carbocation (a tertiary carbocation). This is the reason and source of isomerization in FCC cracking. Since rearrangement can take place before beta-scission,

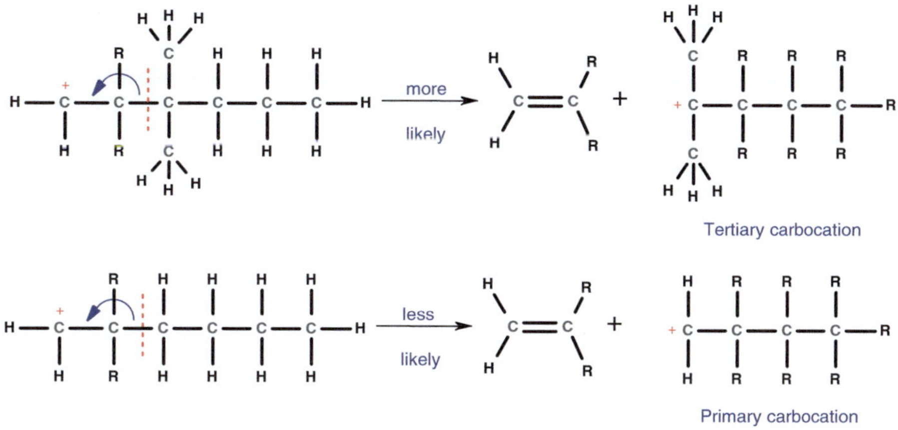

Fig. 8.10 Probability of beta-scission reactions based on product stability

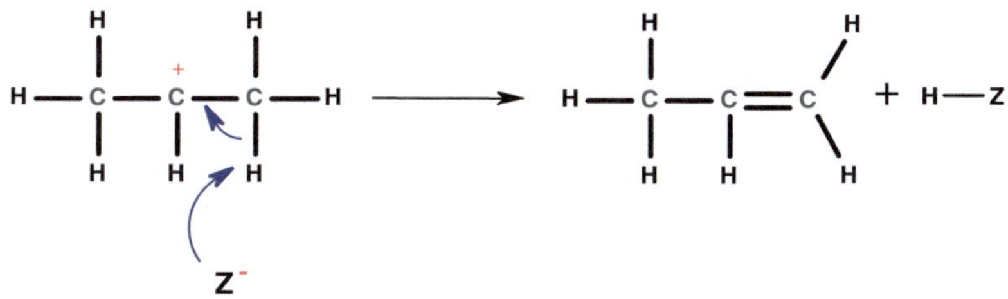

Fig. 8.11 Mechanism of the termination step involving the Brønsted acid

the products in FCC are often highly branched, resulting in isoparaffins instead of normal paraffins (i.e., n-paraffins), which is not the case for thermal cracking reactions.

As a result of isomerization in catalytic cracking, a number of different phenomena are seen in FCC product slates. Isoparaffins have lower boiling points than n-paraffins; since isoparaffins are more branched, they cannot pack as tightly as n-paraffins, lowering the boiling point. In turn, this lowers the cloud point of the diesel fuel fraction. Additionally, isoparaffins have higher octane numbers than n-paraffins, directly resulting in higher octane gasoline.

Hydrogen transfer reactions also play a role in catalytic cracking. Hydrogen transfer can occur when two olefins are activated within close proximity of each other, namely, when two olefins interact with adjacent or proximal acid sites on the FCC catalyst. For this to happen, there needs to be a high number of active sites (high acid site density) in the bulk catalyst so that they are near each other, or the catalyst needs to be rare earth exchanged. In the latter, one rare earth element stabilizes or bridges two or three acid sites on the catalyst. If these acid sites are bridged, the probability of two olefins interacting is greater, and in turn more hydrogen transfer will occur. Hydrogen transfer reactions create paraffins and aromatics, thus depleting the olefins in the product slate. As such, the gasoline cut that is created in the FCCU is more stable (less of the reactive olefins around) and is not over-cracked. However, because there are lower olefins, the gasoline octane number is lower. In addition to impacting the gasoline cut, the LPG cut will also see lower olefins.

Dehydrogenation can also occur during catalytic cracking in an FCC unit producing hydrogen or H_2. However, these reactions are not inherent to the catalyst and only occur when the catalyst is contaminated with metals from the feed such as nickel and vanadium, both of which are active dehydrogenation catalysts themselves.

In summary, the role of active sites on the FCC catalyst is of critical importance, dictating the performance of the catalyst and as a result the whole process. The active sites provide the means for initiating the cracking mechanisms described above, namely, by the formation of a carbocation. Additionally, the active sites are key to the amount of hydrogen transfer reactions occurring in the riser and therefore define the product slate coming out of an FCC unit.

8.6 Understanding the Role of Zeolites in FCC

The greatest breakthrough in catalytic cracking occurred in 1962 when it was demonstrated that zeolite Y can effectively replace amorphous silica–alumina as the main active component of the FCC catalysts [18]. Zeolite Y-based FCC catalysts were commercially introduced in 1964 [19]. The potential of zeolites is probably best illustrated by their history in FCC [20] that led to a step change in the oil refining industry. Their high activity, combined with the increased selectivity toward the gasoline fraction and the lower coke production relative to the previous generation of catalysts (amorphous silica–alumina-based catalysts), allowed refiners to significantly increase gasoline output. This is a dramatic illustration of the leverage of zeolite catalysts [21]. Since then, zeolite Y continues to be the primary component of the FCC catalysts, and all efforts have concentrated on the modification of zeolite Y properties for improving activity, catalytic selectivity, and octane number as well as coke selectivity and stability. Those efforts have already led to significant improvements in the FCC catalyst and process, making the zeolite Y-based catalyst ever more efficient and hard to be replaced by other materials.

Zeolite Y is a large-pore zeolite with the same framework structure as natural zeolite called faujasite (framework code FAU). The composite building units of the Zeolite Y framework are shown in Fig. 8.12. The three building units are the double 6-ring, the sodalite cage which is formed connecting 4- and 6-rings in different arrangements, and a very large cavity with four 12-ring windows known as the supercage. The connectivity of these cages allows molecules to diffuse in the three-dimensional lattice of identical pores, with an opening of around 7.4 Å in size (supercage opening).

Zeolite Y is the main active component in the cracking catalyst and can be prepared in forms that have widely varying performance by manipulating its properties (ion exchange, structural modifications, etc.). The zeolite Y content in the catalyst

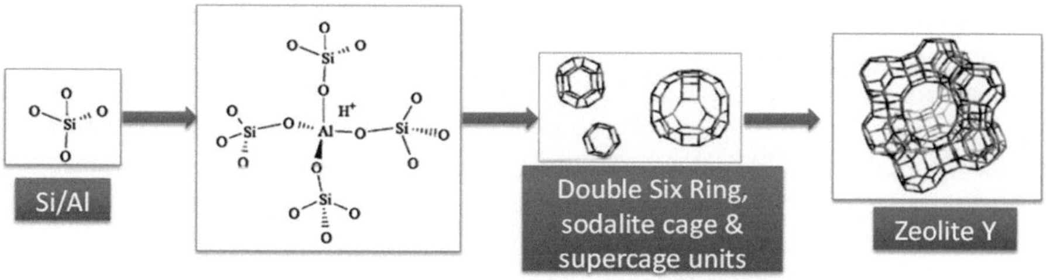

Fig. 8.12 Building blocks and framework structure of Zeolite Y

ranges depending on the catalyst manufacturing technology and can be between 10 and 40 % for incorporated catalysts and can go above 50 % for the in situ manufacturing technology.

Some features of zeolite Y and key factors to its successful use in FCC are [22]:

- Its acidity, which can be tuned easily; the protons which balance the negative charge of a zeolite framework are not strongly bound to the framework and readily react with feed molecules which penetrate into the zeolite pore system. A protonated zeolite thus provides Brønsted acidity.
- The possibility to easily generate mesopores through steaming; together with the matrix porosity, they provide an extra accessibility of bulky feed molecules to the zeolite crystal where the critical cracking and hydrogen transfer reactions can take place efficiently.
- Its structure resists degradation (or adjusts to) at high temperatures allowing their use in the very harsh hydrothermal conditions of the FCC unit.
- Its three-dimensional pore system allows optimal management of the feed and product molecular traffic.
- It has a wide range of composition (Si/Al ratio) and its properties can be tuned by ion exchange.
- It can be easily shaped into microspheres that can be fluidized in the FCC unit.

The properties of the zeolite play a significant role in the overall performance of the catalyst. Understanding these properties increases our ability to predict catalyst response to changes in unit operation. The FCC riser/regenerator environment can cause significant changes in chemical and structural composition of the zeolite. In the regenerator, for instance, the zeolite is subjected to severe thermal and hydrothermal conditions. In the riser, it is exposed to feedstock contaminants such as nickel, vanadium, iron, and sodium. The most important parameters/characteristics governing zeolite behavior through these continual cycles are the following:

- Framework composition
- Unit cell size (UCS)
- Ultra-stabilization by dealumination (USY)
- Sodium content
- Rare earth (RE) exchange

One important parameter controlling the cracking behavior of the zeolite Y is the framework composition which determines, in principle, the amount, strength, and distribution of the Brønsted acid sites associated with framework Al atoms. The product selectivity during gas–oil cracking also depends on the framework composition. Dealuminated Y zeolites with low framework Al content are more selective to diesel in detriment of gasoline, give a higher product olefinicity (and thus a higher research octane number for gasoline), and are more stable under the severe conditions (temperatures above 700 °C or ca. 1300 °F in the presence of steam) used in the regenerator of FCC units. However, it was found that the cracking behavior of dealuminated Y zeolite is also strongly contingent on the nature and amount of extra framework Al species as well as the mesoporosity generated during the dealumination treatments [23, 24]. Therefore, great efforts have been devoted to the optimization of the cracking performance of Y zeolite through the manipulation of both extra framework Al and mesoporosity. Mesoporosity plays an important role in catalytic cracking since the bulky molecules present in commercial feedstocks can hardly access the acid sites located in the micropores and the first cracking events will take place, predominantly, on the external zeolite surface. Therefore, the accessibility of the zeolite within the catalyst particle plays a very important role. In this respect the in situ synthesis route presents a considerable intrinsic advantage since the zeolite is more exposed as it grows directly on the walls of the macropores, providing superior accessibility to the zeolite. Consequently, the activity and selectivity of the zeolite will be mainly determined by the surface composition and the accessibility to the acid sites, which greatly depends on the synthesis method as well as the method and severity of the dealumination treatment.

The elementary building block of the zeolite crystal is a unit cell. The unit cell size (UCS) is the distance between the repeating cells in the zeolite structure. One unit cell in a typical fresh Y zeolite lattice contains 192 framework atomic positions of aluminum and silicon atoms that correspond to silica (SiO_2)-to-alumina (Al_2O_3) molar ratio of 5–8. The UCS is an important parameter in characterizing the zeolite structure. A unit cell of zeolite Y contains 8 large supercages, 8 sodalite cages, and 16 double-six-ring units. Molecules larger than water or ammonia can access only the supercages and cannot pass into the empty space inside sodalite cages. Thus, most of the zeolitic FCC reactions take place in the supercages. The unit cell size is measured in angstroms and is also a measure of the ratio of silicon to aluminum within the zeolite crystal structure. Because aluminum is the larger of the two ions, the unit cell size dimension gets smaller as the aluminum is removed from the zeolite structure. Reducing the aluminum content of the zeolite will increase the silica-to-alumina ratio and result in fewer but stronger and more widely separated acid sites per unit cell. The strength of the acid sites is determined by the extent of their isolation from the neighboring acid sites as well as local site geometry (i.e., site architecture). The close proximity of these acid sites causes destabilization of the zeolite structure. Acid site distribution of the zeolite is a fundamental factor affecting zeolite activity and selectivity. In addition, the UCS measurement can be used to indicate octane number potential of the zeolite-based catalyst. A lower UCS presents fewer active sites per unit cell. The fewer number of acid sites mean that they are farther apart and therefore inhibit hydrogen transfer reactions, which in turn increase gasoline octane number, as well as the production of C3 and lighter components. Thus, the UCS is also an indicator of zeolite acidity, and its changes can help predict changes in catalyst activity and selectivity and ultimately the performance of the FCC unit and the properties of its products.

Introducing a fresh as-synthesized catalyst containing an H–Y zeolite in an FCC regenerator causes rapid, large-scale uncontrolled dealumination of the zeolite framework. The steam partial pressure is too low and the temperature too high for the silicon reinsertion (i.e., defect healing) process to occur ideally, so the equilibrium catalyst's performance would not be optimum. The FCC catalyst manufacturers prefer to carry out this modification, by including high-temperature calcination of the zeolite in the presence of steam. Such Y zeolites are commonly called ultrastable Y (USY) [25]. Removing approximately half of the aluminum and replacing it with silicon is sufficient to stabilize the structure permanently. Associated with this are predominantly three structural changes that have been well documented in the literature. First, destruction of part of the zeolite framework occurs. Electron microscopy studies of steamed Y zeolites have shown the presence of a substantial number of newly generated mesopores [26, 27]. Second, the framework aluminum ions removed subsequently continue to exist as extra framework aluminum species [28]. Finally, new Lewis acid sites and hydroxyl groups appear [29, 30]. The catalytic consequence of steaming is that the zeolite becomes much more active for cracking, sometimes by nearly two orders of magnitude compared to the unmodified catalysts. The enhancement in activity depends on the pretreatment of the sample, such as the method and severity of dealumination procedures, in addition to the concentration of acid sites in the original sample [31, 32]. Thus, post-synthesis modifications of zeolites are technologically critical since they result in frameworks with greater thermal stability and enhanced catalytic properties [32]. Such post-synthesis treatments can also provide another very important advantage since they allow more thorough elimination of sodium if followed by one or more ion exchange(s) with ammonium cations.

The sodium on the catalyst originates from zeolite synthesis during its manufacture. It is important for the fresh zeolite to contain very low amounts of sodium. The residual sodium in zeolite is not desirable since sodium neutralizes the zeolitic acid sites, and in the presence of vanadium in the feed and under the hydrothermal conditions of the regenerator of the FCC, it can promote zeolite destruction [33]. It is therefore advantageous to eliminate sodium totally, but it is not an easy process and is a costly procedure. The sodium cations can be replaced by ion exchange with ammonium or rare earth cations in an aqueous solution. A series of exchanges is needed in the presence of temperature and intermediate calcinations in order to achieve low-sodium zeolite Y. Calcination helps sodium ions to migrate to easily accessible sites where they can be replaced by ion exchange.

Rare earth (RE) elements such as lanthanum and cerium stabilize aluminum atoms in the zeolite structure when they are introduced via ion-exchange procedures. They prevent the aluminum atoms from leaving the zeolite lattice when the catalyst is exposed to harsh hydrothermal conditions in the FCC regenerator. A fully rare earth-exchanged zeolite equilibrates at a high UCS, whereas a non-rare earth zeolite equilibrates at a very low UCS. All intermediate levels of rare earth-exchanged zeolite can be produced. Rare earth cations therefore improve thermal and hydrothermal stability of the zeolite. Rare earth increases zeolite activity and gasoline selectivity with a loss in octane number (Fig. 8.13). The octane number loss is due to the promotion of hydrogen transfer reactions that increase also the coke production. The poorer gasoline quality is largely related to the transformation of olefins into corresponding paraffins to a greater or lesser degree. Paraffins are more stable structures than olefins and so are less likely to undergo consecutive cracking themselves as described in Sect. 8.5. The result is that lower gasoline quality usually goes along with an improvement in gasoline yield. As shown in Fig. 8.13, the decrease in rare earth content means a reduction in the UCS and therefore a decrease in the equilibrium zeolite's acidity. As a result, the catalyst manufacturer can adjust the acidity that the catalyst will develop once it is in equilibrium conditions in the unit by introducing more or less rare earth cations during preparation.

In the field of catalytic cracking, a clear distinction is present between the fresh FCC catalyst as it is supplied by the manufacturer and the equilibrium catalyst that is in use in the FCC unit, as their physicochemical and catalytic properties are drastically different. As soon as it enters the regenerator of the FCC unit, the catalyst undergoes several modifications due to the high temperature of the regenerator (typically ranges between 650 and 750 °C or 1200–1400 °F), the presence of steam, and the contamination by the feed components (Ni, V, Na, Fe, Ca, etc.) and coke. The result is a decline in catalytic performance. If nothing is done, this irreversible degradation results in an unacceptable catalytic performance after a few days, weeks, or months depending on operational conditions/configurations. Therefore, a small fraction of the catalyst inventory circulating in the FCC unit is deliberately

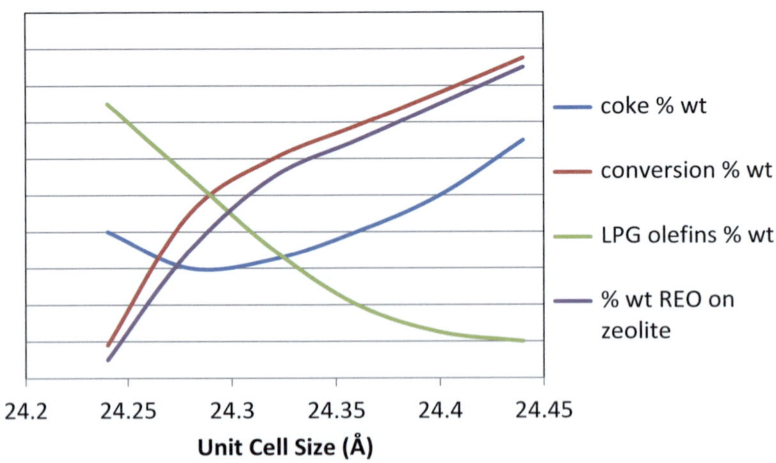

Fig. 8.13 FCC yields and rare earth (RE) correlation to unit cell size (UCS)

withdrawn every day and replaced with an equivalent amount of fresh catalyst. The mixture of this fresh and aged catalyst is called equilibrium catalyst (Ecat).

All catalysts at equilibrium conditions have a reduced unit cell size when compared with their fresh catalyst state. Reducing the unit cell size stabilizes the zeolite while removing aluminum from the zeolite structure. The lower unit cell size of equilibrium catalyst correlates with improvements in research and motor octane number. The unit cell size of steamed or equilibrium FCC catalyst ranges from about 24.25 to 24.35 Å and is significantly lower than the unit cell size measurements of fresh catalysts. In the production of FCC catalysts, most fresh catalysts start with a relatively high unit cell size in the range of 24.50–24.75 Å. It is the hydrothermal environment of the FCC unit regenerator which extracts aluminum from the zeolite structure, thus, reducing the acid site density and unit cell size of the equilibrium catalyst. The final equilibrium level unit cell size is dependent upon the rare earth and sodium levels of the zeolite. The lower the rare earth level of the fresh catalyst, the lower is the resultant equilibrium unit cell size and the greater the octane number.

Besides Y zeolite, the addition of ZSM-5 zeolite to the FCC unit has been recognized as the simplest solution for improving gasoline octane number and increasing the yield of light olefins (propylene and butenes) at the expense of gasoline [18]. ZSM-5 was discovered by Mobil in 1965 (US Patent No. 3702886), and its first scale commercial trial in an FCC unit as a catalyst (mixture with REY) was in 1984. Since 1986 it has been used as a separate FCC catalyst additive [34]. The significant incentives of using ZSM-5 additive provided by the need for increasing the gasoline octane number in the early 1980s were later combined with the considerably more favorable economics for producing light olefins especially in FCC units with propylene maximization as a main target (i.e., max-propylene units) [34].

Since first synthesis of ZSM-5, significant advances have been made to improve its hydrothermal stability (under the regenerator conditions), catalytic activity, selectivity, and metal tolerance [35]. The concept of using ZSM-5 as an additive derived from the effectiveness of ZSM-5 in increasing RON and MON at even small concentrations, the experience in commercializing Pt-based CO promoters, and the cost of manufacturing the ZSM-5 zeolite [34]. A critical question was whether ZSM-5 should be incorporated into a separate additive or whether it was most effective when incorporated in the same particle as the zeolite Y. Testing at Mobil showed that there was no synergism between zeolite Y and ZSM-5, and therefore the separate additive case was considered as more attractive, since it brings added flexibility [34]. Additives with high level of ZSM-5 have become commercially available in the late 1990s. These new ZSM-5 additives produce high levels of light olefins with minimal dilution of the base cracking catalyst [33]. Nowadays, it can be used as an additive in a wide range of amounts (typically 1–25 %) in the unit's total inventory depending on the objective of the refinery.

Zeolite ZSM-5 is an MFI type of zeolite. Its structure is characterized by a three-dimensional channel network with 10-ring cages and an opening of around 5.5 Å in size. The MFI framework topology is depicted in Fig. 8.14.

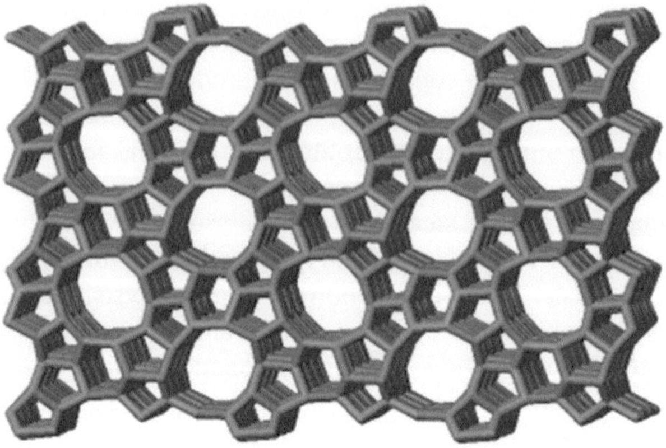

Fig. 8.14 Skeletal diagram of ZSM-5

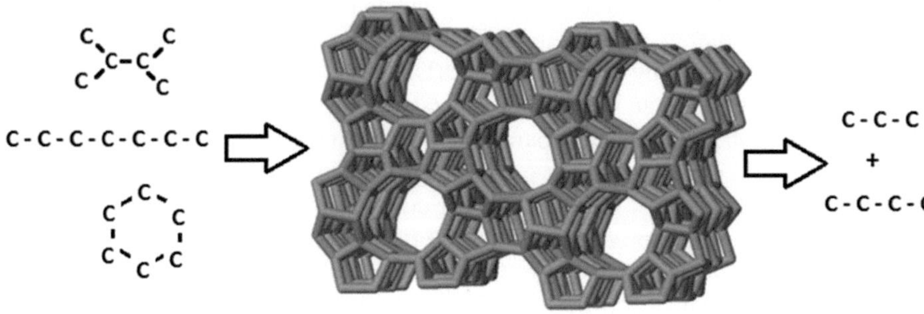

Fig. 8.15 Catalytic cracking in ZSM-5 zeolite

ZSM-5 is well known as a highly shape-selective zeolite in FCC since it works by promoting the cracking of certain gasoline range components, particular linear or single-branched paraffins, and olefins, with poor octane number, eliminating them and producing LPGs (C3 and C4 olefins) at the expense of gasoline yield [36], as illustrated in Fig. 8.15.

In 1987, ZSM-5 additive manufacturers started to add phosphorus to their formulations. The effect was dramatic in improving the efficacy of ZSM-5 and reducing the amount of additive required to achieve the desired effect in octane number and light olefin yields from FCC operations. The benefit of adding phosphorus on the stability and selectivity of ZSM-5 was attributed to its ability to retard aluminum from leaving the zeolite framework. Phosphorus treatment reduces the initial acidity of the zeolite but helps it retain a larger fraction of its acidity after severe steaming produces [37, 38]. In addition, formation of a new type of acid sites with the introduction of phosphorus contributes to the enhanced performance of this new generation of additives.

8.7 FCC Catalyst Deactivation

As fresh catalyst enters and circulates in the FCC unit, deactivation of the catalyst takes place. The presence of steam at elevated temperatures in the FCC regenerator can cause dealumination and the eventual collapse of the zeolite structure [39]. Contaminant metals in the feed can cause further reduction in catalyst activity [33] and undesirable formation of hydrogen and coke during the cracking process [40]. The detrimental effects of the FCC feedstock metals (especially Ni, Na, Fe, and V) have been widely recognized in literature [33, 40–42]. Nickel and vanadium are well-known dehydrogenation catalysts. Nickel has approximately three to four times more dehydrogenation activity than vanadium, but vanadium destroys also the crystalline zeolite framework of FCC catalyst, especially in the presence of sodium [33]. Furthermore, the catalyst circulating in the unit, termed equilibrium catalyst (Ecat) as mentioned above, contains particles of different ages. The freshly added catalyst particles undergo the fewest cycles of reaction/stripping/regeneration and have the highest surface area/activity and lowest level of contaminant metals. On the other hand, the oldest particles go through many cycles and have the lowest surface area/activity and highest contaminant metal levels. Thus, realistically mimicking the Ecat properties via accelerated deactivation in the laboratory is critical for the development of catalyst technologies and the selection of the right catalyst for each FCC unit.

Two types of deactivation approaches are typically used. For low metal FCC operations, steam deactivation without the addition of contaminant metals is employed. For high metal operations, contaminant metals (V, Ni) are deposited onto the catalyst particles using Mitchell impregnation/cyclic propylene steaming (CPS) [41] or cyclic metals deactivation (CMDU) [42].

8.7.1 Steam Deactivation

During steam deactivation, fresh catalyst is hydrothermally treated at high temperature in the absence of metals causing zeolite dealumination and acid site loss. The steam temperature and duration are controlled to match the target Ecat's unit cell size, activity, and, to some extent, selectivity. Hydrothermal deactivation leads to permanent loss of catalytic activity due to zeolite dealumination and zeolitic surface area reduction. To incorporate age distribution into the deactivation and better mimic the Ecat selectivity, a continuous age distribution model has been developed [43].

8.7.2 Mitchell Impregnation/CPS

The deactivation of FCC catalysts in small-scale units in the presence of contaminant metals has been the industry workhorse for a number of years due to the robustness and simplicity of such methods.

The first approach for metalation and deactivation of fresh FCC catalysts was the Mitchell technique [41]. The Mitchell technique consists of incipient wetness impregnation of FCC catalysts with vanadium and nickel naphthenate or octoate solutions followed by a subsequent deactivation performed in small-scale or pilot units by exposing the metallated FCC catalysts to hydrothermal conditions. In this protocol, vanadium is in the +5 state, continuously, where it reacts to form vanadic acid under hydrothermal conditions that has intra-particle and inter-particle mobility, which brings more effective penetration and destruction of the zeolite than in lower oxidation states of vanadium [33, 44–46]. Therefore, this method exaggerates FCC catalyst activity and selectivity.

The cyclic propylene steaming (CPS) method based on the Mitchell technique of adding metals was introduced with the objective of better simulating, on a small or pilot scale, the performance of commercial FCC catalyst compared to Mitchell technique followed by steaming. In the CPS protocol, FCC catalyst is impregnated by the incipient wetness method (Mitchell technique) with vanadium and nickel prior to deactivation in reduction–oxidation cycles using propylene as the reducing medium [44]. This provides significant advantages over traditional methods such as the Mitchell approach. Nevertheless, recent investigations have shown that, in certain situations, the contribution of vanadium to catalyst deactivation is overemphasized in the original CPS-1 method [47, 48], and therefore this method has been adjusted further by applying a number of modifications. These modifications have been made to further address the destructive effects of vanadium on the FCC catalyst during deactivation and reduce its effect to the levels observed on commercial FCC catalysts. The new CPS-3 protocol includes change in the sequence of stripping/reduction/stripping/oxidation, increase of the time the catalyst spends in reducing compared to oxidizing environment, and introduction of pre-stabilization cycles during the heat-up phase. The objective was not only to develop a more realistic and accurate deactivation protocol but also to maintain the productivity of the CPS method, and therefore the total time of the CPS-1 method was not increased. In other words, the catalyst decay in the modified CPS method is governed more by hydrothermal deactivation and less by deactivation from highly oxidized vanadium, especially at high metals levels.

The deactivation procedures for the metal-loaded catalyst in the CPS protocol have been modified continuously over the years [44, 47, 48], while the Mitchell-type impregnation had not been modified substantially. In Mitchell-type methods, the FCC catalysts are artificially metallated with vanadium and nickel naphthenates that are dissolved in an organic solvent. Most of the literature studies concluded that catalyst metallated by the Mitchell method and deactivated by CPS or steaming led to metal concentrations higher than the realistic metal loading, not properly aged and with a misleading distribution (mainly nickel) and inaccurate activity of the metals (both nickel and vanadium) compared to the CMDU method (see below) which leads to lab-deactivated catalyst closer to the commercial equilibrium catalyst [42, 49–51].

More recently, a new method of catalyst metalation was developed using a spray impregnation technique, where the contaminant metals vanadium and nickel are deposited on the outer surface area of the FCC catalyst particles [52]. Using this technique, nickel remains primarily on the surface of the particles under severe hydrothermal conditions, while vanadium migrates into the bulk of the particles and from particle to particle as observed in equilibrium catalysts. This dispersion of the contaminant metals occurs simultaneously with zeolite degradation and thus leads to different effects on the physical and catalytic properties of FCC catalysts compared to a Mitchell-type technique where the contaminant metals already penetrate the FCC catalyst particles during the impregnation step. In commercial FCC unit operation, vanadium and nickel are deposited on the catalyst particles and subsequently migrate inter- and intra-particles depending on their mobility. The spray impregnation method provided a more close to equilibrium catalyst-like migration profile during deactivation compared to the Mitchell method where vanadium and nickel behaved quite differently. However, this spray impregnation technique cannot account for the various degrees of penetration seen in different Ecats and therefore has the risk of oversimplifying the Ni distribution profiles.

To conclude, although there have been significant changes over the years in the CPS protocol, the metal poisoning simulation in the laboratory is still a very complicated and controversial task since:

- The use of metal naphthenate or octoate solutions as the precursor to impregnate the metals instead of in commercial feed type metal molecules like porphyrines.
- The metal–catalyst interaction not properly reflecting the effect of pore volume and matrix.
- The lack of age distribution of the metal, resulting in a deactivated sample with a uniform age instead of sample with age distribution varying from minutes to days mimicking the commercial FCC application.
- The substantial differences in hydrogen and coke yields when comparing catalysts from different technologies in contrast to minor differences when comparing the same catalysts in commercial performance are still not addressed properly, highlighting the need for further improvements in the future.

8.7.3 CMDU

In order to improve the simulation of commercial deactivation in lab testing, cyclic metals deactivation unit, CMDU, was designed where the aging of FCC catalyst is simulated by the deactivation in a fixed fluidized bed through repeated cycles of reaction, stripping, and regeneration. During the cracking step, metals are deposited on the catalyst with metal-spiked FCC feedstock. Volatile hydrocarbons are then stripped from the catalyst with nitrogen and/or steam, and the temperature is raised to the regeneration temperature. The catalyst is regenerated with a gas mixture of steam, oxygen, and nitrogen and optionally other gases, like SO_x. The CO combustion mode (complete or partial) can be controlled with the gas flow and composition. After regeneration, the catalyst temperature is reduced for a new cycle of reaction, stripping, and regeneration. Fresh catalyst addition and/or withdrawal can be done during the run in order to simulate the age distribution of equilibrium catalyst and to collect catalyst samples for the determination of physical and chemical properties and catalytic performance. The metal tolerance can thus be determined easily upon progressive metal buildup and (hydro) thermal deactivation [53, 54].

The key differences of CMDU compared to Mitchell/CPS type of deactivation are:

- Metal deposition in the cracking step at riser temperature by using a metal-spiked feedstock in CMDU versus metal impregnation by using a metal solution at low temperature in Mitchell/CPS
- Improved metals deposition profile in CMDU, which is closer to the actual distribution of metals in refinery versus uniform metal distribution with Mitchell/CPS
- Low-temperature deactivation in stripping followed by high-temperature deactivation during regeneration in CMDU vs. constant high-temperature deactivation throughout CPS

As the metal deposition process in CMDU is more representative of the deposition process in the commercial FCC units, CMDU mimics closer the metal distribution in actual Ecats [42, 53–55]. As shown in Fig. 8.16, in commercial Ecat,

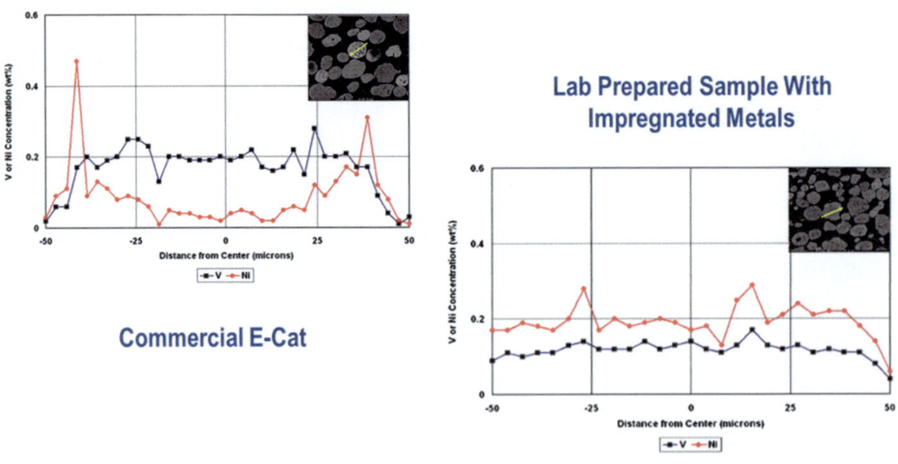

Fig. 8.16 Energy dispersive X-ray spectroscopic line scans of commercial equilibrium catalyst (Ecat) and lab-deactivated catalyst samples using Mitchell impregnation. Mitchell impregnation disperses Ni uniformly across the catalyst particles, whereas Ni is concentrated on the edges of the Ecat

V is distributed uniformly throughout the catalyst due to its mobility under FCC operating conditions. In contrast, Ni is concentrated on the edges of the Ecat particles where it accumulates over time, indicating that Ni mobility is very limited. In the commercial operation, new Ni molecules deposit onto the old Ni molecule and that results in the agglomcration of Ni and a reduction on the activity of Ni with age. Mitchell impregnation disperses Ni uniformly across the catalyst particles and thus exaggerates the dehydrogenation effect of Ni at the same Ecat Ni level. CMDU better mimics the Ni penetration profile and results in relatively higher concentration at the edges and thus more representatively reflect the Ni activity.

It has been demonstrated in the literature that the catalytic performance of catalysts deactivated by CMDU represents commercially equilibrated catalysts much more realistically than the Mitchell/CPS type of deactivation technique [41, 49, 50, 53, 54].

8.8 Conclusions

In no other industrial process is the impact of zeolite catalysts on global economy more clear and visible than fluid catalytic cracking (FCC). Majority of the gasoline used around the world today comes from the FCC process, in which zeolite-based catalyst is the key enabler. The presence of zeolites in FCC catalysts brings a substantial—and much needed—increase in gasoline yield from catalytic cracking, leading to more efficient utilization of crude oil. FCC units worldwide also contribute considerably to the global supply of other fuels. Important basic chemicals such as propylene and butylene are also produced in FCC units in ever-increasing quantities with the help of zeolite-based catalysts and additives. As a result, FCC today represents the largest volume of zeolite catalyst consumption worldwide. It has been more than half a century since the introduction of zeolites into catalytic cracking, and they still are at the core of this key refining process, which fuels our industrialized society.

References

1. Greater Pittsburgh and Allegheny County (1901) The pioneer oil refiner. Original from the New York Public Library: The American Manufacturer and Iron World
2. Bell HS (1945) America petroleum refining. Van Nostrand Co., Princeton
3. Gruse WA, Stevens DR (1960) Chemical technology of petroleum. McGraw-Hill Book Company Inc., New York, pp 346–352
4. Pioneer of Catalytic Cracking: Almer McAfee at Gulf Oil (North American Catalysis Society website)
5. Palucka T (2005) The wizard of octane: Eugene Houdry. Invent Technol 20:36–45
6. Avidan AA, Edwards M, Owen H (1990) Innovative improvements highlight FCC's past and future. Oil & Gas, Journal 88(2): 33–58

7. American Chemical Society National Historic Chemical Landmarks. The Houdry Process for Catalytic Cracking (1996) American Chemical Society, Division of the History of Chemistry and The Office of Public Outreach, Washington, D.C.
8. Avidan AA, Edwards M, Owen H (1989) Paper presented at the sixth international conference on Fluidization, Banff, Alberta, Canada
9. Reichle AD (1992) Fluid catalytic cracking hits 50 year mark on the run. Oil & Gas Journal 90:41–48
10. Cornelis K (2002) Mineral science. Wiley, New York
11. Hosterman JW, Patterson SH (1992) Bentonite and fuller's earth resources of the United States, U.S. geological survey professional paper 1522. U.S. Government Printing Office, Washington, DC
12. Hudec P (2011) FCC catalyst – key element in refinery technology. In: 45th international petroleum conference, Bratislava
13. Bhatia S (1990) Zeolite catalysis: principles and applications. CRC Press, Inc., Boca Raton
14. Yang W (2003) Handbook of fluidization and fluid particle systems. CRC Press, Boca Raton. ISBN 0-8247-0259-X
15. ElzeaKogel, J, Trivedi NC, Barber JM, Krukowsk ST (eds) (2006) Industrial minerals & rocks: commodities, markets and uses, 7th edn. Society of Mining, Metallurgy and Exploration, Littleton
16. Xu M, Stockwell DM (2005) Enhanced FCC catalysts for gas oil and resid applications. US Patent 6942783 B2
17. Sadeghbeigi R (2012) Fluid catalytic cracking handbook. Butterworth-Heinemann., Oxford
18. Corma A, Martinez A (2005) Zeolites in refining and petrochemistry. Stud Surf Sci Catal 157:337–366
19. Plank CJ (1983). The Invention of Zeolite Cracking Catalysts - A Personal Viewpoint. In: Davis BH, Hettuler WP (eds) ACS symposium series 222. ACS, 253, Washington, DC
20. Avidan AA (1993) Origin, development and scope of catalysis. Stud Surf Sci Catal 76:1–39
21. Vermeiren W, Gilson JP (2009) Impact of zeolites on the petroleum and petrochemical industry. Top Catal 52:1131–1161
22. Guisnet M, Gilson J-P (2005) Zeolites for cleaner technologies. Imperial College Press, London
23. Corma A, Fornes V, Martinez A, Melo FV, Pallota O (1988) Effect of the nonuniform dealumination on the acidity and catalytic activity of faujasite. Stud Surf Sci Catal 37:495–499
24. Corma A, Fornes V, Martinez A, Orchilles AV (1988) Perspectives in molecular sieve science. Ser Am Chem Soc, Washington, DC
25. McDaniel CV, Maher PK (1968) Molecular sieves. Society of Chemistry Industry, London
26. Patzelova V, Jaeger NI (1987) Texture of deep bed treated Y zeolites. Zeolites 7:240–242
27. Choi-Feng C, Hall JB, Huggins BJ, Beyerlein RA (1993) Electron-microscope investigation of mesopore formation and aluminum migration in USY catalysts. J Catal 140:395–405
28. Klinowski J, Fyfe CA, Gobbi GC (1985) High-resolution solid-state nuclear magnetic resonance studies of dealuminated zeolite Y. J Chem Soc Faraday Trans 8:3003–3019
29. Janin A, Lavalley JC, Macedo A, Raatz F (1988) In: Flank WH, Whyte TE Jr (eds) Perspectives in molecular sieve science, American chemical society symposium series. American Chemical Society, Washington, DC
30. Auroux A, Ben Taarit Y (1987) Calorimetric investigation of the effect of dealumination on the acidity of zeolites. Thermochim Acta 122:63–70
31. Williams BA, Babitz SM, Miller JT, Snurr RQ, Kung HH (1999) The roles of acid strength and pore diffusion in the enhanced cracking activity of steamed Y zeolites. Appl Catal A Gen 177:161–175
32. Lonyi F, Lunsford JH (1992) The development of strong acidity in hexafluorosilicate-modified Y-type zeolites. J Catal 136:566–577
33. Xu M, Liu X, Madon RJ (2002) Pathways for Y zeolite destruction: the role of sodium and vanadium. J Catal 207:237–246

34. Degnan TF, Chitnis GK, Schipper PH (2000) History of ZSM-5 fluid catalytic cracking additive development at Mobil. Microporous Mesoporous Mater 35–36:245–252
35. Harding RH, Peters AW, Nee JRD (2001) New developments in FCC catalyst technology. Appl Catal A 221:389–396
36. Buchanan JS (2000) The chemistry of olefins production by ZSM-5 addition to catalytic cracking units. Catal Today 55:207–212
37. Blasco T, Corma A, Martínez-Triguero J (2006) Hydrothermal stabilization of ZSM-5 catalytic-cracking additives by phosphorus addition. J Catal 237:267–277
38. Kaeding WW (1981) Selective alkylation toluene with methanol to produce para-xylene. J Catal 67:159–174
39. Corma A et al (1996) Cracking activity and hydrothermal stability of MCM-41 and its comparison with amorphous silica-alumina and a USY zeolite. J Catal 159:375–382
40. Quintana-Solorzano R et al (2011) Cyclic deactivation with steam of metallated cracking catalysts: catalytic testing at the bench scale and the pilot scale. Top Catal 54:547–560
41. Mitchell BR (1980) Ind Eng Chem Prod Res Dev 19:209–213
42. Lappas AA, Nalbandian L, Iatridis DK, Voutetakis SS, Vasalos IA (2001) Effect of metals poisoning on FCC products yields: studies in an FCC short contact time pilot plant unit. Catal Today 65:233–240
43. Stockwell, DM et al (2000) Simulation of FCC equilibrium catalyst age distribution. In: 2nd international symposium on deactivation and testing of catalysts. 219th national ACS meeting, Petroleum Division
44. Boock LT, Petti TF, Rudesill JA (1995) Recent advances in contaminant metal deactivation and metal dehydrogenation effect during cyclic propylene steaming of FCC catalysts. ACS Div Pet Chem 4:421–426
45. Kugler EL, Leta DP (1988) Nickel and vanadium on equilibrium cracking catalysts by imaging secondary ion mass spectrometry. J Catal 109:387–395
46. Wormsbecher RF, Peters AW, Maselli JM (1986) Vanadium poisoning of cracking catalysts: mechanism of poisoning and design of vanadium tolerant catalyst system. J Catal 100:130–137
47. Wallenstein D, Harding RH, Nee JRD, Boock LT (2000) Recent advances in the deactivation of FCC catalysts by cyclic propylene steaming (CPS) in the presence and absence of contaminant metals. Appl Catal A Gen 204:89–106
48. Wallenstein D, Roberie T, Bruhin T (2007) Review on the deactivation of FCC catalysts by cyclic propylene steaming. Catal Today 127:54–69
49. Bendiksen M, Tangstad E, Myrstad T (1995) A comparison of laboratory deactivation methods for FCC catalysts. Appl Catal A Gen 129:21–31
50. Chiranjeevi T, Gokak DT, Ravikumar V, Viswanathan PS (2014) Development of new deactivation method for simulation of fluid catalytic cracking equilibrium catalyst. J Chem Sci 126:353–360
51. Haas A, Suárez W, Young GW (1992) Evaluation of metals contaminated FCC catalysts. AIChE Symp Ser 88:133–142
52. Wallenstein D, Farmer D, Knoell J, Fougret CM, Brandt S (2013) Progress in the deactivation of metals contaminated FCC catalysts by a novel catalyst metallation method. Appl Catal A Gen 462–463:91–99
53. Gerritsen LA, Wijngaards HNJ, Verwoert J, O'Connor P (1991) Cyclic deactivation: a novel technique to simulate the deactivation of FCC catalyst in commercial units. Catal Today 11:61–72
54. O' Connor P, Pouwels AC (1994) Catalyst deactivation 1994. Stud Surf Sci Catal 88:129–144
55. McLean JB et al. Multi stage reaction catalyst: a breakthrough in FCC catalyst technology, NPRA AM-11-01

Chapter 9
Chemistry of the Methanol to Olefin Conversion

Weiguo Song, Yingxu Wei, and Zhongmin Liu

Abstract Methanol conversion to olefin (MTO) has been an intriguing topic for both catalytic research and chemical industry. During the induction time, interesting chemistry occurred for the generation of initial olefin products. After the induction time, the chemistry of MTO proceeds mainly through hydrocarbon pool mechanism, in which carbon pool species are involved in either side-chain growth route or paring route to produce an olefin product. And other minor routes such as direct routes and olefin homologation/cracking route also play certain roles. However, more detailed and extensive research are ahead to fully describe the chemistry of MTO.

The commercialization of methanol conversion technology has been very successful, with several versions designated for certain products, i.e., MTG for gasoline and MTP for propylene. One million ton scale DMTO plant has been in full operation in China. Coupled with coal conversion to methanol, MTO technology is now a new route for major petrochemical raw materials.

Keywords MTO • Zeolite • SAPO-34 • ZSM-5 • Hydrocarbon pool • Coke • Induction • Carbenium cation

W. Song (✉)
Institute of Chemistry, Chinese academy of Sciences,
Beijing 100190, People's Republic of China
e-mail: wsong@iccas.ac.cn

Y. Wei • Z. Liu (✉)
Dalian Institute of Chemical Physics, Chinese Academy of Sciences,
Dalian 116023, People's Republic of China
e-mail: liuzm@dicp.ac.cn

F.-S. Xiao, X. Meng (eds.), *Zeolites in Sustainable Chemistry*,
Green Chemistry and Sustainable Technology,
DOI 10.1007/978-3-662-47395-5_9

9.1 Introduction

9.1.1 Brief History

Methanol to olefin (MTO) conversion is one of the most promising routes to produce useful chemicals from relatively abundant sources such as coal and methane. There are several varied versions of methanol conversion process, including methanol to gasoline (MTG), methanol to hydrocarbon (MTH), or more product-specified methanol to propylene (MTP) or aromatics (MTA). However, all these methanol conversion processes are based on the following overall elementary mass balance and carbon balance:

$$nCH_3OH \rightarrow nCH_2 + n\,H_2O$$

Thus to simplify the nomenclature in this chapter, MTO will be used for the methanol conversion process.

The discovery of MTO was first reported in the early 1970s by Clarence Chang and Tony Silvestri, who were researchers in the Mobil Chemical Research [1–3]. They were trying to produce high-octane gasoline by passing mixture of methanol and isobutene through HZSM-5, hoping that methanol will form the methyl cation (CH_3^+) or carbene ($:CH_2$), which will then add to the isobutene to form longer hydrocarbon chains. What they produced from these experiments were indeed high-octane products that were mainly consisted of aromatics, and gaseous products consisted of light olefins and alkanes. However, careful control experiments suggested that produces with similar product distributions were formed without co-feeding isobutene, and mass balance analysis indicated that isobutene is not consumed when being co-fed with methanol [4]. What they observed was actually a new route to produce hydrocarbon directly from methanol, as known as MTG.

From a broad perspective, syngas (CO and H_2) can be produced from methane or coal, and methanol can be efficiently produced from syngas using $Cu/ZnO/Al_2O_3$-based catalysts and technology. Thus potentially MTO technology fills the gap for an alternative route to produce stock chemicals from C1 sources, which are considerably more abundant than crude oil. With MTO, methanol becomes a carbon source for almost all petroleum-related chemicals and feedstock fine chemicals, especially olefin monomers and polymers. This discovery was a breakthrough in producing aromatics and other hydrocarbons from nonpetroleum sources and was later developed into the methanol-to-gasoline (MTG) process, which was applied in a commercial plant in New Zealand [5].

However, since MTG process produced mainly aromatics as high-octane number fuel, and aromatics were no longer considered as a desirable content in gasoline because of the environmental concerns, MTG technology was not widely commercialized. In recent years, however, with increasing demand for light olefins from polymer industry, various versions of methanol conversion to olefin processes were developed, especially MTP (methanol to propylene), which became an attractive route to produce propylene. In recent years, China has built several commercial MTO or MTP plants. Some of these plants have the annual capacity of one million tons of propylene using MTP package by Dalian Institute of Physical Chemistry (see next chapter for detailed description of commercial MTO plants) [6, 7].

The chemistry of MTO involves almost all aspects of zeolite catalysis. The active site for MTO is a hybrid of inorganic zeolite framework and organic reaction centers, by which hydrocarbon pool mechanism and other parallel mechanisms operate to convert methanol to olefins [3, 8–11]. Typical hydrocarbon chemistry such as alkylation, isomerization, homologation, oligomerization, cracking, cyclization, protonation, deprotonation, and aromatization all occurred during the MTO process. In this regard, studying MTO chemistry will benefit all branches of hydrocarbon chemistry.

Basic research on MTO is still an ongoing subject. The goals for MTO study are to increase the catalyst efficiency to lower the operation cost and to develop new catalyst that has better ability to resist coking. For these goals, mechanism studies are important. Besides its impact on chemical industry, MTO process has also been a puzzling and still debating issue in several aspects of the process [10, 11]. Apparent overall MTO process is quite clear and can be conveniently divided into four steps:

1. The first step is a simple one as the dehydration of methanol, leading to equilibrium among methanol, dimethyl ether, and water. Zeolite catalyst, as solid acid, can readily catalyze such dehydration reaction. In this review, the first step will not be discussed.
2. The second step, usually called the induction time, is for fresh zeolite catalyst to be active for MTO reactions. During induction time, only a few olefins are produced. However, profound change is made inside the zeolite catalyst, as active intermediate species is formed and actively involved in olefin production. There has been an ongoing debate on what happened during induction time. This review will try to offer a balanced account on this subject.
3. During the third step, real MTO chemistry occurs, and high yield of hydrocarbons is observed. Carbon pool-type mechanism is dominating for MTO reactions, while detailed mechanism still waits. And some other mechanisms may also contribute for small part of olefin production.
4. In the fourth step, the catalyst is deactivated by accumulation of coke species. Such process is a typical one in zeolite chemistry and is perhaps not part of MTO as far as olefin production is concerned. There are many studies and reviews from petroleum chemistry on zeolite coking.

The chemistry of MTO has been extensively studied, and there are several outstanding review articles covering almost all aspects of the process [4, 11–19]. This chapter is another attempt to summarize the latest development of this exciting subject.

For MTO mechanism studies, kinetic data analysis has been a powerful method to infer mechanistic insight from chromatographic data. As complex as MTO reaction is, it is difficult to experimentally observe each individual step of the whole process. Kinetic analysis deals with large amount of data. These data provide information such as reaction rates and activation energies during each steps of the MTO process and offer evidence for certain pathways. Kinetic studies on MTO, especially during the induction time, have also contributed to the understanding of hydrocarbon pool mechanism [4, 10, 19–32].

Among many techniques, isotopic labeling is practically useful for mechanistic studies. For MTO study, as primary reactant methanol has one carbon and two kinds of hydrogen atoms (three at methyl group and one at hydroxyl group), it is easy to label methanol with ^{13}C and deuterium. By feeding methanol with specific isotopic atoms, especially ^{13}C, it becomes easier to track the destination of certain species. It is fair to state that isotopic labeling has been the strongest evidence for hydrocarbon pool mechanism to be accepted [21, 22, 33–45].

Another frequently used technique in MTO study is thermal quench of the reactor. By thermal quench, the ongoing MTO reaction can be abruptly halted, and presumably organic intermediate species during the reaction are preserved on the catalyst [46]. By dissolution of the catalyst and extraction of organic species, the nature of the organic intermediate species can be analyzed by GC-MS [39–42, 47–50]. For active species as well as the zeolite catalyst itself, ex situ spectroscopic (solid-state NMR, IR, Raman, UV/Vis, etc.) can provide useful information [51–64].

Computational study of MTO also proved to be an essential partner for better understanding of MTO [10, 29, 65–68]. Computational modeling can find energy barriers for certain pathways and offer insight for the whole reaction pathways [66–68].

For MTO mechanism, two unsolved issues are related to the second step and third step, i.e., how the first hydrocarbon product is produced during the induction time and how olefins are produced in the third step. In this chapter, a review of recent progress in MTO mechanism regards the chemistry in the second and third steps will be offered.

9.1.2 Initial C–C Bond Formation

From methanol to olefin involves C–C bond formation; thus, how initial C–C bonds are formed from C1 species becomes the first question for MTO mechanism and has been a heated debate. As will be discussed in the next section, there are two stages of C–C bond formation. On a working zeolite catalyst, C–C bond formation can be mainly preceded through a carbon pool-type indirect mechanism, as discussed later. However, during the early stage of induction time, when the catalyst is fresh, C–C bond seems to be formed directly from C1 species only. Such direct mechanism can certainly work during the third stage of the MTO process. But the question is how C–C bonds are formed from C1 species.

According to a comprehensive review by Stock [8], there have been many mechanisms proposed to explain the MTO mechanism. Some early studies did not separate induction time from working time of the MTO process, and tried to explain not just the initial C–C bond formation, but rather for all olefin formation. Haw et al. summarized the indirect mechanisms in Fig. 9.1 [4].

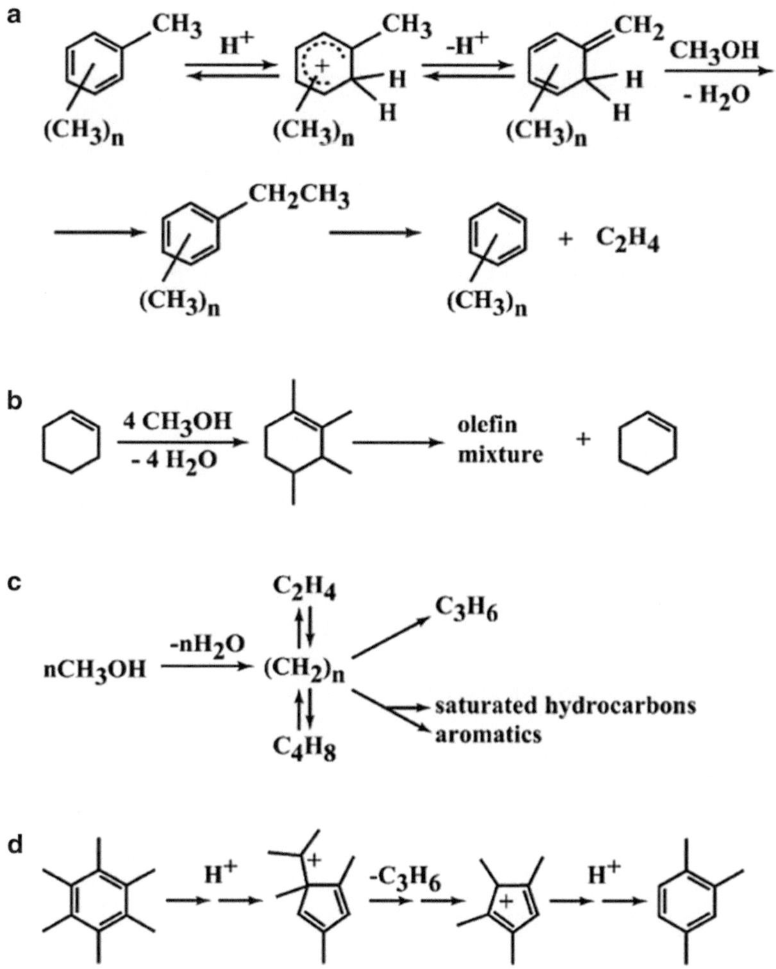

Fig. 9.1 Early proposals that have significantly shaped current thinking about the hydrocarbon pool mechanism in MTO catalysis. (**a**) Mole's 1983 mechanism of methylbenzene side-chain alkylation. (**b**) An abbreviated account of Langner's explanation for the dramatic effect of cyclohexanol and other co-feeds of reducing the kinetic induction period. (**c**) Kolboe's early phenomenological hydrocarbon pool mechanism for MTO catalysis. (**d**) The paring reaction in highly abbreviated form (Reprinted with permission from Ref. [4]. Copyright (2003) American Chemical Society)

It's quite straight forward to consider carbene (:CH_2) as the key intermediate in MTO, especially for initial C–C bond formation, as the following equation is so simple yet elegant:

$$:CH_2 + :CH_2 = CH_2CH_2$$

Chang et al. proposed the carbene mechanism in their earlier report on MTO [1, 3]. Carbene is an active species; thus, direct spectroscopic observation of carbene was rare, and most studies relied on indirect evidences that can be explained by carbene intermediate. Oxonium ylide mechanism and carbocation mechanism were based on the acid nature of zeolite [69–71]. Hunger and Wang considered methoxyl group as key species in C–C bond formation from their in situ NMR studies [16, 72–76].

All these mechanisms are so-called direct mechanisms, i.e., C–C bonds are directly formed from C1 species. The intermediates involved in these mechanisms are usually unstable species and are not easy to be observed by spectroscopic techniques except that methoxy groups are observed by NMR. Several computational researchers have found high energy barriers for reaction routes for olefins being produced from certain direct mechanisms [10, 29, 65–68]. Thus it's likely that the roles of these direct mechanisms become less significant once the MTO reaction is in the third stage. However, the small amounts of olefinic product from these direct mechanisms can oligomerize to form the hydrocarbon pool species during the induction time and thus are key part of the MTO [16].

Closely related issue during the induction time is the induction time itself. Why an induction time is observed on MTO? Autocatalysis is usually related to an induction time in catalysis. It's hard to image any direct mechanism that requires an induction time. In pulse-type experiments, first few pulses of methanol on ZSM-5 or SAPO-34, two of the most used zeolite in MTO research, yield no appreciable amount of hydrocarbon product [41, 46, 52]. Researchers in Mobile, White and Xu, quantitatively measured the methanol conversion against the amount of total methanol feeding [17]. Their experiments were carried out at a relatively mild temperature so that the methanol conversion was less than 25 % at most in order to minimize the interference from secondary reactions. Near bell-shaped conversion curve was observed, which depicts the three stages of MTO quite well: induction, steady state, and coking. The conversion at early stage was very low as the reaction was in the induction time and abruptly increased when the catalyst was in fully functional stage.

In the earlier years of MTO research, the goal of several studies on the induction time was to shorten the induction time. Looking back, we may summarize these studies as foundations of hydrocarbon pool mechanisms, as these work provided solid evidences to show the role of organic intermediates. For example, Mole et al. found that toluene was a cocatalyst in MTO [77].

Haw and Song et al. proposed an unorthodox ideal with a so-called impurity mechanism that further minimizes the role of direct mechanisms [78]. They argued that zeolites such as ZSM-5 and SAPO-34 are not solid super acids and are not able to sustainably produce active C_1 species in abovementioned direct mechanisms. On the other hand, since there are significant amounts of impurity compounds that already contain C–C bonds in any MTO reaction system, i.e., acetone and ethanol in methanol (even the highest-grade methanol is only 99.999 %), and hydrocarbon residual species on zeolite after calcination, these impurities could form hydrocarbon species that can be the starting materials to form the active organic centers for the hydrocarbon pool-type mechanism. This "impurity mechanism" by no means negates any other direct mechanisms. It is indeed impossible to negate the role of any direct mechanism. The "impurity mechanism" is more a practical observation than a conclusion. The evidences for "impurity mechanism" are also mainly circumstantial evidence. Some of the evidences are listed below:

When methanol, carrier gases, and catalysts are intentionally and extensively purified, olefin production from early pulses was significantly reduced comparing to what was observed from normal experimental setups. As shown in Fig. 9.2, olefin production was reduced to near zero when specially purified methanol was pulsed onto SAPO-34 or ZSM-5.

Song et al. found that zeolite with pore sizes of ZSM-22 or smaller was not an active catalyst for MTO [79]. They found that when methanol was pulsed into the ZSM-22 catalyst bed, olefin production was steadily very low after very short time on stream (or methanol feeding). Yet, the seemed inactive catalyst for MTO, ZSM-22 zeolite was still very active for many other reactions, such as butane isomerization, olefin homologation, etc. They claimed that the size of ZSM-22 channel was not big enough for hydrocarbon pool mechanism to work, so that olefin product from hydrocarbon pool was also inhibited [79–84].

Song's claim that ZSM-22 was not active for MTO was debated. Olsbye et al.'s earlier studies showed significant olefin production from ZSM-22 and attributed the apparent activity difference from Song et al.'s observation to the difference experimental setup, especially methanol space velocity [4, 8, 65, 66, 69]. Liu et al.'s work on ZSM-22 also indicated significant MTO activity for ZSM-22 [85–88]. The apparent MTO activity observed on ZSM-22 may be due to several issues, including the experimental setups (especially the flow rate of methanol), composition of the reactant feed, and perhaps the purity of homemade ZSM-22 zeolite used in the research. Another possible explanation for the olefin formation is that they are produced from the methylation/cracking mechanism, as will be covered in the latter part of this chapter [11, 29, 80, 86, 89, 90].

Song et al. recently reported that for ZSM-22 and ZSM-12 both of which have one-dimensional pores, but with small pore size difference (0.57 Å for ZSM-22's ten member ring vs. 6.0 Å for ZSM-12's twelve member ring, see Table 9.1 for comparison), and dramatic MTO activity difference was observed, as shown in Fig. 9.3 [83]. For ZSM-22, essentially no hydrocarbon was formed after several pretreatment methanol pulses. The pretreatment pulses did resulted in significant amount of olefin formation, but the olefin formation quickly reduced to near zero with each additional pulse. Once the olefin formation is reduced to near zero, the catalyst showed little coking, indicating that the ZSM-22 catalyst was not deactivated by coking, and the ZSM-22 catalyst was still a very efficient catalyst for reactions that are typical for such one-dimensional 10-member ring zeolites, including butane isomerization, benzene methylation, olefin homologations, etc., suggesting that the pore of the ZSM-22 was not blocked. However, for ZSM-12, the MTO reactions proceed fast and abundant aromatics as well as alkanes were produced. Once the ZSM-12 was deactivated, the catalyst is totally deactivated for all acid-catalyzed reactions, i.e., even methanol dehydration to dimethyl ether was not able to proceed. And the deactivated ZSM-12 was full of coke species that are fused aromatics not soluble in common organic solvents.

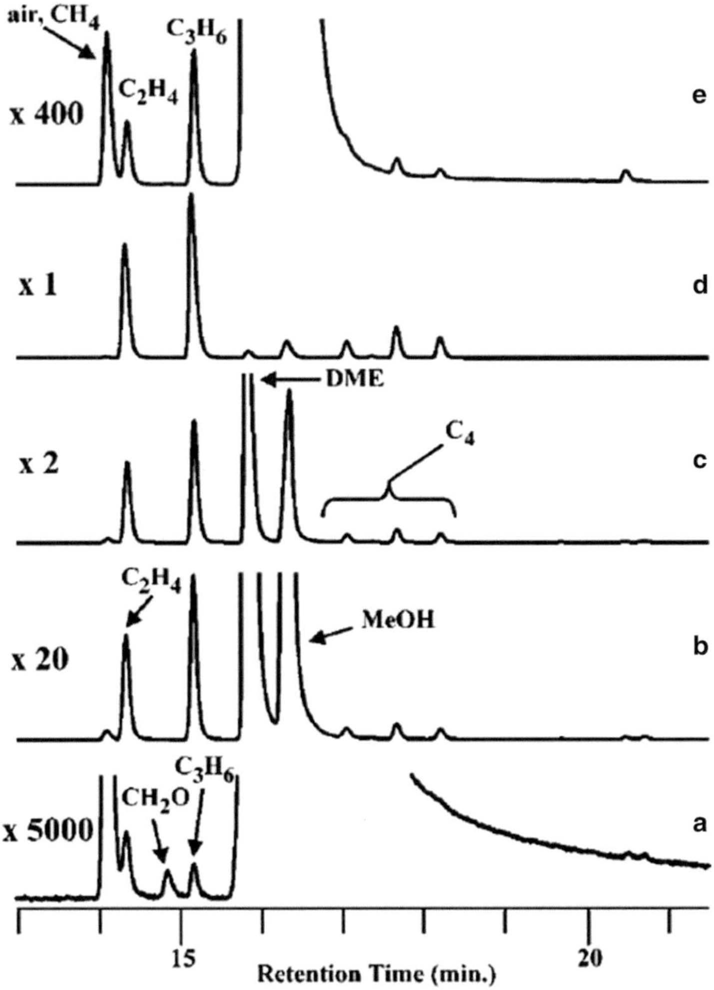

Fig. 9.2 GC (FID detector) analyses of the product streams sampled 2.4 s after pulsing 12.5 íL methanol onto 300 mg beds of HSAPO-34 with purified He flowing at 10 cm^3s^{-1}. (**a–d**) are from a single bed of rigorously calcined HSAPO-34 following a series of pulses of fractionally distilled methanol delivered in 30 min intervals. (**a**) Following the first pulse, the total yield of volatile hydrocarbons was ca. 0.0026 % (26 ppm). (**b**) Following the second, identical methanol pulse, the yield of volatile hydrocarbon products greatly increased to 1.5 %. (**c**) Following the third, identical methanol pulse, the hydrocarbon yield further increased to 10 % as a result of the growing hydrocarbon pool. (**d**) This catalyst bed was reacted with an additional 200 íL of methanol to create a larger hydrocarbon pool. Thirty minutes later 12.5 íL of methanol was pulsed, and the product stream showed nearly complete conversion. (**e**) Following the first pulse of fractionally distilled methanol onto a fresh HSAPO-34 bed prepared using the standard calcination procedure but not the more rigorous second calcination. The volatile hydrocarbon yield, 590 ppm, was substantially higher without rigorous calcination, cf. (**a**). Small amounts of methane and CO are formed on the reactor wall (Reprinted with permission from Ref. [78]. Copyright (2002) American Chemical Society)

Table 9.1 Summary of prepared zeolites H-ZSM-12 and H-ZSM-22

Zeolite	Pore topology	Pore size (Å)	Si/Al [a]
H-ZSM-12	1D, 12 rings	5.6×6.0	35
H-ZSM-22	1D, 10 rings	4.6×5.7	40

1D one-dimensional
[a]ICP analysis

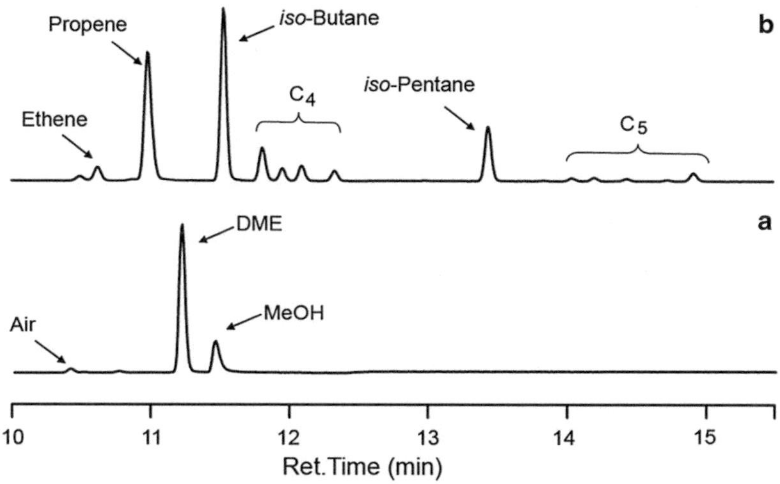

Fig. 9.3 GC profiles of methanol pulsed onto zeolites at 673 K. (**a**) H-ZSM-22; (**b**) H-ZSM-12. Samples were taken 8 s after methanol pulse (Reprinted with permission from Ref. [83]. Copyright 2011, American Chemical Society)

In another study, Song et al. found that for ZSM-22 and ZSM-23, both of which have small one-dimensional pores, the amount of olefin (mainly propene) is higher on smaller zeolite particles, which have higher amount of external surface acid site. And when external acid sites were selectively removed by acid itching, the olefin productions decreased to minimum [84]. External surface acid sites and pore mouth acid sites are important parts of solid acid catalysts, especially when large reaction species are concerned.

In these set of experiments using ZSM-22 as catalyst, regular analytical grade methanol (99.99 %) is used; thus, the small amount of olefin was also perhaps due to the "impurities." For example, propene production was about the same in a wide temperature range from 300 to 500 °C [81], as shown in Fig. 9.4.

This finding does not support any direct mechanism. Higher olefin production is expected at higher temperature if olefins are produced from any C1 species. For example, if a carbene, methoxy group or methyl cation is involved in the C–C bond formation, then the C-H bond on CH_3OH must be broken, and H/D exchange will

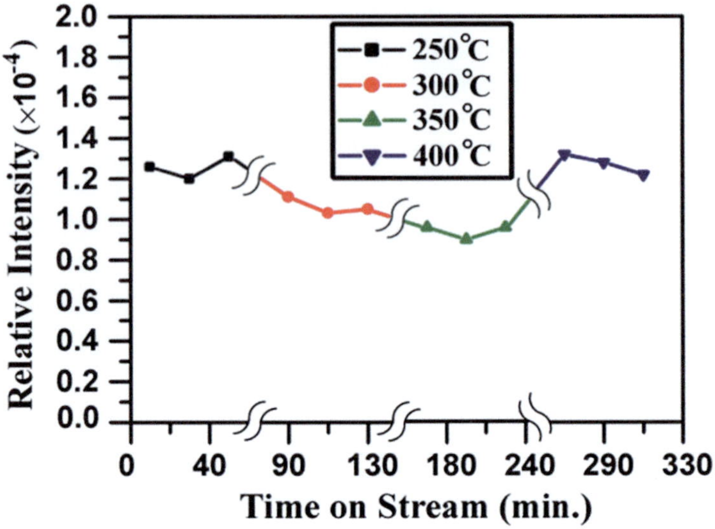

Fig. 9.4 Propene's time on stream relative intensities to DME plus methanol at various temperatures when methanol was flowed onto ZSM-22 constantly (Reprinted with permission from Ref. [81]. Copyright (2008) American Chemical Society)

Table 9.2 Quantitative results of number of D atoms in product DME produced by pulsing CH_3OD onto DZSM-22 at various temperatures

	D_0	D_1	D_2	D_3	D_4	D_5	D_6
200 °C	98.0	1.1	0.8	0.01	0.1	0.01	0.01
300 °C	98.0	1.2	0.7	0.01	0.1	0.01	0.01
400 °C	97.9	0.9	1.0	0.01	0.2	0.01	0.01
500 °C	97.7	1.0	1.1	0.01	0.2	0.01	0.01

Reprinted with permission from Ref. [80]. Copyright (2012) American Chemical Society

be observed if methanol was selectively labeled with deuterium atoms. As shown in Table 9.2, at a temperature range of 200–500 °C, the H/D exchange did not occur. Haw et al. observed similar results on SAPO-34 [45, 91].

These findings on ZSM-22 have one side benefit for MTO research. On ZSM-22, the chemistry of small species can be directly studied without the interference from dominated olefin secondary reactions. The main olefin production route, the hydrocarbon route, is inhibited under certain experimental conditions, so that some other important aspects of the MTO process, such as chain growth [80], scrambling, and double bond shift, may be observed directly. These studies will be covered in the next section.

9.1.3 C_2 and Up

9.1.3.1 Hydrocarbon Pool Mechanism

Chemistry becomes even more interesting in the third stage of MTO reactions, in which the olefins were produced in large quantity. Besides those possible direct mechanisms mentioned in previous chapter, other mechanisms may also contribute, and likely more predominately, to the hydrocarbon formation. Even when direct mechanisms were popular in earlier years, some indirect mechanisms were proposed based on hydrocarbon chemistry in zeolites. For example, Mole et al. proposed a side-chain growth-type mechanism to account for the cocatalyst effect of toluene [77]. In previous MTO studies on shortening the induction time, Mole et al. found that methyl benzenes were a cocatalyst and proposed a side-chain growth mechanism to explain the cocatalyst effect. He was perhaps more close to realizing the importance of methyl benzenes in MTO than anyone at that time. However, since he used HZSM-5 as catalyst, and methyl benzenes are also part of the main product, the true role of methyl benzenes was only considered as cocatalyst. With overwhelming evidence, the hydrocarbon pool mechanism by Kolboe is now considered as the most accurate and dominate one in MTO mechanism.

"Carbon pool" mechanism is at first an empirical description of how methanol is converted into olefins. Using zeolite SAPO-34, which has a cage-like pore structure (eight-member ring opening of 3.8A), Kolboe et al. carried out systematic experiments to show that the organic species trapped inside the SPO-34 cages are vital for olefin formation [33, 34, 92, 93]. In their experiments, labeled methanol (^{13}C-MeOH) was used along unlabeled ethylene or propylene. Such experimental setup is very similar to what Chang et al. did on ZSM-5, when Chang et al. co-feed methanol with isobutene. Kolboe et al. found out that most of the olefin product was from 13C-labeled methanol, and ethylene was almost inactive. This finding was a direct rebuke to direct mechanisms, as ethylene is the primary product from direct mechanisms and key intermediate for longer chain product. While direct mechanisms are unable to account for this result, an indirect route, i.e., a parallel reaction route, was proposed as a "carbon pool" mechanism. The term "carbon pool" was a smart one as exact structure of the $(CH_2)_n$ "pool" species was not known at that time. However, it seems so accurate when one envisions a catalytic cycle, in which methanol or any C1 species are fed into the pool on one side and olefin products are product from another side.

Several groups have vigorously pursued the nature of carbon pools. Their works were summarized in several reviews [3, 4, 9, 11, 16–19, 85, 94–96].

Haw et al. developed a computer-controlled pulse quench micro reactor to study the MTO mechanism as shown in Fig. 9.5 [97]. The pulse quench reactor allows very fast thermal quench of the catalyst bed, and pulse injection of small dose of

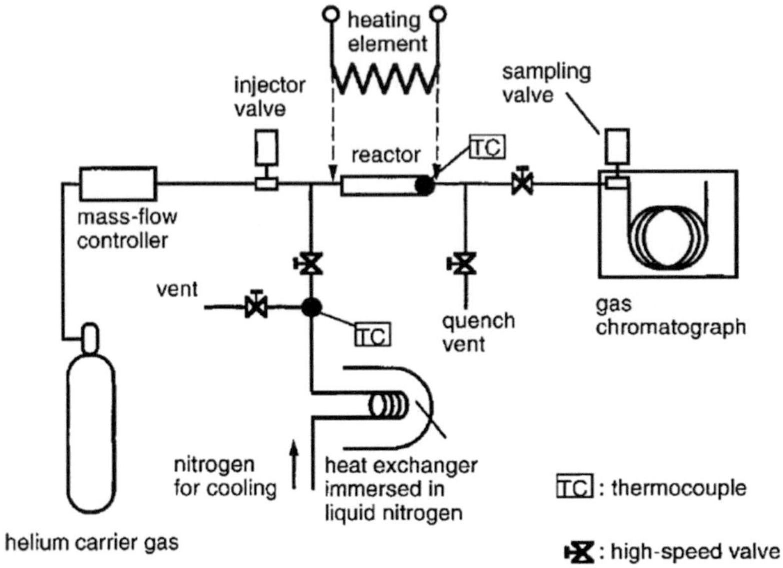

Fig. 9.5 Schematic representation of the pulse ± quench reactor used to prepare samples under standard flow reactor conditions for study by NMR spectroscopy. The *gray* region indicates the catalyst bed. All valves and the heating element are under computer control (Reprinted with permission from Ref. [4]. Copyright 2003, American Chemical Society)

reactant allows only small changes to the catalysts from each pulse. Thus in a pulse quench experiment, a small dose of reactant was injected, then a GC sample was collected several seconds after the injection, and then the catalyst bed was quenched by nitrogen precooled to near cryogenic temperature. The catalyst was then sealed inside the frozen reactor and was later collected for analysis by solid-state NMR or other techniques.

This pulse quench reactor allows essentially simultaneous analysis of the gas phase species in the flow gas and species trapped in the zeolite, assuming that the quench time is short enough and the species on catalyst do not change during and after quench. Several key results relating to carbon pool species were obtained from such pulse quench reactions.

Several stable carbenium cations were observed on ZSM-5 using such protocol by Haw and coworkers. Carbenium cations were considered key intermediates in hydrocarbon chemistry on super acid [69, 98]. However, only a handful of carbenium cations were stable enough on zeolite to be directly observed by solid-state NMR. One reason is the presence of water, which is prone to neutralize the cation. Water is abundant in MTO reactions as water is the by-product of methanol dehydration. Using pulse quench reactor, water was pushed out of the catalyst bed by flow gas, leaving carbenium cations on zeolite. With such design, several carbenium cations were observed in Haw's group, including cyclopentium cations and benzenium cations [46, 52, 97, 99–101] (Fig. 9.6). Some of the carbenium cations were observed

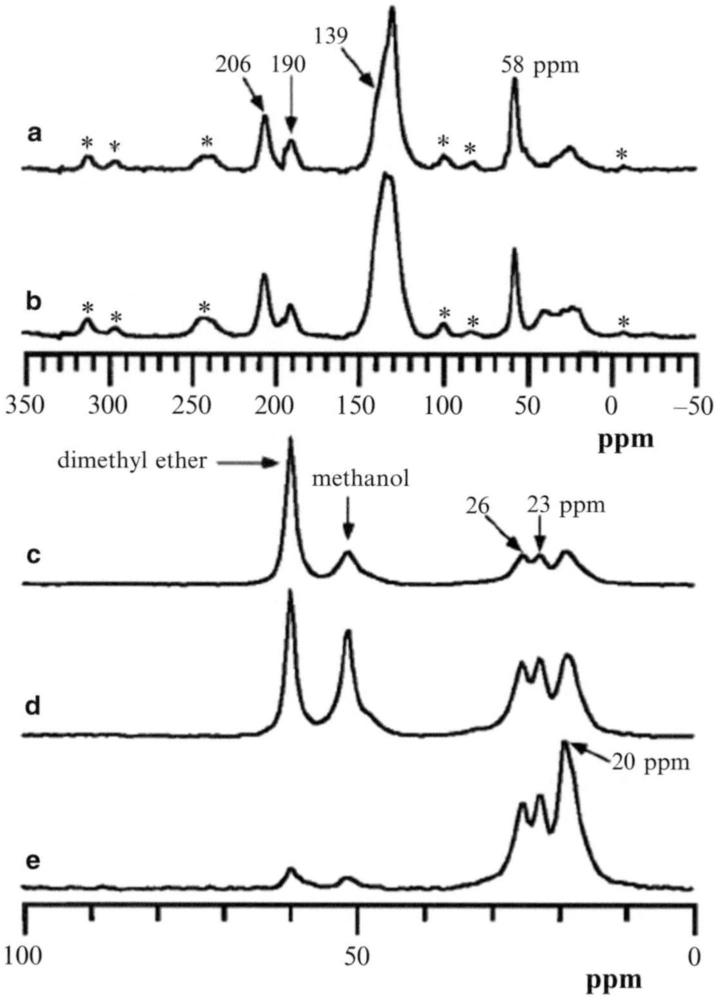

Fig. 9.6 Selected 75.4 MHz ^{13}C MAS spectra of benzenium ion 3 (206, 190, 139, 58, and incompletely resolved signals between 23 and 26 ppm) on zeolite HZSM-5 (Si/Al = 19). Signals near 20 ppm are due to the methyl carbons of neutral aromatic compounds, e.g., toluene and xylenes. Benzene or toluene and methanol were injected into the flow reactor as a pulse and allowed to react for 4 s at 573 K before quench: (**a**) 0.5 equiv. of benzene-$^{13}C6$ and 3 equiv. of methanol-^{12}C, (**b**) 0.5 equiv. of toluene-ring-$^{13}C6$ and 2.5 equiv. of methanol-^{12}C, (**c**) 0.5 equiv. of benzene and 3 equiv. of methanol-^{13}C, (**d**) 0.5 equiv. of toluene and 2.5 equiv. of methanol-^{13}C, and (**e**) 0.5 equiv. of toluene-1-^{13}C and 2.5 equiv. of methanol-^{12}C. 1 equiv. on this catalyst corresponds to 0.58 mmol reactant/g of zeolite. An *asterisk* denotes spinning sideband (Reprinted with permission from Ref. [99]. Copyright (1998) American Chemical Society)

under conditions similar to MTO reactions. For example, Haw et al. observed 1,1,2,3,5-tetra methyl benzenium cation by pulsing methanol and toluene onto the ZSM-5 catalyst [102]. Fingerprint chemical shift peaks were observed by solid

NMR, as shown in Fig. 9.7. Haw and Song et al. later observed a fully methylated benzenium (heptaMB$^+$) on H-beta, whose larger cage allowed the formation of larger cations [102]. Under methylating conditions, hexaMB may be further methylated and form the heptamethyl benzenium ion (heptaMB$^+$), which has the gem methyl groups that are critical for side-chain growth mechanism discussed later. Bjørgen et al. further advanced this carbenium identification route and actually isolated heptaMB$^+$ trapped inside H-beta by co-fed methanol and benzene. The heptaMB$^+$ was isolated as its neutral base form as HMMC (1,2,3,3,4,5-hexamethyl-6-methylene-1,4-cyclohexadiene) and confirmed its structure by NMR spectroscopy [103–105]. The stability of methyl benzenium cations on H-beta is determined by the acid strength of the zeolite framework, i.e., only those benzenium cations that match the acid strength of the zeolite can present as stable cations. IR- and UV-vis spectroscopy studies by Kolboe et al. suggested that trimethyl benzene and tetramethyl benzene are about where the threshold for stable benzenium cations stands. Tetramethyl benzene and higher methylated benzenes can exist as stable cations on H-beta [103, 104]

By pulsing ethylene or propylene onto ZSM-5 at low temperature, another carbenium cation 1,3-dimethyl cyclopentenyl cation was observed by Haw et al. Because of its structure, it was given a nickname of smiley cation [97]. smiley cation is in fact a very stable and common cation on zeolite. Pulsing ethylene or propylene is probably the best choice to produce a relatively large number of the cation. However, pulsing methanol can also produce it. And in pulsing experiments, with one single propene pulse to produce smiley cation on ZSM-5, the induction for methanol conversion was eliminated, proving that smiley cation is a hydrocarbon pool species [52]. Haw et al. reported a fully methylated smiley cation by condensing acetone within the cage of SAPO-34 [100]. This was the first stable cation observed on SAPO-34, which was a considerably weaker acid than ZSM-5. It's the cage-like structure of SAPO-34 and higher stability of the cation due to the methyl substitution that lead to this observation. Liu et al. recently reported that they observed the benzenium cation and smiley cation on SAPO-34, and other MTO catalysts under MTO reaction conditions further solidify the role of cyclopentenium cation in MTO [85, 106–108].

The role of the carbenium cation in MTO is speculated by Haw et al. as the organic reaction center, i.e., one hydrocarbon pool specie with exact structure [52]. Indeed, if the ZSM-5 catalyst was pretreated with ethylene to generate smiley cation (by pulsing ethylene, only smiley cation was left on ZSM-5, all other species were flushed out by carrier gas), then the induction time for MTO was very much shortened. Thus it is fair to conclude that smiley cation is one of the key hydrocarbon pool species in MTO, especially when ZSM-5 was used as catalyst. Liu et al.'s recent finding that smiley cation-type carbenium cations existed on working SAPO-34 further highlighted the role of smiley cation in MTO. As will be shown later, since methyl benzenes are also hydrocarbon pool species. Most studies on hydrocarbon pool mechanism have been focused on methyl benzenes, and smiley cation will be a vital part in the so-called paring mechanism to connect benzene rings and cyclopentenyl rings. For benzenium ion, another possible pathway, the side-chain growth pathway, may require benzenium ion as key intermediate species.

Haw et al.'s work on SAPO-34 proved that methyl benzenes are the hydrocarbon pool species for MTO on SAPO-34 as shown in Fig. 9.7 [41]. With solid-state NMR and online GC to analyze the species trapped inside, the SAPO-34 cage and organic product exited from the reactor, respectively. The key finding is the correlation between the presence of methyl benzenes in SAPO-34 cages and the formation of olefin product. In a series of methanol pulses, the first pulse resulted in DME only for a gas sample taken 4 s after the pulse. Then the catalyst was kept at 400 °C with

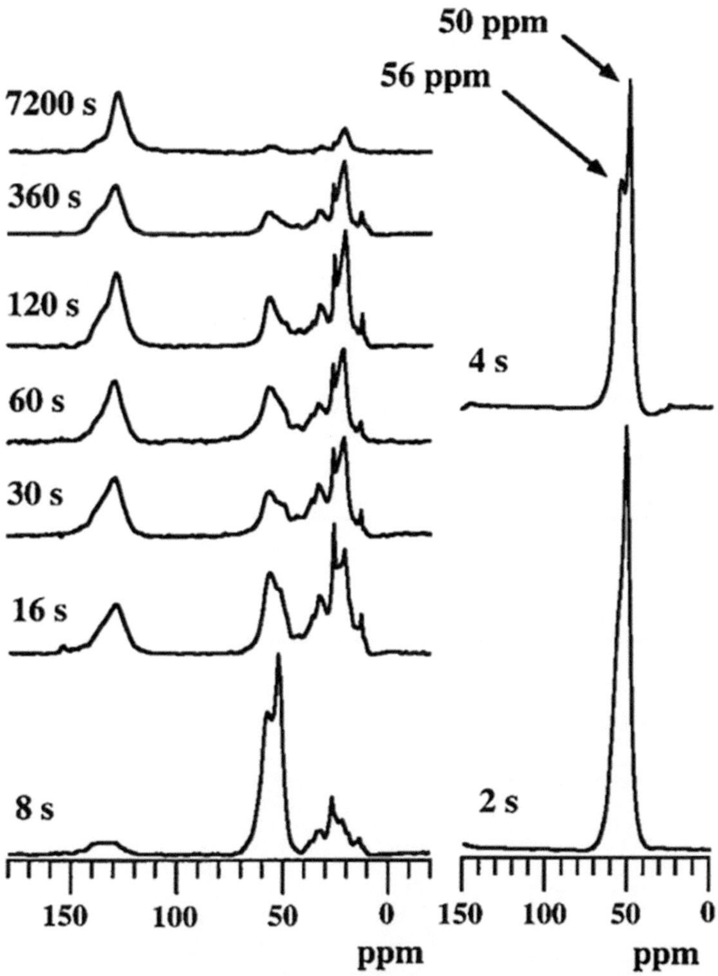

Fig. 9.7 75 MHz 13C CP/MAS NMR spectra of samples from a pulse quench study of methanol conversion on HSAPO-34 at 673 K. Each sample was prepared by injecting 20 íL of methanol-13C onto a freshly activated catalyst bed (0.3 g), while He was flowed at 600 mL min 1, and reaction occurred for the times shown followed by a rapid thermal quench. All spectra (4000 scans) were measured at 298 K using a 2-ms contact time (Reprinted with permission from Ref. [41]. Copyright 2000 American Chemical Society)

helium flow for 5 min in order to flush out all volatile species from the first pulse (as confirmed by a GC sample taken in a control experiment showing almost no organic product in the carrier gas after 5 minutes). Then the second pulse was injected, and a GC sample was taken 4 s after the second methanol pulse. The second pulse resulted in significant amount of olefins (mostly ethylene and propylene), indicating that the catalyst was in working condition. Thus from the induction point of view, the first pulse was able to eliminate the induction time on SAPO-34.

In a parallel solid-state NMR study, series of samples were prepared by quenching the SAPO-34 catalyst at various times after the first pulse to investigate the evolution of organic species trapped inside the SAPO-34 cage after the first pulse. As shown in Fig. 9.8, for catalyst quenched after 2 s, only physically adsorpted

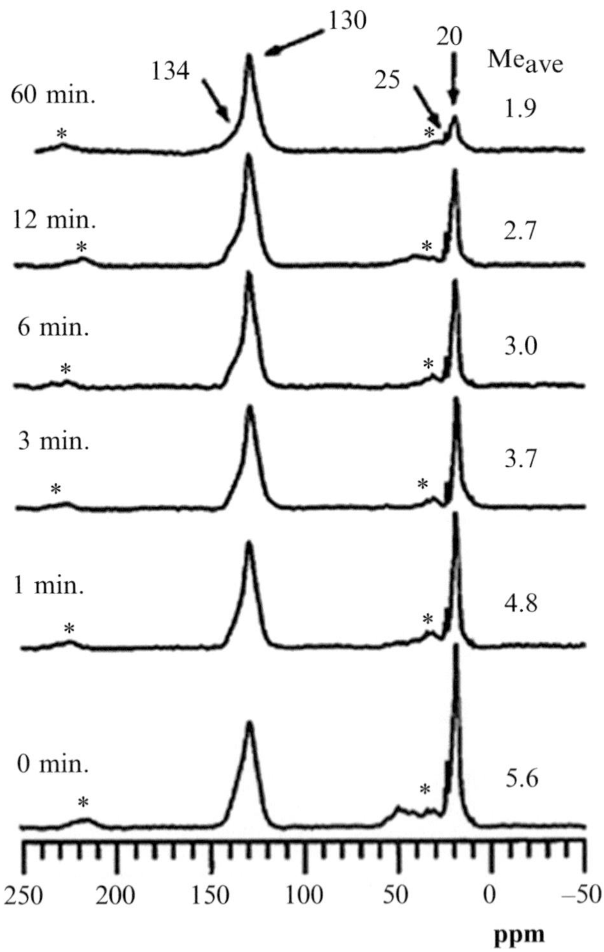

Fig. 9.8 90 MHz ^{13}C CP/MAS NMR spectra showing the loss of methyl groups as a function of time from methylbenzenes trapped in HSAPO-34 at 400 °C (Reprinted with permission from Ref. [137]. Copyright 2001, American Chemical Society)

methanol was detected. Then DME signal appeared after 4 s. Methyl benzene rings started to show after 8 s, with 13C peaks at 25 ppm (methyl group) and 130 ppm (benzene ring), respectively, showing the presence of methyl benzenes (mixture of methyl benzenes). With longer time, the peak intensities for the benzene ring remained the same, while the peaks for DME, methanol, and methyl groups decreased. These findings were proofs that methyl benzenes were hydrocarbon pool species.

In a related study, Haw et al. produced SAPO-34 catalyst that was loaded with only methylnaphthalenes and found that methylnaphthalenes were also active hydrocarbon pool species, though not as active as methyl benzenes [47, 48]. The fact that methylnaphthalenes are active hydrocarbon pool species is quite intriguing in that methylnaphthalenes are usually considered as the early stage coke species. With longer time on stream, methylnaphthalenes will become fused aromatics such as arylene that will eventually deactivate the catalyst by blocking the channel. Thus in MTO process, methyl aromatics are indeed unique. They are carbon pool species as good coke needed for olefin production, and these hydrocarbon pool species will stop the olefin production when they become bad coke.

Because of their cage-like morphology and eight-member pore, it's relatively easy to generate methyl benzenes on SAPO-34 or SAPO-18 to study their role in MTO [43]. However, on other zeolites such as ZSM-5 and H-beta, because methyl benzenes can diffuse out of the zeolites, it is hard to control the nature of organic species on these zeolites. Haw et al. developed a reactor that allowed direct pulse of hexamethyl benzene (HMB) onto H-beta catalyst. Olefin products were product from this experiment, proving that HMB was active for olefin formation on H-beta [42]. Bjørgen et al. investigated the reactivity of polymethylbenzenes over the large pore zeolite H-beta, and the heptamethylbenzenium cation as well as its corresponding base HMMC was identified as a key intermediate [31, 40, 105, 109].

Hunger et al. developed a stop-flow reactor, which combined the flow-type reactor and solid-state NMR to study MTO. Such protocol was able to analyze the effluent gas phase product by GC and organic species on catalyst at working temperature, while in situ solid-state NMR spectra of the catalysts are being acquired. Such protocol requires advanced technology in NMR devices and is ideal for heterogeneous catalysis research. Interesting results were obtained from such reactor, in particular methanol conversion on various zeolites [16, 56, 57, 72–74, 110–119].

Kolboe et al. have been a very productive team in all aspect of MTO research, especially in kinetic analysis and modeling [11, 21, 31, 34–38, 40, 92, 95, 103, 120–134]. For example, with isotope exchange technology, Kolboe et al. obtained similar results as Haw et al. on SAPO-34 in terms of the role of methyl benzene [37, 120, 121]. They studied isotopic scrambling of methyl benzenes in an experiment protocol in which ^{12}C-methanol was first fed and switched to 13C-methanol feed. They observed extensive 12C/13C scrambling on methyl benzenes, especially pentamethyl benzene and HMB. However, the ^{13}C presence in the olefin product increased at a slower pace than methyl benzenes, suggesting that the 13 C labels were scrambling between methyl carbons and benzene ring carbons before the olefins were produced from methyl benzenes. The most rational explanation for such results is that methyl benzenes are hydrocarbon pool species.

Besides academic studies listed above, another main player in MTO study is the oil industry. The ExxonMobil also invested large amount of resources in MTO study. In a recent review, White et al. summarized research work on the chemistry of MTO at the ExxonMobil. Due to the nature of industrial research, their research works were mostly published as patents [135, 136]. In several patents, Xu and White disclosed part of their findings. They used solid-state NMR as the main investigation tool to study the organic species on the catalyst, and ^{13}C-labeled compounds are frequently used to assist determining the chemical structure of carbon pool species. They reached the same conclusion as Haw et al. that methyl benzenes and methylnaphthalenes were carbon pool species on HSAPO-34 for MTO. They proposed an artificial organic/inorganic hybrid structure as the true active catalyst in MTO. Such hybrid structure is almost the same as what Haw et al. proposed, the "reaction center," which is consisted of hydrocarbon pool species (carbenium cation or methyl aromatics) and the adjacent Brønsted acid site in the zeolite channels [13, 66, 91, 137, 138].

These work from academic and industry unequivocally proved that the hydrocarbon pool mechanism is the correct description of the main MTO reaction on frequently used MTO catalysts such as ZSM-5 and SAPO-34. An active hydrocarbon pool species can be benzene, toluene, and naphthalene that are substituted with various numbers of methyl groups. These aromatics, along with the Brønsted acid site, form an active center, which can react with C1 species (methanol, DME, methoxy group, oxonium ion, etc.) by adding a methyl group to the carbon pool species. Then the carbon pool species undergo complicated change to allow the ethyl and propyl group to be formed and detached from the pool. This is how ethylene and propylene are formed as the primary olefin product from MTO.

9.1.3.2 Product Formation

With methyl benzenes as stable hydrocarbon pool species, exactly how olefins are produced from them becomes an actively studied topic. In hydrocarbon pool mechanism, there are two parts of the chemistry: the addition of methanol-related C1 species to the pool and the splitting of olefin from the pool. It may be convenient to separately study them.

Haw et al. monitored the evolution of olefin (ethylene and propene) product on SAPO-34, which was pretreated to load methyl benzenes (with an average number of methyl group per benzene ring of 5.6) [137]. The micro reactor was then kept at reaction temperature of 400 °C with carrier gas continuously flowing. A series of samples were prepared by quenching the SAPO-34 catalyst at a various time after the first pulse to investigate the evolution of organic species trapped inside the SAPO-34 cage after the first pulse. The waiting time before quench ranged from zero seconds to 1 h. The amount and the distribution of ethylene and propylene effluent at different times were then monitored by GC. NMR analysis results are shown in Figs. 9.8 and 9.9. Solid-state NMR results indicated that only methyl benzenes are left on the SAPO-34 catalyst, and the average number of methyl groups on benzene ring decreased steadily with longer waiting time. At first the methylbenzenes are mostly HMB and PMB (with an average number of methyl per ring of 5.6). Such number decreased to 1.9 after 60 min. However, even after 60 min, very small amounts of olefins, mostly ethylene and propene with trace amount of butane, were diffusing out of the catalyst, suggesting that these methyl benzenes were producing olefins. Since there was no methanol, methoxyl, and DME signal from Fig. 9.9, it is safe to conclude that these C1 species were flushed out and were not involved in the evolution of the methyl benzenes. These results seem to suggest that both ethylene and propylene are primary olefin product in MTO.

Exact mechanism of methyl benzene transformation is not elucidated in this work. However, the distribution of ethylene and propene as a function of average methyl group per benzene ring (M_{ave}) was an interesting result. Olefin selectivity in MTO is quite important for industry, as the demand for ethylene and propylene does not necessary agree with the olefin selectivity of the catalysts for MTO. From Fig. 9.9, apparently ethylene is preferred with lower M_{ave} value, while propylene will dominate with HMB and PMB. A selectivity switch occurred at M_{ave} value of about 2.8. This finding offers a rational approach to control the olefin selectivity on SAPO-34 by controlling the number of methyl groups on benzene ring. For example, by limiting the available space inside the SAPO-34 cage, Haw et al. developed a ship in a bottle approach to add enhancement to the ethylene selectivity [49].

Thus it is clear that methylbenzenes or other related cyclic species are direct reaction intermediates during the MTO catalytic cycles for olefin formation. However, the exact mechanism of how the alkenes are formed from such species remains an ongoing pursuit. So far two distinct hypotheses have been proposed to explain the olefin formation. They are called exocyclic methylation mechanism (side-chain growth mechanism) and paring mechanism (ring expansion and contraction mechanism).

The exocyclic methylation route is also called side-chain growth mechanism. In earlier MTG studies, Mole et al. proposed such route to explain the cocatalyst role of toluene [77]. Later on, many researchers found more evidences for this route. For example, Haw and Kolboe suggested that ethylene may well be formed from an ethyl group, which is formed from methylation of methyl benzenes [37, 38, 42, 50, 91, 92, 95, 103, 123–125, 139]. A typical exocyclic methylation route involves a benzenium cation such as heptaMB$^+$, which is associated with HMMC by protonation and is deprotonated. For HMMC, its exocyclic double bond provides ideal target for methylation agents (presumably methoxyl groups on the catalyst or nearby methyl groups on the same benzene ring) for homologation reactions, which result in an ethyl group and consequently ethyl benzene. Once ethyl benzene is formed, ethylene can be readily produced from dealkylation (Fig. 9.11). Further methylation of ethyl groups to propyl groups may lead to propene formation and possibly even butane. However, propene and butane formation from this exocyclic methylation route is unlikely a prominent route, as adding two or even three methyl groups to the same target seems difficult in terms of steric restriction.

The side-chain growth route goes through two gem methyl benzene species, which forms ethyl benzene species. Kolbe et al. observed HMMC by mass spectrum

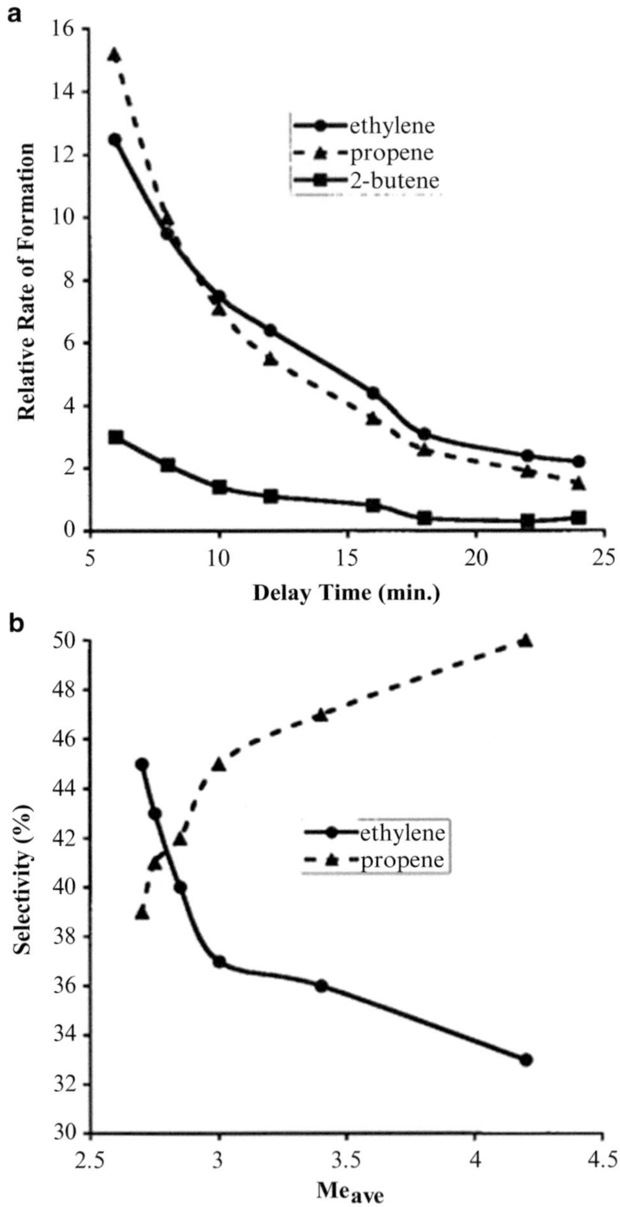

Fig. 9.9 (**a**) The rates of formation of ethylene, propene, and 2-butene as a function of time from a single experiment similar to those used for Fig. 9.9. For these measurements, multiple gas samples were analyzed from a single catalyst bed as it evolved over time after cessation of methanol flow. Ethylene and propene are major primary products of MTO chemistry. (**b**) Ethylene and propene selectivity as a function of the average number of methyl groups per ring, Me$_{ave}$. The data in part (**b**) are replotted by using an abscissa derived by empirically fitting the time evolution of Me$_{ave}$ in Fig. 9.9 to a smooth curve generated by Excel. This mapping shows that ethylene is favored by methylbenzenes with 2 or 3 methyl groups, while propene is favored with 4 or more methyl groups (Reprinted with permission from Ref. [137]. Copyright 2001, American Chemical Society)

[11, 39, 122]. Hunger et al. used in situ for MTO study, and their results also led to support for side-chain growth mode under their experimental conditions [113, 114, 116]. Haw et al.'s observation that lower number of methyl groups on benzene ring leading to higher ethylene yield was an evidence supporting such gem methyl benzene intermediate [137].

Another equally appealing mechanism is the so-called "paring" mechanism, which was proposed by Sullivan et al. for olefin formation in hydrocarbon conversion during petrochemical processes [140]. Thus the paring mechanism was not a new proposal for MTO, but rather a borrowed one for MTO. MTO process is indeed very much like petrochemical process, except that methanol and other C1 species are involved. The hydrocarbon evolution chemistry may well be the same. The essential part of pairing mechanism is the ring contraction and expansion cycle, in which alkyl groups, presumably propyl groups, are generated as shown in Fig. 9.11. Thus the major product from paring mechanism should be propene. With the fact that, in MTO reactions, propene is indeed a major product, pairing mechanism received support from many researchers.

And there are quite some experimental supports for this mechanism, as shown in many review articles [4, 11, 13, 16–18]. The most convincing evidence has been the 13C label scrambling between ring carbons and methyl group carbons on polymethyl benzenes. The 13C labels can interchange between the methyl groups and ring. Such label scrambling becomes more extensive as the number of methyl group on benzene ring increases [37, 38, 42, 47, 85, 87, 91, 95, 139, 141]. And the olefin products contain mixed 13 C labels when 13C methanol and 12C benzenes are co-fed into the zeolite catalysts. Such label scrambling is the main evidence for Kolbe et al. to propose "carbon pool" mechanism [33, 92, 93], and pairing type mechanism is the most logical explanation for these observations.

9.1.3.3 Parallel Reactions

Once there are olefins in the zeolite channel, they will also undergo methylation reactions by methanol-related species. Such methylation reactions result in the addition of methyl groups to the original olefin, i.e., the increase of the carbon chain. Thus such process can be regarded as homologation reactions. Before the carbon pool mechanism, olefin homologations were considered as an important part of the olefin evolution within ZEM-5. Dessau et al. proposed that the MTH reaction proceeds as sequential increase of the olefin chain length by methylation. Longer chain olefins such as octene will then crack into smaller olefins, which will then undergo methylation again [142, 143]. By such repeated methylation and cracking, methanol is converted into olefins [21, 22, 39, 40, 95, 105, 122, 127, 134].

Direct observation of the homologation of olefins is difficult with ZSM-5 and H-beta catalyst because fast oligomerization and cracking of olefins make it hard to tell the origin of the chain length change [31]. Isotope exchange has again been the most useful tool in this subject. For example, Svelle et al. co-fed 13C methanol and 12C olefins (ethylene, propylene, and butane) over ZSM-5 and SAPO-34 [21].

By choosing appropriate reaction conditions such as short contact time, and with the presence of abundant methanol and water molecules, the olefin oligomerization and cracking were minimized, making it possible to obtain kinetic data on olefin methylation. They observed the first order for olefins and zero order for methanol. And the activity for methylation increased with increased olefin chain length (ethylene < propene < butane).

Song et al. used ZSM-22 as catalyst and pulsed mixture of 13C methanol and ethanol, propanol, or butanol (alcohols were precursors for corresponding olefins). With the one-dimensional pore of ZSM-22, and pulse-type reaction mode that resulted in short contact time, step-by-step methylation of olefins was directly observed [80]. For the mixture of 13C methanol and ethanol, ethylene has essentially no ^{13}C as ethylene was produced from ethanol dehydration. Propene has one 13C atoms, and butenes have two 13C atoms, while pentenes have three 13C atoms. For the mixture of 13C methanol and propanol, butenes have one 13C atoms, while pentenes have two ^{13}C atoms. And the methylation rate increased with similar order of ethylene < propene < butane. And on ZSM-22, the butane products are a mixture of 1-butene, 2-butene (E and Z confirmation), and isobutene, indicating full equilibrium of the butane species inside the ZSM-22 pores. Such results also showed that double bond shift for butane is facile inside ZSM-22 channel.

With olefin methylation and cracking being a viable route for olefin formation, MTO can be considered as a parallel process, in which hydrocarbon pool-type mechanism is the major route, while the olefin methylation and cracking route is the minor route. The relative contribution of each route certainly depends on the structure of the zeolite. Olsbye et al. proposed that on ZSM-5 ethylene was predominately formed from hydrocarbon pool mechanism, and ethylene did not involve in the methylation route, since ethylene is inert to methylation. Propene and other olefins are produced by both hydrocarbon pool mechanism and methylation/cracking route [38, 39, 105, 109]. Such parallel olefin formation routes were referred as the dual-cycle concept [39].

Since propene and higher olefins can be produced from both hydrocarbon pool mechanism and methylation/cracking mechanism, while ethylene is produced solely from hydrocarbon pool mechanism, it is envisioned that the olefin selectivity from MTO might be manipulated by promoting or suppressing one of the routes [109]. Such approach is rational and promising.

In summary of the olefin formation, the pore size and pore structure of the zeolite catalyst play control role in determining which mechanism operates for MTO. For limiting pores such as 10-ring one-dimensional pores, i.e., ZSM-22, cyclic hydrocarbon pool mechanism is inhibited due to the spatial restriction, while olefin methylation/cracking route may work. If the pores are larger such as the pores on zeolite Y and beta, or there is pore intersection such as ZSM-5, or cage-like structures such as SAPO-34 and ferrierite, the space will be large enough for hydrocarbon pool mechanism, and these catalysts are active catalysts for MTO.

9.1.3.4 Theoretical Computation

The detailed mechanism of MTO, including side-chain growth route, paring mechanism, and methylation/cracking mechanism, involves many transition state species and intermediate species that are virtually impossible to identify and be observed by experimental means. Such subjects are what computational chemistry is suited for. Computational chemistry has been an important tool in elucidating detailed mechanism of MTO. Many publications have covered about all aspects of MTO process, especially the relative energy barrier for side-chain growth of paring mechanism. Several outstanding computational works have also provided the complete catalytic cycle of MTO [29, 65–68, 144–146].

9.1.4 Catalyst Deactivation

Like another hydrocarbon chemistry inside zeolite catalysts, in MTO process, the zeolite catalyst will be deactivated after a certain amount of methanol is being fed. Besides mechanical degradation of the zeolite powder granule and consequently higher pressure drop leading to the blockage of the reaction flow, two major reasons may lead to the zeolite catalyst's deactivation: the loss of acid sites due to ion exchange and the block of internal pores due to the cokes.

Acid sites can be lost if methanol and water were co-fed into the zeolite catalyst bed. Such co-feeding sometimes is used to enhance the yield of a certain olefin product. However, if the water contains even a small amount of sodium or other alkali metal ions, ion exchange between these metal ions and proton may occur, resulting in the loss of acid sites. This problem may be more serious in industry, though, since in academy lab, the quality of water is mostly well controlled.

Coking, on the other hand, is a universal problem in essentially all catalysts, but especially important for zeolite catalysts in MTO, where aromatics are important intermediates for olefin products and yet are also precursors for coke species [47]. For SAPO-34, coking is inevitable as the aromatics are trapped inside its cages. Methylbenzenes will eventually be converted to polycyclic aromatics and finally coke species with long time onstream. In general, coke formation is affected by the pore structure of the zeolites, acid strength, and the acid site density. Since the acid strength of the most frequently used zeolites is roughly about the same (though SAPO-34 was considerably a weaker acidic zeolite than ZSM-5), acid site density was the most influential factor in terms of coke formation [12, 14, 40, 58, 67, 84, 87, 96, 124, 147–162].

Though coke formation seems unavoidable for SAPO-34, it can be minimized for ZSM-5. The pore structures of ZSM-5 allow some methylbenzenes to diffuse out of its pores; thus, the accumulation of coke species is much slower on ZSM-5 than that on SAPO-34. Further reducing the diffusion length of the methylbenzenes will shorten the resident time of coke precursor specie, and thus less coke formation. Ryoo et al. reported that using H-ZSM-5 nanosheets, which has very thin layer

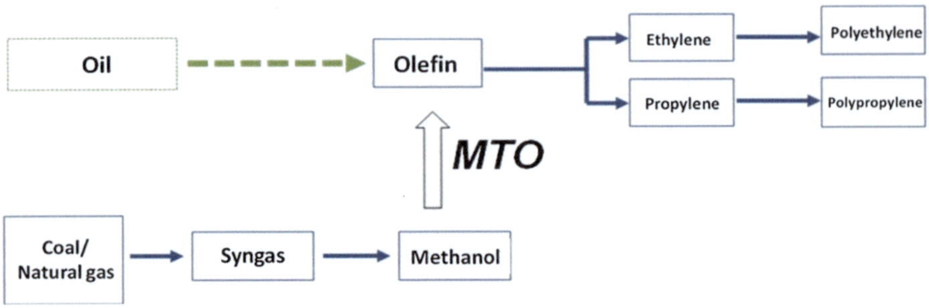

Fig. 9.10 MTO process linking petrochemical industry and coal or natural gas chemical industry

structures, will considerably reduce the coke formation [154, 158, 163–165]. With similar approach, mesoporous zeolites, which have hierarchical pore structures, can also reduce the coke formation rate [166–168].

9.2 MTO Technology

Research and development of coal or natural gas to olefin technology was inspired by the serious impact of the first and second oil crises on the world economy in the 1970s. Many countries have put their research effort on the technology development of producing light olefins from non-oil resources. Up to date, methanol to light olefin technology has become the bridge linking coal or natural gas chemical industry with petrochemical industry (Fig. 9.10). The non-oil route light olefin production from coal or natural gas contains several steps, including syngas preparation, manufacture of methanol, methanol to olefins, and olefin polymerization. Except the MTO process, the other processes have already been commercialized, so MTO process development attracts more and more attention from academy and science all over the world.

9.2.1 Characteristics of MTO Reaction

1. Acid catalysis. Methanol, due to its active chemical property, is easily converted to dimethylether and olefins over solid acid catalyst. Dimethylether can be prepared from methanol using the catalysts of Al_2O_3 and SiO_2-Al_2O_3. However olefin selectivity over these amorphous solid acids is much lower. Selection of suitable zeolite catalyst for methanol transformation can improve the selectivity of light olefins notably.
2. High conversion. Catalyzed by zeolites at above 400 °C, methanol or dimethylether can be completely transformed to hydrocarbons.

3. Low pressure. Olefins and water generate in MTO reaction. Due to the molecular number that tends to increase when light hydrocarbon products and water are generated, low pressure is in favor of light olefin generation.
4. Strong exothermic reaction. In the temperature range of 200–300 °C, the heat of methanol to dimethylether reaction is −10.9–10.4 kJ/mol. At 400–500 °C, the heat of −22.4 to −22.1 kJ/mol is released in MTO reaction.
5. Fast reaction. Experimental studies have confirmed that MTO reaction is a very quick reaction and can occur in a few milliseconds. Under general reaction condition, 100 % conversion can be achieved with contact time of 0.04 s. When the reaction is performed in a short contact time, the secondary reactions of olefin products can be suppressed, and olefin selectivity can be improved.
6. Shape selectivity and coking on zeolite catalyst. By the usage of shape-selective zeolite catalyst, high light olefin selectivity can be achieved. However, coke deposition from the secondary reactions of olefin products leads to the loss of catalytic activity and affects the product selectivity.

9.2.2 Development of MTO Technologies

9.2.2.1 MTO Process of UOP/Hydro and Combined MTO and Olefin Cracking Process

UOP company developed MTO-100 catalyst based on SAPO-34 molecular sieve for MTO process and cooperated with Norsk Hydro company to develop the procedures of synthesizing methanol from natural gas and subsequent methanol conversion to olefins. A fluidized-bed methanol conversion, dealing with 0.75 ton/day methanol at Prosgrann in Norway, had been conducted to verify the MTO catalyst and process. In 1995, pilot test results were published, and it was claimed that MTO industrial device could be designed for producing 0.3–0.5 million tons of light olefins every year. MTO-100 catalyst was reported to show higher attrition resistance performance than the standard FCC catalyst. The MTO-100 catalyst also presented good stability and the conversion and selectivity kept after 450 times of regeneration. During 90 days continuous methanol reaction, conversion kept close to 100 %, and total selectivity of ethylene and propylene (based on carbon number) was about 75–80 % [169]. The ratio of ethylene to propylene could be varied from 0.75 to 1.25 by modifying the reaction conditions, and the total selectivity of ethylene and propylene reached its maximum when the selectivity of ethylene was close to that of propylene [169, 170]. The flow diagram of MTO technology developed by UOP/Hydro is shown in Fig. 9.11 [171].

For improving the selectivity of ethylene and propylene, olefin cracking process (OCP) was suggested by the UOP company for converting higher olefins to smaller parts, ethylene and propylene. OCP process was proved in 1988 and further developed and demonstrated by Total Petrochemicals. It was reported that MTO process combined with the OCP process could further improve the selectivity of light olefins to

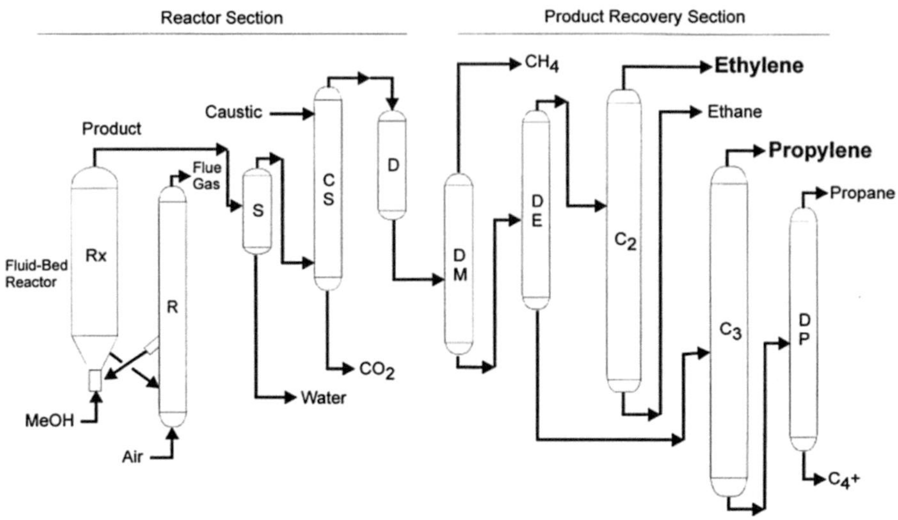

Fig. 9.11 Flow diagram of MTO technology developed by UOP/Hydro. *Rx* reactor, *R* regenerator, *S* separator, *CS* caustic scrubber, *D* dryer, *DM* demethanizer, *DE* deethanizer, C_2 C_2 splitter, C_3 C_3 splitter, *DP* depropanizer (Reprinted from Ref. [171], Copyright 1999, with permission from Elsevier)

about 85–90 %, and the ratio of propylene to ethylene could be varied to 2.1. The integration of MTO and OCP utilizes the C_4^+ products of the MTO process for more ethylene and propylene production. Catalytic cracking is usually in favor of producing propylene, which in turn enhanced the propylene to ethylene ratios [170] (Fig. 9.12). In OCP process, fixed-bed reactors and zeolite catalyst were adopted for the reaction of $C_4 \sim C_7$ mixtures to produce light olefins, under the reaction condition of high space velocity, low pressure (0.1–0.3 MPa), and high reaction temperature (500–600 °C).

A unit of integrated MTO (UOP/Hydro) and OCP (Total/UOP) was established in the research and development center of Total at Feluy in Belgium. In 2010, it was claimed that the demonstration device has run well since it was built in October 2008.

9.2.2.2 MTO Technology of ExxonMobil

A patent of ExxonMobil demonstrated a fluidized-bed process for methanol conversion to olefins catalyzed by SAPO-34, which to some extent represented its MTO technology development and scale. The MTO process of ExxonMobil adopted riser reactor as indicated in Fig. 9.13 [172]. In the device, 300 kg catalyst was loaded, methanol of 550 kg/h was fed, and reaction and regeneration temperature was 490 and 685 °C, respectively, and the published results are given in Table 9.3. According to this patent, increasing the ratio of catalyst inventory in reaction zone to that in circulation zone would improve the quality of products.

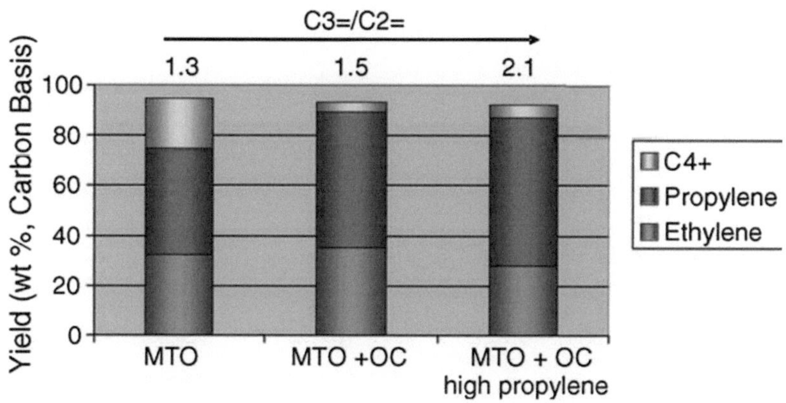

Fig. 9.12 Combination of optimized MTO catalyst and integrated MTO and olefin cracking process for improved propylene-to-ethylene ratios (over 2.0) and reduced by-product formation (Reprinted from Ref. [170], Copyright 2005, with permission from Elsevier)

9.2.2.3 MTP Technology of Lurgi

Lurgi Company developed methanol-to-propylene (MTP) process based on a modified ZSM-5 catalyst with high propylene selectivity. In MTP process, methanol feedstock was firstly converted in pre-reactor containing acidic catalysts into an equilibrium mixture including methanol, dimethyl ether, and water vapor, and then the mixture was converted into olefins (mainly propylene) in fixed-bed reactors installed in series at a reaction temperature of 450–500 °C. Finally, propylene was separated, and the other products were recycled to reactors for further propylene yield improvement. The conversion of methanol and dimethyl ether was higher than 99 %, with 71 % propylene yield calculated on the carbon basis. Based on the microscale tests in the laboratory, the Demo Unit of MTP process was built by Lurgi in Statoil methanol plant at Tjeldbergodden, Norway, in 2001 and started in January 2002. The demonstration test was finished in May 2004, which confirmed that the catalyst lifetime met the commercial target of 8000 h with regeneration period of 500–600 h. Propylene selectivity was higher than 60 % and propylene purity attained to the polymerization grade quality. At the same time, high-quality gasoline was also produced as by-products. The product yield on large-scale MTP devices was announced by Lurgi as follows: for MTP device with the processing capacity of 5000 tons methanol per day (1.67 million tons annually), approximately 519,000 of tons propylene annually, 143,000 tons of gasoline annually, 54,000 tons of LPG annually, 15,000 tons of self-use fuel gas, and 936,000 tons of water annually could be produced [173]. The recommended MTP process flow diagram was shown in Fig. 9.14 [173].

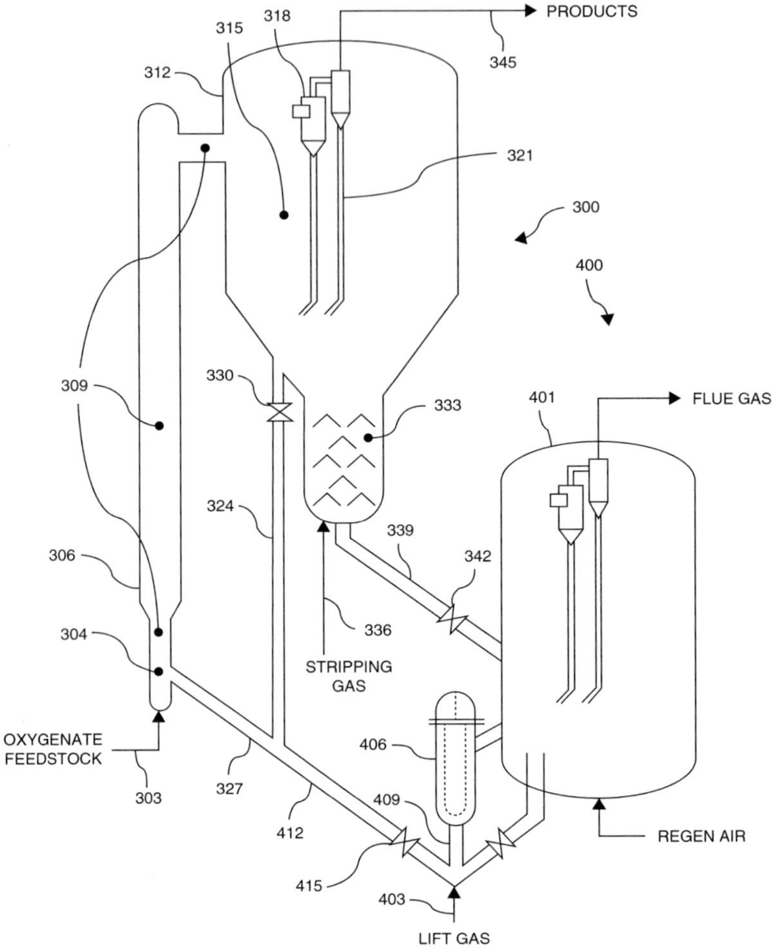

Fig. 9.13 Flow diagram of reaction and regeneration process of ExxonMobil [172] (303, 324, 327, 339, 345, 409, 412-line, 300-reactor, 304-inlet zone, 306-reactor vessel, 309-reaction zone, 312-termination vessel, 315-termination vessel volume, 318-cyclone separator device, 321-dipleg, 324-recirculation line, 333-catalyst stripper, 336-stripping gas, 330, 342, 415-control valve, 400-regenerator, 401-catalyst regenerator vessel, 403-lift gas, 406-catalyst cooler)

Table 9.3 The results of MTO process by ExxonMobil

Sample Point	a	b	c
Rate hydrocarbon plus unconverted methanol (kg/h)	235	14.2	249
Hydrocarbon composition (selectivity wt. %)			
Ethylene	36.7	26.2	35.6
Propylene	39.8	31.3	39.4
Methane + ethane	2.04	5.67	2.37
Propane	2.77	8.43	3.46
C4s	12.5	14.8	12.8
C5+	6.19	13.6	6.37

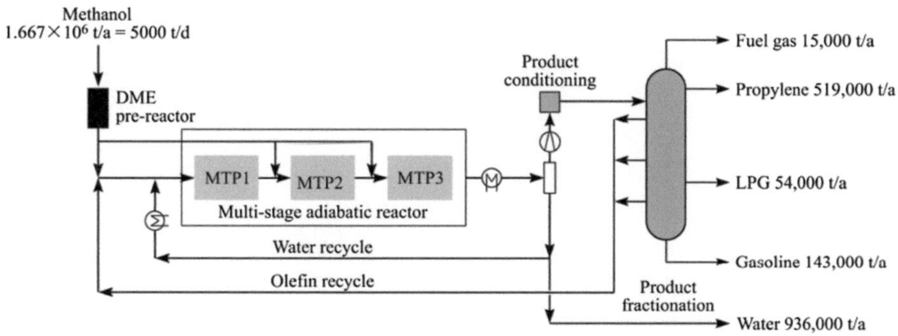

Fig. 9.14 Lurgi MTP simplified process flow diagram (Reprinted with permission from Ref. [173])

9.2.3 MTO Technology of DICP

In China, the study of methanol to olefins in Dalian Institute of Chemical Physics (DICP), Chinese Academy of Sciences, started from the early 1980s. DMTO process (dimethyl ether/methanol to olefin, DMTO) has been successfully developed and applied in the World's first industrial MTO plant [174]. In recent years, Tsinghua University developed methanol-to-propylene process (FMTP) based on a small pore SAPO molecular sieve and fluidized-bed technique and also completed the industrial test with 100 tons of methanol feed per day [175]. China Petrochemical Corporation also developed methanol-to-olefin (SMTO) process and completed the industrial test based on small pore molecular sieve catalyst. It is reported that the SMTO technology has been applied in ethylene plant expansion and reconstruction in Zhongyuan Oilfield [176]. This section will mainly introduce the development of DMTO technology.

9.2.3.1 Research History and MTO Process Development in DICP

9.2.3.1.1 Fixed-Bed MTO Technology of DICP

Since the early 1980s, DICP has been engaged in the MTO process development and non-petrochemical route for the production of light olefins using coal or natural gas as the starting materials. Based on modified ZSM-5 catalysts, DICP developed the catalyst (5200 series) with higher ethylene selectivity (~30 %) and the catalyst (M792 series) with higher propylene selectivity (50–60 %).

The pilot test of fixed-bed MTO reaction was carried out with 300 tons of methanol feeding annually. The flow diagram of fixed-bed process was shown in Fig. 9.15. To avoid the big temperature rise of reactor bed and remove the reaction heat quickly, the dilute methanol (30 wt%) was used as the reaction feedstock and two reactor systems were applied. Methanol was partly dehydrated in the first reactor loaded with Al_2O_3 catalyst to get the mixture of methanol, dimethyl ether, and

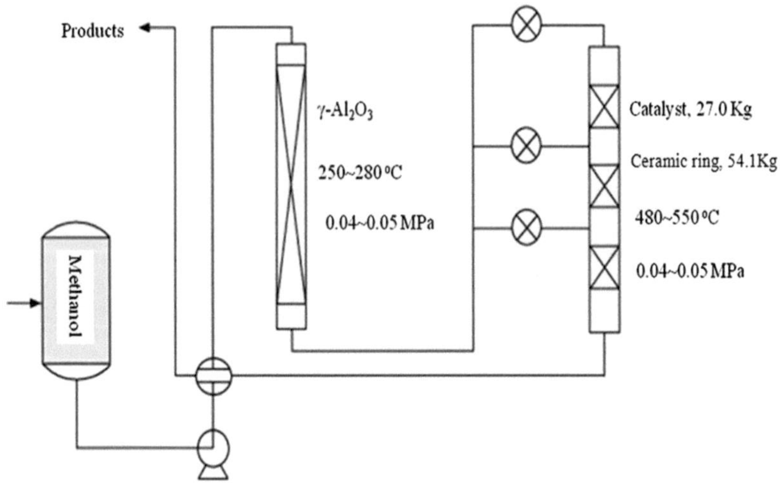

Fig. 9.15 Flow diagram of 300 t/a fixed-bed MTO pilot plant of DICP

water, and then in the second reactor, the mixture was transformed to hydrocarbon products. In this way, severe reaction heat release could be avoided. Furthermore, the catalyst loaded segmentedly in the second reactor to achieve an appropriate bed temperature distribution. The pilot test of fixed-bed methanol conversion verified catalyst performance. 1000 h stability test was completed with seven regeneration cycles. The representative results are shown in Table 9.4.

The selectivity of ethylene is not very satisfactory in the fixed-bed MTO technology based on modified ZSM-5 catalysts, however, the selectivity of propylene could reach 40 wt% at high temperature. These results has laid a good foundation for the propylene production technology via varying the reaction conditions combined with the improvements in techniques.

9.2.3.1.2 Fluidized-Bed MTO Technology of DICP

The previous fixed-bed MTO technology was based on modified ZSM-5 catalyst, although proven to be successful, the selectivity of ethylene and the overall selectivity of both ethylene and propylene was not very high. From the point of shape-selective principle of molecular sieve, it is very difficult for further substantial increase in the selectivity toward light olefins, especially for ethylene based on the modification of ZSM-5 zeolite. The application of novel zeolites with small pore made an important breakthrough in the MTO technology.

In the 1980s, the researchers of Union Carbide company (now UOP) synthesized the silicoaluminophosphate (SAPO) molecular sieves and applied for United State patent in 1984 [176]. The discovery of SAPOs' molecular sieves has been a milestone in the field of zeolite and related catalysis. The researcher of DICP realized

Table 9.4 The pilot test results of 300 t/a fixed-bed MTO process of DICP

Cycle	TOS/h	Methanol WHSV/h^{-1}	Methanol solution /wt%	Product distribution/wt%							$C_2^=$ ~ $C_4^=$ selectivity / wt%
				CH_4	$C_2^=$	$C_3^=$	$C_4^=$	C_2^0 ~ C_4^0	C_5^+	CO_x	
1	162	1.54	34.5	1.75	24.8	39.6	20.7	5.90	5.46	1.61	85.2
2	324	1.49	36.4	1.69	23.8	39.2	21.7	5.34	6.77	1.49	84.6
3	486	1.55	33.9	1.95	24.3	39.6	20.9	5.44	5.87	1.79	84.8
4	638	1.58	36.9	2.07	23.8	40.2	20.9	5.07	6.05	1.86	84.9
5	744	1.56	44.3	1.82	23.3	39.3	22.0	5.51	6.26	1.72	84.6
6	890	1.56	34.6	2.37	24.2	40.3	20.7	5.50	4.70	1.91	85.2
7	1022	1.52	34.5	2.04	23.3	40.0	21.3	5.38	6.24	1.59	84.6

the special significance of SAPO molecular sieve with unique structure and mild acidity on the catalytic conversion of methanol. They reported for the first time the application of SAPO-34 in the MTO reaction [177]: the selectivity of C_2~C_4 alkenes could reach 89 % with ethylene selectivity of 57–59 % at methanol conversion of 100 %.

In the 1990s, researchers in DICP have developed microsphere SAPO molecular sieve catalyst (DO123) prepared via an inexpensive route. When it was applied in the lab-scale MTO reaction at atmospheric pressure, 500–550 °C, and methanol WHSV of 6 h^{-1}, methanol conversion was ~100 % with $C_2^=$~$C_4^=$ selectivity of 85–90 %. Among the products, ethene selectivity was 50–60 %, and ethylene and propylene selectivity attained to 80 %. The industrial scale-up of SAPO-34 molecular sieve and DO123 catalysts were also successfully realized. The scale-up catalyst was tested in a fluidized-bed reactor and presented the performance with DME conversion of 100 %, $C_2^=$~$C_4^=$ selectivity of 90 %, and $C_2^=$ selectivity of ~60 % at 550 °C with DME WHSV of 6 h^{-1}, confirming the scale-up method for SAPO-34 synthesis and catalyst preparation.

As for fluidized-bed reaction technique, DICP constructed the downflow type dilute phase and dense phase fluidized-bed reaction devices and investigated the fluidization mode. Finally, dense phase fluidized-bed reaction system was adopted for the research of MTO process based on the comprehensive consideration of the reaction characteristics, the difficulty of process scale-up, and the helpful experiences of FCC. Dense phase fluidized-bed reactor tests optimize the reaction conditions and lay the foundation for further scale-up of MTO process.

9.2.3.1.3 DMTO Industrial Test

In order to verify and optimize the MTO process and provide basic data for designing and constructing a large-scale industrial plant, the world's first DMTO (Dalian methanol-to-olefins) industrial test plant of 10,000 tons per year (methanol feeding: 50–75 ton/day) stared to be built from August 2004 jointly by the Dalian Institute of Chemical Physics (DICP), Chinese Academy of Sciences, Shanxi Xinxing Coal Chemical Science and Technology Development Co., Ltd., and Luoyang Petrochemical Engineering Corporation Ltd/SINOPEC. The industrial test plant includes the reaction-regeneration part, heat removal part, quenching-stripping part, air compressors, and power station. The plant flow diagram is shown in Fig. 9.16. The industrial test was completed in June 2006 after running for nearly 1200 hours, and the basic data for designing large-scale industrial plant were obtained, and the feasibility of catalysts and technologies was verified [178]. The results under typical operation conditions are shown in Fig. 9.17.

9.2.3.1.4 DMTO-II Industrial Test

To further improve the economics of the DMTO process, DICP developed the new-generation DMTO technology (named DMTO-II) which combines MTO conversion with further conversion of C4+ products into ethylene and propylene.

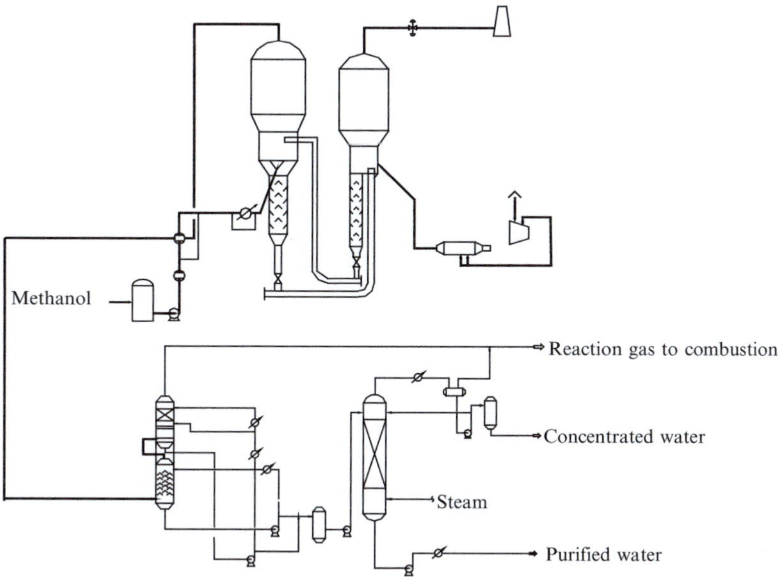

Fig. 9.16 Flow diagram of DMTO industrial test plant

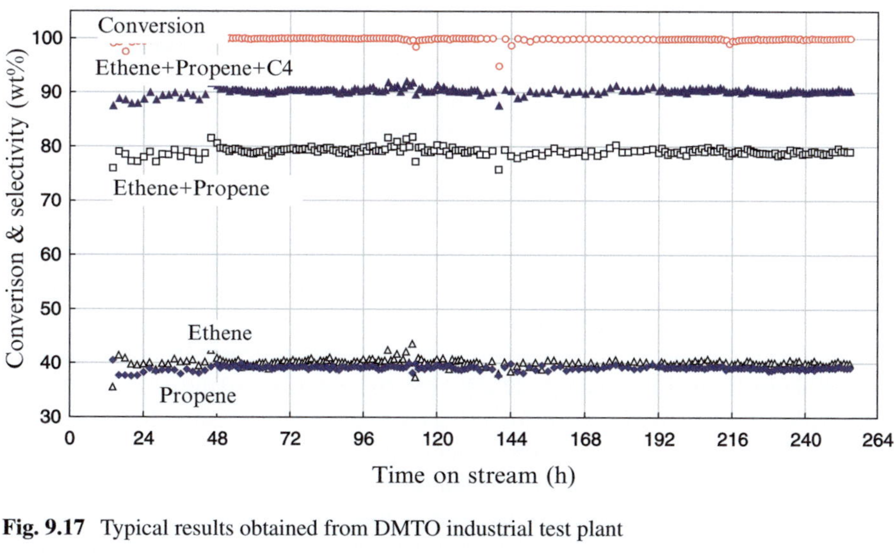

Fig. 9.17 Typical results obtained from DMTO industrial test plant

Although conversions of methanol to olefins and C4+ catalytic cracking are two different types of reactions with distinct reaction mechanisms, both of them can use molecular sieves as solid acid catalysts for the catalytic conversions. Based on the consideration of the principles and technical feasibility, DICP has developed the DMTO-II process based on DMTO technology, especially taking into account

Table 9.5 Comparison of reaction principles and technological characteristics of MTO conversion and C4+ catalytic cracking

		MTO	C_4^+ catalytic cracking
Reaction principles	Reactants	Methanol (C1 reactant)	C_4–C_6 (mainly straight-chain alkenes)
	Mechanism	Hydrocarbon pool, C–C bond propagation	Carbonium ion, C–C bond cracking
	Induction period	Yes	No
	Reaction heat	Strong exothermic	Strong endothermic
	Catalyst	Solid acid	Solid acid
	Pressure effect	Low pressure beneficial	Low pressure beneficial
Technological characteristics	Feed	Gasification feed	Gasification feed
	Main product	Light olefins	Light olefins
	Product distribution	C_1–C_6	C_1–C_6
	Conversion	Nearly 100 %	<100 %
	Catalyst to methanol/oil ratio	0.1–0.5	>5
	Reaction temperature	400–550 °C	550–650 °C
	Reaction pressure	0.15 Mpa	0.15 Mpa
	Regeneration temperature	550–700	600–700
	Coke	1–2 %	2–3 %
	Heat balance	Self balance, heat removal from reactor and regenerator	Heat supply
	Reactor	Bed reactor	Riser

the C4+ conversion performance on the catalyst for DMTO process. The technique characteristic of methanol-to-olefin conversion and C4+ hydrocarbon catalytic cracking reaction are compared in Table 9.5.

The DMTO-II industrial test setup with the handling capacity of 50 tons of methanol per day was based on the DMTO process but included the new constructed the reaction-regeneration system for C4+ conversion, compressing-drying system, and C4+ separation system. Accomplishment of DMTO-II industrial test for 1000 h verified the feasibility of DMTO-II technology and the excellent performance of catalysts manufactured at commercial scale. The test also obtained reliable data for industrial design, laying the foundations for construction of large-scale industrial unit of DMTO-II process. The 72 h accurate evaluation results showed that methanol conversion reaches 99.97 % and raw material consumption per ton of ethene and propene production is 2.67 tons of methanol. Compared with the DMTO technology, the methanol consumption per ton of olefin production was reduced in DMTO-II by more than 10 %.

9.2.3.2 Technological Characteristics of DMTO and DMTO-II

9.2.3.2.1 Technological Characteristics of DMTO

1. Continuous reaction-regeneration dense phase circulation fluidized-bed reaction adapts to the catalytic conversion of methanol over molecular sieve.

By the usage of microporous molecular sieves, SAPO-34, methanol-to-olefin conversion presents very high selectivity of ethylene and propylene due to the acidity and shape selectivity of the catalyst. At the same time, the catalysts also deactivate fast due to coke generation in the cage of the catalyst. It is necessary to regenerate the deactivated catalyst frequently. Taking into account all the reaction factors, including catalysts and reaction characteristics, the fluidized-bed reaction is proved to be a suitable reaction mode.

The characteristics of circulation fluidized-bed reaction for DMTO technology are listed below:

1. Realize continuous reaction and regeneration processes.
2. Beneficial to heat removal in time and well resolving the temperature distribution uniformity of reactor bed (temperature gradient can be controlled in less than 1 °C in the dense phase zone of fluidized-bed reactor).
3. Appropriately controlling the reaction and regeneration conditions for conveniently achieving the heat balance of the reaction system.
4. Methanol conversion performed at high space velocity to reduce the reactor volume.
5. Appropriately setting linear velocity of reactants, controlling the reaction contact time close to an ideal range.
6. The reaction raw materials can be crude methanol, methanol, dimethyl ether, or the mixer of the abovementioned materials.
7. The reaction temperature is 400–550 °C, and the regeneration temperature is 550–700 °C. No rigorous limitation for the materials of reactor and regenerator.

2. Special catalyst for circulation fluidized-bed reaction technique.

The MTO catalyst is developed specially for DMTO technology. It not only possesses the excellent catalytic properties, good thermal stability, and hydrothermal stability, suitable for various raw materials including methanol, dimethyl ether, and their mixtures, but also has appropriate physical properties. Especially, the physical properties, particle size distribution, and fluidization property of the microspherical catalyst used in the fluidized-bed MTO reaction are close to the catalyst used in the fluidized-bed catalytic cracking reaction, so DMTO technology can draw experiences from mature fluidization technology (e.g., FCC). Table 9.6 gives the comparisons of DMTO and FCC technological characteristics.

3. Ethylene/propylene ratio can be adjusted in a proper range.

Without changing the catalyst, the ethylene/propylene ratio can be properly regulated by changing the reaction and regeneration conditions, to adapt to market requirements. According to the industrial test, the ethylene/propylene ratio can be

Table 9.6 Comparisons of DMTO and FCC technological characteristics

No.	Contents	DMTO technology	FCC technology
1	Raw materials	Methanol or DME	Wax oil
			Atmospheric residue
			Vacuum residue
2	Feeding temperature	185–320 °C	180–250 °C
3	Feedstock state	Gas phase	Liquid phase
4	Reaction heat	Exothermic reaction	Endothermic reaction
5	Ratio of catalyst to methanol or oil	0.1–0.5	6–8
6	Reaction temperature	400–550 °C	500–520 °C
7	Reaction pressure	0.15 MPa	0.2–0.3 MPa
8	Regeneration temperature	550–700 °C	600–700 °C
9	Coke yield	1–2 %	5–9 %
10	Thermal equilibrium	Heat removal	Heat balance
11	Reactor	Dense phase fluidized bed	Riser
12	Regenerator	Dense phase fluidized bed	Dense phase fluidized bed
13	Feeding distribution	Gas grid	Nozzle
14	Catalyst transport	Standpipe, sloped pipe, slide valve	Standpipe, sloped pipe, slide valve
15	Main air blower outlet pressure	0.22 MPa	0.3–0.4 MPa
16	Main air blower quantity	Small	Large
17	Catalyst particle type	A type	A type
18	Fluidizing medium in reactor	Lower viscosity	Higher viscosity

adjusted in the range of 0.8–1.2 considering the operation elasticity of a large industrial plant.

4. Special requirements on raw materials and equipment of DMTO technology.

DMTO technology uses an acid molecular sieve catalyst. In order to ensure the long-term stability of the catalyst performance, impurity in raw material of methanol, which may cause the poisoning and permanent deactivation of the catalyst, should be avoided. In addition, due to the active chemical properties of methanol, some metals (such as nickel) can cause the decomposition of methanol to syngas. Therefore, the material which may cause the methanol decomposition should not be used at high-temperature parts of the reactor which are in contact with the feeding of methanol.

9.2.3.2.2 Technological Characteristics of DMTO-II

DMTO-II technology realizes methanol conversion and C_4^+ conversion on the same catalyst for the target olefin product generation. Both the two conversions adopt the fluidized-bed technique. In addition to DMTO technical characteristics, DMTO-II has new features as follows:

1. The reaction of methanol conversion and C_4^+ conversion uses the same catalyst (DMTO catalyst). Beside the effective methanol conversion, catalytic cracking of C_4^+ can be realized, which significantly improves the selectivity of light olefins.
2. Methanol conversion and C_4^+ conversion adopt the fluidized-bed reaction mode. They are performed separately in different reaction zones. Considering the strong endothermic reaction of C_4^+ catalytic cracking, C_4^+ conversion is performed in the high-temperature zone for heat coupling.
3. Methanol conversion and C_4^+ conversion can share one set of separation system, due to the close product distribution.
4. The ratio of ethylene/propylene in DMTO-II is 0.7–1.1.

DMTO-II technology can be not only used for the new MTO setup, but also can be used in technology reforming and upgrading of DMTO industrial equipment.

9.2.3.3 Industrial Application of DMTO Technology

After the industrial test of DMTO technology completed in 2006, the National Development and Reform Commission of People's Republic of China approved coal to the olefin project of Shenhua Group in Baotou, China, using DMTO technology. The project was the world's first industrialization attempt of using coal resources to produce basic chemical products, such as ethylene and propylene, and it was also the first example of industrialization of the methanol-to-olefin technology. The coal to olefin plant includes coal-based methanol joint chemical plant with an annual output of 1.8 million tons of methanol, MTO plant with an annual output of 0.6 million tons of methanol-based polyolefins, power station, public equipments, auxiliary production facilities, etc. The main products are 0.3 million tons of polyethylene and 0.3 million tons of polypropylene per year. The by-product is about 0.15 million tons per year, including 22,000 tons of sulfur, 0.1 million tons of C_4 mixture, and 25,000 tons of C_5+ mixture. The main process of coal to olefin project is shown in Fig. 9.18, and the main technologies used are listed in Table 9.7. The main reaction steps from coal to polymerized olefins include the processes of coal to syngas, syngas to methanol, and methanol to olefins, olefin separation, and olefin polymerization. Most of these steps are realized with the mature technologies from international companies. Only methanol-to-olefin step is realized with innovative DMTO technology. The development of DMTO technology has successfully solved the technical bottleneck of coal or natural gas to olefin process and makes the non-petrochemical route for olefins production a reality.

In August 8, 2010, methanol was fed to MTO plant. In August 13, 2010, qualified ethylene and propylene were the output from olefin separation device. The qualified products of polypropylene pellets and polyethylene pellets were produced since August 15 and August 21, 2010, respectively. After a long period operation, in May, 2011, a 72 h calibration of Shenhua Baotou coal to olefin plant was carried out with the results about methanol-to-olefin unit as follows: methanol conversion 99.99 %, 1 ton olefin production with consumption of 2.96 tons of methanol, and all other

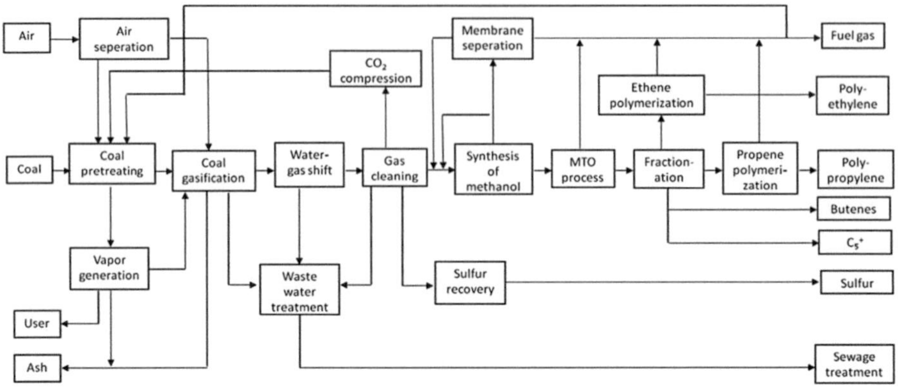

Fig. 9.18 Coal to olefin process

Table 9.7 Main technologies adopted by Shenhua Baotou coal to olefin plant

Unit	Technology source
Coal gasification	GE company, USA
Purification	Linde company, Germany
Methanol synthesis	Davy company, G.B.
Methanol to olefins	DICP, CAS, China
	SYN energy technology Co. ltd
	Luoyang Petrochemical Engineering Corporation ltd, SINOPEC
Olefin separation	Lummus company, USA
Polyethylene	Univation company, USA
Polypropene	Dow company, USA

parameters in the designed range. The long period operation of the device proves that industrial technical feasibility of the DMTO and the financial data show the economic feasibility. Up to present, 18 DMTO technology licensing contracts for 20 MTO plant construction have been signed, and six of them have been industrialized successfully. DMTO-II technology has been industrialized by the Shanxi Pucheng new energy chemical industry limited company.

9.3 Outlook

With the increasing demand for MTO technology, basic study for MTO chemistry will be an active field for the years to come. One important direction for future MTO research is to develop zeolite catalysts with new pore structures for designated product, such as light olefins or aromatics. The size and shape of the opening windows as well as the cage are two basic forms of pores, but the combinations of them lead to numerous possibilities. With better understanding of MTO mechanism, fine-tune of pore parameters may fine-tune the product selectivity, especially by promoting or inhibiting certain mechanisms. Some recent results in this direction have shown promising ability to control the olefin product [160, 162, 179–193].

Another direction is to prolong the life of the zeolite catalyst by enhancing the stability of the catalyst, especially against coking. Coking not only deactivates the catalyst, but also consumes valuable methanol resource. Regenerating deactivated also consumed energy. It's not acceptable that too much valuable methanol feed ends up in cokes. To reduce the coke formation rate, nano-structured zeolite crystals are promising because of their shortened diffusion length. Another approach is to design zeolite pore structures that allow aromatics to diffuse readily so that they don't accumulate in zeolite channels, since coke precursors such as methyl benzenes are useful chemicals. With the goal of less coke for longer catalyst life and the market need for all kinds of chemical raw materials, maybe it is wise to go another direction in terms of product selectivity. This direction is more of technology concern than chemistry.

MTO process industrialization has been successfully realized in recent years. As fluidized-bed catalytic cracking, from its first industrialization to date, its technology innovation never stops. MTO process development still requires continuous innovation and optimization. For the catalyst development, new-generation catalyst development depends on the new material invention and application. Molecular sieves with small size or mesoporous-microporous complex structure will become promising catalysts due to the conversion improvement and prolongation of catalyst life. Industrialization also calls for the precise control of the large-scale synthesis of molecular sieves and more cheap synthesis methods. For the engineering aspect, how to develop new model reactors and optimize the operation condition will be a long-term mission for more efficient methanol conversion and olefin selectivity.

References

1. Chang CD, Silvestri AJ (1977) The conversion of methanol and other O-compounds to hydrocarbons over zeolite catalysts. J Catal 47:249–259
2. Chang CD (1981) A reply to Kagi: mechanism of conversion of methanol over ZSM-5 catalyst. J Catal 69:244–245
3. Chang CD (1983) Hydrocarbons from methanol. Catal Rev Sci Eng 25:1–118
4. Haw JF, Song W, Marcus DM, Nicholas JB (2003) The mechanism of methanol to hydrocarbon catalysis. Acc Chem Res 36:317–326
5. Chang CD (1992) The New Zealand Gas-to-Gasoline plant: An engineering tour de force. Catal Today 13:103–111
6. Ondrey G (2011) Coal-to-chemicals. Chem Eng 118:16–20
7. Methanol-to-Olefins unit starts up in China Oil & Gas Journal, 2010, 108
8. Stocker M (1999) Methanol-to-hydrocarbons: catalytic materials and their behavior. Microporous Mesoporous Mater 29:3–48
9. Wang W, Jiang YJ, Hunger M (2006) Mechanistic investigations of the methanol-to-olefin (MTO) process on acidic zeolite catalysts by in situ solid-state NMR spectroscopy. Catal Today 113:102–114

10. Lesthaeghe D, Van Speybroeck V, Waroquier M (2009) Theoretical evaluation of zeolite confinement effects on the reactivity of bulky intermediates. Phys Chem Chem Phys 11:5222–5226
11. Olsbye U, Svelle S, Bjorgen M, Beato P, Janssens TV, Joensen F, Bordiga S, Lillerud KP (2012) Conversion of methanol to hydrocarbons: how zeolite cavity and pore size controls product selectivity. Angew Chem Int Ed Engl 51:5810–5831
12. Guisnet M (2002) "Coke" molecules trapped in the micropores of zeolites as active species in hydrocarbon transformations. J Mol Catal A Chem 182:367–382
13. Haw JF, Marcus DM (2005) Well-defined (supra) molecular structures in zeolite methanol-to-olefin catalysis. Top Catal 34:41–48
14. Froment G (2008) Kinetic modeling of hydrocarbon processing and the effect of catalyst deactivation by coke formation. Catal Rev 50:1–18
15. Smit B, Maesen TL (2008) Molecular simulations of zeolites: adsorption, diffusion, and shape selectivity. Chem Rev 108:4125–4184
16. Wang W, Hunger M (2008) Reactivity of surface alkoxy species on acidic zeolite catalysts. Acc Chem Res 41:895–904
17. White JL (2011) Methanol-to-hydrocarbon chemistry: the carbon pool revolution. Catal Sci Technol 1:1630–1635
18. Ilias S, Bhan A (2013) Mechanism of the catalytic conversion of methanol to hydrocarbons. ACS Catal 3:18–31
19. Ahmadi SMA, Askari S, Halladj R (2013) A review on kinetic modeling of deactivation of SAPO-34 catalyst during Methanol to Olefins (MTO) process. Afinidad 70:130–138
20. Wang W, Buchholz A, Seiler M, Hunger M (2003) Evidence for an initiation of the methanol-to-olefin process by reactive surface methoxy groups on acidic zeolite catalysts. J Am Chem Soc 125:15260–15267
21. Svelle S, Rønning PO, Kolboe S (2004) Kinetic studies of zeolite-catalyzed methylation reactions 1. Coreaction of [C-12]ethene and [C-13]methanol. J Catal 224:115–123
22. Svelle S, Ronning P, Olsbye U, Kolboe S (2005) Kinetic studies of zeolite-catalyzed methylation reactions. Part 2. Co-reaction of [C-12]propene or [C-12]n-butene and [C-13] methanol. J Catal 234:385–400
23. Gayubo A, Aguayo A, Alonso A, Atutxa A, Bilbao J (2005) Reaction scheme and kinetic modelling for the MTO process over a SAPO-18 catalyst. Catal Today 106:112–117
24. Chen D, Grønvold A, Moljord K, Holmen A (2007) Methanol conversion to light olefins over SAPO-34: reaction network and deactivation kinetics. Ind Eng Chem Res 46:4116–4123
25. Gayubo AG, Aguayo AT, Alonso A, Bilbao J (2007) Kinetic modeling of the methanol-to-olefins process on a silicoaluminophosphate (SAPO-18) catalyst by considering deactivation and the formation of individual olefins. Ind Eng Chem Res 46:1981–1989
26. Zhou H, Wang Y, Wei F, Wang D, Wang Z (2008) Kinetics of the reactions of the light alkenes over SAPO-34. Appl Catal A Gen 348:135–141
27. Froment GF (2008) Kinetic modeling of hydrocarbon processing and the effect of catalyst deactivation by coke formation. Catal Rev Sci Eng 50:1–18
28. Mier D, Gayubo AG, Aguayo AT, Olazar M, Bilbao J (2011) Olefin production by cofeeding methanol and n-butane: kinetic modeling considering the deactivation of HZSM-5 zeolite. AIChE J 57:2841–2853
29. Van Speybroeck V, Van der Mynsbrugge J, Vandichel M, Hemelsoet K, Lesthaeghe D, Ghysels A, Marin GB, Waroquier M (2011) First principle kinetic studies of zeolite-catalyzed methylation reactions. J Am Chem Soc 133:888–899
30. Hill IM, Hashimi SA, Bhan A (2012) Kinetics and mechanism of olefin methylation reactions on zeolites. J Catal 285:115–123
31. Van der Mynsbrugge J, Visur M, Olsbye U, Beato P, Bjørgen M, Van Speybroeck V, Svelle S (2012) Methylation of benzene by methanol: single-site kinetics over H-ZSM-5 and H-beta zeolite catalysts. J Catal 292:201–212
32. Taheri Najafabadi A, Fatemi S, Sohrabi M, Salmasi M (2012) Kinetic modeling and optimization of the operating condition of MTO process on SAPO-34 catalyst. J Ind Eng Chem 18:29–37

33. Dahl IM, Kolboe S (1993) On the reaction mechanism for propene formation in the MTO reaction over SAPO-34. Catal Lett 20:329–336
34. Dahl IM, Kolboe S (1996) On the reaction mechanism for hydrocarbon formation from methanol over SAPO-34.2: isotopic labeling studies of the co-reaction of propene and methanol. J Catal 161:304–309
35. Mikkelsen O, Ronning PO, Kolboe S (2000) Use of isotopic labeling for mechanistic studies of the methanol-to-hydrocarbons reaction: methylation of toluene with methanol over H-ZSM-5, H-mordenite and H-beta. Microporous Mesoporous Mater 40:95–113
36. Bjorgen M, Olsbye U, Kolboe S (2003) Coke precursor formation and zeolite deactivation: mechanistic insights from hexamethylbenzene conversion. J Catal 215:30–44
37. Bjorgen M, Olsbye U, Petersen D, Kolboe S (2004) The methanol-to-hydrocarbons reaction: insight into the reaction mechanism from [C-12]benzene and [C-13]methanol coreactions over zeolite H-beta. J Catal 221:1–10
38. Svelle S, Joensen F, Nerlov J, Olsbye U, Lillerud KP, Kolboe S, Bjorgen M (2006) Conversion of methanol into hydrocarbons over zeolite H-ZSM-5: ethene formation is mechanistically separated from the formation of higher alkenes. J Am Chem Soc 128:14770–14771
39. Bjorgen M, Svelle S, Joensen F, Nerlov J, Kolboe S, Bonino F, Palumbo L, Bordiga S, Olsbye U (2007) Conversion of methanol to hydrocarbons over zeolite H-ZSM-5: on the origin of the olefinic species. J Catal 249:195–207
40. Bjørgen M, Akyalcin S, Olsbye U, Benard S, Kolboe S, Svelle S (2010) Methanol to hydrocarbons over large cavity zeolites: toward a unified description of catalyst deactivation and the reaction mechanism. J Catal 275:170–180
41. Song WG, Haw JF, Nicholas JB, Heneghan CS (2000) Methylbenzenes are the organic reaction centers for methanol-to-olefin catalysis on HSAPO-34. J Am Chem Soc 122:10726–10727
42. Sassi A, Wildman MA, Ahn HJ, Prasad P, Nicholas JB, Haw JF (2002) Methylbenzene chemistry on zeolite HBeta: multiple insights into methanol-to-olefin catalysis. J Phys Chem B 106:2294–2303
43. Marcus DM, Song WG, Ng LL, Haw JF (2002) Aromatic hydrocarbon formation in HSAPO-18 catalysts: cage topology and acid site density. Langmuir 18:8386–8391
44. Marcus DM, Hayman MJ, Blau YM, Guenther DR, Ehresmann JO, Kletnieks PW, Haw JF (2006) Mechanistically significant details of the H/D exchange reactions of propene over acidic zeolite catalysts. Angew Chem Int Ed 45:1933–1935
45. Marcus DM, McLachlan KA, Wildman MA, Ehresmann JO, Kletnieks PW, Haw JF (2006) Experimental evidence from H/D exchange studies for the failure of direct C-C coupling mechanisms in the methanol-to-olefin process catalyzed by HSAPO-34. Angew Chem Int Ed 45:3133–3136
46. Goguen PW, Xu T, Barich DH, Skloss TW, Song W, Wang Z, Nicholas JB, Haw JF (1998) Pulse-quench catalytic reactor studies reveal a carbon-pool mechanism in methanol-to-gasoline chemistry on zeolite HZSM-5. J Am Chem Soc 120:2650–2651
47. Song W, Fu H, Haw JF (2001) Selective synthesis of methylnaphthalenes in HSAPO-34 cages and their function as reaction centers in methanol-to-olefin catalysis. J Phys Chem B 105:12839–12843
48. Fu H, Song W, Haw JF (2001) Polycyclic aromatics formation in HSAPO-34 during methanol-to-olefin catalysis: ex situ characterization after cryogenic grinding. Catal Lett 76:89–94
49. Fu H, Song W, Marcus DM, Haw JF (2002) Ship-in-a-bottle synthesis of methylphenols in HSAPO-34 cages from methanol and air. J Phys Chem B 106:5648–5652
50. Svelle S, Bjorgen M, Kolboe S, Kuck D, Letzel M, Olsbye U, Sekiguchi O, Uggerud E (2006) Intermediates in the methanol-to-hydrocarbons (MTH) reaction: a gas phase study of the unimolecular reactivity of multiply methylated benzenium cations. Catal Lett 109:25–35
51. Barich DH, Xu T, Song W, Wang Z, Deng F, Haw JF (1998) N-alkylnitrilium cations in zeolites: a study using theoretical chemistry and in situ NMR with the pulse-quench reactor. J Phys Chem B 102:7163–7168
52. Haw JF, Nicholas JB, Song W, Deng F, Wang Z, Xu T, Heneghan CS (2000) Roles for cyclopentenyl cations in the synthesis of hydrocarbons from methanol on zeolite catalyst HZSM-5. J Am Chem Soc 122:4763–4775

53. Chua YT, Stair PC, Nicholas JB, Song W, Haw JF (2003) UV Raman spectrum of 1,3-dimethylcyclopentenyl cation adsorbed in zeolite H-MFI. J Am Chem Soc 125:866–867
54. Salehirad F, Anderson MW (1998) NMR studies of methanol-to-hydrocarbon chemistry. Part 1 – Primary products and mechanistic considerations using a wide-pore catalyst. J Chem Soc Faraday Trans 94:1911–1918
55. Salehirad F, Anderson MW (1998) Solid-state NMR study of methanol conversion over ZSM-23, SAPO-11 and SAPO-5 molecular sieves: Part 2. J Chem Soc Faraday Trans 94:2857–2866
56. Hunger M, Freude D, Pfeifer H (1991) Magic-angle spinning nuclear magnetic resonance studies of water molecules adsorbed on Brønsted- and Lewis-acid sites in zeolites and amorphous silica–aluminas. J Chem Soc Faraday Trans 87:657–662
57. Hunger M (2008) In situ flow MAS NMR spectroscopy: state of the art and applications in heterogeneous catalysis. Prog Nucl Magn Reson Spectrosc 53:105–127
58. Huang J, Jiang Y, Marthala VRR, Bressel A, Frey J, Hunger M (2009) Effect of pore size and acidity on the coke formation during ethylbenzene conversion on zeolite catalysts. J Catal 263:277–283
59. Dai W, Wang X, Wu G, Guan N, Hunger M, Li L (2011) Methanol-to-olefin conversion on silicoaluminophosphate catalysts: effect of bronsted acid sites and framework structures. ACS Catal 1:292–299
60. Dai W, Li N, Li L, Guan N, Hunger M (2011) Unexpected methanol-to-olefin conversion activity of low-silica aluminophosphate molecular sieves. Catal Commun 16:124–127
61. Dai W, Scheibe M, Guan N, Li L, Hunger M (2011) Fate of bronsted acid sites and benzene-based carbenium ions during methanol-to-olefin conversion on SAPO-34. ChemCatChem 3:1130–1133
62. Jiang Y, Huang J, Dai W, Hunger M (2011) Solid-state nuclear magnetic resonance investigations of the nature, property, and activity of acid sites on solid catalysts. Solid State Nucl Magn Reson 39:116–141
63. Dai W, Scheibe M, Li L, Guan N, Hunger M (2012) Effect of the methanol-to-olefin conversion on the PFG NMR self-diffusivities of ethane and ethene in large-crystalline SAPO-34. J Phys Chem C 116:2469–2476
64. Dai W, Wu G, Li L, Guan N, Hunger M (2013) Mechanisms of the deactivation of SAPO-34 materials with different crystal sizes applied as MTO catalysts. ACS Catal 3:588–596
65. Lesthaeghe D, Van Speybroeck V, Marin GB, Waroquier M (2006) Understanding the failure of direct C-C coupling in the zeolite-catalyzed methanol-to-olefin process. Angew Chem Int Ed 45:1714–1719
66. McCann DM, Lesthaeghe D, Kletnieks PW, Guenther DR, Hayman MJ, Van Speybroeck V, Waroquier M, Haw JF (2008) A complete catalytic cycle for supramolecular methanol-to-olefins conversion by linking theory with experiment. Angew Chem Int Ed 47:5179–5182
67. Lesthaeghe D, Horré A, Waroquier M, Marin GB, Van Speybroeck V (2009) Theoretical insights on methylbenzene side-chain growth in ZSM-5 zeolites for methanol-to-olefin conversion. Chem Eur J 15:10803–10808
68. Lesthaeghe D, Van der Mynsbrugge J, Vandichel M, Waroquier M, Van Speybroeck V (2011) Full theoretical cycle for both ethene and propene formation during methanol-to-olefin conversion in H-ZSM-5. ChemCatChem 3:208–212
69. Olah GA (1981) Higher coordinate (hypercarbon containing) carbocations and their role in electrophilic reactions of hydrocarbons. Pure Appl Chem 53:201–207
70. Hutchings GJ, Gottschalk F, Hall MVM, Hunter R (1987) Hydrocarbon formation from methylating agents over the zeolite catalyst ZSM-5: comments on the mechanism of carbon–carbon bond and methane formation. J Chem Soc Faraday Trans I 83:571–583
71. Hutchings GJ, Watson GW, Willock DJ (1999) Methanol conversion to hydrocarbons over zeolite catalysts: comments on the reaction mechanism for the formation of the first carbon-carbon bond. Microporous Mesoporous Mater 29:67–77
72. Wang W, Seiler M, Hunger M (2001) Role of surface methoxy species in the conversion of methanol to dimethyl ether on acidic zeolites investigated by in situ stopped-flow MAS NMR spectroscopy. J Phys Chem B 105:12553–12558

73. Seiler M, Wang W, Buchholz A, Hunger M (2003) Direct evidence for a catalytically active role of the hydrocarbon pool formed on zeolite H-ZSM-5 during the methanol-to-olefin conversion. Catal Lett 88:187–191
74. Wang W, Buchholz A, Arnold A, Xu M, Hunger M (2003) Effect of surface methoxy groups on the Al-27 quadrupole parameters of framework aluminum atoms in calcined zeolite H-Y. Chem Phys Lett 370:88–93
75. Wang W, Jiao J, Jiang Y, Ray SS, Hunger M (2005) Formation and decomposition of surface ethoxy species on acidic zeolite Y. ChemPhysChem 6:1467–1469
76. Jiang Y, Hunger M, Wang W (2006) On the reactivity of surface methoxy species in acidic zeolites. J Am Chem Soc 128:11679–11692
77. Mole T, Whiteside JA, Seddon D (1983) Aromatic co-catalysis of methanol conversion over zeolite catalysts. J Catal 82:261–266
78. Song W, Marcus DM, Fu H, Ehresmann JO, Haw JF (2002) An oft-studied reaction that may never have been: direct catalytic conversion of methanol or dimethyl ether to hydrocarbons on the solid acids HZSM-5 or HSAPO-34. J Am Chem Soc 124:3844–3845
79. Cui ZM, Liu Q, Song WG, Wan LJ (2006) Insights into the mechanism of methanol-to-olefin conversion at zeolites with systematically selected framework structures. Angew Chem Int Ed 45:6512–6515
80. Cui ZM, Liu Q, Ma Z, Bian SW, Song WG (2008) Direct observation of olefin homologations on zeolite ZSM-22 and its implications to methanol to olefin conversion. J Catal 258:83–86
81. Cui ZM, Liu Q, Baint SW, Ma Z, Song WG (2008) The role of methoxy groups in methanol to olefin conversion. J Phys Chem C 112:2685–2688
82. Song ZX, Takahashi A, Mimura N, Fujitani T (2009) Production of propylene from ethanol over ZSM-5 zeolites. Catal Lett 131:364–369
83. Wang Q, Cui Z-M, Cao C-Y, Song W-G (2011) 0.3 angstrom makes the difference: dramatic changes in methanol-to-olefin activities between H-ZSM-12 and H-ZSM-22 zeolites. J Phys Chem C 115:24987–24992
84. Wei F-F, Cui Z-M, Meng X-J, Cao C-Y, Xiao F-S, Song W-G (2014) Origin of the low olefin production over HZSM-22 and HZSM-23 zeolites: external acid sites and pore mouth catalysis. ACS Catal 4:529–534
85. Liu Z, Liang J (1999) Methanol to olefin conversion catalysts. Curr Opin Solid State Mater Sci 4:80–84
86. Li J, Qi Y, Liu Z, Liu G, Zhang D (2008) Co-reaction of ethene and methylation agents over SAPO-34 and ZSM-22. Catal Lett 121:303–310
87. Li J, Wei Y, Qi Y, Tian P, Li B, He Y, Chang F, Sun X, Liu Z (2011) Conversion of methanol over H-ZSM-22: the reaction mechanism and deactivation. Catal Today 164:288–292
88. Li JZ, Wei YX, Liu GY, Qi Y, Tian P, Li B, He YL, Liu ZM (2011) Comparative study of MTO conversion over SAPO-34, H-ZSM-5 and H-ZSM-22: correlating catalytic performance and reaction mechanism to zeolite topology. Catal Today 171:221–228
89. Maihom T, Boekfa B, Sirijaraensre J, Nanok T, Probst M, Limtrakul J (2009) Reaction mechanisms of the methylation of ethene with methanol and dimethyl ether over H-ZSM-5: an ONIOM study. J Phys Chem C 113:6654–6662
90. Simonetti DA, Ahn JH, Iglesia E (2011) Mechanistic details of acid-catalyzed reactions and their role in the selective synthesis of triptane and isobutane from dimethyl ether. J Catal 277:173–195
91. Xu T, Haw JF (1994) NMR observation of indanyl carbenium ion intermediates in the reactions of hydrocarbons on acidic zeolites. J Am Chem Soc 116:10188–10195
92. Kolboe S (1986) Methanol reactions on ZSM-5 and other zeolite catalysts: autocatalysis and reaction mechanism. Acta Chem Scand 40:711–713
93. Dahl IM, Kolboe S (1994) On the reaction mechanism for hydrocarbon formation from methanol over SAPO-34: I. Isotopic labeling studies of the co-reaction of ethene and methanol. J Catal 149:458–464

94. Hunger M (1997) Bronsted acid sites in zeolites characterized by multinuclear solid-state NMR spectroscopy. Catal Rev 39:345–393
95. Olsbye U, Bjørgen M, Svelle S, Lillerud K-P, Kolboe S (2005) Mechanistic insight into the methanol-to-hydrocarbons reaction. Catal Today 106:108–111
96. Chen D, Moljord K, Holmen A (2012) A methanol to olefins review: diffusion, coke formation and deactivation SAPO type catalysts. Microporous Mesoporous Mater 164:239–250
97. Haw JF, Goguen PW, Xu T, Skloss TW, Song W, Wang Z (1998) In situ NMR investigations of heterogeneous catalysis with samples prepared under standard reaction conditions. Angew Chem Int Ed 37:948–949
98. Olah GA, Gupta B, Farina M, Felberg JD, Ip WM, Husain A, Karpeles R, Lammertsma K, Melhotra AK, Trivedi NJ (1985) Electrophilic reactions at single bonds. 20. Selective monohalogenation of methane over supported acidic or platinum metal catalysts and hydrolysis of methyl halides over.gamma.-alumina-supported metal oxide/hydroxide catalysts: a feasible path for the oxidative conversion of methane into methyl alcohol/dimethyl ether. J Am Chem Soc 107:7097–7105
99. Xu T, Barich DH, Goguen PW, Song W, Wang Z, Nicholas JB, Haw JF (1998) Synthesis of a benzenium ion in a zeolite with use of a catalytic flow reactor. J Am Chem Soc 120:4025–4026
100. Song W, Nicholas JB, Haw JF (2001) A persistent carbenium ion on the methanol-to-olefin catalyst HSAPO-34: acetone shows the way. J Phys Chem B 105:4317–4323
101. Song W, Nicholas JB, Haw JF (2001) Acid-base chemistry of a carbenium ion in a zeolite under equilibrium conditions: verification of a theoretical explanation of carbenium ion stability. J Am Chem Soc 123:121–129
102. Song W, Nicholas JB, Sassi A, Haw JF (2002) Synthesis of the heptamethylbenzenium cation in zeolite-beta: in situ NMR and theory. Catal Lett 81:49–53
103. Bjørgen M, Bonino F, Kolboe S, Lillerud K-P, Zecchina A, Bordiga S (2003) Spectroscopic evidence for a persistent benzenium cation in zeolite H-beta. J Am Chem Soc 125:15863–15868
104. Bjørgen M (2004) The methanol-to-hydrocarbons reaction: insight into the reaction mechanism from [C-12]benzene and [C-13]methanol coreactions over zeolite H-beta. J Catal 221:1–10
105. Bjørgen M, Joensen F, Lillerud KP, Olsbye U, Svelle S (2009) The mechanisms of ethene and propene formation from methanol over high silica H-ZSM-5 and H-beta. Catal Today 142:90–97
106. Li J, Wei Y, Chen J, Tian P, Su X, Xu S, Qi Y, Wang Q, Zhou Y, He Y, Liu Z (2012) Observation of heptamethylbenzenium cation over SAPO-type molecular sieve DNL-6 under real MTO conversion conditions. J Am Chem Soc 134:836–839
107. Xu S, Zheng A, Wei Y, Chen J, Li J, Chu Y, Zhang M, Wang Q, Zhou Y, Wang J, Deng F, Liu Z (2013) Direct observation of cyclic carbenium ions and their role in the catalytic cycle of the methanol-to-olefin reaction over chabazite zeolites. Angew Chem Int Ed 52:11564–11568
108. Li J, Wei Y, Xu S, Tian P, Chen J, Liu Z (2014) Heptamethylbenzenium cation formation and the correlated reaction pathway during methanol-to-olefins conversion over DNL-6. Catal Today 226:47–51
109. Svelle S, Olsbye U, Joensen F, Bjørgen M (2007) Conversion of methanol to alkenes over medium- and large-pore acidic zeolites: steric manipulation of the reaction intermediates governs the ethene/propene product selectivity. J Phys Chem C 111:17981–17984
110. Hunger M, Horvath T (1995) A new MAS NMR probe for in situ investigations of hydrocarbon conversion on solid catalysts under continuous-flow conditions. J Chem Soc Chem Commun 14:1423–1424
111. Hunger M (1996) Multinuclear solid-state NMR studies of acidic and non-acidic hydroxyl protons in zeolites. Solid State Nucl Magn Reson 6:1–29
112. Hunger M, Horvath T (1996) Adsorption of methanol on bronsted acid sites in zeolite H-ZSM-5 investigated by multinuclear solid-state NMR spectroscopy. J Am Chem Soc 118:12302–12308
113. Seiler M, Schenk U, Hunger M (1999) Conversion of methanol to hydrocarbons on zeolite HZSM-5 investigated by in situ MAS NMR spectroscopy under flow conditions and on-line gas chromatography. Catal Lett 62:139–145

114. Wang W, Seiler M, Weitkamp J, Hunger M, Ivanova II (2001) In situ stopped-flow (SF) MAS NMR spectroscopy: a novel NMR technique applied for the study of aniline methylation on a solid base catalyst. Chem Commun 15:1362–1363
115. Seiler M, Wang W, Hunger M (2001) Local structure of framework aluminum in zeolite H-ZSM-5 during conversion of methanol investigated by in situ NMR spectroscopy. J Phys Chem B 105:8143–8148
116. Hunger M, Seiler M, Buchholz A (2001) In situ MAS NMR spectroscopic investigation of the conversion of methanol to olefins on silicoaluminophosphates SAPO-34 and SAPO-18 under continuous flow conditions. Catal Lett 74:61–68
117. Wang W, Seiler M, Ivanova II, Sternberg U, Weitkamp J, Hunger M (2002) Formation and decomposition of N, N, N-trimethylanilinium cations on zeolite H-Y investigated by in situ stopped-flow MAS NMR spectroscopy. J Am Chem Soc 124:7548–7554
118. Buchholz A, Wang W, Xu M, Arnold A, Hunger M (2002) Thermal stability and dehydroxylation of Bronsted acid sites in silicoaluminophosphates H-SAPO-11, H-SAPO-81 H-SAPO-31, and H-SAPO-34 investigated by multi-nuclear solid-state NMR spectroscopy. Microporous Mesoporous Mater 56:267–278
119. Buchholz A, Wang W, Arnold A, Xu M, Hunger M (2003) Successive steps of hydration and dehydration of silicoaluminophosphates H-SAPO-34 and H-SAPO-37 investigated by in situ CF MAS NMR spectroscopy. Microporous Mesoporous Mater 57:157–168
120. Arstad B, Kolboe S (2001) The reactivity of molecules trapped within the SAPO-34 cavities in the methanol-to-hydrocarbons reaction. J Am Chem Soc 123:8137–8138
121. Arstad B, Kolboe S (2001) Methanol-to-hydrocarbons reaction over SAPO-34: molecules confined in the catalyst cavities at short time on stream. Catal Lett 71:209–212
122. Svelle S, Kolboe S, Olsbye U, Swang O (2003) A theoretical investigation of the methylation of methylbenzenes and alkenes by halomethanes over acidic zeolites. J Phys Chem B 107:5251–5260
123. Arstad B, Kolboe S, Swang O (2004) Theoretical study of protonated xylenes: ethene elimination and H, C-scrambling reactions. J Phys Org Chem 17:1023–1032
124. Bjorgen M, Olsbye U, Svelle S, Kolboe S (2004) Conversion of methanol to hydrocarbons: the reactions of the heptamethylbenzenium cation over zeolite H-beta. Catal Lett 93:37–40
125. Arstad B, Kolboe S, Swang O (2005) Theoretical study of the heptamethylbenzenium ion: intramolecular isomerizations and C2, C3, C4 alkene elimination. J Phys Chem A 109:8914–8922
126. Svelle S, Aravinthan S, Bjorgen M, Lillerud KP, Kolboe S, Dahl IM, Olsbye U (2006) The methyl halide to hydrocarbon reaction over H-SAPO-34. J Catal 241:243–254
127. Teketel S, Skistad W, Benard S, Olsbye U, Lillerud KP, Beato P, Svelle S (2012) Shape selectivity in the conversion of methanol to hydrocarbons: the catalytic performance of one-dimensional 10-ring zeolites: ZSM-22, ZSM-23, ZSM-48, and EU-1. ACS Catal 2:26–37
128. Bleken F, Skistad W, Barbera K, Kustova M, Bordiga S, Beato P, Lillerud KP, Svelle S, Olsbye U (2011) Conversion of methanol over 10-ring zeolites with differing volumes at channel intersections: comparison of TNU-9, IM-5, ZSM-11 and ZSM-5. Phys Chem Chem Phys 13:2539–2549
129. Olsbye U, Saure OV, Muddada NB, Bordiga S, Lamberti C, Nilsen MH, Lillerud KP, Svelle S (2011) Methane conversion to light olefins – how does the methyl halide route differ from the methanol to olefins (MTO) route? Catal Today 171:211–220
130. Sommer L, Krivokapić A, Svelle S, Lillerud KP, Stöcker M, Olsbye U (2011) Enhanced catalyst performance of zeolite SSZ-13 in the methanol to olefin reaction after neutron irradiation. J Phys Chem C 115:6521–6530
131. Svelle S, Sommer L, Barbera K, Vennestrom PNR, Olsbye U, Lillerud KP, Bordiga S, Pan YH, Beato P (2011) How defects and crystal morphology control the effects of desilication. Catal Today 168:38–47
132. Olsbye U, Svelle S, Bjørgen M, Beato P, Janssens TVW, Joensen F, Bordiga S, Lillerud KP (2012) Conversion of methanol to hydrocarbons: how zeolite cavity and pore size controls product selectivity. Angew Chem Int Ed 51:5810–5831

133. Kumar P, Thybaut JW, Svelle S, Olsbye U, Marin GB (2013) Single-event microkinetics for methanol to olefins on H-ZSM-5. Ind Eng Chem Res 52:1491–1507
134. Westgård Erichsen M, Svelle S, Olsbye U (2013) H-SAPO-5 as methanol-to-olefins (MTO) model catalyst: towards elucidating the effects of acid strength. J Catal 298:94–101
135. Xu T, White JL (2004) Crystalline silicoaluminophosphate molecular sieve as catalyst in converting oxygenate feedstock to olefin compositions, includes integrated hydrocarbon co-catalyst. US Patent 6,743,747, 1 Jun 2004
136. Xu T, White JL (2004) Crystalline silicoaluminophosphate molecular sieve as catalyst in converting oxygenate feedstock to olefin compositions, includes integrated hydrocarbon co-catalyst. US Patent 6,734,330, 11 May 2004
137. Song WG, Fu H, Haw JF (2001) Supramolecular origins of product selectivity for methanol-to-olefin catalysis on HSAPO-34. J Am Chem Soc 123:4749–4754
138. Ehresmann JO, Wang W, Herreros B, Luigi D-P, Venkatraman T, Song W, Nicholas JB, Haw JF (2002) Theoretical and experimental investigation of the effect of proton transfer on the Al-27 MAS NMR line shapes of zeolite-adsorbate complexes: an independent measure of solid acid strength. J Am Chem Soc 124:10868–10874
139. Sassi A, Wildman MA, Haw JF (2002) Reactions of butylbenzene isomers on zeolite HBeta: methanol-to-olefins hydrocarbon pool chemistry and secondary reactions of olefins. J Phys Chem B 106:8768–8773
140. Sullivan RF, Sieg RP, Langlois GE, Egan CJ (1961) A new reaction that occurs in the hydrocracking of certain aromatic hydrocarbons. J Am Chem Soc 83:1156–1160
141. Wei YX, Zhang DZ, Liu ZM, Su BL (2007) Mechanistic elucidation of chloromethane transformation over SAPO-34 using deuterated probe molecule: a FTIR study on the surface evolution of catalyst. Chem Phys Lett 444:197–201
142. Dessau RM, Lapierre RB (1982) On the mechanism of methanol conversion to hydrocarbons over HZSM-5. J Catal 78:136–141
143. Dessau RM (1986) On the H-ZSM-5 catalyzed formation of ethylene from methanol or higher olefins. J Catal 99:111–116
144. Lesthaeghe D, De Sterck B, Van Speybroeck V, Marin GB, Waroquier M (2007) Zeolite shape-selectivity in the gem-methylation of aromatic hydrocarbons. Angew Chem Int Ed 46:1311–1314
145. Hemelsoet K, Nollet A, Vandichel M, Lesthaeghe D, VanSpeybroeck V, Waroquier M (2009) The effect of confined space on the growth of naphthalenic species in a chabazite-type catalyst: a molecular modeling study. ChemCatChem 1:373–378
146. Vandichel M, Lesthaeghe D, Mynsbrugge JV, Waroquier M, Van Speybroeck V (2010) Assembly of cyclic hydrocarbons from ethene and propene in acid zeolite catalysis to produce active catalytic sites for MTO conversion. J Catal 271:67–78
147. Chen D, Rebo H, Moljord K, Holmen A (1997) Influence of coke deposition on selectivity in zeolite catalysis. Ind Eng Chem Res 36:3473–3479
148. Chen D, Moljord K, Fuglerud T, Holmen A (1999) The effect of crystal size of SAPO-34 on the selectivity and deactivation of the MTO reaction. Microporous Mesoporous Mater 29:191–203
149. Bjørgen M (2003) Coke precursor formation and zeolite deactivation: mechanistic insights from hexamethylbenzene conversion. J Catal 215:30–44
150. Jiang Y, Huang J, Reddy Marthala V, Ooi YS, Weitkamp J, Hunger M (2007) In situ MAS NMR-UV/Vis investigation of H-SAPO-34 catalysts partially coked in the methanol-to-olefin conversion under continuous-flow conditions and of their regeneration. Microporous Mesoporous Mater 105:132–139
151. Qi G, Xie Z, Yang W, Zhong S, Liu H, Zhang C, Chen Q (2007) Behaviors of coke deposition on SAPO-34 catalyst during methanol conversion to light olefins. Fuel Process Technol 88:437–441
152. Mores D, Stavitski E, Kox MH, Kornatowski J, Olsbye U, Weckhuysen BM (2008) Space- and time-resolved in-situ spectroscopy on the coke formation in molecular sieves: methanol-to-olefin conversion over H-ZSM-5 and H-SAPO-34. Chem Eur J 14:11320–11327

153. Kustova M, Holm MS, Christensen CH, Pan YH, Beato P, Janssens TVW, Joensen F, Nerlov J (2008) Synthesis and characterization of mesoporous ZSM-5 core-shell particles for improved catalytic properties. In: Gedeon A, Massiani P, Babonneau F, (eds) Zeolites and related materials: trends, targets and challenges, proceedings of the 4th international feza conference, Paris (France). vol 174, pp. 117–122
154. Choi M, Na K, Kim J, Sakamoto Y, Terasaki O, Ryoo R (2009) Stable single-unit-cell nanosheets of zeolite MFI as active and long-lived catalysts. Nature 461:246–249
155. Guisnet M, Costa L, Ribeiro FR (2009) Prevention of zeolite deactivation by coking. J Mol Catal A Chem 305:69–83
156. Wragg DS, Johnsen RE, Balasundaram M, Norby P, Fjellvag H, Gronvold A, Fuglerud T, Hafizovic J, Vistad OB, Akporiaye D (2009) SAPO-34 methanol-to-olefin catalysts under working conditions: a combined in situ powder X-ray diffraction, mass spectrometry and Raman study. J Catal 268:290–296
157. Schulz H (2010) "Coking" of zeolites during methanol conversion: basic reactions of the MTO-, MTP- and MTG processes. Catal Today 154:183–194
158. Kim J, Choi M, Ryoo R (2010) Effect of mesoporosity against the deactivation of MFI zeolite catalyst during the methanol-to-hydrocarbon conversion process. J Catal 269:219–228
159. Mores D, Kornatowski J, Olsbye U, Weckhuysen BM (2011) Coke formation during the methanol-to-olefin conversion: in situ microspectroscopy on individual H-ZSM-5 crystals with different bronsted acidity. Chem Eur J 17:2874–2884
160. Mores D, Stavitski E, Verkleij SP, Lombard A, Cabiac A, Rouleau L, Patarin J, Simon-Masseron A, Weckhuysen BM (2011) Core-shell H-ZSM-5/silicalite-1 composites: bronsted acidity and catalyst deactivation at the individual particle level. Phys Chem Chem Phys 13:15985–15994
161. Wang F, Luo M, Xiao WD, Cheng XW, Long YC (2011) Coking behavior of a submicron MFI catalyst during ethanol dehydration to ethylene in a pilot-scale fixed-bed reactor. Appl Catal A Gen 393:161–170
162. Inagaki S, Shinoda S, Kaneko Y, Takechi K, Komatsu R, Tsuboi Y, Yamazaki H, Kondo JN, Kubota Y (2013) Facile fabrication of ZSM-5 zeolite catalyst with high durability to coke formation during catalytic cracking of paraffins. ACS Catal 3:74–78
163. Na K, Choi M, Park W, Sakamoto Y, Terasaki O, Ryoo R (2010) Pillared MFI zeolite nanosheets of a single-unit-cell thickness. J Am Chem Soc 132:4169–4177
164. Verheyen E, Jo C, Kurttepeli M, Vanbutsele G, Gobechiya E, Korányi TI, Bals S, Van Tendeloo G, Ryoo R, Kirschhock CEA, Martens JA (2013) Molecular shape-selectivity of MFI zeolite nanosheets in n-decane isomerization and hydrocracking. J Catal 300:70–80
165. Cho K, Na K, Kim J, Terasaki O, Ryoo R (2012) Zeolite synthesis using hierarchical structure-directing surfactants: retaining porous structure of initial synthesis gel and precursors. Chem Mater 24:2733–2738
166. Cho HS, Ryoo R (2012) Synthesis of ordered mesoporous MFI zeolite using CMK carbon templates. Microporous Mesoporous Mater 151:107–112
167. Na K, Choi M, Ryoo R (2013) Recent advances in the synthesis of hierarchically nanoporous zeolites. Microporous Mesoporous Mater 166:3–19
168. Vora BV, Marker TL, Barger PT, Nilson HR, Kvisle S, Fuglerud T (1997) Economic route for natural gas conversion to ethylene and propylene. Stud Surf Sci Catal 107:87–98
169. Chen JQ, Bozzano A, Glover B, Fuglerud T, Kvisle S (2005) Recent advancements in ethylene and propylene production using the UOP/Hydro MTO process. Catal Today 106:103–107
170. Keil FJ (1999) Methanol-to-hydrocarbons: process technology. Microporous Mesoporous Mater 29:49–66
171. Coute NP, Kuechler KH, Chrisholm PN, Vaughn SN, Lattner JR, Kuechler WL (2004) US Patent 6673978, 2004
172. Kioes S, Liebner W (2004) Methane-the promising career of a humble molecule. J Nat Gas Chem 13:71–78
173. Tian P, Wei Y, Ye M, Liu Z (2015) Methanol to olefins (MTO): from fundamentals to commercialization. ACS Catal 5:1922–1938

174. Zhang H (2008) Advances in process research of methanol to light olefins. Chem React Eng Technol 24:178–182
175. Lok BM, Messina CA, Patton RL, Gajek RT, Cannan TR, Flanigen EM (1984) New family of silico-alumino-phosphate molecular sieves – prepd. by hydrothermal crystallisation are useful for sepn. and as catalysts US Patent 4440871, 3 Apr 1984
176. Liang J, Li HY, Zhao SQ, Guo WG, Wang RH, Ying ML (1990) Characteristics and performance of SAPO-34 catalyst for methanol-to-olefin conversion. Appl Catal 64:31–40
177. Liu Z, Qi Y (2006) The industrial experimental research of methanol to olefins technology. Bull Chin Acad Sci 21:406–408
178. Xin Y, Qi P, Duan X, Lin H, Yuan Y (2013) Enhanced performance of Zn-Sn/HZSM-5 catalyst for the conversion of methanol to aromatics. Catal Lett 143:798–806
179. Wang P, Yang D, Hu J, Xu JA, Lu G (2013) Synthesis of SAPO-34 with small and tunable crystallite size by two-step hydrothermal crystallization and its catalytic performance for MTO reaction. Catal Today 212:62.e1–62.e8
180. Nesterenko N, Donk SV, Minoux D, Dath JP (2013) US Patent 2013, vol 0204061 A1
181. Martins J, Birot E, Guillon E, Lemos F, Ribeiro FR, Magnoux P, Laforge S (2013) Sodium exchange over H-EU-1. zeolite. Part II: Catalytic properties. Microporous Mesoporous Mater 171:238–245
182. Pushparaj H, Mani G, Muthiahpillai P, Velayutham M, Park Y-K, Choi WC, Jang HT (2013) Effects of crystallinity of ZSM-5 zeolite on para-selective tert-butylation of ethylbenzene. Chin J Catal 34:294–304
183. Yang S-T, Kim J-Y, Chae H-J, Kim M, Jeong S-Y, Ahn W-S (2012) Microwave synthesis of mesoporous SAPO-34 with a hierarchical pore structure. Mater Res Bull 47:3888–3892
184. Viswanadham N, Saxena SK, Kumar J, Sreenivasulu P, Nandan D (2012) Catalytic performance of nano crystalline H-ZSM-5 in ethanol to gasoline (ETG) reaction. Fuel 95:298–304
185. Bleken FL, Chavan S, Olsbye U, Boltz M, Ocampo F, Louis B (2012) Conversion of methanol into light olefins over ZSM-5 zeolite: strategy to enhance propene selectivity. Appl Catal A 447:178–185
186. Álvaro-Muñoz T, Márquez-Álvarez C, Sastre E (2012) Use of different templates on SAPO-34 synthesis: effect on the acidity and catalytic activity in the MTO reaction. Catal Today 179:27–34
187. Wang P, Lv A, Hu J, Xu JA, Lu G (2011) In situ synthesis of SAPO-34 grown onto fully calcined kaolin microspheres and its catalytic properties for the MTO reaction. Ind Eng Chem Res 50:9989–9997
188. Salmasi M, Fatemi S, Taheri Najafabadi A (2011) Improvement of light olefins selectivity and catalyst lifetime in MTO reaction; using Ni and Mg-modified SAPO-34 synthesized by combination of two templates. J Ind Eng Chem 17:755–761
189. Möller K, Bein T (2011) Crystallization and porosity of ZSM-23. Microporous Mesoporous Mater 143:253–262
190. Dimitrov L, Mihaylov M, Hadjiivanov K, Mavrodinova V (2011) Catalytic properties and acidity of ZSM-12 zeolite with different textures. Microporous Mesoporous Mater 143:291–301
191. Wang B, Tian Z, Li P, Wang L, Xu Y, Qu W, He Y, Ma H, Xu Z, Lin L (2010) A novel approach to synthesize ZSM-23 zeolite involving N, N-dimethylformamide. Microporous Mesoporous Mater 134:203–209
192. Teketel S, Olsbye U, Lillerud K-P, Beato P, Svelle S (2010) Selectivity control through fundamental mechanistic insight in the conversion of methanol to hydrocarbons over zeolites. Microporous Mesoporous Mater 136:33–41
193. Lee JH, Park MB, Lee JK, Min H-K, Song MK, Hong SB (2010) Synthesis and characterization of ERI-type UZM-12 zeolites and their methanol-to-olefin performance. J Am Chem Soc 132:12971–12982

Chapter 10
Zeolite Catalysis for Biomass Conversion

William N.P. van der Graaff, Evgeny A. Pidko, and Emiel J.M. Hensen

Abstract In this contribution, opportunities for the application of zeolites as catalysts in the conversion of biomass into fuels and chemicals are discussed. The constitution of various types of biomass in terms of composition and potential uses is briefly introduced. This contribution will focus on selective transformation of lignocellulosic biomass. Relevant mechanistic insights in the conversion of biomass constituents as well as an overview of the state-of-the-art in the application of zeolites in biomass conversion reactions are discussed. This chapter highlights the unique features of zeolites which make them an appealing choice as catalysts in future chemical biorefinery approaches.

Keywords Catalysis • Zeolites • Biomass • Platform chemicals • Carbohydrate valorization • Isomerization • Dehydration • Acidity

10.1 Introduction

The development of novel catalytic routes to convert biomass to chemicals and fuels is becoming increasingly important to industry spurred by sustainability and feedstock diversification targets [1, 2]. Biomass is the only renewable source of carbon and it has accordingly been argued to play a pivotal role in transition technologies towards a sustainable society. Whereas traditionally biomass was the main source of energy before the industrial revolution, its current share in meeting the

W.N.P. van der Graaff • E.J.M. Hensen (✉)
Inorganic Materials Chemistry Group, Eindhoven University of Technology, PO Box 513, 5600 MB Eindhoven, The Netherlands
e-mail: e.j.m.hensen@tue.nl

E.A. Pidko
Inorganic Materials Chemistry Group, Eindhoven University of Technology, PO Box 513, 5600 MB Eindhoven, The Netherlands

Institute of Complex Molecular Systems, Eindhoven University of Technology, PO Box 513, 5600 MB Eindhoven, The Netherlands

F.-S. Xiao, X. Meng (eds.), *Zeolites in Sustainable Chemistry*, Green Chemistry and Sustainable Technology, DOI 10.1007/978-3-662-47395-5_10 total global energy demand is small, mainly contributing with thermal combustion approaches in underdeveloped countries. The development of novel practical technologies to produce tailor-made fuels and chemicals from biomass in an efficient, clean and economically viable manner is an important topic in academia and industry. Current industrial processes utilizing biomass as a raw material for fuels and chemicals are still in their infancy. A notable example of an established biobased process is the conversion of xylose to furfural, which can be used as lubricant and herbicide [3, 4]. Catalysis is recognized as a key ingredient in the development of sustainable, efficient and economical routes for the transformation of biomass constituents to value-added products.

The emergence of catalysis as a scientific discipline in the last century has been mostly driven by the establishment of an oil-centered chemical and fuels industry. The vast knowledge about catalytic chemistry of current industrial processes have focused on the upgrading of apolar, low reactive and non-functionalized hydrocarbon streams derived from crude oil. Many operations exist in which C-H and C-C bonds need to be activated in order to upgrade these streams into high-quality transportation fuels. Examples of such process include hydrocarbon cracking to shift the boiling point (fluid catalytic cracking and hydrocracking) and isomerization to improve the octane number, for which zeolites are extensively used. The chemical industry typically makes use of the high reactivity of compounds with C-C double bonds. In particular, aromatics derived from oil feedstock directly or via other refinery processes such as catalytic reforming as well as olefins, usually obtained by thermal cracking of naphtha, are important platform chemicals to a range of intermediate and final products. Local trends such as the availability of cheap shale gas might shift this balance in the near future, so that it might become interesting to focus on the ethane fraction in natural gas instead of naphtha as a feedstock for cracking, and also to consider the lignin fraction in biomass as a source of aromatics [5–7]. In China, technology to convert cheap coal via syngas and methanol into olefins has been commercialized. Zeolites are used in the final methanol conversion step to olefins.

The transition to new chemical technologies utilizing biomass as the primary feedstock represents a number of significant fundamental challenges to catalysis science and technology. Indeed, from a chemical perspective oil can be regarded as a quite homogeneous substance with its main components represented by non-polar, low-reactive, non-functionalized hydrocarbon molecules. On the contrary, biomass is a highly heterogeneous substrate and its chemical structure and composition largely depends on the source. Furthermore, biomass is composed of polar, extensively functionalized and highly reactive substances. This implies that completely new chemocatalytic pathways have to be engineered for the efficient and selective conversion of biofeedstock to commodity chemicals and fuels. Therefore, the transition to new chemical technologies utilizing biomass as one of the primary feedstock represents a number of significant fundamental challenges to catalysis science and technology.

Zeolites are crystalline microporous aluminosilicates and are broadly applied as catalysts in important industrial chemical processes. Their widespread application is facilitated by their favorable physicochemical properties including high surface area, a three-dimensional system of pores in the sub-nanometer range endowing zeolites shape-selective properties and high chemical and hydrothermal stability. The strong Brønsted acidity in zeolites resides in the Al^{3+} substitutions for tetrahedral Si^{4+} in the framework renders. Acidity is for instance crucial for the large-scale oil upgrading technology operated in refineries. The versatile properties of zeolites render them also promising for the transformation of other feedstock including biomass as a whole or product streams derived from biomass. The industrial application of zeolites in biomass processing, however, is still very limited.

In this contribution, we will discuss possible applications of zeolites as catalysts for the conversion of biomass into fuels and chemicals. We will first briefly introduce the various types of biomass, in terms of composition and potential uses. Relevant mechanistic insights in the conversion of biomass constituents as well as an overview of the state-of-the-art in the application of zeolites in biomass conversion reactions will be discussed.

10.2 Biomass Feedstock

Throughout the history of the mankind biomass was extensively utilized as a fuel. Initial industrial efforts to utilize biomass for sustainability purposes follow these historical approaches and are mainly focused on either burning or application of crop-based feedstock (*e.g.*, sugar cane, palm oil) for the production of biofuels. Such fuels produced from edible biomass are commonly referred to as the first generation biofuels. Current examples are the production of green diesel from palm and waste oils by catalytic hydrogenation [8, 9]. There is general consensus that non-food biomass should form the basis for sustainable production of fuels and chemicals so as to avoid competition with food production as much as possible. Furthermore, serious environmental issues are entwined with the production of first-generation biofuels, such as the burning of crop residues, causing considerable pollution [10]. It is therefore a key interest to develop sustainable routes using the complete spectrum of non-edible biomass including agricultural waste streams as feedstock.

In analogy with today's petroleum refineries, biorefineries are facilities that integrate conversion of, in this case, biomass feedstock and separation processes for the production of fuels, chemicals and power [11]. Given the nature and availability of biomass, it is likely that the size of biorefineries will be smaller than that of current petroleum refineries [12]. An important aspect in the transition to a biobased economy may also be integration of biomass as a feedstock in current fuel and chemicals production facilities. Despite the complexity of petroleum refineries, its chemistry starts from a number of platform molecules (BTX, light olefins and methanol). It has been proposed that analogously a limited number of platform molecules obtained from lignocellulosic biomass will serve as starting chemicals in a biorefinery. A recent discussion on this issue is given by Bozell and Petersen [13].

Biomass is an abundant source of renewable carbon that comes in a wide variety of chemical structures containing many diverse functional groups of different reactivity.

Its functionalized nature brings opportunities for tailor-made (catalytic) conversion processes that preserve or make use of these functionalities to chemicals and fuels. An inherent drawback of the reactive nature of especially the carbohydrate fraction of biomass is its tendency to undergo side-reactions that often lead to high-molecular weight products (caramelization reactions, humin formation).

Lignocellulose is the most abundant non-edible biomass resource. It is the major component of cell walls of plants and with an estimated global annual production of 180 billion tons [14], holds a great promise to be used as a feedstock for chemicals and fuel components. Lignocellulosic biomass is composed mainly of three types of biopolymers: cellulose, hemicellulose and lignin. The exact composition of the biomass is highly dependent on its origin. Table 10.1 presents an overview of the typical composition of common types of lignocellulosic biomass.

Lignin is the most recalcitrant constituent in lignocellulose. It is characterized by a highly heterogeneous structure (see Fig. 10.1) representing interconnecting *p-coumaryl*, coniferyl and sinapyl alcohol units resulting in an amorphous biopolymer. The poorly defined structure of lignin and its low reactivity limit its applicability for the selective production of bulk chemicals. Lignin conversion is typically carried out under harsh conditions resulting in the formation of a wide range of compounds [17].

The structure and composition of the carbohydrate part of biomass, namely cellulose and hemi-cellulose, are much more uniform (see Fig. 10.1). This renders carbohydrates much more attractive candidates for the development of targeted conversion routes under relatively mild conditions. Cellulose is a crystalline homopolymer of glucose monomeric units, linked by β-1,4 glycosidic bonds. Hemicellulose is an amorphous heteropolymer of different sugar monomers, predominantly pentoses and hexoses.

A possible approach to deal with the highly complex nature of biomass feedstock for the production of biofuels is to employ gasification or thermochemical conversion (pyrolysis) as the initial step. Gasification as a thermal, non-catalytic process will not be discussed in this chapter. Inspired by the successful application of zeolites as acidic catalysts for the cracking of heavy fractions of oil feedstock (fluid catalytic cracking, FCC) for the cleavage of large hydrocarbon molecules, efforts have been

Table 10.1 Composition of various types of biomass

Biomass type	Cellulose	Hemicellulose	Lignin
Hardwood stem	40–55	24–40	18–25
Softwood stem	45–50	25–35	25–35
Switch grass	31–45	24–31	12–23
Rice straw	35–38	25–26	13–25
Wheat straw	35–40	21–27	15–25
Bagasse	41–50	25–30	18–25
Corn stover	27–48	13–27	14–31
Corn cob	34	34	18
Beet pulp	23	36	1
Coconut husk	44	12	33

Adapted from Ref. [15]

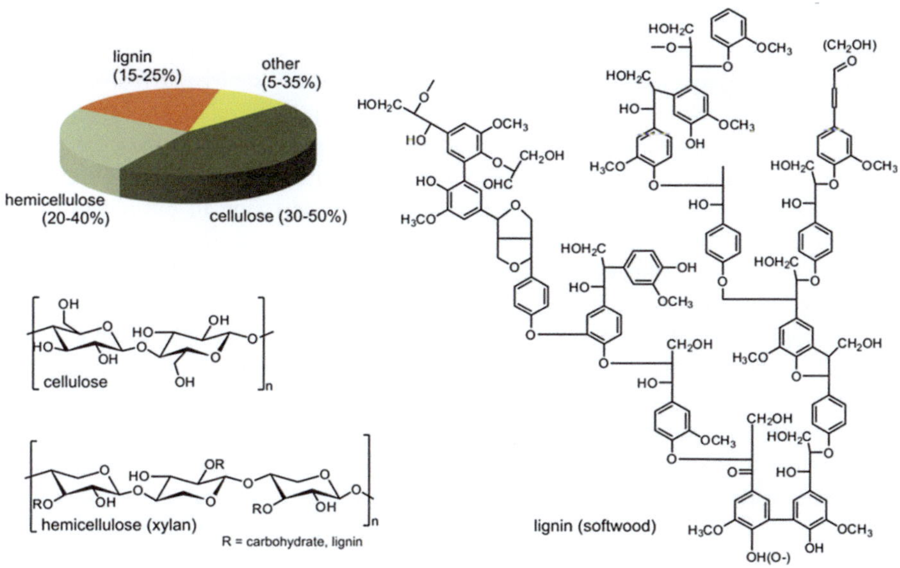

Fig. 10.1 Composition of lignocellulosic biomass and chemical structure of its main constituents cellulose, hemicellulose and lignin. The lignin structure is adapted from Ref. [16]

made to investigate the possibility of catalytically cracking lignocellulosic biomass using zeolites [18–21]. As catalysts, acidic zeolites for common FCC processing of crude oil, mainly Y-zeolite and to a lesser extent H-ZSM-5 have been used. In its embodiment for molecular weight reduction of biomass, the zeolites will need to catalyze cleavage of C-C and C-O bonds. The interested reader is referred to a recent review by Liu et al. for further details [22].

Another important option for direct complete biomass conversion is pyrolysis, which yields pyrolysis oils that are not directly usable due to their high acidity, oxygen content and, related, low heating values [23]. Pyrolysis oil is a complex mixture of acids (formic, acetic and propionic acid), aldehydes, alcohols, furans, phenolic and other aromatic hydrocarbons. During processing, also gases (small hydrocarbons, H_2, CO, CO_2) are formed. Utilization of these oils as fuels requires further upgrading processes. Huber and co-workers have explored catalytic fast pyrolysis of biomass and found that zeolites can improve the overall product yield [24–26]. More recently, interesting chemistry is developed based on Diels-Alder reactions of common pyrolysis oil constituents, such as furans and alkenes, to improve the yield of aromatics [5–7]. For instance, such an approach allowed obtaining relatively high yields of *p*-xylene, which is a desirable precursor for terephthalic acid production.

More selective processes can be established when biomass is pretreated to separate lignin and (hemi-)cellulose prior to further conversion. (Hemi-)cellulose depolymerization to monomeric sugars can be achieved via Brønsted acid-catalyzed hydrolysis [27] that requires much milder conditions than lignin depolymerization.

The applicability of zeolites as catalysts in this process is not straightforward. Zeolite Brønsted acid sites are located inside the micropores that are inaccessible to the large biopolymers. Therefore, cellulose hydrolysis is usually carried out in the presence of mineral acids. However, zeolites have recently found an important niche application in this area. By catalytic depolymerization of intermediate oligomeric species they can decrease the concentration of mineral acid needed to efficiently cleave the sugar biopolymers. A good example can be found in the recent work describing the conversion of cellulose to sorbitol [28]. In this work, Ru nanoparticles were combined with mineral and solid zeolite acids to convert cellulose efficiently to sorbitol in a single-step process.

The application of microporous zeolitic materials in the field of biomass conversion is inherently limited to the transformation of small biomass-derived molecules to value-added products. In such reactions, selectivity is the key aspect that can be finely tailored by modifying the chemical composition of the zeolite catalysts as well as their topological properties. Below we will discuss in more details selected examples in this field and highlight the unique features of zeolite catalysis that allow establishing highly efficient conversion routes towards bulk chemicals.

10.3 The Concept of Platform Chemicals

The implementation of biomass as the primary feedstock in bulk chemical synthesis requires not only the compatibility of new processes with the current fossil-based technology, but also the compatibility of the final products with established markets. The chemical conversion routes have to be engineered such that the biomass-derived products at least bear significant resemblance with the valuable chemicals originally produced from fossil feedstock. A promising approach is to establish a biomass conversion process towards value-added chemicals with a wide range of applications in the current chemical industry. Such compounds are commonly referred to as platform chemicals. As a result of a careful analysis of the applicability and feasibility of mass production of distinct chemical compounds from lignocellulosic biomass, the US Department of Energy proposed several highly promising top value-added chemicals, which can lead towards market-ready products [29]. Bozell and Petersen updated this analysis recently [13]. Different pathways for valorization of carbohydrate biomass have been the subject of a number of excellent recent reviews. In particular, van Putten et al. provided a comprehensive review on the 5-hydroxymethylfurfural platform [30], whereas Dusselier et al. explored the potential of lactic acid and lactates [31].

Figure 10.2 presents an overview of possible compounds that can be produced from glucose. The fact that many of these compounds are currently produced from fossil feedstock highlights the important role that biomass can serve in establishing more sustainable chemical technologies. Zeolite catalysis was shown to hold a great promise in carbohydrate biomass processing. Below we will discuss three major valorization pathway, namely sugar conversion to 5-hydroxymethylfurfural (HMF), γ-valerolactone (GVL) and lactates. Special emphasis will be placed on the role of zeolites in the respective conversion processes.

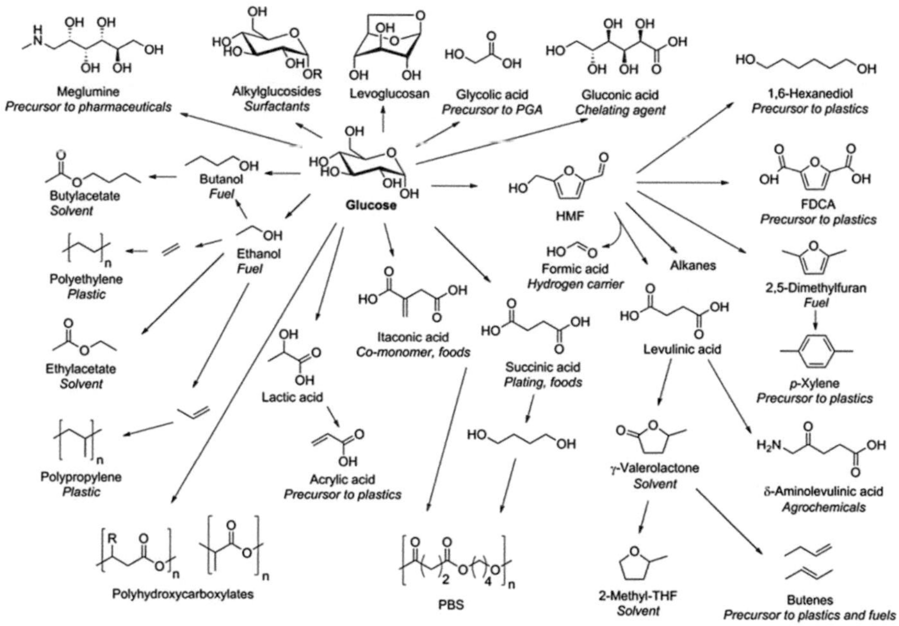

Fig. 10.2 Proposed routes for glucose valorization towards different market-ready bulk chemicals (Reproduced from Ref. [15] with permission from The Royal Society of Chemistry)

10.4 Carbohydrate Chemistry

We will focus on three promising carbohydrate-derived chemical platforms (HMF, GVL and lactates), the chemistry involved and the role of zeolites as catalysts in the associated transformation steps. The selective conversion of glucose, the most abundant monosaccharide, into other compounds is difficult to achieve with chemocatalysts. An essential reaction towards dehydration products from sugars is the isomerization of glucose to its more reactive isomer fructose. Accordingly, most glucose conversion routes start with glucose to fructose isomerization. It can be readily achieved via the high-temperature alkaline treatment of glucose. This reaction proceeds via the Lobry-de Bruyn – Alberda – van Ekenstein (LdB-AvE) mechanism [32].

The mechanism involves deprotonation of the sugar, followed by a rearrangement towards ene-diol intermediates (see Fig. 10.3). The proton of the OH-group on the carbon atom adjacent to the C-O$^-$ can be abstracted by the enolate oxygen. Subsequent reprotonation leads to a carbonyl group formed at the C_2-carbon position, essentially forming the ketose sugar. The presence of a base is necessary to promote deprotonation and subsequent rearrangement of the sugar. However, the alkaline conditions required for this reaction limit the possibility of its integration into a sustainable chemocatalytic process for glucose valorization. At higher temperatures, this gives rise to numerous undesired reactions (*e.g.*, polymerization of sugar intermediates). This complicates work-up procedures of the reaction mixtures

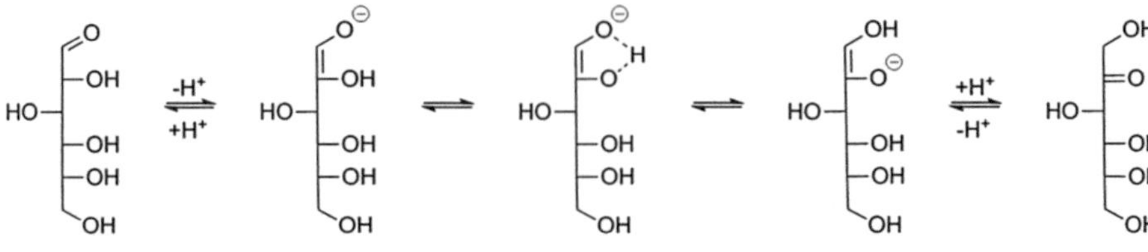

Fig. 10.3 The Lobry-de Bruyn-Alberda-van Ekenstein mechanism of base-catalyzed isomerization of glucose to fructose

generated by the base-catalyzed glucose isomerization, causing the generation of large waste streams. Furthermore, the use of base is not compatible with other steps of the tandem sequence envisaged for an integrated process for the conversion of carbohydrate biomass to the desired compounds, for instance because these often require the presence of Brønsted acids. In addition to either mineral, like lye [33], or solid bases (*e.g.*, hydrotalcites) [34, 35], strong Lewis acids can also catalyze the isomerization of carbohydrates. There has been much interest in the development of stable, solid Lewis acid catalysts for this reaction. To date, the most efficient catalyst for glucose isomerization is Sn-BEA zeolite. This material, first reported in 1999 [36], turned out to be highly efficient for the Bayer-Villiger oxidation as well as the Meerwein-Ponndorf-Verley reaction [37]. In the latter reaction, the catalyst acts as a transfer hydrogenation catalyst. As the chemistry in these transformations is similar to sugar isomerization, Sn-BEA was found to be also highly active for the isomerization of carbohydrates under relatively mild conditions [38]. This reaction, however, is not selective. Sn-BEA promotes both the aldose-ketose isomerization reaction to fructose and the epimerization of glucose resulting in mannose.

In epimerization reactions, the stereochemistry of one of the asymmetric carbon atoms is altered. This involves an intramolecular hydride shift via ene-diol intermediates (see Fig. 10.4). Selective epimerization that is not accompanied by the aldose-ketose isomerization was first observed by Bílik and coworkers in 1972 in the presence of molybdenic acid as catalyst [39]. Epimerization reactions give access to rare sugars, which can be used as chiral building blocks for *e.g.* pharmaceuticals [40]. However, it was found that for molybdenic acid-mediated epimerization, the mechanism involves a 1,2-intramolecular carbon shift rather than a LdB-AvE-type hydride shift (see Fig. 10.4) [41].

Many efforts have been made to elucidate the mechanism involved in the transformation of aldoses to ketoses. Davis and coworkers have studied the catalytic role of both framework and extraframework Sn [42]. They critically assessed the behavior of different Sn-containing materials including SnO_2, SnO_2 supported on BEA zeolite, and well-defined Sn sites incorporated in the framework of Sn-BEA zeolite. Extraframework Sn-sites in the form of tin oxide deposits are inactive in water, but can isomerize sugars in methanol. By studying the H/D scrambling in C_2-deuterated glucose during the glucose to fructose isomerization by Sn-containing BEA zeolites, it has been demonstrated that the kinetic behavior using SnO_2-containing Sn-BEA is very different from that of the well-defined Sn sites in

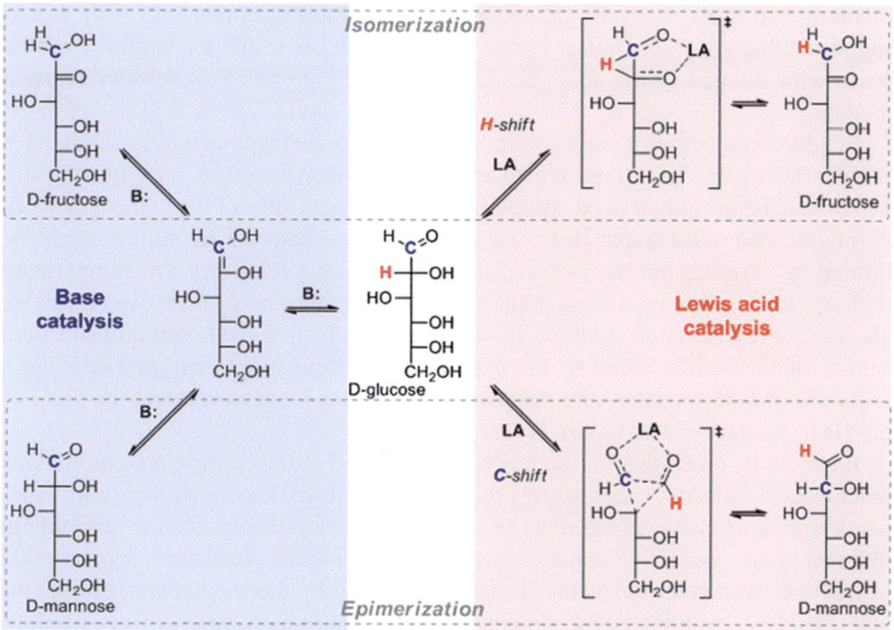

Fig. 10.4 Simplified mechanism for the isomerization and epimerization of glucose in the presence of a Brønsted base (*left*) and Lewis acid (*right*) catalyst

Sn-BEA. In the presence of extraframework Sn-sites, both C_1 and C_2 sites of glucose present in the reaction mixtures contained the deuterium label. This implies that extraframework Sn-species act as base sites and catalyze the isomerization reaction following the LdB-AvE mechanism (base-catalyzed path, see Fig. 10.4). A very different catalytic behavior of isolated Sn sites embedded into the framework of BEA was observed. The position of the deuterium label in the carbohydrate molecules did not change when the isomerization reaction was carried using a well-defined Sn-BEA catalyst with isolated lattice Sn sites. This pointed to the direct Lewis acid-catalysed hydride shift mechanism (Lewis acid-catalyzed path, Fig. 10.4).

Well-defined Sn-BEA mainly produced mannose when methanol was used as solvent, via epimerization with fructose as a secondary product. No H/D scrambling was observed in such case. Reactions employing $^{13}C_1$-labled glucose as reactant yielded mannose with the label at the C_2-position, proving that the epimerization reaction occurs *via* the Bílik mechanism (Fig. 10.4, Lewis acid-catalyzed C-shift) rather than reversible enolization according to the LdB-AvE-mechanism (see Base-catalyzed epimerization, Fig. 10.4). The ^{13}C label in the minor fructose product was primarily located at the C_1 position indicating that it was exclusively formed via direct glucose isomerization. Extra framework Sn catalyzed the isomerization of glucose to fructose in methanol without formation of mannose. It was concluded that Lewis acidic framework Sn sites are responsible for the epimerization. Other studies employing Sn-BEA, however, reported the formation of mannose in water, implying that in this case epimerization can also take place [43]. Selectivity towards mannose can furthermore be enhanced by the addition of borate salts [44]. It was suggested that the formation of borate-sugar complexes inside the confined zeolite pores with isolated Lewis acid sites are of critical importance to selectively yield mannose.

In addition to strong Lewis acidic Sn-substituted zeolites, also ultrastabilized Y zeolite (USY) has been reported to exhibit good isomerization activity [45, 46], which may be presumed to be caused by the Lewis acid sites of the extraframework Al-phase. The subsequent Brønsted acid-catalyzed sugar dehydration could be limited by carrying out the isomerization reaction at a relatively low temperature (below 120 °C). These findings point to the USY zeolites as an attractive option for glucose isomerization to fructose. Potentially, by adjusting the Lewis and Brønsted acidity of the zeolite catalyst, the cascade conversion of glucose to HMF via a sequence of isomerization and dehydration steps can be achieved and the yield of the HMF product can be tailored (*vide infra*).

It should be noted that all the reactions discussed in this section are equilibrium reactions. In industry, mixtures of glucose and fructose, known as high-corn fructose syrup, are extensively used as sweetener in *e.g.* soft drinks and are available in different grades, generally ranging from 42 to 90 % fructose. The selective production of fructose from glucose/fructose mixtures is bound by thermodynamic limitations and, therefore, requires energy-intensive and costly process engineering solutions [47]. For bulk commercial applications, full conversion of glucose can only be achieved by, for instance, oxidation to gluconic acid or hydrogenation to sorbitol. If the ketose-aldose equilibrium is involved, one needs to further convert fructose to useful products to obtain full conversion. Therefore, there is a great interest in the selective conversion of glucose and fructose to value-added chemicals. Several examples of representative processes are given in the course of this chapter.

Recent theoretical studies on glucose isomerization with Sn-BEA have pointed out conceptual similarities between heterogeneous, homogeneous and enzymatic systems for aldose-ketose isomerization [48–50]. It has been identified that the high efficiency usually stems from the cooperative action between the different functionalities at the active site determine a low-energy path for the difficult 1,2-hydride shift step [51]. For enzymes, the reaction involves a highly complex transformation promoted by two metal cofactors and a water molecule confined at the active center. Transient formation of a binuclear metal site has been proposed as one of the key features of the ionic liquid mediated homogeneous Lewis acid catalysis by chromium ions [48–50, 52]. The active site in Sn-BEA most probably consists of an isolated Sn site, as Sn-content is generally low. Bare et al. [53] have shown that Sn-centers in isomorphously substituted Sn-BEA may be positioned in the opposite vertices of the 6-membered ring. This would cause the Sn-centers to be too far apart for direct cooperative interaction. DFT calculations by Yang et al. [54] evidence that binuclear substitution of Ti, Sn or Zr in the 6MR results in higher Lewis acidity than having a single-atom substitution in the 6MR. The DFT calculations predict higher barriers for isomerization for mono-substituted models, so that one may speculate that other cooperativity effects play a role here. One such effect may be the presence of vicinal hydroxyl groups [55, 56]. These may come in the form of co-adsorbed water, nearby framework defects or partially hydrolyzed Sn framework sites. Experimental evidence for the presence of the latter is very difficult. A similar intensive discussion on the nature of the active framework Ti-species in TS-1 has not been unequivocally resolved yet, although in this case experimental evidence points to partially hydrolyzed Ti-OH being the active site rather than $Ti(OSi)_4$-sites [57]. FTIR spectroscopy of adsorbed acetonitrile has been used to indirectly probe partially hydrolyzed Sn-OH lattice sites [58, 59]. The possible formation of partially hydrolyzed Sn sites has also been the subject of several theoretical studies [51, 56]. It was shown that the partial hydrolysis of the Sn-site, yielding a vicinal silanol, lowers the activation energy barrier for the hydride shift step. These computational studies also focused on the analysis of the effect of the framework topology on the glucose isomerization activity of Sn-containing zeolites. In a study by Li et al. [55], it was shown that the variation in shape of the micropores with zeolite topology affects the mode and strength of carbohydrate adsorption, which is dominated by van der Waals forces. Their influence on the intrinsic reactivity of intrazeolite Sn sites is, however, small. Therefore, it is proposed that different reactivities of zeolite catalysts with different topologies stem from intrachannel diffusion limitations. There is a need for more extensive research to unravel the origin of the different catalytic behavior of various zeolite catalysts.

10.5 5-Hydroxymethylfurfural

One of the most interesting biomass-derived platform molecules is 5-hydroxymethyl furfural (HMF), often considered a ‘sleeping giant’ because of its wide potential applicability in chemical processes. It is formed by the triple dehydration of fructose. The dehydration of sugars to obtain furfural-type compounds has been extensively studied in the last decade. The industrial potential of HMF and the chemistry underlying its synthesis and further functionalization have been very recently reviewed by van Putten et al. [30] (Fig. 10.5).

It is generally accepted that the dehydration of carbohydrates is catalyzed by Brønsted acids. Apart from mineral acids such as HCl or H_2SO_4 [60], also solid acids such as sulfonated resins (*e.g.*, Amberlyst) [61], sulfonated organosilica [61, 62] and Brønsted acidic zeolites [63, 64] can readily convert fructose to HMF. Although water is the preferred green solvent, conversion of carbohydrates in water

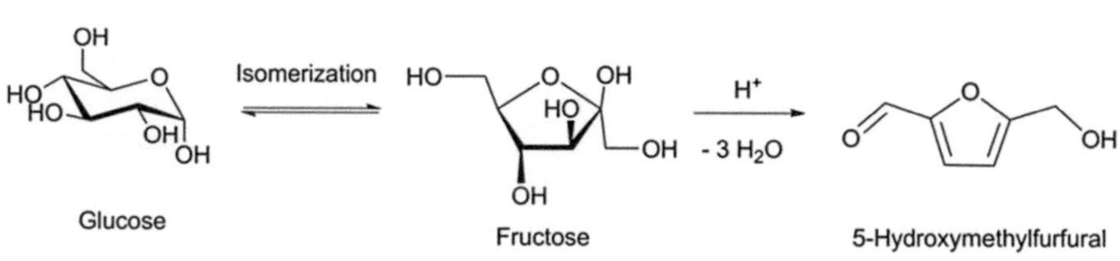

Fig. 10.5 Hexose dehydration to HMF

typically leads to formation of by-products. These can be useful rehydration products of HMF, such as levulinic acid and formic acid (vide *infra*), but also more complex transformations to undesirable humins [65, 66]. Humins are ill-defined carbonaceous heteropolymers of furans and sugars. Although currently efforts are undertaken to valorize humins to improve overall process economics [67], it would be more efficient to avoid their formation as much as possible. Approaches to achieve this include the in-situ extraction of HMF by organic solvents [30, 63, 64]. Solvents play a critical role in the dehydration of fructose. Whereas in water harsh conditions are required, for other solvents and most notably for DMSO [61], THF [62] and ionic liquids [65] dehydration of fructose is a facile reaction and near-quantitative yields of HMF are obtained. In DMSO and some ionic liquids, the dehydration can also take place in the absence of Brønsted acids [68]. The exact reason for this reactivity is unclear. It has been suggested that it can either be due to the stabilization of the furanic ring [69], small amounts of contaminants in the DMSO [70] or the role the solvent itself can fulfill as catalyst [68].

As part of efforts to valorize the hemicellulose fraction of biomass, the dehydration of other five-membered ring containing sugars, such as xylose, towards furfural is being investigated. Xylose is obtained by hydrolysis of the so-called xylan backbone, xylan being the generic term for polysaccharides that contain a β-(1 → 4) xylopyranose backbone with different side chains [71]. The depolymerization of this backbone, isomerization to xylulose (to form the 5-ring) and the dehydration of xylose can be achieved in the presence of Brønsted acids. Many studies have been devoted to the dehydration of xylose, which has been comprehensively reviewed by Mäki-Arvela et al. [14]. The dehydration of xylose has been investigated extensively as well, although yields are often moderate, especially when starting directly from xylan or high-hemicellulose. Good yield (~70 %) was obtained from xylan using $FeCl_3.6H_2O$ at 170 °C in γ-valerolactone as the solvent [72]. At low conversion, however, high selectivities have been reported from xylose using H-faujasite and H-mordenite zeolites in aqueous media [73]. A maximum of 96 % selectivity at 27 % conversion was obtained using H-mordenite as the catalyst in toluene/water at 170 °C. Furans hold promise as platform for the future generation of olefins and aromatics, as for instance has been shown by Huber and coworkers using Ga-ZSM-5 [5].

The direct conversion of glucose to HMF is met with more difficulties due to the required isomerization step from glucose to fructose before the dehydration can take place. The main challenge here is that isomerization and dehydration occur under different optimum reaction conditions. Combining these reactions in one pot usually results in by-product formation. Pioneering work in this field was performed by Zhao et al. [74], who found that Lewis acidic $CrCl_2$ in ionic imidazolium-based liquids is not only active for the isomerization of glucose to fructose, but also efficiently catalyzes the dehydration of fructose. The finding that zeolites containing Lewis acidic metal centers and most prominently Sn can also isomerize sugars efficiently is a breakthrough, because zeolites are relatively stable in aqueous media [75, 76] compared to, for instance, functionalized hydrotalcites. This considerably expanded the possibilities to use zeolites in biomass conversion.

Relatively high yields of HMF from glucose were obtained by combining mineral acid (0.1 M HCl) with Lewis-acidic zeolites [77]. Promising yields can also be obtained using solid Brønsted acids instead of mineral acids. For a system employing a combination of Sn-BEA with Amberlyst-131 in ethanol at temperatures as low as 90 °C, ethoxymethylfurfural was obtained in yields up to 31 % from glucose [78]. A significant factor contributing to this moderate yield is the competing retro-aldol reaction, which will be discussed later in this chapter. For a monophasic THF:H_2O-system employing Sn-BEA and Amberlyst-70, 63 % yield of HMF was obtained from glucose in only 30 min at 130 °C [79]. This example shows that careful choice of solvent is crucial. A very elegant approach combining both Lewis and Brønsted acidity in one multifunctional catalyst was reported by Wang et al., who reacted glucose in a THF/DMSO mixture at 160 °C with montmorillonite clay ion-exchanged with Sn [80]. Otomo et al. reported on a bifunctional BEA-zeolite catalyst prepared by partial dealumination, achieving a yield of 43 % in THF at 180 °C in 3 h [81]. A comprehensive overview of the catalytic systems used in the production of HMF can be seen in Table 10.2.

Table 10.2 Most promising examples of the conversion of carbohydrates to HMF

Substrate	Catalyst	Solvent	Extraction	Conditions	$Y/(S)_{HMF}$[a] (%)	Ref
Fructose	HCl (0.1 M)	H_2O/DMSO 4:1	MIBK[b]/2-BuOH 4:1	185 °C, 1 min	82(86)	[60]
Fructose	H-Mordenite	H_2O	MIBK	165 °C, 30 min	35(96)	[63]
Fructose	None	DMSO	No	150 °C, 36 min	88(88)	[68]
Glucose	Steamed BEA	THF	No	180 °C, 180 min	43(55)	[81]
Glucose	Sn-Mont	THF/DMSO 7:3	No	160 °C, 180 min	54(54)	[80]
Glucose	$CrCl_2$	EMImCl	No	100 °C, 180 min	70(74)	[74]
Glucose	Sn-BEA/HCl	3.5 % NaCl in H_2O (pH = 1)	THF	180 °C, 70 min	57(72)	[77]
Glucose	Ti-BEA/HCl	3.5 % NaCl in H_2O (pH = 1)	THF	180 °C, 105 min	53(70)	[77]
Starch	Sn-BEA/HCl	3.5 % NaCl in H_2O (pH = 1)	THF	40 min hydrolysis, 60 min dehydration, 180 °C	52(69)	[77]
Glucose	Sn-BEA/ Amberlyst-131	Ethanol	No	90 °C, 24 h	31(31)	[78]
Glucose	Sn-BEA/ Amberlyst-70	H_2O/THF 1:9	No	130 °C, 30 min	63(70)	[79]
Cellulose	NHC[c] -$CrCl_2$/ H-Y zeolite	BMImCl	No	120 °C, 12 h	47.5	[82]

[a] *$Y/(S)_{HMF}$* yield and selectivity of HMF, respectively
[b] *MIBK* methyl isobutyl ketone (4-methyl-2-pentanone)
[c] *NHC* N-heterocyclic carbene (here 1,3-bis(2,6-diisopropylphenyl)imidazole)

10.6 Levulinic Acid and γ-Valerolactone as Intermediates to Fuel and Chemicals

In aqueous acidic media, rehydration of HMF to levulinic acid and formic acid is one of the potential side reactions. Levulinic acid was already recognized as a potential key resource for the chemicals industry in the 1950s [83]. The acid-catalyzed conversion of carbohydrates to levulinic acid is considered to be one of the key steps in future biorefineries [84, 85] (Fig. 10.6).

A key application of levulinic acid is its hydrogenation to γ-hydroxyvaleric acid, and its dehydration product, γ-valerolactone (GVL). This compound is currently under investigation as a key intermediate for future chemical processes, as GVL is already directly applicable as solvent and can serve as gasoline blender. Its usability as a solvent is very promising due to its low toxicity, low volatility, stability and high solvation capacity. As such it is even able to dissolve very recalcitrant biopolymers like lignin, while at the same time being miscible with water. It is seen as an important intermediate for the production of fuels and several pathways have been devised to further increase its energy density, prominently featuring decarboxylation reactions like ketonization of pentanoic acid or direct decarboxylation of GVL to give butenes [84]. A considerable advantage of the manufacture of GVL starting from hexoses is the formation of levulinic acid and formic acid in equimolar amounts, which can be considered an indirect decarboxylation as formic acid is a potential hydrogen carrier.

Most reports on the hydrogenation of levulinic acid to GVL focused on catalytic systems such as homogeneous transition metal complexes, supported noble metals and mixed oxides. The use of zeolites for the production of GVL directly from

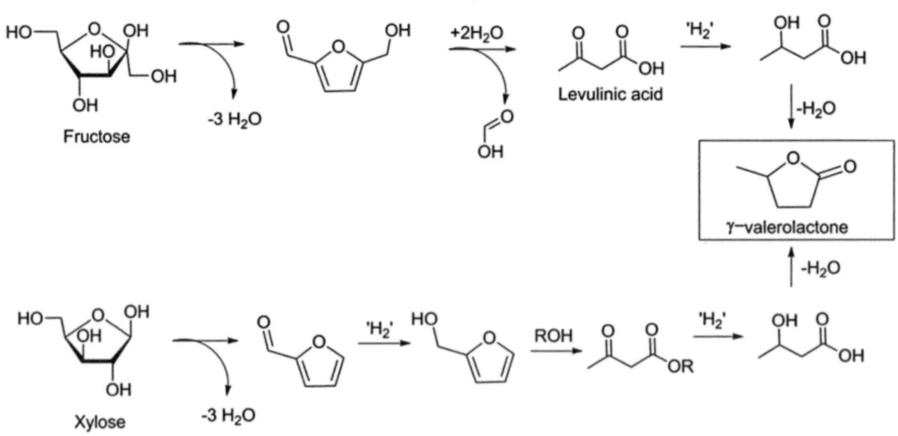

Fig. 10.6 Pathways towards GVL from fructose (cellulose) and xylose (hemicellulose)

furfural was explored by Bui et al. [86] who reported on a system which makes use of a combination of Zr-BEA and Al-MFI nanosheets in a packed bed flow reactor to selectively yield GVL using 2-butanol as hydrogen source for the two (transfer) hydrogenation steps involved. Near quantitative yield was obtained at 120 °C. The strength of this approach can be attributed to the combination of Lewis acid sites (transfer hydrogenation) and Brønsted acidity. The near-quantitative selectivities obtained with this catalyst reinforce the notion that strongly Lewis acidic zeolites can also facilitate benign transfer hydrogenation for the hydrogenation of levulinates.

10.7 Lactic Acid/Alkyl Lactates

Lactic acid is a compound with a wide range of applications. It is not only extensively used in the food industry, but also as a monomer for biodegradable plastics and bulk chemicals such as acrylic acid [31]. The production of lactic acid from carbohydrates is nowadays almost exclusively done by tedious enzymatic processes, which require the use of neutralizing agents to retain enzyme activity. Recently, new chemocatalytic routes towards lactates have been gaining attention. This opens the possibility for non-glycolytic pathways towards lactates from glucose, which is significant as lactic acid is not only a monomer and monomer precursor for e.g. acrylic acid, but it can also be converted to ethanol (fuel, solvent) by decarboxylation and further to ethylene via dehydration of ethanol. High lactic acid yields up to 90 % have already been obtained from cellulose using lead(II) ions [87] and lanthanide triflates [88]. In zeolite chemistry, the direct conversion of cellulose is rather difficult as the large cellulose molecules cannot enter the micropores and reactions of large oligomers and polymers can only take place on the outer surface. However, recently a major breakthrough was reported in the use of strongly Lewis acidic metal-containing (Ti, Zr, Sn) zeolites regarding the conversion of hexose mono- and oligomers to alkyl lactates at 160 °C [89]. Usually, high temperatures are required for this reaction, although this particular reaction was observed in ethanol at temperatures as low as 90 °C when reacting glucose over Sn-Mont [80].

The pivotal factor for the selective production of lactates is the necessity of the correct position of the sugar carbonyl group (see Fig. 10.7). More specifically, for hexose sugars, this means that if glucose is used as starting compound, isomerization to fructose is necessary before the retro-aldol condensation can occur. Usually, this does not happen and apart from lactates, often products originating from glycolaldehyde and aldotetroses are detected due to direct retro-aldol condensation of aldohexoses [90]. In Fig. 10.8, the reaction pathway for the conversion of glucose to lactates is shown.

A maximum yield of 64 % of methyl lactate was obtained from sucrose in methanol with Sn-BEA as a catalyst [89]. Indeed, the reaction products do not only comprise lactates formed from the ketose fructose, but also other compounds due to retro-aldol condensation of aldoses like glucose are typically observed.

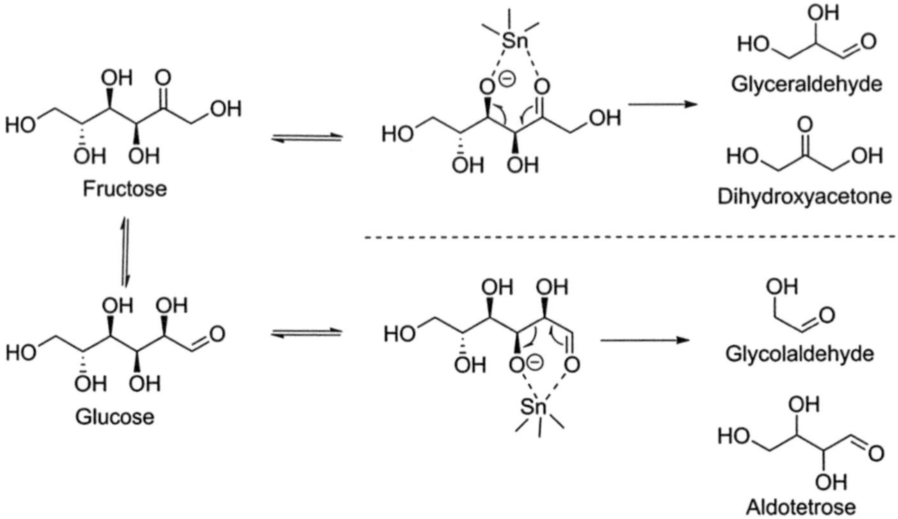

Fig. 10.7 Retro-aldol condensation of glucose (aldose) and fructose (ketose) over Sn-silicates

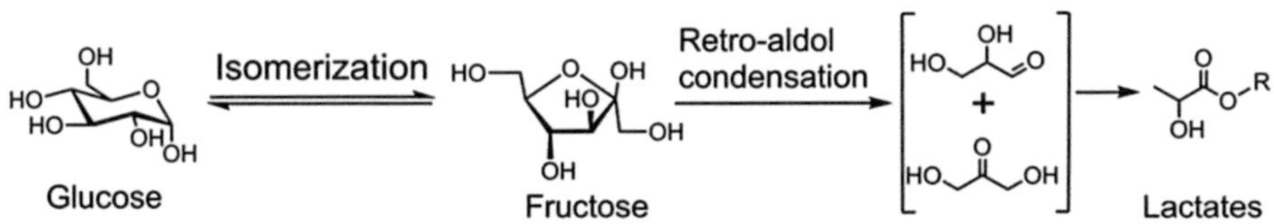

Fig. 10.8 The conversion of hexose sugars to lactates *via* retro-aldol condensation

Reusing the catalyst often results in only marginal loss of performance and this is often hailed as an argument for its stability. Although at lower temperature the stability of the Sn-centers has been demonstrated [76], Osmundsen et al. showed that considerable amounts of Sn are leached from the zeolite at elevated temperatures [59]. This implies that the Sn-content of the materials is of secondary importance and that activity is dominated by other effects. This possibility has already been pointed out before [91], and the exact nature of the active site is currently a hot research topic (Table 10.3).

The retro-aldol condensation reaction of sugars remains a fairly unselective reaction, and therefore there is special interest in the conversion of C_3-sugars such as glyceraldehyde and 1,3-dihydroxyacetone to lactates. The conversion of dihydroxyacetone to alkyl lactates is in general a facile process, readily catalyzed by strongly Lewis acidic solid acids as the key step is a 1,2-hydride shift similar to that from glucose to fructose. The presence of Brønsted acid sites shifts the selectivity towards the dialkylacetal species (see Fig. 10.9) [59]. NMR spectroscopy studies on Sn-MFI zeolite and USY in the conversion of glyoxal to glycolic acid suggest that only the Lewis acid sites catalyze the hydride shift. Recently, Dapsens et al. showed that by ion exchanging parent H-forms of zeolites with Ga^{3+}-ions the selectivity in

Table 10.3 Conversion of carbohydrates to the corresponding lactates

Substrate	Catalyst	Solvent	Conditions	$Y_{lactate}$ (%)	Ref
Glucose	Sn-BEA	Methanol	160 °C, 20 h	43	[89]
Fructose	Sn-BEA	Methanol	160 °C, 20 h	44	[89]
Sucrose	Sn-BEA	Methanol	160 °C, 20 h	64	[89]
DHA	Sn-BEA	Methanol	40 °C, 24 h	95	[59]
DHA	Sn-BEA	Water	125 °C, 24 h	90	[92]
Glucose	Sn-MWW	Methanol	160 °C, 20 h	44	[93]
DHA/GLA	Sn-MWW	Methanol	120 °C, 24 h	>99	[93]
DHA/GLA	Sn-MWW	Water	120 °C, 6 h	96	[93]
Glucose	Sn-C-MCM-41	Methanol	155 °C, 20 h	17	[94]
Sucrose	Sn-C-MCM-41	Methanol	155 °C, 20 h	45	[94]

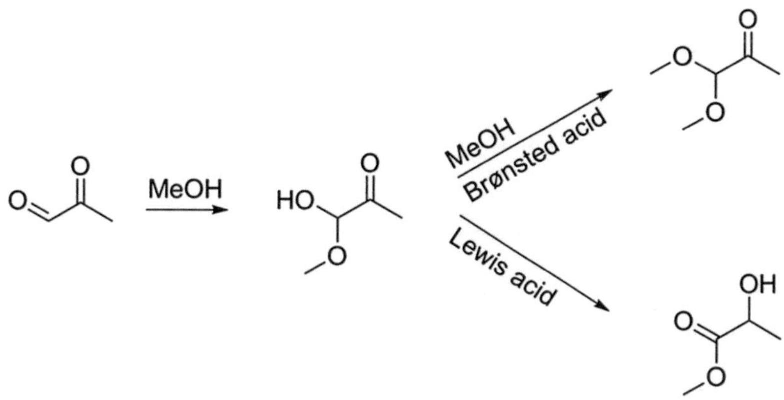

Fig. 10.9 Influence of Lewis and Brønsted acid sites on the conversion of pyruvaldehyde in methanol

the reaction can be decisively altered towards the alkyl lactates [95]. This shows that significantly more facile methods than hydrothermal synthesis can be applied to obtain active and selective zeolite catalysts.

10.8 State-of-the-Art of the Development of Zeotype Materials for Biomass Conversion Reactions

Generally, there are two types of single-site heterogeneous catalysts (SSHC), as defined by Thomas [96]. The first one, which comprises the immobilization of well-defined ensembles (metal nanoparticles, organic functionalities) onto not necessarily well-defined carrier materials fall beyond the scope of this chapter. The second class of SSHCs comprises zeotype materials. The use of zeolites is particularly appealing for industrial applications due to their high thermal and mechanical stability. Diffusion limitations, although needed for some applications, are often unwanted and therefore large particles should be avoided. Furthermore, the incorporation of significant amounts of large heteroatoms in a selective manner is often rather difficult.

Since the development of titanium silicalite-1 (TS-1) in the early 1980s [97], a wide variety of heteroatom-containing zeolites has been developed. The introduction of heteroatoms into zeolite frameworks was a revolution in zeolite chemistry as well as catalysis as it paved the way for the development of tailor-made Lewis or Brønsted acid catalyst. The development of Lewis acidic zeolites, specifically when incorporating large heteroatoms such as tin or titanium is often problematic. Especially the direct incorporation of large atoms like Sn is difficult and a solution developed in the past is based on applying fluoride media to increase crystallization times. The use of fluorides is, however, undesirable due to its extreme toxicity and also because large zeolite crystallites are formed and only small amounts of Sn can be incorporated into the framework. This approach has also met with reproducibility issues. Although efforts have been undertaken to reduce synthesis times by *e.g.* seeding [98] and steam-assisted conversion [99] methods, still toxic fluoride media are extremely difficult to circumvent in order to introduce large heteroatoms by direct synthesis. Alternatively, methods to incorporate metal centers selectively into zeolite frameworks, post-synthesis methods are subject of many investigations. Post-synthesis incorporation of metals such as titanium or tin is usually performed by dealuminating or deboronating the parent zeolite followed by grafting the generated silanol nests with a suitable source of the metal of choice, see Fig. 10.10.

Already several examples of post-synthesis metal incorporation have been demonstrated. Solid-state ion exchange [100], in which $Sn(OAc)_2$ is physically ground with dealuminated BEA zeolite and subsequently calcined, is a promising method to introduce large amounts of active sites into the framework. A similar approach has been reported for $Zr(OEt)_4$ [101]. Also chemical vapor deposition using an appropriate, *i.e.* volatile metal source is a method that can be applied. However, as the temperatures used during grafting are often quite high, the control over the formation of extraframework species in such methods is rather poor. This was shown by Li et al. [102],

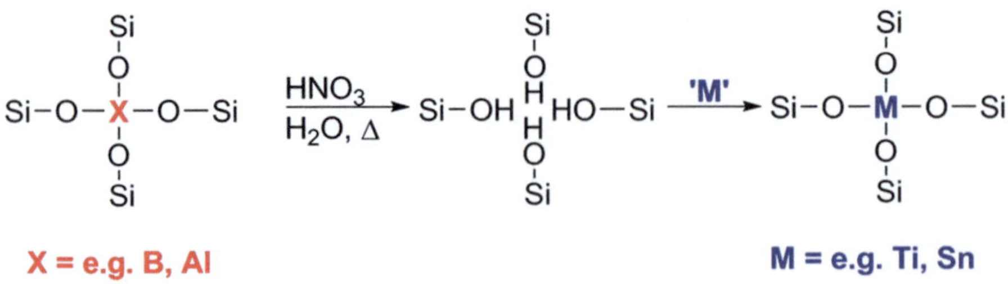

Fig. 10.10 A generalized scheme for the post-synthetic incorporation of metal sites into zeolite framework positions

who applied $SnCl_4$ as grafting agent at high temperatures. Significant amounts of extraframework Sn sites were formed, although high amounts of Sn could be incorporated into the zeolite framework. A factor of considerable influence seems to be the reactivity of the applied metal precursor. Whereas for $TiCl_4$ many extraframework sites and even titania nanoclusters were formed [103], the use of Cp_2TiCl_2 under similar conditions yielded almost exclusively framework Ti species [104]. Also other methods, such as solution impregnation of dealuminated zeolites followed by calcination can yield zeolites with relatively low amounts of extraframework metal centers [43, 93]. Another way of introducing additional Lewis acidity is based on ion exchange upon which charge balancing cations of an Al-containing zeolite are substituted by another metal, *e.g.* gallium. Such an approach was shown to yield promising catalysts for the conversion of C_3-carbohydrates to lactates [95].

Due to the microporous nature of zeolites, much efforts are being made on either decreasing the particle size or on introducing mesopores into the zeolite framework in order to shorten diffusion limitations and avoid deactivation. Especially at elevated temperatures, the micropores of the zeolites become blocked by the formations of cokes. It has been shown for several reactions already that the introduction of mesopores into the zeolite matrix enhances the catalytic stability of the material. An elegant example was given by Wu et al. [105], who showed that the activity of the zeolite in the methanol-to-olefins reaction is maintained for much longer when a hierarchical meso/microporous SSZ-13 is used instead of common microporous SSZ-13. In connection to the introduction of mesopores, also many efforts are undertaken to develop ultra large-pore zeolites by introducing germanium in the lattice, leading to new zeolite topologies [106]. Large-scale application of such materials is hampered by the high cost of germanium, as well as their low thermal/hydrothermal stability, which is associated with their low framework density [107].

10.9 Stability of Zeolites

A major drawback of zeolites for large-scale processing of biomass is their relatively low tolerance towards water at elevated temperatures. Often this instability arises from the defects present in the zeolite, which facilitate the hydrolysis of the Si-O-Si-bonds [108]. Several pathways can be distinguished to improve the stability of the zeolites under hydrothermal conditions. A method that is in use on the industrial scale is steaming [109], which is done to obtain ultrastable Y-zeolite (USY). Furthermore increasing the hydrophobicity of the zeolite, by increasing the Si/Al ratio [110], grafting of hydrophobic moieties onto the surface [111] or adjusted synthesis procedures such as dry-gel conversion [112] can lead to improved stability. The formation of defects during synthesis can be largely avoided by the use of fluoride media [36]. However, as it has already been emphasized above, the use of fluoride media is generally not desirable.

The presence of defects in the zeolite has a detrimental effect on the zeolite stability. This, however, is not necessarily a negative property as it allows for facile modification of the zeolite to meet particular requirements. The manner in which

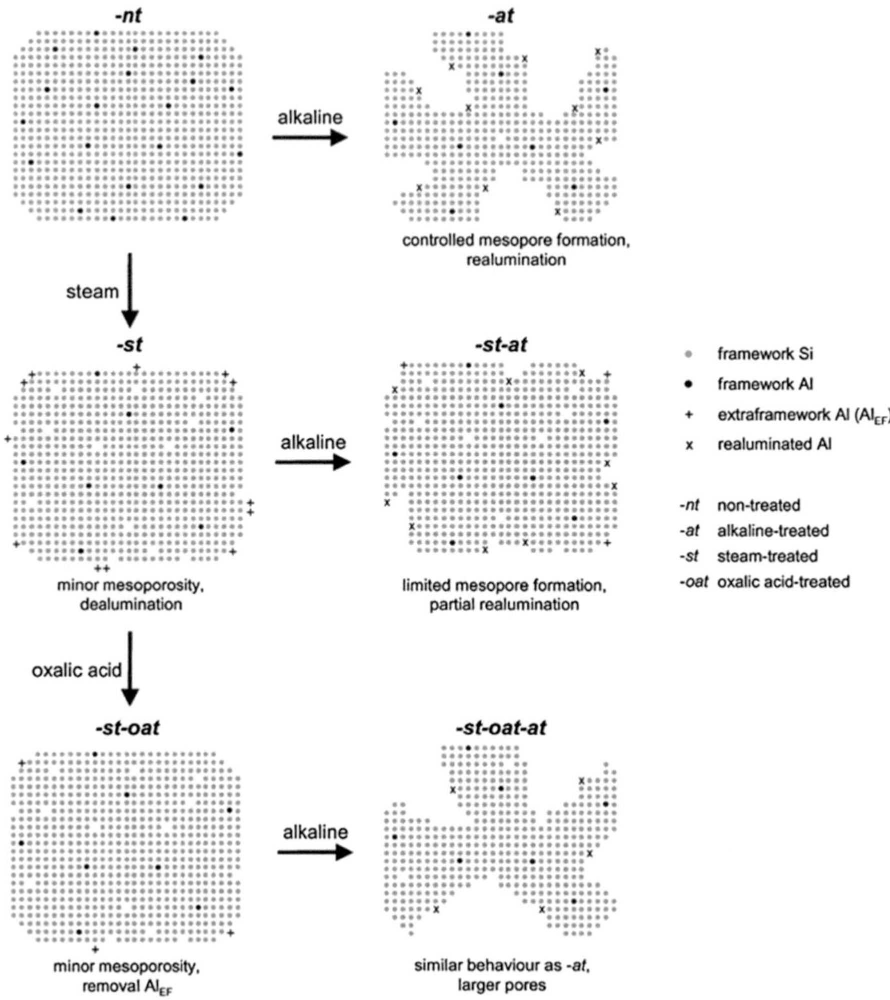

Fig. 10.11 Schematic proposed hierarchical evolution of mesoporosity upon various treatments and their combinations (Reprinted from Ref. [113] by permission of John Wiley & Sons, Inc.)

the characteristics of the zeolite change is especially dependent on the pH of the medium the zeolite is in contact with, as well as on the chemical composition of the zeolite itself (Fig. 10.11) [114, 115]. This fact is extensively used to introduce mesoporosity in the zeolites or change their elemental composition by removing silicon or aluminum atoms from the framework.

A frequently applied method for generating mesopores is the base treatment of zeolites. Stability issues play an important role in steering the functionalization of the zeolite. It was shown for MFI zeolite that the presence of aluminum as a pore directing agent is of critical importance for the optimal generation of mesopores [116]. High-alumina zeolites tend to be more stable in basic media, whereas the absence of the stabilizing influence of aluminum for high-silica zeolites causes extensive damage to the zeolite framework at high pH. Therefore, it was assumed that framework aluminum acts as pore-directing agent. Later it was found that additions of external aluminum or other pore directing agents (such as tetrapropylammonium cations) also can direct the desilication process in a controlled manner. Also lower defect and intergrowth density of the zeolites has a positive effect on the generation of mesopores [117].

The behavior of heterogeneous catalysts in presence of water is very important when considering the effects of biomass processing conditions on the activity and stability of the catalysts [118]. One could think of acidic pyrolysis oils or application in reactions in which large amounts of water are generated such as HMF synthesis. Boiling water was shown by Ravanelle et al. to have a great impact on the crystallinity of zeolites [119]. Such stability studies are instrumental in improving understanding of the implications of biomass processing conditions on catalyst performance. In this respect, theoretical studies on the behavior of zeolites in the presence of water are necessary to develop further understanding of the underlying processes taking place at the molecular level. Malola et al. have investigated the influence of water on desilication and dealumination processes in MFI. They found that desilication is dominant under steaming conditions [120].

Classical application of zeolites in chemical processes relies mainly on the thermal stability of the zeolites. In order to effectuate the transition to the large-scale application of zeolites in biomass conversion reactions, more research has to be devoted to their stability under hydrothermal conditions. By identifying trends in the stability and activity of zeolites after being exposed to hydrothermal conditions, routes to improve their long-term stability can be developed. Theoretical studies can play an important supporting role in identifying pathways for improving the performance of zeolite catalysts in biomass processing.

10.10 Concluding Remarks

The application of zeolites for biomass conversion reactions has undergone a significant development. Traditional cracking processes using biomass, resembling the current oil-based petrochemical industrial processes, are gradually being substituted by more selective processes for the conversion of biomass constituents. In this way, lignocellulosic biomass can be developed into a competitive feedstock for the future generation of bulk chemicals and fuels. The use of zeolites for this kind of processes is interesting due to their abundance and tunability.

Several promising pathways towards the selective formation of platform chemicals have been identified and are currently intensively investigated. This plays a vital role in the development of renewable bulk chemicals with a wide range of applications, as well as reducing environmental footprint of the chemical industry.

Great progress has been made in understanding the role of topologies and functional groups on the performance of zeolites in catalysis. However, the understanding of the role of confinement space effects on activity and selectivity of zeolites, especially in the presence of solvents, is still a matter of debate. The use of theoretical studies in this field could prove a vital tool for the further development of new catalytic systems.

References

1. Huber G, Corma A (2007) Synergies between bio- and oil refineries for the production of fuels from biomass. Angew Chem Int Ed 46:7184–7201
2. Huber G, Iborra S et al (2006) Synthesis of transportation fuels from biomass: chemistry, catalysts, and engineering. Chem Rev 106:4044–4098
3. Karinen R, Vilonen K et al (2011) Biorefining: heterogeneously catalyzed reactions of carbohydrates for the production of furfural and hydroxymethylfurfural. ChemSusChem 4:1002–1016
4. Lange J, van der Heide E et al (2012) Furfural – a promising platform for lignocellulosic biofuels. ChemSusChem 5:150–166
5. Cheng Y, Jae J (2012) Production of renewable aromatic compounds by catalytic fast pyrolysis of lignocellulosic biomass with bifunctional Ga/ZSM-5 catalysts. Angew Chem Int Ed 51:1387–1390
6. Vispute T, Zhang H et al (2010) Renewable chemical commodity feedstocks from integrated catalytic processing of pyrolysis oils. Science 330:1222–1227
7. Zhang H, Cheng Y-T et al (2011) Catalytic conversion of biomass-derived feedstocks into olefins and aromatics with ZSM-5: the hydrogen to carbon effective ratio. Energy Environ Sci 4:2297
8. Furimsky E (2013) Hydroprocessing challenges in biofuels production. Catal Today 217:13–56
9. Gosselink R, Hollak SA (2013) Reaction pathways for the deoxygenation of vegetable oils and related model compounds. ChemSusChem 6:1576–1594
10. Crutzen PJ, Andreae MO (1990) Biomass burning in the tropics: impact on atmospheric chemistry and biogeochemical cycles. Science 250:1669–1678
11. Kamm B, Kamm M, Gruber PR, Kromus S (2005) Biorefinery systems – an overview. In: Kamm B, Gruber PR, Kamm M (eds) Biorefineries-industrial processes and products: status quo and future directions. Wiley-VCH Verlag GmbH, Weinheim
12. Krieger K (2014) Renewable energy: biofuels heat up. Nature 508:448–449
13. Bozell JJ, Petersen GR (2010) Technology development for the production of biobased products from biorefinery carbohydrates – the US Department of Energy's "Top 10" revisited. Green Chem 12:539–554
14. Mäki-Arvela P, Salmi T et al (2011) Synthesis of sugars by hydrolysis of hemicelluloses – a review. Chem Rev 111:5638–5666
15. Kobayashi H, Fukuoka A (2013) Synthesis and utilisation of sugar compounds derived from lignocellulosic biomass. Green Chem 15:1740–1763
16. Adler E (1977) Lignin chemistry -past, present and future. Wood Sci Technol 11:169–218
17. Zakzeski J, Bruijnincx PCA et al (2010) The catalytic valorization of lignin for the production of renewable chemicals. Chem Rev 110:3552–3599
18. Sharma RK, Bakhshi NN (1993) Catalytic upgrading of pyrolysis oil. Energy Fuel 7:306–314
19. Mante OD, Agblevor FA et al (2011) Fluid catalytic cracking of biomass pyrolysis vapors. Biomass Convers Biorefin 1:189–201
20. Adjaye JD, Bakhshi NN (1995) Production of hydrocarbons by catalytic upgrading of a fast pyrolysis bio-oil. Part I: Conversion over various catalysts. Fuel Process Technol 45:161–183
21. Adjaye JD, Bakhshi NN (1995) Production of hydrocarbons by catalytic upgrading of a fast pyrolysis bio-oil. Part II: Comparative catalyst performance and reaction pathways. Fuel Process Technol 45:185–202

22. Liu C, Wang H et al (2014) Catalytic fast pyrolysis of lignocellulosic biomass. Chem Soc Rev 43:7594–7623
23. Dickerson T, Soria J (2013) Catalytic fast pyrolysis: a review. Energies 6:514–538
24. Jae J, Tompsett GA (2010) Depolymerization of lignocellulosic biomass to fuel precursors: maximizing carbon efficiency by combining hydrolysis with pyrolysis. Energy Environ Sci 3:358
25. Carlson TR, Tompsett GA (2009) Aromatic production from catalytic fast pyrolysis of biomass-derived feedstocks. Top Catal 52:241–252
26. Carlson TR, Jae J et al (2010) Catalytic fast pyrolysis of glucose with HZSM-5: the combined homogeneous and heterogeneous reactions. J Catal 270:110–124
27. Luterbacher JS, Rand JM et al (2014) Nonenzymatic sugar production from biomass using biomass-derived gamma-valerolactone. Science 343:277–280
28. Geboers J, Van de Vyver S et al (2011) Efficient hydrolytic hydrogenation of cellulose in the presence of Ru-loaded zeolites and trace amounts of mineral acid. Chem Commun 47:5590–5592
29. Werpy T, Petersen G (2004) Top value added chemicals from biomass volume I—results of screening for potential candidates from sugars and synthesis gas. US Department of Energy (DOE) Report (DOE) DOE/GO-102004-1992), Golden
30. van Putten RJ, van der Waal JC et al (2013) Hydroxymethylfurfural, a versatile platform chemical made from renewable resources. Chem Rev 113:1499–1597
31. Dusselier M, Van Wouwe P et al (2013) Lactic acid as a platform chemical in the biobased economy: the role of chemocatalysis. Energy Environ Sci 6:1415
32. Speck JC Jr (1958) The Lobry de Bruyn-Alberda van Ekenstein transformation. Adv Carbohydr Chem 13:63–103
33. Kooyman C, Vellenga K et al (1977) The isomerization of d-glucose into d-fructose in aqueous alkaline solutions. Carbohydr Res 54:33–44
34. Takagaki A, Ohara M et al (2009) A one-pot reaction for biorefinery: combination of solid acid and base catalysts for direct production of 5-hydroxymethylfurfural from saccharides. Chem Commun 41:6276–6278
35. Lecomte J, Finiels A et al (2002) Kinetic study of the isomerization of glucose into fructose in the presence of anion-modified hydrotalcites. Starch Stärke 54:75–79
36. Valencia S, Corma A (1999) Three-dimensional framework of silica, titania, germania, and tin oxide; selective oxidation catalysts, US5968473A
37. Corma A, Domine ME et al (2003) Water-resistant solid Lewis acid catalysts: Meerwein–Ponndorf–Verley and Oppenauer reactions catalyzed by tin-beta zeolite. J Catal 215:294–304
38. Moliner M, Román-Leshkov Y et al (2010) Tin-containing zeolites are highly active catalysts for the isomerization of glucose in water. Proc Natl Acad Sci U S A 107:6164–6168
39. Bílik V (1972) Reactions of saccharides catalyzed by molybdate ions: II: Epimerization of D-glucose and D-mannose. Chem Zvesti 26:183–186
40. Moliner M (2014) State of the art of Lewis acid-containing zeolites: lessons from fine chemistry to new biomass transformation processes. Dalton Trans 43:4197–4208
41. Hayes ML, Pennings NJ et al (1982) Epimerization of aldoses by molybdate involving a novel rearrangement of the carbon skeleton. J Am Chem Soc 104:6764–6769
42. Bermejo-Deval R, Gounder R et al (2012) Framework and extraframework tin sites in zeolite beta react glucose differently. ACS Catal 2:2705–2713
43. Dijkmans J, Gabriëls D et al (2013) Productive sugar isomerization with highly active Sn in dealuminated β zeolites. Green Chem 15:2777–2785
44. Gunther WR, Wang Y et al (2012) Sn-Beta zeolites with borate salts catalyse the epimerization of carbohydrates via an intramolecular carbon shift. Nat Commun 3:1109
45. West RM, Holm MS et al (2010) Zeolite H-USY for the production of lactic acid and methyl lactate from C3-sugars. J Catal 269:122–130
46. Saravanamurugan S, Paniagua M et al (2013) Efficient isomerization of glucose to fructose over zeolites in consecutive reactions in alcohol and aqueous media. J Am Chem Soc 135:5246–5249
47. Zhang Y, Hidajat K et al (2004) Optimal design and operation of SMB bioreactor: production of high fructose syrup by isomerization of glucose. Biochem Eng J 21:111–121

48. Zhang Y, Pidko EA et al (2011) Molecular aspects of glucose dehydration by chromium chlorides in ionic liquids. Chemistry 17:5281–5288
49. Pidko EA, Degirmenci V et al (2010) Glucose activation by transient Cr2+ dimers. Angew Chem 122:2584–2588
50. Pidko EA, Degirmenci V et al (2014) On the mechanism of Lewis acid catalyzed glucose transformations in ionic liquids. ChemCatChem 4:1263–1271
51. Assary RS, Curtiss LA (2011) Theoretical study of 1,2-hydride shift associated with the isomerization of glyceraldehyde to dihydroxy acetone by Lewis acid active site models. J Phys Chem A 115:8754–8760
52. Degirmenci V, Pidko EA et al (2011) Towards a selective heterogeneous catalyst for glucose dehydration to 5-hydroxymethylfurfural in water: CrCl2 catalysis in a thin immobilized ionic liquid layer. ChemCatChem 3:969–972
53. Bare SR, Kelly SD et al (2005) Uniform catalytic site in Sn-beta-zeolite determined using X-ray absorption fine structure. J Am Chem Soc 127:12924–12932
54. Yang G, Pidko EA et al (2013) Structure, stability, and Lewis acidity of mono and double Ti, Zr, and Sn framework substitutions in BEA zeolites: a periodic density functional theory study. J Phys Chem C 117:3976–3986
55. Li G, Pidko EA et al (2014) Synergy between Lewis acid sites and hydroxyl groups for the isomerization of glucose to fructose over Sn-containing zeolites: a theoretical perspective. Catal Sci Technol 4:2241–2250
56. Rai N, Caratzoulas S et al (2013) Role of silanol group in Sn-Beta zeolite for glucose isomerization and epimerization reactions. ACS Catal 3:2294–2298
57. Khouw C, Davis ME (1995) Catalytic activity of titanium silicates synthesized in the presence of alkali-metal and alkaline-earth ions. J Catal 151:77–86
58. Boronat M, Concepción P et al (2005) Determination of the catalytically active oxidation Lewis acid sites in Sn-beta zeolites, and their optimisation by the combination of theoretical and experimental studies. J Catal 234:111–118
59. Osmundsen CM, Holm MS et al (2012) Tin-containing silicates: structure–activity relations. Proc R Soc A 468:2000–2016
60. Tuercke T, Panic S et al (2009) Microreactor process for the optimized synthesis of 5-hydroxymethylfurfural: a promising building block obtained by catalytic dehydration of fructose. Chem Eng Technol 32:1815–1822
61. van der Graaff WNP, Garrido Olvera K et al (2014) Stability and catalytic properties of porous acidic (organo)silica materials for conversion of carbohydrates. J Mol Catal A Chem 388–389:81–89
62. Tucker MH, Crisci AJ et al (2012) Acid-functionalized SBA-15-type periodic mesoporous organosilicas and their use in the continuous production of 5-hydroxymethylfurfural. ACS Catal 2:1865–1876
63. Rivalier P, Duhamet J et al (1995) Development of a continuous catalytic heterogeneous column reactor with simultaneous extraction of an intermediate product by an organic solvent circulating in countercurrent manner with the aqueous phase. Catal Today 24:165–171
64. Moreau C, Durand R et al (1996) Dehydration of fructose to 5-hydroxymethylfurfural over H-mordenites. Appl Catal A Gen 145:211–224
65. Zakrzewska ME, Bogel-Łukasik E et al (2011) Ionic liquid-mediated formation of 5-hydroxymethylfurfural-a promising biomass-derived building block. Chem Rev 111:397–417
66. Blanksma JJ, Egmond G (2010) Humin from hydroxymethylfurfuraldehyde. Recueil des Travaux Chimiques des Pays-Bas 65:309–310
67. Hoang TM, Lefferts L et al (2013) Valorization of humin-based byproducts from biomass processing-a route to sustainable hydrogen. ChemSusChem 6:1651–1658
68. Amarasekara AS, Williams LD et al (2008) Mechanism of the dehydration of D-fructose to 5-hydroxymethylfurfural in dimethyl sulfoxide at 150 degrees C: an NMR study. Carbohydr Res 343:3021–3024

69. Caratzoulas S, Vlachos DG (2011) Converting fructose to 5-hydroxymethylfurfural: a quantum mechanics/molecular mechanics study of the mechanism and energetics. Carbohydr Res 346:664–672
70. Shimizu K-I, Uozumi R et al (2009) Enhanced production of hydroxymethylfurfural from fructose with solid acid catalysts by simple water removal methods. Catal Commun 10:1849–1853
71. Wyman CE, Decker SR et al (2005) Hydrolysis of cellulose and hemicellulose. In: Dumitriu S (ed) Chapter 43: Polysaccharides: structural diversity and functional versatility, 2nd edn. Marcel Dekker, Inc., New York, pp 995–1033
72. Zhang L, Yu H et al (2014) Production of furfural from xylose, xylan and corncob in gamma-valerolactone using $FeCl_3.6H_2O$ as catalyst. Bioresour Technol 151:355–360
73. Moreau C, Durand R et al (1998) Selective preparation of furfural from xylose over microporous solid acid catalysts. Ind Crops Prod 7:95–99
74. Zhao H, Holladay JE et al (2007) Metal chlorides in ionic liquid solvents convert sugars to 5-hydroxymethylfurfural. Science 316:1597–1600
75. Li L, Stroobants C et al (2011) Selective conversion of trioses to lactates over Lewis acid heterogeneous catalysts. Green Chem 13:1175–1181
76. Corma A, Renz M (2004) Sn-Beta zeolite as diastereoselective water-resistant heterogeneous Lewis-acid catalyst for carbon-carbon bond formation in the intramolecular carbonyl-ene reaction. Chem Commun 5:550–551
77. Nikolla E, Román-Leshkov Y et al (2011) "One-pot" synthesis of 5-(Hydroxymethyl)furfural from carbohydrates using tin-Beta zeolite. ACS Catal 1:408–410
78. Lew CM, Rajabbeigi N et al (2012) One-pot synthesis of 5-(Ethoxymethyl)furfural from glucose using Sn-BEA and amberlyst catalysts. Ind Eng Chem Res 51:5364–5366
79. Gallo JMR, Alonso DM et al (2013) Production and upgrading of 5-hydroxymethylfurfural using heterogeneous catalysts and biomass-derived solvents. Green Chem 15:85–90
80. Wang J, Ren J et al (2012) Direct conversion of carbohydrates to 5-hydroxymethylfurfural using Sn-Mont catalyst. Green Chem 14:2506–2512
81. Otomo R, Yokoi T et al (2014) Dealuminated Beta zeolite as effective bifunctional catalyst for direct transformation of glucose to 5-hydroxymethylfurfural. Appl Catal A Gen 470:318–326
82. Tan M, Zhao L et al (2011) Production of 5-hydroxymethyl furfural from cellulose in CrCl2/Zeolite/BMIMCl system. Biomass Bioenergy 35:1367–1370
83. Leonard RH (1956) Levulinic acid as a basic chemical raw material. Ind Eng Chem 48:1330–1341
84. Bond JQ, Alonso DM et al (2010) Integrated catalytic conversion of gamma-valerolactone to liquid alkenes for transportation fuels. Science 327:1110–1114
85. Lange JP, Price R et al (2010) Valeric biofuels: a platform of cellulosic transportation fuels. Angew Chem 49:4479–4483
86. Bui L, Luo H et al (2013) Domino reaction catalyzed by zeolites with Brønsted and Lewis acid sites for the production of gamma-valerolactone from furfural. Angew Chem 52:8022–8025
87. Wang Y, Deng W et al (2013) Chemical synthesis of lactic acid from cellulose catalysed by lead(II) ions in water. Nat Commun 4:2141
88. Wang F-F, Liu C-L et al (2013) Highly efficient production of lactic acid from cellulose using lanthanide triflate catalysts. Green Chem 15:2091–2095
89. Holm MS, Saravanamurugan S et al (2010) Conversion of sugars to lactic acid derivatives using heterogeneous zeotype catalysts. Science 328:602–605
90. Holm MS, Pagán-Torres YJ et al (2012) Sn-Beta catalysed conversion of hemicellulosic sugars. Green Chem 14:702–706
91. Sheldon RA, Wallau M et al (1998) Heterogeneous catalysts for liquid-phase oxidations: philosophers' stones or Trojan horses? Acc Chem Res 31:485–493
92. Taarning E, Saravanamurugan S et al (2009) Zeolite-catalyzed isomerization of triose sugars. ChemSusChem 2:625–627
93. Guo Q, Fan F et al (2013) Highly active and recyclable Sn-MWW zeolite catalyst for sugar conversion to methyl lactate and lactic acid. ChemSusChem 6:1352–1356

94. de Clippel F, Dusselier M et al (2012) Fast and selective sugar conversion to alkyl lactate and lactic acid with bifunctional carbon-silica catalysts. J Am Chem Soc 134:10089–10101
95. Dapsens PY, Kusema BT et al (2014) Gallium-modified zeolites for the selective conversion of bio-based dihydroxyacetone into C1–C4 alkyl lactates. J Mol Catal A Chem 388–389:141–147
96. Thomas JM (2012) The societal significance of catalysis and the growing practical importance of single-site heterogeneous catalysts. Proc R Soc A Math Phys Eng Sci 468:1884–1903
97. Taramasso M, Perego G et al (1983) Preparation of porous crystalline synthetic material comprised of silicon and titanium oxides, US4410501A
98. Chang C-C, Wang Z et al (2012) Rapid synthesis of Sn-Beta for the isomerization of cellulosic sugars. RSC Adv 2:10475–10477
99. Kang Z, Zhang X et al (2013) Factors affecting the formation of Sn-Beta zeolites by steam-assisted conversion method. Mater Chem Phys 141:519–529
100. Hammond C, Conrad S et al (2012) Simple and scalable preparation of highly active Lewis acidic Sn-beta. Angew Chem 51:11736–11739
101. Wolf P, Hammond C et al (2014) Post-synthetic preparation of Sn-, Ti- and Zr-beta: a facile route to water tolerant, highly active Lewis acidic zeolites. Dalton Trans 43:4514–4519
102. Li P, Liu G et al (2011) Postsynthesis and selective oxidation properties of nanosized Sn-Beta zeolite. J Phys Chem C 115:3663–3670
103. Rigutto MS, de Ruiter R et al (1994) Titanium-containing large pore molecular sieves from boron-beta: preparation, characterization and properties. J Catal 84:2245–2252
104. Tang B, Dai W et al (2014) A procedure for the preparation of Ti-Beta zeolites for catalytic epoxidation with hydrogen peroxide. Green Chem 16:2281–2291
105. Wu L, Degirmenci V et al (2013) Mesoporous SSZ-13 zeolite prepared by a dual-template method with improved performance in the methanol-to-olefins reaction. J Catal 298:27–40
106. Davis ME (2002) Ordered porous materials for emerging applications. Nature 417:813–821
107. Bellussi G, Carati A, Millini R (2010) Industrial potential of zeolites. In: Čejka J, Corma A, Zones S (eds) Zeolites and catalysis: synthesis, reactions and applications. Wiley-VCH Verlag GmbH & Co. KGaA, Weinheim
108. Okuhara T (2002) Water-tolerant solid acid catalysts. Chem Rev 102:3641–3666
109. Beyerlein RA, Choi-Feng C, Hall JB, Huggins BJ, Ray GJ (1994) Investigation of mesopore formation. In: Occelli ML, O'Connor P (eds) Fluid catalytic cracking III. ACS Publications, Washington, DC
110. Hayakawa K, Morita T et al (1996) Adsorption of cationic surfactants on hydrophobic mordenites of different Si/Al ratio. J Colloid Interface Sci 177:621–627
111. Zapata PA, Faria J et al (2012) Hydrophobic zeolites for biofuel upgrading reactions at the liquid-liquid interface in water/oil emulsions. J Am Chem Soc 134:8570–8578
112. Tatsumi T, Jappar N (1998) Properties of Ti-beta zeolites synthesized by dry-gel conversion and hydrothermal methods. J Phys Chem B 102:7126–7131
113. Groen JC, Peffer LAA et al (2005) Mechanism of hierarchical porosity development in MFI zeolites by desilication: the role of aluminium as a pore-directing agent. Chemistry 11:4983–4994
114. Groen JC, Jansen JC et al (2004) Optimal aluminum-assisted mesoporosity development in MFI zeolites by desilication. J Phys Chem B 108:13062–13065
115. Groen JC, Peffer LAA et al (2004) On the introduction of intracrystalline mesoporosity in zeolites upon desilication in alkaline medium. Microporous Mesoporous Mater 69:29–34
116. Groen JC, Moulijn JA et al (2006) Desilication: on the controlled generation of mesoporosity in MFI zeolites. J Mater Chem 16:2121–2131
117. Svelle S, Sommer L et al (2011) How defects and crystal morphology control the effects of desilication. Catal Today 168:38–47
118. Van Pelt AH, Simakova OA et al (2014) Stability of functionalized activated carbon in hot liquid water. Carbon 77:143–154
119. Ravenelle RM, Schüβler F et al (2010) Stability of zeolites in hot liquid water. J Phys Chem C 114:19582–19595
120. Malola S, Svelle S et al (2012) Detailed reaction paths for zeolite dealumination and desilication from density functional calculations. Angew Chem Int Ed 51:652–655

Chapter 11
Catalytic Oxidations Over Titanosilicate Zeolites

Liang Wang and Feng-Shou Xiao

Abstract An important topic in the field of zeolite is the titanosilicate with tetrahedral Ti sites, which have already been regarded as efficient catalyst for the oxidation of various organic substrates by hydrogen peroxide. Since the first report of TS-1 zeolite in 1983, a variety of titanosilicates with various morphologies have been synthesized and applied in the oxidation reactions. The physical parameters of structures, porosities, Ti loadings, and wettabilities strongly determine the catalytic properties of the titanosilicate catalysts. The TS-1 zeolite is efficient for the hydroxylation of benzene and phenol, as well as the epoxidation and ammoxidation of olefins and ketons with small molecule size. Ti-Beta with larger micropores than TS-1 exhibits excellent catalytic performances in the oxidation of bulky molecules. Ti-MWW zeolites with a unique two-dimensional structure are also used in the oxidation of olefins, ammoxidation of ketons, and oxidative desulfurization.

Keywords Titanosilicate zeolite • Oxidation • Hydrogen peroxide • Catalytic activity • Selectivity

11.1 Introduction

The titanosilicate zeolites with isolated Ti(IV) sites such as TS-1 are very successful catalysts for the liquid-phase oxidation of various organic substrates under mild reaction conditions by hydrogen peroxide in industrial processes, where the hydrogen peroxide is regarded as a "green" oxidant with water as the sole product [1–5].

Although TS-1 zeolite has exhibited excellent catalytic properties in the oxidation of a series of organic compounds with small molecule sizes (*e.g.* phenol, propylene,

F.-S. Xiao (✉)
Institute of Catalysis, Zhejiang University, Hangzhou 310028, China
e-mail: fsxiao@zju.edu.cn

L. Wang
Department of Chemistry, Zhejiang University, Hangzhou 310028, China

F.-S. Xiao, X. Meng (eds.), *Zeolites in Sustainable Chemistry*,
Green Chemistry and Sustainable Technology,
DOI 10.1007/978-3-662-47395-5_11

1-hexene, hexane, and ethylene), it failed to catalyze the oxidation of relatively bulky molecules (*e.g.* cyclohexene and cyclooctene), because the ten-membered ring micropores of TS-1 hinders the transfer of bulky organic molecules. To overcome this problem, great efforts have been made to synthesize larger pore titanosilicate zeolites. The typical examples are 12-membered ring zeolites of Ti-Beta [6–9], Ti-MOR [10], Ti-ITQ-7 [11], 12- and 10-membered ring zeolites of Ti-MWW [12–14] and Ti-MCM-68 [15]. Particularly, these large-pore titanosilicate zeolites exhibited an enhanced catalytic performance in the oxidation of bulky molecules. In this chapter, several titanosilicate zeolites and their applications in various oxidations will be systemically illustrated.

11.2 TS-1 Zeolite

TS-1 zeolite with Ti(IV) sites on the MFI-type framework is the first titanosilicate zeolite synthesized by Enichem Company in Italy in 1983 [1]. In the framework of TS-1 zeolite, the tetracoordinate Ti sites are highly isolated with $Ti(OSi)_4$ units. The unique tetrahedral Ti sites make the TS-1 zeolite highly efficient for various oxidations with hydrogen peroxide, including the hydroxylation of benzene and phenol, oxidation of olefins, alcohols, alkanes, and ammoxidation of ketones.

11.2.1 Hydroxylation of Benzene and Phenol Over TS-1 Zeolite

Although the hydroxylation of phenol over TS-1 was extensively investigated by Enichem Company in the early of 1980s, the details on the reaction had been reported until in the early of 1990s. After that, great efforts have been focused on systemical study on the hydroxylation of phenol over TS-1 zeolite.

In 1991, Thangaraj et al. [16] found that various reaction parameters (*e.g.* the Ti content on TS-1, catalyst concentration, reaction time, solvent, and temperature) strongly influence the phenol conversion in the hydroxylation over TS-1 zeolite by hydrogen peroxide. For example, increasing the Ti content on TS-1 zeolite with Ti/(Ti + Si) molar ratio from 0 to 0.056, or increasing the concentration of TS-1 zeolite in the reaction system with catalyst/phenol weight ratio from 0 to 10, could obviously improve the conversion of phenol and hydrogen peroxide.

In the same year, Tuel et al. [17] studied the contribution of Ti sites on the different position of TS-1 zeolite to the hydroxylation. They found that the formation of hydroquinone from the hydroxylation of phenol is strongly related to the effect of the micropore constraints. Therefore, the internal Ti sites in the microporous structure of TS-1 play an important role for the formation of hydroquinone. In contrast, due to the lack of micropore constraint effect, the Ti sites on the external surface could catalyze the formation of catechol and tar compounds immediately at the start of the reaction. The bulky tar molecules could solely poison and deactivate the external Ti sites, without any effect on the internal Ti sites. Later, van der Pol et al. found that the catalytic activity of TS-1 in the hydroxylation of phenol strongly depends on the crystal size of TS-1. The TS-1 crystals smaller than 300 nm were regarded as efficient catalyst as a result of less micropore diffusion limitation.

Bhaumik et al. [18] firstly reported the hydroxylation of benzene over TS-1 under solvent-free and triphase conditions (solid catalyst, organic substrate, and H_2O_2), where the TS-1 catalyst exhibited even 20 times higher benzene conversion than that under conventional biphase conditions (solid catalyst and homogeneous reactants). This phenomenon is attributed to the hydrophobic nature of TS-1 zeolite, which is favorable to adsorb the organic benzene reactant [19, 20]. In contrast, the adsorption of benzene over TS-1 was hindered in biphase system because of the competitive adsorption with organic solvent, leading to lower catalytic activity than that in triphase system. Bianchi et al. [21–24] reported that the post-treatment with NH_4HF_2 and H_2O_2 could efficiently improve the catalytic activity of TS-1 zeolite in the hydroxylation of benzene, which is probably due to its F-modified surface and/or the formation of new non-tetrahedral titanium species generated by the post-synthesis treatment.

Despite TS-1 is very active for the hydroxylation of phenol, it is almost inactive for the hydroxylation of trimethylphenol, which is due to that the TS-1 micropores are too small for access by bulky reactants. To solve this problem, ordered mesoporous titanosilicates assembled from TS-1 nanoclusters was rationally synthesized [25], designed as MTS-9 (Fig. 11.1). The MTS-9 sample has the combined advantages of both mesoporous materials and zeolites, including open mesopores and high thermal stability. Very importantly, the MTS-9 sample exhibited comparably catalytic activity and product selectivity in the hydroxylation of phenol. Furthermore, in the hydroxylation of bulky molecule of trimethylphenol, MTS-9 exhibited much higher catalytic activity than TS-1, which is attributed to the mesoporous structure of MTS-9 for the transfer of bulky reactants. Later, Xin et al. [26] introduced mesopores into the TS-1 crystals by using carbon black as a hard template. Compared with the conventional TS-1 zeolite, the mesoporous TS-1 exhibited much higher catalytic activity than the conventional TS-1 in the hydroxylation of phenol. The improved catalytic activity of mesoporous TS-1 is attributed to the rapid diffusion rate to and out of the micropores conferred by the mesopores.

11.2.2 Epoxidation of Olefins Over TS-1 Zeolite

The epoxidation of olefin plays key roles in the synthesis of relevant building blocks for the synthesis of polyols and polymers. For example, the propylene epoxide from the epoxidation of propylene is an important intermediate for synthesizing propylene glycol, which is important for production of paints and adhesives [27–34]. Conventionally, the epoxides are synthesized from chlorohydrin and hydroperoxidation routes, which have disadvantages such as low selectivity and the formation of environmentally unfriendly wastes [35].

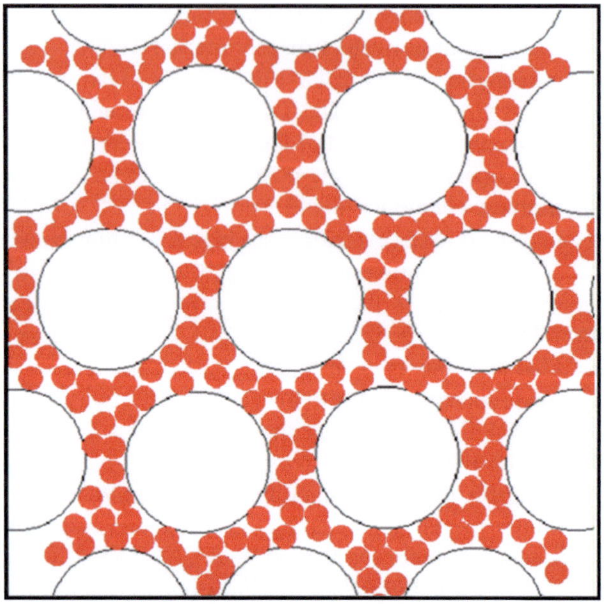

Fig. 11.1 Ordered mesoporous titanosilicates with high stability and catalytic activity assembled from TS-1 crystal (Reprinted with permission from Ref. [25], Copyright 2014, American Chemical Society)

Recently, BASF showed a sustainable route for the epoxidation of propylene using hydrogen peroxide over TS-1 zeolite, which has been one of the most successful commercial processes. In this case, the selectivity to propylene epoxide could reach as high as 98 %.

Compared with hydrogen peroxide, the use of relatively low cost oxidant of molecular oxygen led to a significant reduction of the epoxide selectivity in the reaction [36]. To overcome this shortcoming, Haruta and coworkers opens new route for selective epoxidation of propylene in the presence of hydrogen and oxygen over TS-1 supported Au nanoparticles (Au/TS-1) [28, 29, 37–47]. They found that the activities and stabilities of Au/TS-1 catalysts are strongly dependent on the Au and Ti loading, where the dilute Au and Ti species are favorable for obtaining highly active and stable Au/TS-1 catalysts. The TS-1 with different crystal size at 170 and 519 nm supported Au nanoparticles have similar reaction rates, indicating that the active Au-Ti sites for the reaction are not exclusively the Au nanoparticle-external Ti sites.

Apart from propylene, TS-1 catalyst was also used in the epoxidation of 1-hexene to epoxide product. Fan et al. [48] reported that poisoning the external Ti sites of TS-1 could not decrease the catalytic activity in 1-hexene oxidation, indicating that the internal Ti atoms are the active sites for this reaction, because the 1-hexene molecule with linear structure could transfer in the micropores of TS-1 zeolite. An important feature for the highly catalytic activity of TS-1 zeolite is the hydrophobicity, which is favorable for adsorbing and enriching the organic substrate. According to the rate equation for the chemical reaction over various reactants, increasing concentration of one reactant around the active sites (*e.g.* the Ti sites on TS-1) under the premise of similar concentration of other reactants should improve reaction rates.

Recently, considering that the first step in the oxidations with H_2O_2 over TS-1 zeolite is the activation of H_2O_2 by Ti species [49–52], it is proposed that the preparation of TS-1 zeolite with hydrophilic hydroxyl groups for enriching the reactants of H_2O_2 could also significantly enhance the catalytic activities. As example, Wang et al. [53] rationally synthesized TS-1 zeolite with controllable hydroxyl groups by using dimethyldimethoxysilane in the starting gel. Interestingly, the hydroxyl-rich TS-1 zeolite synthesized from suitable dimethyldimethoxysilane amount exhibits a significantly higher reaction rate at 4.9 mmol L^{-1} min^{-1} than that (0.61 mmol L^{-1} min^{-1}) of conventional TS-1 zeolite in the oxidation of 1-hexene with an extremely low concentration of H_2O_2 (0.083 mol L^{-1}). This phenomenon is strongly related to the hydrophilicity of hydroxyl-rich TS-1 zeolite, which could efficiently adsorb the H_2O_2 reactant to form high H_2O_2 concentration in the zeolite micropores (Fig. 11.2).

Notably, in the oxidation of propylene and 1-hexene by H_2O_2, the TS-1 zeolite generally exhibited high selectivity to epoxide product. However, in the oxidation of styrene by H_2O_2, TS-1 gives only a small amount of epoxide product and phenylacetaldehyde is a major product, which is related to synergistic effect of water and TiOOH species [54–58]. In this case, the TiOOH species interacts with H_2O to form a five-membered cyclic structure, which could generate acid site (Fig. 11.3) that is favorable for the formation of phenylacetaldehyde. To solve this

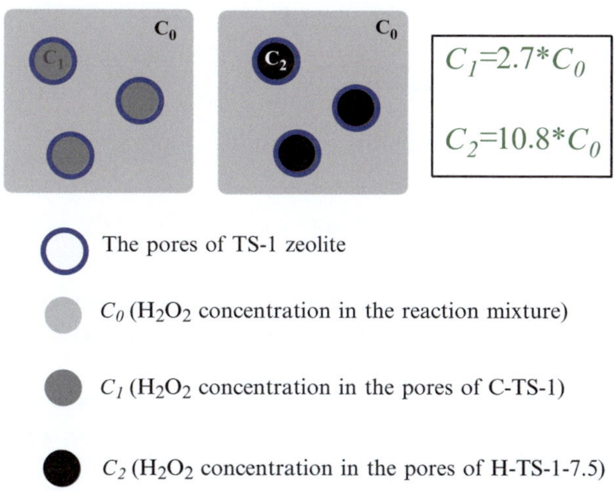

Fig. 11.2 Enrichment of H_2O_2 concentration in conventional TS-1 (C-TS-1) and hydrophilic TS-1 (H-TS-1-7.5) catalysts. The H_2O_2 concentration in the reaction mixture (C_0) was measured to be 1.61 mol L^{-1} by the $KMnO_4$ titration method. The H_2O_2 concentrations in C-TS-1 ($C_1 \approx 4.35$ mol L^{-1}) and H-TS-1-7.5 ($C_2 \approx 17.4$ mol L^{-1}) were calculated by the pore volume of TS-1 catalysts and the change of H_2O_2 concentration in the mixture before and after adding the solid samples (Reproduced from Ref. [53], by permission of The Royal Society of Chemistry)

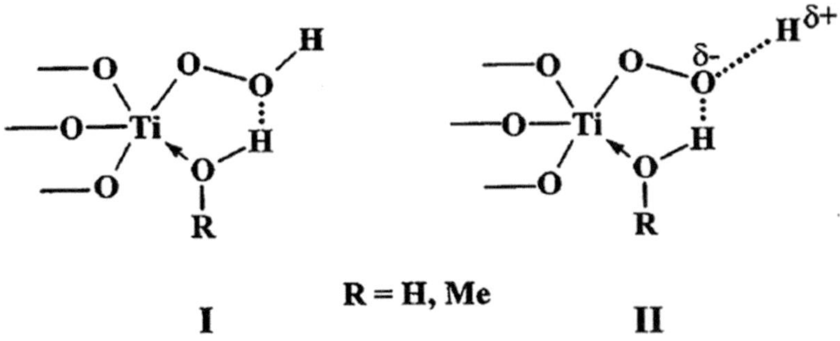

Fig. 11.3 A stable five-membered cyclic structure with a donor hydroxyl moiety coordinated on Ti-OOH generated from Ti site and H_2O_2 (species I). The enhanced acid strength may be generated in charge-separated species (species II) due to greater hydrogen bonding (Reprinted from Ref. [57], Copyright 2001, with permission from Elsevier)

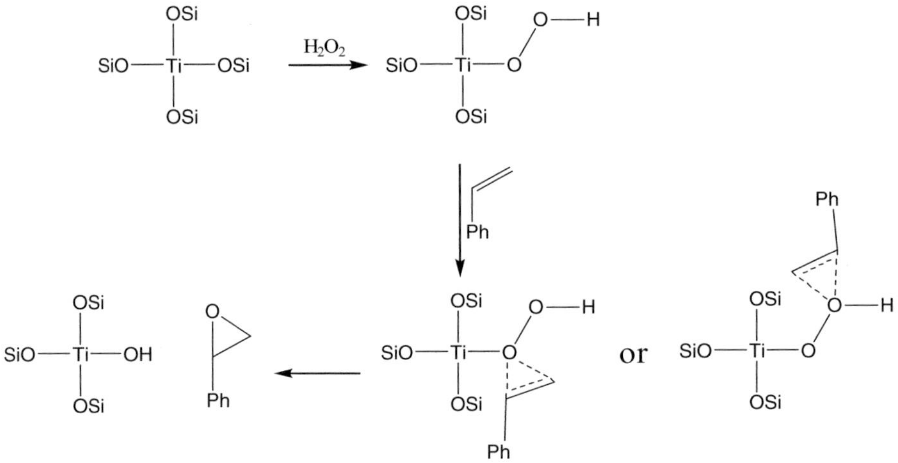

Fig. 11.4 The proposed mechanism for the formation of TiOOH and styrene epoxide

issue, Laha et al. [57] used urea to neutralize the acid site formed in the oxidation, remarkably improving epoxide selectivity in the epoxidation of styrene (85 %), compared to that (5–15 %) in the conventional system.

With *in situ* nuclear magnetic resonance (NMR) spectroscopy, Zhuang et al. [49] systematically studied that the mechanisms on the oxidation of styrene with urea-H2O2 over TS-1 zeolite. It is found the Ti-superoxo radical (TiOO·) is responsible for the formation of hemiacetal species, which is stable on the TS-1 framework in the absence of any Brønsted acid species (Fig. 11.4). In contrast, the presence of Brønsted acid species could catalyze the transformation of the hemiacetal species to phenylacetaldehyde. By combining with the *in situ* electron paramagnetic resonance (EPR) technique, it is definitely indicated that the formation of phenylacetaldehyde and styrene epoxide are directly from the oxidation of styrene as two competitive processes.

Although TS-1 is very active for the oxidation of propylene, 1-hexene, and styrene, it is almost inactive for the oxidation of bulky olefins with molecules such as cyclohexene and cyclooctene, because the mass transfer of the bulky molecules are hindered by the relatively small micropores (~0.56 nm). Therefore, to improve the catalytic performances of TS-1 in the oxidation of bulky olefins, Cundy et al. [59] synthesized nanosized TS-1 crystals (166 nm) using a colloidal dispersion of 43 nm crystals of TPA-silicalite-1 as seeds. By adjusting these conditions, Wang et al. [60] synthesized TS-1 with crystal size at 100 nm. As expected, the nanosized TS-1 crystals are highly active in the oxidation of various bulky olefins. Furthermore, Schmidt et al. [61] synthesized hierarchical TS-1 with both zeolitic micropores and mesopores by using carbon black as a hard template in the starting gel for synthesizing zeolite. In the oxidation of bulky cyclohexene molecule with H_2O_2, the hierarchical TS-1 exhibited nearly ten times higher cyclohexene conversion than conventional TS-1; Serrano et al. [62] synthesized hierarchical TS-1 with rich supermicro/mesopores by a unique method using silanization agent functionalized zeolitic seeds, which could partially hinder the zeolite growth and create rich defects in the crystals. Interestingly, in the epoxidation of 1-octene with *tert*-butyl hydroperoxide (TBHP), the hierarchical TS-1 exhibits 1-octene conversion at 42.1 %, which is much higher than that at 5.6 % of conventional TS-1.

More recently, Wang et al. [63] synthesized a lamellar TS-1 by employing a bifunctional surfactant as the structure-directing agent. The obtained lamellar TS-1has a multilayer structure constructed from a collection of ~2-nm zeolite nanosheets in thickness and ~3.2-nm mesopores between the nanosheets (Fig. 11.5). In the oxidation of various bulky olefins (*e.g.* cyclohexene, cycloheptene, cyclooctene, and cyclododecene) with TBHP, and in the oxidation of small olefins (*e.g.* propylene and 1-hexene) with bulky cumene hydroperoxide, the lamellar TS-1 always exhibits superior catalytic performances compared with the conventional TS-1. For example, in the epoxidation of cyclohexene with TBHP, the lamellar TS-1 gives cyclohexene conversion at 16.7 %, while conventional TS-1 is nearly inactive; in the epoxidation of propylene with bulky cumene hydroperoxide, the lamellar TS-1 exhibited propylene conversion at 20.0 %, while the conventional TS-1 showed the conversion at only 0.8 %. The unique catalytic behavior of the lamellar TS-1 is attributed to the mesoporosity and ultrathin crystal morphology, which improves the accessibility of the Ti sites to the bulky substrate molecules.

11.2.3 Ammoxidation Over TS-1 Zeolite

The oximes are important intermediates in the synthesis of lactam products, such as ε-caprolactam, which has been widely used in the production of degradable plastics [64, 65]. Conventionally, the synthesis of lactam products from ketone ammoxidation involves in several steps, giving very low selectivity to the oxime products. This synthesis is costly and environmentally unfriendly [66]. In 1989, Martin et al. [67] found that TS-1 could catalyze the ammoxidation of ketones with H_2O_2 and NH_3 with very high selectivity for oxime products under mild conditions. During this

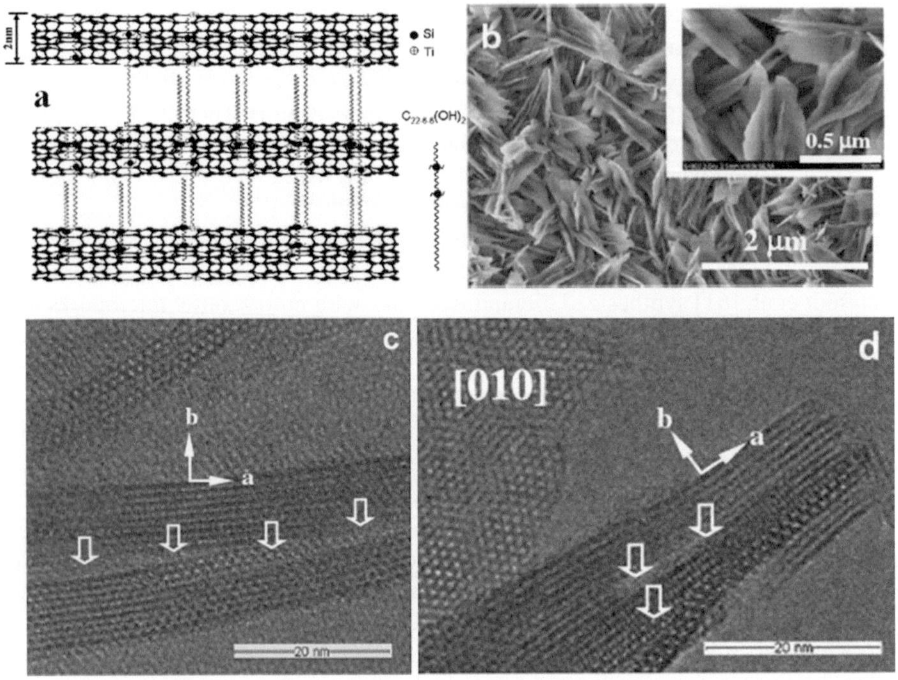

Fig. 11.5 (**a**) Lamellar MFI titanosilicate directed by bifunctional structure-directing agent; (**b**) SEM and (**c** and **d**) TEM images of lamellar TS-1 (Reprinted from Ref. [63], Copyright, 2012, with permissions from Elsevier)

reaction, several parameters could strongly influence the yields of ε-caprolactam, including the suitable Ti content, reaction temperature, NH_3/H_2O_2 molar ratio, high catalyst concentration, method of H_2O_2 addition, and solvent. By adjusting these reaction conditions, high conversions (>94 %) and excellent selectivities (>99.0 %) could be obtained [67].

Additionally, TS-1 could also catalyze the full conversion of methyl ethyl ketone by ammoxidation in *t*-BuOH/water solvent. However, the selectivity to methyl ethyl ketone oxime is only 80.0 % with 2-nitrobutane as byproduct [68]. This is due to the over-oxidation of methyl ethyl ketone oxime product.

11.2.4 Oxidation of Alkanes Over Ti-Rich TS-1 (YNU) Zeolite

Because of industrial importance of TS-1 zeolite, improving the catalytic activity of TS-1 has paid much attention for a long time [69–72]. To increase the TS-1 activities, it is suggested to enhance the content of tetrahedral Ti sites in the TS-1 framework. Fan et al. [73] developed a simple method to synthesize tetrahedral Ti-rich TS-1 (denoted as YNU, the Si/Ti ratio could reach as low as 34, Fig. 11.6) using $(NH_4)_2CO_3$ as a crystallization mediating agent. In contrast, the Si/Ti ratio in TS-1 synthesized

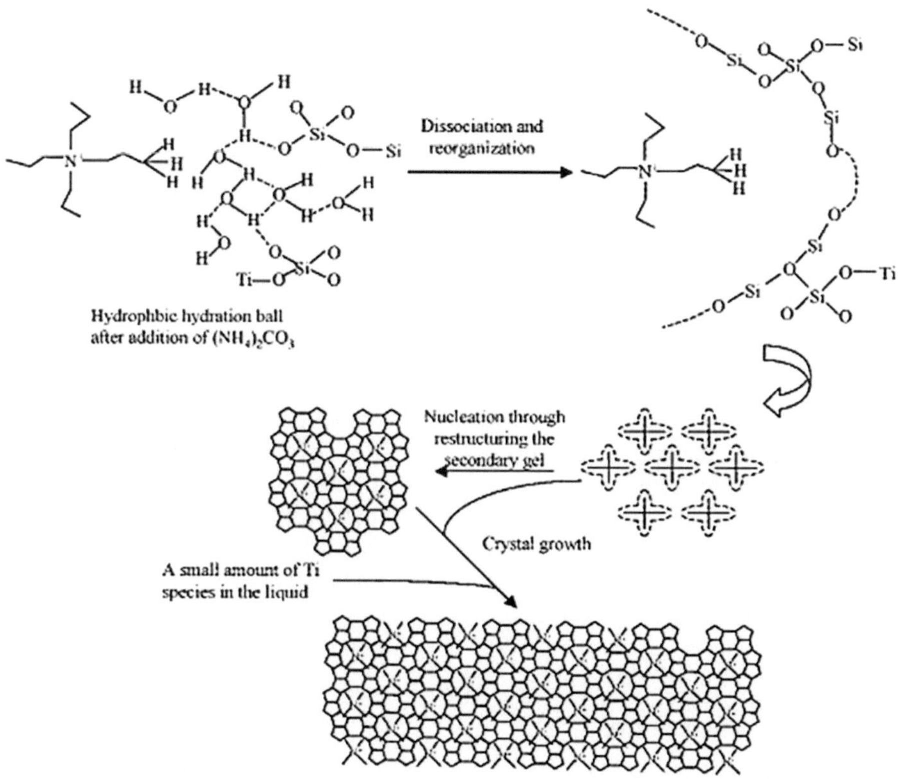

Fig. 11.6 The proposed mechanism for the formation of Ti-rich TS-1 under YNU conditions (Reprinted with permission from Ref. [73], Copyright 2008, American Chemical Society)

in conventional system by Enichem Co. could reach only 58 [74]. In the oxidation of *n*-hexane, the YNU catalyst with Si/Ti ratio at 34.2 exhibited *n*-hexane conversion at 50.8 %, while conventional TS-1 with Si/Ti ratio at 58.3 gave *n-hexane* conversion at only 30.6 %. This phenomenon should be mainly owing to the high content of active Ti sites on the YNU catalyst. Even with similar Si/Ti ratios, the YNU catalysts are still slightly more active than the conversional TS-1 catalyst. This result should be attributed to less defect sites and hydroxyl groups in the YNU catalyst than the conventional TS-1, leading to strong hydrophobicity on YNU. This feature is favorable for the catalytic conversion of organic compounds.

11.3 TS-2 Zeolite

Since the discovery of TS-1 zeolite, it has become an interesting topic to synthesize more novel titanosilicates with the expectation of unique catalytic performances. TS-2 with MEL structure is a typical example, which is synthesized from similar procedures to TS-1 but using tetrabutylammonium as a template [17, 75–81].

11.3.1 Hydroxylation of Phenol Over TS-2 Zeolite

Tuel et al. [75] carefully compared the catalytic properties of TS-1 and TS-2 zeolite in the hydroxylation of phenol with H_2O_2, and it was found that TS-2 (~500 nm) is less active than TS-1 (<300 nm) due to the difficulty for the synthesis of smaller TS-2 crystals. Because TS-2 and TS-1 have very similar framework structure, the difference in the catalytic activities from the structure contribution is not observed.

11.3.2 Epoxidation of Olefins Over TS-2 Zeolite

Compared with the hydroxylation of phenol, which is sensitive with the crystal size of titanosilicate, the epoxidation of olefins is not so strongly related to the crystal sizes ranged at 200–500 nm. Therefore, it is expected that the TS-2 zeolite could be effective in the epoxidation of olefins. Laha et al. [82] reported that TS-2 zeolite is highly efficient in the epoxidation of various olefins with H_2O_2 oxidant. The mechanism on the reaction monitored by UV-visible and EPR spectra is in good agreement with that of TS-1.

11.4 Ti-Beta Zeolite

TS-1 zeolite has already been confirmed to be a great breakthrough in oxidations. However, both TS-1 and TS-2 cannot catalyze the oxidations of bulky molecules, because of their narrow 10-MR channels. To solve this limitation, Camblor et al. [83] incorporated Ti atoms into the 12-MR channels of Beta zeolite, obtaining a novel titanosilicate, denoted as Ti-Beta. Due to the advantages of high stability, controllable wettability, and relatively large micropores, Ti-Beta exhibited excellent catalytic properties in the oxidation of bulky molecules of olefins, alkanes, and sulfur-containing compounds.

11.4.1 Oxidation of Olefins and Alkanes Over Ti-Beta Zeolite

Corma et al. [84] firstly reported the oxidation of various olefins and alkanes over Ti-Beta zeolite. For small molecules without steric restrictions to diffuse inside the pores of Ti-Beta and TS-1, the TS-1 catalyst is generally more active than Ti-Beta. For example, in the oxidation of 1-hexene and *n*-hexane, TS-1 gave turnover numbers (TONs) at 50 and 48.5, which are higher than those of 12 and 0.5 over Ti-Beta, respectively. However, for relatively bulky molecules with diffusion problems on TS-1, Ti-Beta is more active than TS-1. For example, TS-1 is inactive for the oxidation of methylcyclohexane, while Ti-Beta is very active, giving TONs at 20.

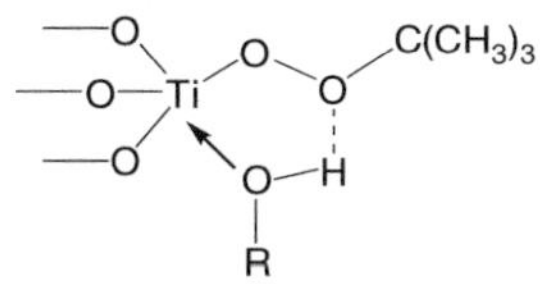

Fig. 11.7 Ti-OO-C$(CH_3)_3$ intermediate generated from interaction between Ti sites and TBHP

Later, they systemically studied the oxidation of a series of linear, branched, and cyclic olefins catalyzed by Ti-Beta zeolite with TBHP or H_2O_2 as oxidants. It is proposed that both olefin and oxidant structures strongly influence the catalytic results [85]. For example, when the TBHP was used, the steric factor is pronounced, because of the relatively bulky Ti-OO-C$(CH_3)_3$ intermediate, which would be formed in the micropores of titanosilicate zeolite (Fig. 11.7). As a result, the activity in olefin oxidation with TBHP is lower than that with H_2O_2 over Ti-Beta.

It is worth noting that the Ti-Beta zeolite exhibited higher epoxide selectivity with TBHP oxidant than that with H_2O_2, owing to less acidity of Ti-OO-C$(CH_3)_3$ intermediate, where the acid sites could catalyze the side-reaction to form ketone by-products.

Furthermore, they found that aluminum species in the Ti-Beta would strongly reduce the epoxide selectivity in the oxidation of 1-hexene. Comblor et al. [7] found that when the Si/Al ratio was increased to as high as 700, the selectivity to epoxide could be effectively improved. Based on this principle, Blasco et al. [86] further synthesized Al-free Ti-Beta zeolite in F^- medium, which exhibited 1-hexene epoxide selectivity as high as 76.6 %.

In addition to the epoxidation of olefins, Ti-Beta also works efficiently in the epoxidation of chiral allylic alcohols, giving excellent diastereoselectivities [87].

11.4.2 Oxidation of Organic Sulfides Over Ti-Beta Zeolite

Sulfur in transportation fuels is a major source of air pollution from automobiles. Oxidative desulfurization has been regarded as an effective method to remove the sulfur-containing compounds under mild reaction conditions. Corma et al. [88] demonstrated that Ti-Beta is more active than TS-1 and mesoporous Ti-MCM-41 in the oxidation of a methyl phenyl sulfide with TBHP, because of the combined advantages of Ti-Beta including larger micropores than TS-1 and crystalline framework with more perfect tetrahedral Ti sites than MCM-41. However, in the oxidation of larger isopentyl phenyl sulfide, which diffuses more slowly in 12-MR channel, Ti-Beta is less active than Ti-MCM-41.

11.5 Ti-MWW Zeolite

Although many titanosilicate with 12-membered (12-M) ring (*e.g.* Ti-Beta, Ti-MOR, and Ti-ITQ-7) have been synthesized and applied in the oxidation of bulky molecules, it is worth mentioning that these materials still do not satisfy the demands from various oxidations in fine chemical synthesis and petrochemical industry. Therefore, a new type of titanosilicate zeolites with more accessibility is required. A new type of zeolite with MWW topology has a unique structure with two-dimensional sinusoidal channels of 10-MR running throughout the framework, presenting unique pore system with large supercages (0.7*0.7*1.8 nm) in the crystal and pocket (0.7*0.7 nm) at the external surface [12], which act as an open space for the conversion of bulky molecules.

Incorporation of Ti atoms into the MWW framework has attracted much attention in order to synthesize MWW-type titanosilicate. For example, Corma and coworkers [89] grafted titanocene onto the surface of pure silica MCM-22 to synthesize ITQ-2; Mobil Co. [90] synthesized Ti-MCM-22 by using $TiCl_4$ to introduce Ti atoms into the framework of dealuminated MCM-22 to get Ti-MCM-22. However, these titanosilicates from post-synthesis exhibited the disadvantage of Ti leaching during the liquid phase reaction. Therefore, one-pot synthesis of Ti-MWW is strongly desirable. Wu et al. [12] firstly reported directly hydrothermal synthesis of Ti-MWW under base-free conditions using piperidine or hexamethyleneimine as a structure-directing agent and using boric acid as a structure-supporting agent. Since then, the synthesis and catalytic application of Ti-MWW have attracted much interest.

11.5.1 Oxidation of Olefins Over Ti-MWW Zeolite

Because of the 12-MR pocket structure on the surface of MWW topology, the Ti-MWW exhibited superior catalytic activities in the oxidation of olefins with bulky molecule sizes, compared with TS-1 and Ti-Beta zeolites. Wu et al. [91] extensively studied the catalytic properties of three-dimension (3D) Ti-MWW and Del-Ti-MWW from delamination of 3D Ti-MWW (Fig. 11.8). The Del-Ti-MWW showed much higher catalytic activity than 3D Ti-MWW. For example, Del-Ti-MWW gives the turnover number (TON) values as high as 300 and 57 in the oxidation of cyclopentene and cyclododecene, which are much higher than those at 89 and 9 over 3D Ti-MWW catalyst. This phenomenon is attributed to higher external surface area of Del-Ti-MWW (421 m^2/g) than 3D Ti-MWW (163 m^2/g). Particularly, it is noted that the Del-Ti-MWW exhibits higher catalytic activities than 3D Ti-MWW even in the oxidations of small molecules such as 1-hexene and 2-hexene. This phenomenon is due to that the Del-Ti-MWW would shorten the length of 10-MR channels, making more rapid transfer of substrates/products in the channels. Very interestingly, in the epoxidation of 2-hexene, the Ti-MWW based catalysts exhibit much higher selectivity to *trans* product (81–84 %) than TS-1 and Ti-Beta catalysts (35–36 %), which is because that the unique sinusoidal channels of Ti-MWW are favorable for the accessibility of *trans* molecules to Ti sites [91].

Notably, the synthesis of Del-Ti-MWW requires expensive reactants (cetyl trimethyl ammonium salts and tetrapropyl ammonium hydroxide), which strongly hinders its scale-up synthesis. To overcome this drawback, Wang et al. [92] developed a new method by silylation of Ti-MWW lamellar precursor and subsequent calcination to

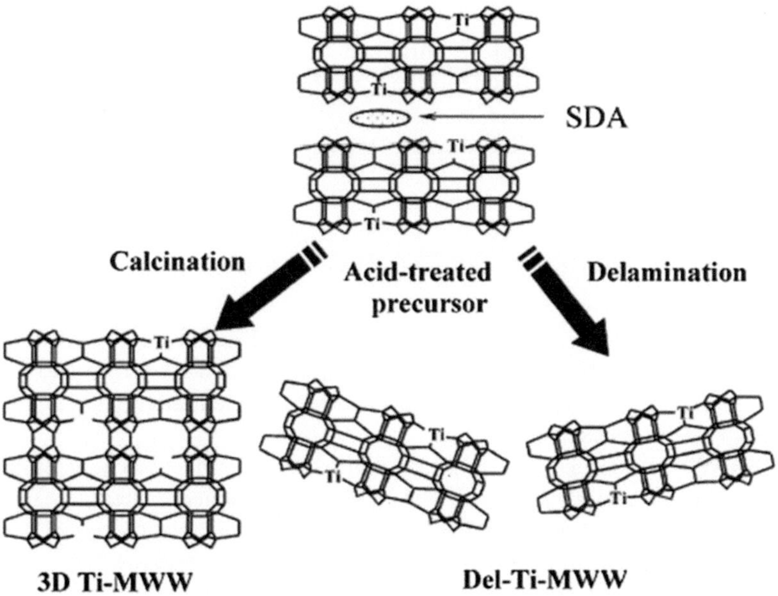

Fig. 11.8 Possible structures of 3D Ti-MWW and Del-Ti-MWW obtained from acid-treated precursor (Reprinted with permission from Ref. [91], Copyright 2004, American Chemical Society)

construct a novel interlayer expanded structure. In this case, the diethoxydimethylsilane, trimethylethoxysilane, and triethoxymethylsilane were employed as silylating agents, whose methyl groups inhibited the intermolecular condensation of silanes effectively, leading to the formation of interlayer expanded Ti-MWW zeolite (IEZ-Ti-MWW) with rich large 12-MR pores. In the epoxidation of bulky cyclohexene, IEZ-Ti-MWW always exhibited higher catalytic activities than 3D Ti-MWW. For example, with suitable Ti content (Ti/Si+Ti at 0.010~0.015), the conversion of cyclohexene could reach more than 20 % over IEZ-Ti-MWW with rich large pores of 12-MR, while the cyclohexene conversion is no more than 5 % over 3D Ti-MWW with only medium pores of 10-MR.

The Ti-MWW zeolite also works well in the epoxidation of functionalized olefins with H_2O_2 [93], such as the epoxidation of allyl alcohol, an important reaction to produce glycidol [50, 94, 95]. Conventionally, the allyl alcohol oxidations were performed over tungstate-based catalysts, which suffer from problems in activity and recyclability. Additionally, the TS-1 zeolite has acceptable catalytic activity and recyclability in the allyl alcohol epoxidation. However, its selectivity to glycidol is extremely poor, because the acidity of TS-1 could catalyze the further hydrolysis or solvolysis of glycidol to glycerol and others. Controlling the crystal sizes and poisoning the acids sites have been employed to enhance the glycidol selectivity, but both approaches remarkably reduces the activities [96–99]. Interestingly, it was found that the Ti-MWW could catalyze the epoxidation of allyl alcohol more efficiently than TS-1 and Ti-Beta catalysts, where the allyl alcohol conversion and glycidol could reach as high as 92.6 and 99.5 %, respectively. The excellent catalytic properties of Ti-MWW should be attributed to the sinusoidal 10-MR channels, which make the Ti sites accessible to the C=C bonds of allyl alcohol. In addition, the hydrophobic environment of Ti sites on Ti-MWW is simultaneously important for improving the catalytic activity and selectivity.

11.5.2 Ammoxidation Over Ti-MWW Zeolite

Although the ammoxidation of cyclohexone over TS-1 have been regarded to be an environmentally friendly process, the use of expensive TS-1 catalyst and *t*-BuOH solvent are costly. Therefore, development of more efficient ammoxidation process in aqueous media is desirable.

Ti-MWW has been confirmed to be highly active, selective, and reusable in ammoxidation of cyclohexene with NH_3 and H_2O_2 in aqueous media [70], giving high cyclohexone conversion (97.0–99.4 %) and oxime selectivity (99.9 %). In contrast, TS-1 exhibits relatively low cyclohexone conversion at 16.2 % and oxime selectivity at 72.8 %. The origin of this phenomenon is similar to that in the oxidation of olefins, which is explained by that the unique pore system of Ti-MWW is favorable for the access of substrate molecules to Ti sites. In this reaction, the dropwise addition of H_2O_2 into the reaction system, suitable reaction temperature, suitable catalyst/cyclohexone/H_2O_2/NH_3 ratios, and Ti content could significantly influence the conversion of cyclohexone. Very importantly, Ti-MWW catalyst is stable and recyclable in the ammoxidation of cyclohexone. Apart from the ammoxidation of cyclohexone, Ti-MWW is also active and selective in the ammoxidation of *p*-tolualdehyde and benzaldehyde, giving substrate conversion at 42.4–55.6 % and oxime selectivity at 99.2–99.7 %.

It is well known that the ammoxidation of methyl ethyl ketone is a more challengeable reaction than ammoxidation of cyclohexone, because of the easy formation of byproduct of 2-nitrobutane. Ti-MWW is highly efficient in this reaction, giving methyl ethyl ketone conversion and oxime selectivity at >99.0 % in the presence of water, where the water could efficiently hinder the side-reactions, and the unique hydrophobic/hydrophilic environment of Ti sites on Ti-MWW could promote the conversion of methyl ethyl ketone [68]. These results indicate that the Ti-MWW is an ideal "green" catalyst for the production of oximes.

11.5.3 Oxidative Desulfurization Over Ti-MCM-56

In the oxidative desulfurization of dibenzothiophene, TS-1 exhibits low activity with dibenzothiophene at only ~5 % due to the limitation of its small micropores. The Ti-Beta and Ti-MWW with relatively large pores are active for this reaction (conversion at 83 %), but it still does not satisfy the requirement of industrial applications. Therefore, developing alternative titanosilicates with large external surface and rich exposed Ti sites to dibenzothiophene molecules still has a challenge.

Wang et al. [100] developed a post-synthesis methodology to prepare a titanosilicate catalyst from acid-treating the 3D Ti-MWW precursors at low temperatures, leading to high crystalline Ti-MCM-56 analogues after calcination. Compared with 3D Ti-MWW, the Ti-MCM-56 analogues has a much larger external surface, which mitigated effectively the steric restrictions to bulky molecules. As expected, in the oxidative desulfurization, Ti-MCM-56 analogues give dibenzothiophene conversion as high as 93.0 %. This work creates a simple route to synthesize novel titanosilicates for the conversion of bulky molecules.

11.6 Summary and Perspective

In this chapter, a series of titanosilicates and their catalytic applications in various oxidations have been described. Notably, there are still challenges for applying these catalysts. For example, TS-1, although it has been applied in industry successfully, is one of the most expensive zeolites because of costly TPA^+ template; the synthesis of Ti-Beta use relatively cheap TEA^+ template, but crystallization for a long time in F^- medium is still an environmentally unfriendly process. Therefore, developing novel green routes to synthesize new microporous titanosilicate zeolites with excellent catalytic properties is still on its way to industrial application.

References

1. Taramasso M, Perego G et al (1983) Preparation of porous crystalline synthetic material comprised of silicon and titanium oxides. US Patent 4,410,501
2. Thangaraj A, Kumar R et al (1991) Catalytic properties of caystalline titanium silicalites 1. Synthesis and characterization of titanium-rich zeolites with MFI structure. J Catal 130:1–8
3. Corma A (2003) State of the art and future challenges of zeolites as catalysts. J Catal 216:298–312
4. Millini R, Massara E et al (1992) Framework composition of titanium silicate-1. J Catal 137:497–503
5. Thangaraj A, Sivasanker S et al (1991) Catakytic properties of crystalline titanium silicates. 3. ammoximation of cyclohexone. J Catal 131:394–400
6. Camblor M, Corma A et al (1993) Synthesis of titanoaluminosilicates isomorphous to zeolite Beta, active as oxidation catalysts. Zeolite 13:82–87
7. Camblor M, Constantini M et al (1996) Synthesis and catalytic activity of aluminium-free zeolite Ti-beta oxidation catalysts. Chem Commun 11:1339–1440
8. Blasco T, Camblor M et al (1993) The state of Ti in titanoaluminosilicates isomorphous with zeolite-Beta. J Am Chem Soc 115:11806–11813
9. Corma A, Esteve P et al (1995) Oxidation of olefins with hydrogen-peroxide and *tert*-butyl hydroperoxide on Ti-Beta catalyst. J Catal 152:18–24
10. Wu P, Komatsu T et al (1996) Characterization of titanium species incorporated into dealuminated mordenites by means of IR soectroscopy and O-18-exchange technique. J Phys Chem 100:10316–10322

11. Diaz-Cabanas M, Villaescusa LA et al (2000) Synthesis and catalytic activity of Ti-ITQ-7: a new oxidation catalyst with a three-dimensional system of large pore channels. Chem Commun 9:761–762
12. Wu P, Tatsumi T et al (2001) A novel titanosilicate with MWW structure. I. Hydrothermal synthesis, elimination of extraframework titanium, and characterizations. J Phys Chem B 105:2897–2905
13. Wu P, Tatsumi T et al (2001) A novel titanosilicate with MWW structure: II. Catalytic properties in the selective oxidation of alkenes. J Catal 202:245–255
14. Wu P, Liu Y et al (2004) A novel titanosilicate with MWW structure: catalytic properties in selective epoxidation of diallyl ether with hydrogen peroxide. J Catal 228:183–191
15. Kubota Y, Koyama Y et al (2008) Synthesis and catalytic performance of Ti-MCM-68 for effective oxidation reactions. Chem Commun 46:6224–6226
16. Thangaraj A, Kumar R et al (1991) Catalytic properties of crystalline titanium silicates. 2. Hydroxylation of phenol with hydrogen-peroxide over TS-1 zeolites. J Catal 131:294–297
17. Tuel A, Moussa-Khouzami et al (1991) Hydroxylation of phenol over TS-1: surface and solvent effects. J Mol Catal 68:45–52
18. Bhaumik A, Mukherjee P et al (1998) Triphase catalysis over titanium–silicate molecular sieves under solvent-free conditions: I. Direct hydroxylation of benzene. J Catal 178:101–107
19. Thangaraj A, Kumar R et al (1990) Direct catalytic hydroxylation of benzene with hydrogen–peroxide over titanium silicate zeolites. Appl Catal 57:L1–L3
20. Bhaumik A, Kumar R et al (1995) Titanium silicate molecular sieve (TS-1)/H2O2 induced triphase catalysis in the oxidation of hydrophobic organic compounds with significant enhancement of activity and Para-selectivity. J Chem Soc Chem Commun 3:349–350
21. Bianchi D, Bortolo R et al (2004) Preparing phenol for use as intermediate, e.g. in production of polycarbonates, by synthesis process of phenol with hydrodeoxygenation section of by-products which are selectively transformed into phenol and recycled in process streams. US Patent 7038093-B2
22. Bianchi D, Bortolo R et al (2006) Continuous process to prepare phenol involves feeding to fixed bed reactor containing catalyst based on titanium silicalite, stream having hydrogen peroxide, benzene, sulfolane and water at specific operative conditions. US Patent 2010184593-A1
23. Bianchi D, D'Aloisio R et al (2007) Oxidation of mono- and bicyclic aromatic compounds with hydrogen peroxide catalyzed by titanium silicalites TS-1 and TS-1B. Appl Catal A Gen 327:295–299
24. Bianchi D, Balducci L et al (2007) Oxidation of benzene to phenol with hydrogen peroxide catalyzed by a modified titanium silicalite (TS-1B). Adv Synth Catal 349:979–986
25. Xiao F-S, Han Y et al (2002) Hydrothermally stable ordered mesoporous titanosilicates with highly active catalytic sites. J Am Chem Soc 124:888–889
26. Xin H, Zhao J et al (2010) Enhanced catalytic oxidation by hierarchically structured TS-1 zeolite. J Phys Chem C 114:6553–6559
27. Trent DL (1996) Kirk-Othmer encyclopedia of chemical technology. Wiley, New York
28. Haruta M, Date M (2001) Advances in the catalysis of Au nanoparticles. Appl Catal A Gen 222:427–437
29. Hayashi T, Tanaka K et al (1998) Selective vapor-phase epoxidation of propylene over Au/TiO_2 catalysts in the presence of oxygen and hydrogen. J Catal 178:566–575
30. Grasselli RK, Burrington JD et al (1981) Selective oxiaiton and ammoxidation of propylene by heterogeneous catalysis. Adv Catal 30:133–163
31. Kamata K, Yonehara K et al (2003) Efficient epoxidation of olefins with >= 99% selectivity and use of hydrogen peroxide. Science 300:964–966
32. Bettahar MM, Costentin G et al (1996) On the partial oxidation of propane and propylene on mixed metal oxide catalysts. Appl Catal A Gen 145:1–48
33. Adams CR, Jennings TJ (1963) Investigation of the mechanism of catalytic oxidation of propylene to acrolein and acrylonitrile. J Catal 2:63–68
34. Reitz JN, Solomon EI (1998) Propylene oxidation on copper oxide surfaces: electronic and geometric contributions to reactivity and selectivity. J Am Chem Soc 120:11467–11678

35. Weissermel K, Arpe HJ (1993) Industrial organic chemistry. VCH, New York
36. Yap N, Andres RP et al (2004) Reactivity and stability of Au in and on TS-1 for epoxidation of propylene with H_2 and O_2. J Catal 226:156–170
37. Hayashi T, Tanaka K et al (1996) Selective partial oxidation of hydrocarbons over Au/TiO_2 catalysts. Am Chem Soc Div Petrol Chem 41:71–74
38. Haruta M, Uphadc BS et al (1998) Selective oxidation of propylene over gold deposited on titanium-based oxides. Res Chem Intermed 24:329–336
39. Kalvachev YA, Hayashi T et al (1997) Selective partial oxidation of propylene to propylene oxide on Au/Ti-MCM catalysts in the presence of hydrogen and oxygen. Stud Surf Sci Catal 110:965–972
40. Kalvachev YA, Hayashi T et al (1999) Vapor-phase selective oxidation of aliphatic hydrocarbons over gold deposited on mesoporous titanium silicates in the co-presence of oxygen and hydrogen. J Catal 186:228–233
41. Uphade BS, Tsubota S et al (1998) Selective oxidation of propylene to propylene oxide or propionaldehyde over au supported on titanosilicates in the presence of H_2 and O_2. Chem Lett 12:1277–1278
42. Uphade BS, Okumura M et al (2000) Effect of physical mixing of CsCl with Au/Ti-MCM-41 on the gas-phase epoxidation of propene using H_2 and O_2: drastic depression of H_2 consumption. Appl Catal A Gen 190:43–50
43. Uphade BS, Okumura M et al (2000) Vapor-phase epoxidation of propene using H_2 and O_2 over Au/Ti-MCM-41 and Au/Ti-MCM-48. Stud Surf Sci Catal 130:833–838
44. Uphade BS, Akita T et al (2002) Vapor-phase epoxidation of propene using H_2 and O_2 over Au/Ti–MCM-48. J Catal 209:331–340
45. Stangland EE, Stavens KB et al (2000) Characterization of gold–titania catalysts via oxidation of propylene to propylene oxide. J Catal 191: 332–347
46. Stangland EE, Stavens KB et al (2000) Propylene epoxidation over gold-titania catalysts. Stud Surf Sci Catal 130: 827–832
47. Haruta M, Yamada N et al (1989) Gold catalysts prepared by coprecipitation for low-temperature oxidation of hydrogen and of carbon monoxide. J Catal 115:301–309
48. Fan W, Wu P et al (2008) Unique solvent effect of microporous crystalline titanosilicates in the oxidation of 1-hexene and cyclohexene. J Catal 256:62–73
49. Zhuang J, Yang G et al (2004) In situ magnetic resonance investigation of styrene oxidation over TS-1 zeolites. Angew Chem Int Ed 43:6377–6381
50. Clerici M, Ingallina P (1993) Epoxidation of lower olefins with hydrogen peroxide and titanium silicalite. J Catal 140:71–83
51. Bordiga S, Damin A et al (2002) The structure of the peroxo species in the TS-1 catalyst as investigated by resonant Raman spectroscopy. Angew Chem Int Ed 41:4734–4737
52. Notari B (1996) Microporous crystalline titanium silicates. Adv Catal 41:253–334
53. Wang L, Sun J et al (2014) A significant enhancement of catalytic activities in oxidation with H_2O_2 over the TS-1 zeolite by adjusting the catalyst wettability. Chem Commun 50:2012–2014
54. Lin W, Frei H (2002) Photochemical and FT-IR probing of the active site of hydrogen peroxide in Ti silicalite sieve. J Am Chem Soc 124:9292–9298
55. Kumar S, Mirajkar S et al (1995) Epoxidation of styrene over a titanium silicate molecular sieve TS-1 using dilute H_2O_2 as oxidizing agent. J Catal 156:163–166
56. Uguina M, Serrano D et al (2000) Preliminary study on the TS-1 deactivation during styrene oxidation with H_2O_2. Catal Today 61:263–270
57. Laha SL, Kumar R (2001) Selective epoxidation of styrene to styrene oxide over TS-1 using urea–hydrogen peroxide as oxidizing agent. J Catal 2001(204):64–70
58. Bellussi G, Carati A, Clerici MG, Maddinelli G, Millini R (1992) J Catal 133:220
59. Cundy CS, Forrest JO et al (2003) Some observations on the preparation and properties of colloidal silicalites. Part I: synthesis of colloidal silicalite-1 and titanosilicalite-1 (TS-1). Microporous Mesoporous Mater 66:143–156
60. Wang X, Guo X et al (2014) Quick synthesis of nano-scale TS-1 and its catalytic properties. Stud Surf Sci Technol 154:2589–2595

61. Schmidt I, Krogh A et al (2000) Catalytic epoxidation of alkenes with hydrogen peroxide over first mesoporous titanium-containing zeolite. Chem Commun 21:2157–2158
62. Serrano D, Sanz R et al (2009) Turning TS-1 zeolite into a highly active catalyst for olefin epoxidation with organic hydroperoxides. Chem Commun 11:1407–1409
63. Wang J, Xu L et al (2012) Multilayer structured MFI-type titanosilicate: synthesis and catalytic properties in selective epoxidation of bulky molecules. J Catal 288:16–23
64. Ichihashi H, Sato H (2001) The development of new heterogeneous catalytic processes for the production of ε-caprolactam. Appl Catal A Gen 221:359–366
65. Armor JN (1981) Ammoximation: I. A direct route to cyclohexanone oxime and caprolactam from NH_3, O_2, and cyclohexanone. J Catal 70:72–83
66. Forni L, Stanga M (1979) Pyridines by propylene ammoxidation over Te oxide/silica-alumina. J Catal 59:148–151
67. Martin A, Lucke B et al (1989) Ammoxidation of picolines on vanadium phosphate catalysts. Appl Catal 49:205–211
68. Song F, Liu Y et al (2007) Highly selective synthesis of methyl ethyl ketone oxime through ammoximation over Ti-MWW. Appl Catal A Gen 327:22–31
69. Wu P, Komatsu T et al (1997) Ammoximation of ketones over titanium mordenite. J Catal 168:400–411
70. Song F, Liu Y et al (2006) A novel titanosilicate with MWW structure: highly effective liquid-phase ammoximation of cyclohexanone. J Catal 237:359–367
71. Roffia P, Leofanti G et al (1990) Cyclohexanone ammoximation: a break through in the 6-caprolactam production. Process Stud Surf Sci Catal 55:43–50
72. Zecchina A, Spoto G et al (1993) Ammoximation of cyclohexanone on titanium silicalite: investigation of the reaction mechanism. Stud Surf Sci Catal 75:719–729
73. Fan WB, Duan R-G et al (2008) Synthesis, crystallization mechanism, and catalytic properties of titanium-rich TS-1 free of extraframework titanium species. J Am Chem Soc 130:10150–10164
74. Clerici MG, Bellussi G et al (1991) Synthesis of propylene oxide from propylene and hydrogen peroxide catalyzed by titanium silicalite. J Catal 129:159–167
75. Tuel A, Ben TY (1993) Comparison between TS-1 and TS-2 in the hydroxylation of phenol with hydrogen peroxide. Appl Catal A Gen 102:69–77
76. Pauls DL, Cohen DJ et al (1981) Familial pattern and transmission of gilles-delatourette syndrome and multiple tics. Arch Gen Psychiatry 38:1091–1093
77. Reddy JS, Kumar R et al (1990) Titanium silicalite-2: synthesis, characterization and catalytic properties. Appl Catal 58:Ll–L4
78. Reddy JS, Kumar R (1991) Synthesis, characterzation, and catalytic properties of a titanium silicate, TS-2, with mel structure. J Catal 130:440–446
79. Reddy JS, Sivasanker S et al (1992) Hydroxylation of phenol over ts-2, a titanium silicate molecular sieve. J Mol Catal 71:373–381
80. Kraushaar B, Van Hoof JHC (1988) A new method for the preparation of titanium-silicalite (TS-1). Catal Lett 1:81–84
81. Reddy JS, Sivasanker S et al (1992) Crystallization kinetics of a new titanium silicate with MEL structure (TS-2). Zeolites 12:95–100
82. Laha SC, Kumar R (2002) Highly selective epoxidation of olefinic compounds over TS-1 and TS-2 redox molecular sieves using anhydrous urea–hydrogen peroxide as oxidizing agent. J Catal 208:339–344
83. Camblor MA, Corma A et al (1992) Synthesis of a titaniumsilicoaluminate isomorphous to zeolite beta and its application as a catalyst for the selective oxidation of large organic molecules. J Chem Soc Chem Commun 8:589–590
84. Corma A, Camblor MA et al (1994) Activity of Ti-Beta catalyst for the selective oxidation of alkenes and alkanes. J Catal 145:151–158
85. Camblor MA, Costantini M et al (1996) The roots of organic development. Elsevier, Amsterdam
86. Blasco T, Camblor MA et al (1998) Direct synthesis and characterization of hydrophobic aluminum-free Ti-Beta zeolite. J Phys Chem B 102:75–88

87. Adam W, Corma A et al (1997) Diastereoselective catalytic epoxidation of chiral allylic alcohols by the TS-1 and Ti-â zeolites: evidence for a hydrogen-bonded, peroxy-type loaded complex as oxidizing species. J Org Chem 62:3631–3637
88. Corma A, Iglesia M et al (1996) Large pore Ti-zeolites and mesoporous Ti-silicalites as catalysts for selective oxidation of organic sulfides. Catal Lett 39:153–156
89. Corma A, Diaz U et al (1999) Ti/ITQ-2, a new material highly active and selective for the epoxidation of olefins with organic hydroperoxides. Chem Commun 9:779–780
90. Levin D, Chang CD et al (2000) Crystalline zeolite for separation and catalytic conversion of organic compounds. US Patent 6,114,551
91. Wu P, Nuntasri D et al (2004) Delamination of Ti-MWW and high efficiency in epoxidation of alkenes with various molecular sizes. J Phys Chem B 108:19126–19131
92. Wang L, Wang Y et al (2009) Alkoxysilylation of Ti-MWW lamellar precursors into interlayer pore-expanded titanosilicates. J Mater Chem 19:8594–8602
93. Wu P, Tatsumi T (2003) A novel titanosilicate with MWW structure III. Highly efficient and selective production of glycidol through epoxidation of allyl alcohol with H_2O_2. J Catal 214:317–326
94. Weissermel K, Arpe HJ (1994) Industrielle organische chemie. VCH, Weinheim
95. Wittcoff HA, Reuben BG (1996) Industrial organic chemicals. Wiley, New York
96. Taramasso M, Perego G et al (1983) Heat-stable modified crystalline silicic acid catalyst useful as catalyst and adsorbent. US Patent 4,410,501
97. Hutchings GJ, Lee DF (1994) Catalytic heterogeneous aziridination of alkenes using microporous materials. Chem Commun 15:1601–1602
98. Hutchings GJ, Lee DF et al (1995) Epoxdatopn of allyl alcohol to glycidol using titanium silicalite TS-1 effect of the method of preparation. Catal Lett 33:369–385
99. Hutchings GJ, Lee DF et al (1996) Epoxidation of allyl alcohol to glycidol using titanium silicalite TS-1: effect of the reaction conditions and catalyst acidity. Catal Lett 39:83–90
100. Wang L, Wang Y et al (2008) Post-transformation of MWW-type lamellar precursors into MCM-56 analogues. Microporous Mesoporous Mater 113:435–444

Chapter 12
Emerging Applications of Environmentally Friendly Zeolites in the Selective Catalytic Reduction of Nitrogen Oxides

Fudong Liu, Lijuan Xie, Xiaoyan Shi, and Hong He

Abstract The catalytic removal of nitrogen oxides (NO_x) from diesel exhaust under oxygen-rich conditions remains a major challenge in the field of environmental catalysis, and the selective catalytic reduction of NO_x with NH_3 (NH_3-SCR), especially over zeolite catalysts, is a well-proven technique for the catalytic deNO_x process for diesel vehicles. The comprehensive understanding of the structure-activity relationship of zeolite catalysts in the NH_3-SCR reaction and the elucidation of the detailed reaction mechanism are very important for the practical use of these catalytic materials. In this chapter, using the environmentally friendly Fe- and Cu-based zeolite catalysts (i.e. Fe-ZSM-5, Cu-ZSM-5, Cu-SSZ-13 and Cu-SAPO-34, etc.) as examples, the influence of preparation methods on NH_3-SCR performance, the role of metal active sites/surface acid sites, the impact of NO_2 and co-existing pollutants, the hydrothermal stability and also the possible SCR reaction pathways over these materials are discussed in detail, which can provide theoretical and empirical guidance for the redesign and activity improvement of these catalysts in their practical applications.

Keywords Selective catalytic reduction • Nitrogen oxides • NH_3-SCR • Diesel exhaust • Fe- and Cu-based zeolite catalysts • Hydrothermal stability • Reaction pathway • Practical applications

12.1 Introduction

Nitrogen oxides (NO_x), which mainly come from mobile sources and industrial combustion of fossil fuels, are a major source of air pollution and can cause a series of environmental problems such as acid rain, photochemical smog and fine-particle

F. Liu • L. Xie • X. Shi • H. He (✉)
Research Center for Eco-Environmental Sciences, Chinese Academy of Sciences,
Beijing 100085, People's Republic of China
e-mail: honghe@rcees.ac.cn

F.-S. Xiao, X. Meng (eds.), *Zeolites in Sustainable Chemistry*,
Green Chemistry and Sustainable Technology,
DOI 10.1007/978-3-662-47395-5_12

(haze) pollution. Owing to their high fuel efficiency and durability, the number of diesel engines worldwide has increased greatly in recent years; however, the emission of NO_x and particulate matter (PM) from diesel engines remains a major constraint for their wider practical use. Through the adjustment of heavy-duty diesel engines, the PM emission from diesel engines can be effectively reduced to meet the Euro IV and Euro V standards. However, due to the well-known "trade-off" relationship between PM and NO_x emitted from diesel engines, the increasing demand to eliminate NO_x, especially in oxygen-rich conditions, has drawn great attention from both academia and industry [1, 2]. The selective catalytic reduction of NO_x with urea/NH_3 (NH_3-SCR) is one of the most promising technologies for NO_x emission control from diesel engine exhaust, although it remains one of the major challenges in the field of environmental catalysis. In particular, to meet higher emission standards such as Euro VI, a typical diesel emission control system using NH_3-SCR technology also includes a diesel particulate filter (DPF), which may expose the SCR catalysts to high temperatures (>650 °C) with high moisture during its regeneration process. Thus, excellent hydrothermal stability is very important for NH_3-SCR catalysts used for the NO_x emission control process from diesel exhaust [3]. As a result, in recent years, many researchers have tried to develop catalysts exhibiting both excellent NH_3-SCR activity and hydrothermal stability.

The first commercial NH_3-SCR catalysts were vanadium-based materials, which have been applied in stationary deNO_x systems [4]. However, many problems exist when vanadium-based catalysts are used for diesel emission control, mainly owing to the toxicity of the V_2O_5 active phase, the poor thermal stability of TiO_2 and the easy sublimation of V_2O_5 [5]. Therefore, much effort has been dedicated to develop environmentally benign NH_3-SCR catalysts with high deNO_x efficiency, excellent stability and also outstanding poisoning resistance that can meet the emerging pollutant regulations for diesel vehicles. Metal oxide catalysts and zeolitic catalysts are two well-studied types of vanadium-free catalysts for the NH_3-SCR process in heavy-duty diesel engines, some of which show great potential in practical use. The improvement of hydrothermal stability is highly needed for metal oxide NH_3-SCR catalysts and is a main factor restricting their wide application in the deNO_x process of diesel engines. Zeolite-based catalysts promoted by transition metals, which show higher hydrothermal stability, are being considered for application in diesel vehicle SCR systems [6, 7].

Fe-exchanged zeolites have been proven to be the active catalysts for the NH_3-SCR of NO_x and received much attention because of their high deNO_x activity, good N_2 selectivity and the innocuous nature of iron. A series of Fe-exchanged zeolites (such as MFI, MOR, BEA and FER) have been studied as catalysts for SCR reaction. Among the numerous Fe-based zeolite catalysts, the application of the MFI structure (Fe-ZSM-5) in the NH_3-SCR reaction has attracted much attention from researchers since the end of the last century. In 1997, Feng and Hall [8] first reported that "over-exchanged" Fe-ZSM-5 prepared by ion-exchange of Na-ZSM-5 with an aqueous solution of ferrous oxalate showed high activity and good resistance to H_2O and SO_2

in the reduction of NO by *i*-C_4H_{10} (HC-SCR). In 1999, Ma and Grünert [9], Long and Yang [10] both reported that Fe-ZSM-5 catalysts prepared by over-exchange of Fe into H-ZSM-5 through $FeCl_3$ sublimation and ion-exchange of Fe into NH_4-ZSM-5 using $FeCl_2$ as precursor, respectively, showed excellent NH_3-SCR activity even under high GHSV and in the presence of SO_2. Thereafter, more and more researchers focused on the study of Fe-exchanged zeolite catalysts, such as the influence of the pore framework of zeolites, the preparation methods, the Fe exchange level and the Si/Al ratio *etc.* on NH_3-SCR performance, which will be discussed in detail in this chapter.

Cu-zeolite catalysts also attracted much attention over the past two decades. These Cu-zeolite catalysts are mainly prepared on the basis of large- or medium-pore zeolites, such as Y (FAU framework), BETA (BEA framework) and ZSM-5 (MFI framework) [11–13]. Cu-ZSM-5 and Cu-BETA, especially Cu-ZSM-5, are the most-studied catalysts due to their high NH_3-SCR activity at low temperatures. Many improvements have been made in order to further improve their NH_3-SCR performance, including the optimization of preparation methods and changing the acidity of the zeolite supports. However, the poor hydrothermal stability of these zeolite catalysts with large or medium pores still limits their wide application for the catalytic deNO_x process from diesel engines [14]. Recently, Cu-zeolite catalysts, especially the Cu-CHA catalysts with small pore structure, attracted much attention from researchers, mainly due to their excellent NH_3-SCR performance and extraordinary hydrothermal stability, and are promising candidates as NH_3-SCR catalysts for the NO_x abatement from diesel vehicles [15, 16]. Moreover, the small pore structure of these Cu-zeolite catalysts contributes to high resistance to hydrocarbon poisoning at the same time, which is also very important for their practical application [17, 18].

Conventional post-synthetic methods and one-pot synthesis methods are both effective for the preparation of Cu-CHA catalysts. Wet ion-exchange and chemical vapor deposition methods are typical conventional post-synthetic methods, which are based on the hydrothermal synthesis of the parent zeolite first. Then, many steps are followed, such as the calcination process for the removal of organic templates, metal incorporation and thermal treatment for activation. Although the IE method is widely used, it is not easy to control the Cu loading due to the limitation of the ion-exchange capacity. Additionally, some structure-directing agents (SDA) have very high cost (*e.g. N,N,N*-trimethyl-1-adamantammonium hydroxide (TMAdaOH) for the synthesis of SSZ-13 zeolite), which also limits their wide application [19]. A new synthesis strategy, called the one-pot synthesis method, can greatly decrease the cost of Cu-CHA catalysts together with allowing controllable metal loading. More importantly, these one-pot synthesized Cu-CHA catalysts can exhibit excellent NH_3-SCR performance, high resistance to space velocity and good hydrothermal stability simultaneously. Thus, it is believed that the one-pot synthesis method could promote the wide application of Cu-CHA catalysts significantly in the catalytic deNO_x process from diesel engine exhaust, which will also be discussed in detail in this chapter.

12.2 Fe-Zeolite Catalysts for NH_3-SCR of NO_x

12.2.1 Synthesis Strategies of Fe-Zeolite Catalysts

Fe-exchanged zeolites with different framework structures have been studied as NH_3-SCR catalysts. Long and Yang [20] found that Fe-ZSM-5 and Fe-MOR catalysts were highly active for the NH_3-SCR reaction under a relatively high space velocity, while Fe-Y and Fe-MCM-41 catalysts showed lower activity under the same experimental conditions. Comparing the activity of various fresh and hydrothermally aged metal-exchanged zeolites (ZSM-5, MOR, BETA, FER, Y), Tolonen et al. [21] concluded that the optimal zeolite catalysts for the NH_3-SCR reaction are Fe-BETA, Cu-FER, Cu-MOR and Fe-ZSM-5. In the study by Iwasaki et al. [22], the NH_3-SCR activity of various Fe-exchanged zeolites decreased in the sequence of MFI > BEA > FER > LTL > MOR. So far, among the different zeolite framework types, MFI (ZSM-5) and BETA are the most attractive ones [7], and thus a large amount of literature on the preparation methods of Fe-ZSM-5 catalysts with high NH_3-SCR activity can be found, as discussed below.

12.2.1.1 Liquid Ion Exchange (IE)

In the conventional aqueous ion exchange method, the H^+, Na^+ or $NH4^+$ cations in the zeolites are replaced by Fe^{3+} or Fe^{2+} ions in a slurry of iron salt solution, such as $FeCl_2 \cdot 4H_2O$ or FeC_2O_4 [8, 23]. Long and Yang [10, 20] prepared a series of Fe-exchanged zeolites including Fe-ZSM-5, Fe-MOR, Fe-Y and Fe-MCM-41 by the conventional ion exchange method and tested them in NH_3-SCR of NO. Under the same aqueous ion exchange conditions, the Fe loading was greatly dependent on the Si/Al ratio of the parent zeolites. A higher exchange degree can be achieved using a parent zeolite with lower Si/Al ratio [24, 25]. Fe-ZSM-5 with the same Si/Al ratio but with different ion exchange degree can be obtained by adding ammonium chloride into the solution of $FeCl_2 \cdot 4H_2O$ [26]. However, it is difficult to realize the full exchange level using the conventional ion exchange procedure. This is due to the fact that one Fe^{3+} ion must compensate three Brønsted acid sites to balance the spatially separated negative charges in the parent zeolite. The reproducibility of the preparation procedure of Fe-zeolite catalysts by conventional aqueous ion exchange is quite low due to the difficulty of accurate control of the pH value inside the zeolite pores [25]. Long and Yang [10] reported an improved aqueous ion-exchange method, by which an Fe-ZSM-5 catalysts (Si/Al = 10) with ion exchange level of 130 % can be obtained by exchanging H-ZSM-5 with iron powder mixed in HCl solution at 50 °C for 10 days in flowing He.

12.2.1.2 Chemical Vapor Deposition (CVD)

An over-exchanged Fe-ZSM-5 catalyst prepared by the CVD method was first reported by Chen and Sachtler [27]. With this method, Fe/ZSM-5 with an iron loading as high as Fe/Al = 1 can be easily achieved. During the preparation process, $FeCl_3$ is sublimed into the cavities of H/ZSM-5, where it reacts chemically with the acid sites of the zeolite according to the following reaction [27–29]:

$$FeCl_3(g) + H^+_{zeol} \rightarrow [FeCl_2]^+_{zeol} + HCl(g)$$

A typical CVD preparation process involves the steps of $FeCl_3$ sublimation, washing and calcination. Each step in the CVD preparation process could affect the NH_3-SCR performance of the final Fe-ZSM-5 catalysts. For instance, it was reported that the NO conversion over Fe-ZSM-5 catalysts prepared by CVD increased with the sublimation temperature, and Fe-ZSM-5 prepared with a sublimation temperature of $FeCl_3$ at 700 °C was more active in the SCR reaction than those prepared at lower temperatures [28, 29]. K. Krishna et al. [28] suggested that $FeCl_3$ would be present as dimer (Fe_2Cl_6) at 320 °C while as monomeric $FeCl_3$ around 700 °C. When the ion exchange is performed, monomeric $FeCl_3$ can access all the protonic sites equally. However, only those protons at a particular distance can be accessed by the dimer (Fe_2Cl_6).

Various techniques, including FTIR, H_2-TPR, XANES, EXAFS, Mössbauer spectroscopy, EPR and UV/Vis DRS, were used for the characterization of iron species during the CVD preparation process of Fe-ZSM-5 catalysts [30–32]. After $FeCl_3$ sublimation, iron was introduced into the zeolite, exchanging with the Brønsted protons in the form of isolated $FeCl^{2+}$ species [30–32]. The EXAFS fitting results revealed that the majority of Fe species formed after $FeCl_3$ sublimation and reaction with the Brønsted acid sites were mononuclear, and consisted of isolated tetrahedral Fe^{3+} species attaching to the zeolite matrix *via* two oxygen atoms at the same distance. The charge of the Fe^{3+} species was further balanced by two Cl^- anions [31].

Besides removing chlorine from the Fe local environment, washing is the key step for the formation of the binuclear Fe complexes. The results of Marturano et al. [32] showed that during the washing process of the zeolite, most of the isolated $FeCl^{2+}$ species formed during sublimation could aggregate into small binuclear hydroxo-iron clusters with a structure similar to that of the binuclear units in α-FeOOH. The extent of washing was found to influence the distribution of Fe species in the final product. For example, the formation of large Fe_xO_y clusters in Fe-ZSM-5 was slightly diminished after washing with 10 L water compared to a sample washed with 1 L water, and the well-washed Fe-ZSM-5 catalyst showed higher normalized reaction rates for the NH_3-SCR reaction [30].

In conclusion, the formation mechanism of the Fe species in Fe-ZSM-5 catalyst prepared by the CVD method can be summarized as follows: after the sublimation of $FeCl_3$ into zeolite pores, iron is stabilized in the form of isolated mononuclear complexes located at Brønsted acid sites; during the hydrolysis and washing step to remove residual chlorine, the mononuclear iron species are displaced and subsequently stabilized in the framework in the form of octahedrally coordinated binuclear Fe-complexes; finally, the calcination process results in the aggregation of iron species removed from Brønsted acid sites. In the above-mentioned processes, a low Si/Al ratio of the parent H-ZSM-5, high washing extent and low heating rate during calcination can suppress the formation of Fe_xO_y clusters, which is beneficial to the promotion of Fe dispersion and thus the improvement of NH_3-SCR activity.

12.2.1.3 Solid-State Ion Exchange (SSIE)

This method is more or less similar to CVD but the required experimental setup is quite simple, and it is an easy method to apply on an industrial scale [7]. Typically, a mechanical mixture of appropriate amounts of the parent zeolite and Fe precursor is heated (usually above the sublimation temperature of the Fe source) under an inert gas flow in an oven for a desired time followed by washing with H_2O [23, 33]. For the Fe-zeolites used as NH_3-SCR catalysts, the H-form zeolites and iron chloride precursors ($FeCl_3$ or $FeCl_2 \cdot 4H_2O$) are commonly used [23, 34].

It was reported that an Fe-ZSM-5 catalyst prepared by the SSIE method and coated on cordierite monoliths showed high activity in the NH_3-SCR reaction [34]. Yet, a decrease of pore volume during the SSIE process can be observed due to the blockage of some pores by the formed iron oxide particles [34]. Compared to the CVD method, the pore blocking induced by the $FeCl_3$ precursor might be more severe for the SSIE method. For example, M.S. Kumar et al. [30] showed that the pore volume of Fe-ZSM-5 prepared by the SSIE method decreased *ca.* 57 % compared with the parent H-ZSM-5, while the pore volume of Fe-ZSM-5 prepared by the CVD method decreased *ca.* 31 %.

12.2.1.4 Other Preparation Methods

Besides the IE, CVD and SSIE methods as mentioned above, other preparation techniques for Fe-zeolite catalysts applied in the NH_3-SCR reaction were also reported. A two-step incipient wetness impregnation method developed by Yang et al. [35, 36] was in effect a combination of conventional impregnation and SSIE processes, but the gas atmosphere for calcination was not inert. The Fe-ZSM-5 catalysts prepared by this method showed high activity in the NH_3-SCR reaction and also in the selective catalytic oxidation (SCO) of NH_3 [35, 36]. More importantly, this preparation process can be very useful for large-scale production of Fe-ZSM-5 catalyst because the preparation conditions are more facile and easy to implement. Furthermore, a novel mechano-chemical route (MR) was reported by M. S. Kumar et al. [30] to produce Fe-ZSM-5 catalysts, during which the intense grinding of the parent H-ZSM-5 with $FeCl_3 \cdot 6H_2O$ was followed by 2–3 short-duration washing steps. They found that an Fe-ZSM-5 catalyst with very high normalized reaction rates in the NH_3-SCR reaction could be obtained *via* such an MR preparation [30].

12.2.1.5 Summary of Preparation Methods

Previously, the preparation methods were believed to influence the NH_3-SCR activity of the final Fe-zeolite catalysts. The results of Long and Yang [23] showed that the NH_3-SCR activity of Fe-ZSM-5 catalysts prepared by different methods mainly followed the sequence as follows: improved aqueous ion-exchange>conventional aqueous ion-exchange>SSIE>CVD. Delahay et al. [37] prepared Fe-ZSM-5 catalysts by $FeCl_3$ sublimation, $Fe(acac)_3$ reaction with HZSM-5 in toluene solution and conventional ion exchange, and found that the Fe-ZSM-5 catalyst with high Fe/Al ratio prepared by $FeCl_3$ sublimation showed the best NH_3-SCR performance. Iwasaki et al. [38] also found that for the same preparation method, the NO_x conversion over Fe-ZSM-5 catalysts was improved with the increase of Fe/Al ratio. However, they further pointed out that the SCR activity was more dependent on the preparation method than on the Fe loading, according to the sequence of CVD>reductive SSIE>impregnation [38]. Shi et al. [39] concluded that the SCR activities of Fe-ZSM-5 catalysts prepared by different methods with similar Fe loadings were quite different (Fig. 12.1). The NO_x conversion over Fe-ZSM-5 catalysts with *ca.* 1.2 % Fe loading decreased in the sequence IE>SSIE>incipient wetness impregnation (IWI) at temperatures ranging from 200 to 350 °C. However, at temperatures above 350 °C, the NO_x conversion over the three catalysts showed the opposite sequence and Fe-ZSM-5 prepared by IWI showed the best SCR performance. Although different research groups reported a diversity of optimal methods for the preparation of Fe-ZSM-5 catalysts with high apparent SCR activity, such as the improved aqueous ion-exchange method by Long and Yang [23], the SSIE method by Schwidder et al. [40], and the CVD method by Iwasaki et al. [38], recently the researchers eventually concluded that the preparation method was actually not a decisive factor in determining the intrinsic catalytic activity of Fe-ZSM-5 catalysts in the NH_3-SCR reaction (*i.e.* the turnover frequency), but the microstructure of Fe species was [7, 38, 40].

12.2.2 Active Sites of Fe-zeolite Catalysts in the NH_3-SCR Reaction

A large number of papers have been devoted to the understanding of the nature of active sites in Fe-zeolite catalysts for the NH_3-SCR reaction, but the scientific debates on this subject are still ongoing. The complexity of various types of Fe species coexisting in Fe-zeolite catalysts is the main reason for the different conclusions about the role of active sites in the SCR process.

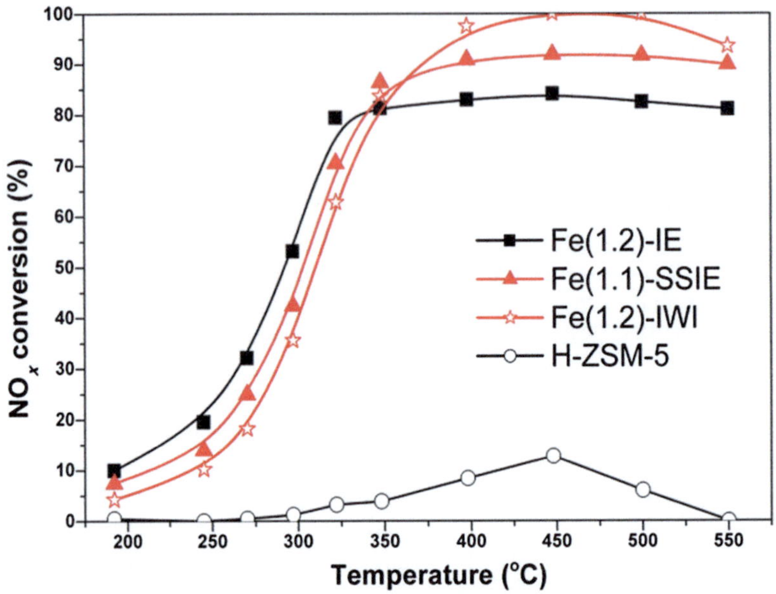

Fig. 12.1 NO_x conversion in NH_3-SCR reaction over H-ZSM-5 and Fe-ZSM-5 catalysts prepared by different methods with the same Fe loading (Reprinted with permission from Ref. [39]. Copyright 2012, CAS/DICP)

In the NH_3-SCR reaction, the catalytic role of Fe is usually confined to the oxidation of NO, resulting in the occurrence of a subsequent "fast SCR" process exclusively on Brønsted sites of the zeolite [24, 40]. Delahay et al. [37] found a good correlation between the NO oxidation activity and the NH_3-SCR activity over their Fe-ZSM-5 catalysts and proposed that an Fe^{2+}/Fe^{3+} redox cycle was involved in the deNO_x process. Based on an ESR study of Fe-ZSM-5 catalysts prepared by different methods, Long and Yang [23] concluded that iron oxides and Fe^{3+} ions with tetrahedral and distorted tetrahedral coordinations were all present in the catalysts, but only Fe^{3+} ions with tetrahedral coordination were the real active sites for the SCR reaction. In contrast, Schwidder et al. [30, 41] studied the structure of Fe species in Fe-ZSM-5 catalysts using UV-Vis DRS, EPR, XAS, etc. and concluded that the SCR rates did not correlate with the concentration of oligomeric Fe species alone. They considered that all Fe species in Fe-ZSM-5 were accessible during the SCR process, including the surfaces of iron oxide aggregates serving as part of the active sites for the NH_3-SCR reaction, and at the same time, the NH_3 oxidation might also be catalyzed by the clustered iron species (oligomers and particles) [30, 41]. According to the characterization results of XRD, UV-Vis DRS and Mössbauer spectroscopy, Iwasaki et al. [38] classified the Fe species in Fe-ZSM-5 catalysts into aggregated α-Fe_2O_3 particles, Fe_xO_y oligomers in the extra-framework and oxo-Fe^{3+} at ion-exchanged sites. They proposed that the ion-exchanged oxo-Fe^{3+}species were the active sites for the NH_3-SCR reaction, which determined the turnover frequency of NO_x.

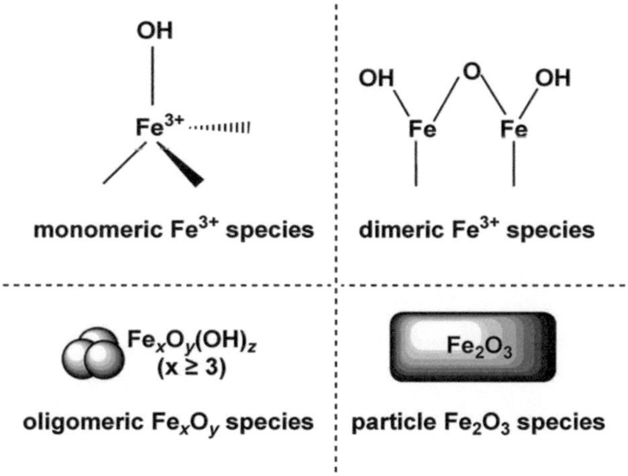

Fig. 12.2 The Fe species probably existing in Fe-zeolite NH_3-SCR catalysts

Recently, Brandenberger et al. [42] systematically studied the different Fe sites located in Fe-ZSM-5 catalysts including the monomeric, dimeric, clustered and oligomeric species and correlated them with the measured NH_3-SCR activity (Fig. 12.2). They reasonably concluded that actually all Fe species in Fe-ZSM-5 catalysts were active, although they exhibited different temperature dependencies in the SCR reaction. Below 300 °C, only monomeric Fe species contributed to the SCR reaction, and this type of Fe species did not catalyze the NH_3 unselective oxidation below 500 °C, which was beneficial to high N_2 selectivity [42]. The important role of monomeric Fe^{3+} sites was also emphasized by Høj et al. [43] in their Fe-BEA catalysts. The UV-Vis DRS, EPR and XAS results by Høj et al. [43] suggested that low Fe loadings (≤1.2 wt.% Fe) could result in mostly iron monomers in Fe-BEA catalysts, especially for the ion-exchanged samples. However, a mixture of monomers, oligomers and Fe_2O_3 particles would form at medium to high loadings (such as 2.5–5.1 wt.% Fe). Above 300, 400 and 500 °C, respectively, the dimeric Fe species, the oligomeric species (e.g. trimeric and tetrameric Fe species) and partially uncoordinated Fe sites in the outermost layer of Fe_xO_y particles could also contribute to the NH_3-SCR reaction, and at high temperatures (> 500 °C), the contribution of dimeric Fe species dominated [42]. However, the NH_3 unselective oxidation would occur on Fe_xO_y particles above 350 °C, resulting in low N_2 selectivity, and the dimeric Fe species governed the NH_3 oxidation up to 500 °C [42]. A similar effect of highly clustered Fe_xO_y species on the unselective consumption of reducing agent NH_3 in SCR reaction was also proposed by Devadas et al. [34] and Schwidder et al. [44].

Therefore, it is concluded that to obtain excellent NH_3-SCR activity at low temperatures together with good N_2 selectivity at high temperatures, an Fe-zeolite catalyst with a maximum quantity of monomeric Fe^{3+} sites should be prepared and the formation of Fe_xO_y clusters should be largely avoided. Although some other factors may also influence the apparent SCR activity of Fe-zeolite catalysts [45], the conclusions drawn by Brandenberger et al. [42] in the above-mentioned study are generally useful for the design, synthesis and application of efficient Fe-zeolite catalysts.

12.2.3 Acid Sites of Fe-Zeolite Catalysts in the NH_3-SCR Reaction

The Brønsted acid sites in Fe-ZSM-5 catalysts are mainly associated with the framework aluminum and terminal silanol groups present in the channels and at the external surfaces (Si(OH)Al). FTIR spectroscopy has been extensively used to study the surface acidity of Fe-zeolite catalysts. For instance, Schwidder et al. [46] investigated the acidic properties of Fe-ZSM-5 catalysts that were prepared by an improved liquid ion exchange method, using IR spectra of pyridine adsorption. Their study indicated that in the H-ZSM-5 sample the Brønsted acid sites were dominant, although some Lewis acid sites resulting from Al^{3+}were also available. Upon the introduction of Fe species, the ratio between Brønsted acid sites and Lewis acid sites dramatically changed. The Brønsted acid sites decreased with increasing exchange degree, and at the same time, an increase of Lewis acid sites could be observed. In this process, the stronger Lewis acid sites of Al^{3+} might be covered by FeO_x species, creating Lewis acid sites of Fe^{3+} with intermediate strength [46]. After the saturation of NH_3 adsorption over Fe-ZSM-5 catalysts, an obvious negative band at around 3610 cm^{-1} could be observed by *in situ* DRIFTS. This could be assigned to the OH stretching vibration mode, confirming the presence of abundant Brønsted acidic sites in these samples, and the negative band was mainly due to the relatively strong interaction between surface hydroxyls and NH_3 [26, 39].

In the NH_3-SCR reaction, the acidity of catalysts is generally considered to be favourable, because the acidic surfaces can adsorb NH_3 in large quantities and thus provide a reservoir of the NH_3 reductant in the vicinity of reduction sites [46]. Therefore, many researchers paid close attention to the investigation on the role of acidity in zeolite materials for the NH_3-SCR reaction. On Fe-based zeolite catalysts with the same type of framework structure, the SCR activity usually increases with decreasing Si/Al ratio [22, 47], although the hydrothermal stability of the zeolites will deteriorate to a certain extent. Previously, Long and Yang [20, 24] studied the influence of Si/Al ratio on SCR activity over Fe-ZSM-5 catalysts and found that the NO conversion was increased with the decrease of the Si/Al ratio. They considered that the increase of SCR activity of Fe-ZSM-5 with low Si/Al ratio corresponded to the enhancement of Brønsted acidity induced by aluminium sites (Al-OH), since the NH_4^+ species adsorbed on Brønsted acid sites directly participated in the SCR reaction to reduce NO_x [20, 24]. However, recent studies reached some different conclusions on the roles of acid sites of Fe-ZSM-5 catalysts for the NH_3-SCR reaction. After extensive study, the researchers finally realized that the Brønsted acidity is actually not a decisive factor for the high NH_3-SCR activity of Fe-zeolite catalysts. For example, Schwidder et al. [46] found that some zeolite materials without acidic sites, such as Fe-silicalite, could still exhibit high activity in the NH_3-SCR reaction with high NO_x conversion rates. However, Schwidder et al. [46] also suggested that a promotion effect of Brønsted acidity on the NH_3-SCR reaction could be observed if the catalysts contained the most favorable Fe species. They further concluded that this was probably due to the obvious promotion of an acid-catalyzed intermediate step in the NH_3-SCR reaction (*e.g.* the decomposition of NH_4NO_2) similar to the results reported by Li et al. [48] and Savara et al. [49], which might be the rate-determining step for the SCR reaction at low temperatures.

In addition, the Brønsted acid sites could play an important role in the distribution of Fe species in Fe-zeolite catalysts. Brandenberger et al. [50] found that the first-order rate constants for the NH_3-SCR reaction were increased with the decreasing of Brønsted acidity of Fe-ZSM-5 catalysts with different Fe/Al ratios. Similar results were also reported by Iwasaki et al. [22], showing that there was actually no correlation between the amount of NH_3 desorption and the NH_3-SCR activity of Fe-ZSM-5 catalysts. They proposed that the Brønsted acid sites might not be required directly in the SCR process for adsorbing or activating NH_3, but they were necessary to bind and disperse the reactive Fe^{3+} ions in the preparation process of Fe-based zeolite catalysts [50]. Rivallan et al. [25] prepared Fe-ZSM-5 catalysts by a liquid ion exchange method using a ferric oxalate precursor, and the Fe-exchange degree of the Fe-ZSM-5 catalysts could be estimated by measuring the intensity decrease of the Brønsted acid band at 3610 cm^{-1} before and after the liquid ion exchange process. They concluded that a low Al content in the parent zeolite could result in Fe-zeolite catalysts with very low Fe loading and relatively large clustering, mainly due to the lack of sufficient Brønsted acid sites to bind and disperse Fe species. Based on the results of ESR and UV/Vis DRS *etc.*, Schwidder et al. [30] also found that more Fe_xO_y clusters were formed in Fe-ZSM-5 catalysts with higher Si/Al ratio when the catalysts were prepared *via* the CVD method. Furthermore, the zeolite support with Brønsted acidity could play a role as NH_3 reservoir no matter what form of NH_3 was stored, and in the NH_3-SCR reaction, the stored NH_3 could migrate to the active sites so as to undergo a reaction with NO [50]. A similar spill-over effect of NH_3 on the zeolite support to the active Fe^{3+} sites in the NH_3-SCR reaction was also reported by Klukowski et al. [51] on their Fe-HBEA catalyst.

Therefore, in brief, the researchers nowadays basically agree that in the standard NH_3-SCR reaction the oxidation function of Fe-based zeolite catalysts and the amount of active Fe species, but not the acidic properties of the zeolite supports, are the main factors controlling the deNO_x efficiency [22, 50].

12.2.4 *Hydrothermal Stability of Fe-Zeolite Catalysts for NH_3-SCR Reaction*

The hydrothermal stability is an important factor for the practical use of Fe-based zeolite catalysts, to which the researchers also paid great attention. The SCR system may be installed at the downstream side of the DPF in order to reduce NO_x and PM simultaneously [6]. Because of the thermal regeneration of the DPF, the temperatures of the diesel exhaust containing *ca.* 10 % water vapor are greatly increased, resulting in the formation of a high temperature hydrothermal condition and thus leading to the deactivation of SCR catalysts. For instance, Kröcher et al. [52] investigated the hydrothermal stability of an Fe-ZSM-5 catalyst coated on cordierite and observed 5–15 % loss in the $deNO_x$ activity after hydrothermal aging at 650 °C in 10 % H_2O for 50 h. The hydrothermal treatment could result in the dealumination of zeolite supports, leading to an decrease of Brønsted acid sites or breakdown of the framework structure, and more severely, the migration of active Fe^{3+} species to form clustered Fe_xO_y/Fe_2O_3 species, leading to the decline of NH_3-SCR activity and N_2 selectivity [26, 39, 53, 54]. Brandenberger et al. [26] concluded that the hydrothermal breakdown of Fe-ZSM-5 catalyst was dominated by three parallel processes (Fig. 12.3): (1) the rapid dealumination of Al-OH-Si sites, (2) the rapid depletion of dimeric iron species and (3) the slow migration of isolated Fe ions followed by the dealumination of Al sites. The remaining NH_3-SCR activity after hydrothermal aging under different conditions was mainly attributed to the residual monomeric Fe^{3+} species located at ion exchange sites [26], and the loss of active Fe sites with high redox ability, rather than the loss of Brønsted acidity, was responsible for the decreased activity [50]. Besides, hydrothermal deactivation could result in a slight increase in the high temperature NH_3-SCR activity of Fe-ZSM-5 catalysts, mainly due to a decrease in over-oxidation of the reducing agent NH_3 at high temperatures [26, 39].

Methods to improve the hydrothermal stability of Fe-zeolite catalysts have been studied extensively by many researchers. The presence of acidic protons was

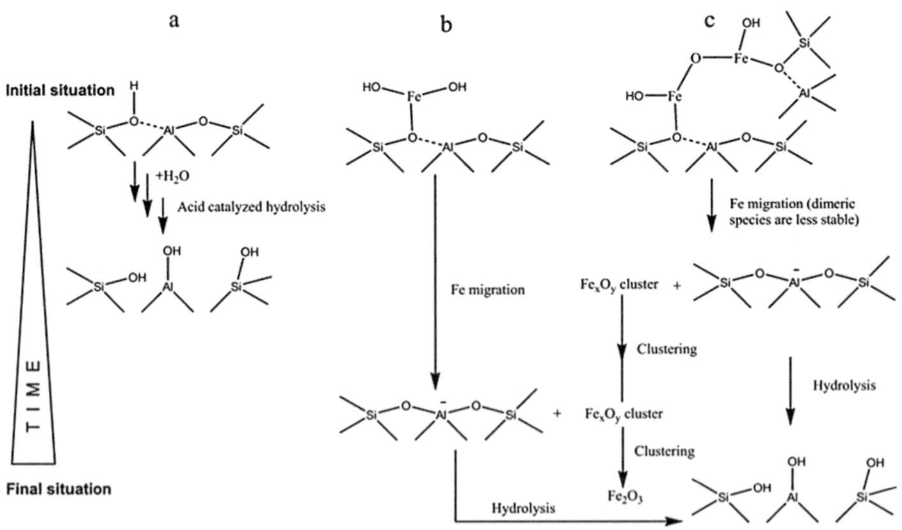

Fig. 12.3 Scheme of hydrothermal aging of (**a**) Brønsted-acid sites, (**b**) isolated iron sites and (**c**) dimeric iron sites involving dealumination, migration and clustering (Reprinted from Ref. [26], Copyright 2011, with permission from Elsevier)

commonly considered to accelerate the dealumination process of zeolites [25, 26, 55]. In the study by Pieterse et al. [55], an Fe-ZSM-5 catalyst prepared by a wet ion-exchange method was exchanged once more in $NaNO_3$ solution, and the hydrothermal stability could be improved by partially replacing protons by sodium [55]. The high temperature stability of an Fe-ZSM-5 catalyst prepared from a ferric oxalate precursor could also be improved through using parent zeolites with high Si/Al ratios [25]. Lee and Rhee [56] found that the secondary sublimation of iron chloride could increase the Fe loading and decrease the concentration of remaining protons, consequently improving the stability of the final Fe-ZSM-5catalyst remarkably [56]. Generally speaking, the application of zeolite supports with high Si/Al ratio is expected to obtain acceptable hydrothermal stability, but the NH_3-SCR activity may be negatively affected owing to the presence of fewer Brønsted acid sites to exchange, bind and disperse the active Fe^{3+} sites.

The pore structure of zeolites can also affect the hydrothermal stability of Fe-zeolite catalysts. From the perspective of hydrothermal stability, Rahkamaa-Tolonen et al. [21] suggested that the best zeolite catalysts for the NH_3-SCR reaction should be Fe-BETA, Cu-FER, Cu-MOR and Fe-ZSM-5. From the test results of zeolites with different Si/Al ratios and pore structures, Iwasaki et al. [22] concluded that the NH_3-SCR activity of hydrothermally aged Fe-zeolite catalysts decreased in the sequence MFI > BEA > FER > LTL > MOR. They suggested that the dealumination process during hydrothermal aging should be the main reason for the deterioration of Fe-zeolite catalysts. Furthermore, the crystal size of zeolites was also an important factor influencing their hydrothermal stability, and larger crystal size of zeolites could result in higher hydrothermal stability for Fe-zeolite catalysts [22].

Brandenberger et al. [45] showed that the hydrothermal stability of Fe-ZSM-5 prepared by the CVD method was higher than those prepared by IE and SSIE methods, but the reason for this phenomenon was still unclear. Recently, the data from Shi et al. [39] showed that the difference in NO_x conversion over Fe-ZSM-5 catalysts prepared from IE, SSIE and IWI methods became less distinct after hydrothermal aging at 700 °C for 8 h and was more significant over the fresh catalysts. Therefore, it seems that the influence of preparation methods on the hydrothermal stability of Fe-ZSM-5 catalysts is actually insignificant.

Lately, Iwasaki and Shinjoh [57] successfully improved the hydrothermal stability of Fe-BEA catalysts by sequential ion-exchange of rare earth (RE) metals and concluded that the improvement was dependent on the ionic radius of the exchanged metals. When using RE metals with radii of 1.05–1.15 Å such as Ce, Nd, Sm, Gd and Tb, the dealumination of aged Fe-BEA catalysts was markedly reduced. After hydrothermal aging at 700 °C for 5 h under 3 % H_2O/air conditions, the Ce-exchanged sample showed higher NH_3-SCR activity than the unmodified Fe-BEA sample. This method is noteworthy for potential use with other Fe-based zeolite catalysts for the effective improvement of hydrothermal stability in practical use.

12.2.5 *Effects of H_2O, SO_2 and Hydrocarbons (HC) on the NH_3-SCR Reaction Over Fe-Zeolite Catalysts*

The effects of H_2O and SO_2 on the NH_3-SCR reaction over Fe-zeolite catalysts are important for practical application, due to the permanent presence of H_2O and SO_2 in diesel engine exhaust. Ma and Grünert [9] observed that the presence of H_2O could promote the NH_3-SCR activity of Fe-ZSM-5 catalysts to a certain extent in the temperature range of 300–550 °C. The impact of SO_2 on the NH_3-SCR activity of Fe-ZSM-5 catalysts was different in the low and high temperature ranges [9, 20, 58, 59]. An SCR activity enhancement of Fe-ZSM-5 catalysts in the presence of H_2O and SO_2 was observed at high temperatures (above 350 °C) by Yang et al. [10, 20, 35], similar to the results from Ma and Grünert [9]. According to an FTIR study, Long and Yang [58] concluded that the treatment of Fe-ZSM-5 in flowing SO_2 and O_2 at 400 °C could result in the formation of sulfate species on Fe^{3+} sites, and a structural transformation from "organic" sulfate to inorganic sulfate with decreasing temperature could take place. The sulfate species on Fe-ZSM-5 catalysts increased the surface Brønsted acidity, and thus resulted in the enhancement of NH_3 adsorption and also high temperature NH_3-SCR activity [58]. However, at low temperatures, the formation of ammonia sulfite/sulfate might occur and cover some active Fe sites, resulting in the decrease of deNO_x activity [58]. Carja et al. [59] reported that an Fe-Ce-ZSM-5 catalyst prepared from an organic phase showed high NO conversion in the temperature range of 250–550 °C and was stable even in the presence of SO_2 and H_2O.

Unburnt hydrocarbons (HCs) are always present in real diesel engine exhaust, and zeolite SCR catalysts can be gradually deactivated due to the poisoning of active sites and pore blockage. Thus, it is very important to understand how the HCs inhibit the activity of Fe-zeolite catalysts for the NH_3-SCR reaction [53, 60–63]. The HC poisoning of Fe-based zeolite catalysts is closely related to the dimensional structure of the zeolite supports. For example, on Fe-ZSM-5 and Fe-BEA catalysts with three-dimensional structure, after HC poisoning the NH_3-SCR activity was obviously decreased, especially at low temperatures, mainly due to carbonaceous deposition resulting in the decline of surface area and pore volume together with suppressed NO oxidation ability for blocked Fe^{3+} sites, or the partial reduction of Fe^{3+} species to Fe^{2+} species by HC, or the competitive adsorption of HC with NH_3/NO onto the catalyst surface [53, 60, 61]. The deactivation effect of HC on NH_3-SCR activity was less on Fe-MOR catalysts with one-dimensional structure due to the difficulty in HC diffusion into the structure, and using this feature Ma et al. [62] effectively improved the HC resistance of a modified Fe-BEA monolith catalyst by coating another layer of MOR zeolite on the outer surface. This useful modification strategy is worthy of application in the preparation of other efficient NH_3-SCR monolith catalysts for emission control from diesel engines.

The HC poisoning effect on Fe-zeolite catalysts is also dependent on the types of hydrocarbons. For example, Malpartida et al. [63] studied the impact of HCs on the NH_3-SCR performance of Fe-ZSM-5 catalysts using propene, toluene and decane as representatives. They proposed that the negative effect of HCs on the deNO$_x$ activity of Fe-ZSM-5 catalyst was mainly due to the competitive adsorption between HCs and NH_3, and this competition was stronger when $C_{10}H_{22}$ was present but almost negligible when $C_3H_6$was present [63]. Similar results were also reported by Li et al. [61] that the NH_3 adsorption was not significantly hindered in the presence of C_3H_6 and the blockage of Fe^{3+} sites was the major reason for C_3H_6 poisoning.

12.2.6 Effects of Inorganic Components on Fe-Zeolite Catalysts for the NH_3-SCR Reaction

The deactivation effects on Fe-based zeolite NH_3-SCR catalysts by inorganic components contained in the diesel exhaust are also worthy of investigation for their practical utilization, including the combustion products of lubricant oil additives (*i.e.* Ca, Mg, Zn, P, B, Mo) and the impurities of biodiesel fuel/urea solution (*i.e.* K, Ca) together with components of aerosol particulates from intake air (*i.e.* Na, Cl) [64]. The predominant deactivation mechanism of Fe-zeolite catalysts should be pore plugging by poisoning components and/or micropore narrowing due to the uptake of metals with alkaline character on cationic positions. Reduced surface acidity and NH_3 storage capacity could also be observed after the deactivation of Fe-zeolite catalysts by alkali metals. Silver et al. [65] concluded that P, originating mainly from lube oil, could deactivate the Fe-zeolite monolith catalysts by physically blocking the adsorption sites for NH_3. Shwan et al. [66] also investigated the chemical deactivation mechanism of Fe-BEA by P using phosphoric acid as poison, and they proposed that the deactivation proceeded in two steps: (1) short duration phosphorous exposure resulted in the information of phosphorous pentoxides (P_2O_5) with a relatively low amount of metaphosphates (PO^{3-}), which did not affect the NO$_x$ conversion but resulted in an increased amount of strongly adsorbed NH_3 not active in the SCR reaction; and (2) longer duration phosphorous exposure resulted in the continuous increase of PO^{3-} replacing the hydroxyl groups on the active Fe species, which led to the significantly decreased NO$_x$ conversion, especially in the low-temperature NH_3-SCR process.

12.2.7 Modification of Fe-Zeolite Catalysts for the NH_3-SCR Reaction by Secondary Metals

In practical use, the low temperature NH_3-SCR activity of Fe-based zeolite catalysts still needs to be improved to meet the diesel emission standards during cold-start and idle speed processes [67]. The enhancement of the NH_3-SCR activity of Fe-zeolite catalysts by introducing secondary metals has been tried by some researchers. Carja et al. [59] reported that the addition of a small amount of Ce to Fe-ZSM-5 could markedly increase its NH_3-SCR activity, and Fe-Ce-ZSM-5 catalysts exhibited very high NO conversion (75–100 %) in a wide temperature window (250–550 °C), even in the presence of H_2O and SO_2. Fe-ZSM-5 containing a small amount of noble metal (such as 0.1 wt.% Pt, Pd or Rh) also showed higher low-temperature NH_3-SCR activity (below 300 °C) than that of unmodified Fe-ZSM-5 [68]. Mn-Fe/ZSM-5 catalysts reported by Nam et al. [69] exhibited high deNO$_x$ activity and N_2 selectivity, but the Mn-Fe/ZSM-5 catalyst deactivated by SO_2 was not easily regenerated by post-treatment at high temperature in air mainly due to the formation of stable $MnSO_4$ on the catalyst surface [69]. Recently, Sultana et al. [70] found that the presence of a small amount of Cu species in Cu-Fe/ZSM-5 catalysts could apparently improve the low temperature NO_x conversion without affecting the high temperature SCR performance significantly, and the higher NO_x conversion over Cu-Fe/ZSM-5 catalysts could be correlated to the facile reduction of metal species. In brief, through changing the composition or preparation method of M-Fe/ZSM-5 catalysts (M=Ce, Pt, Cu, etc.), the redox ability, the acidic properties and thus the NH_3-SCR activity of Fe-zeolite catalysts can be effectively tuned, meeting different practical utilization conditions for the deNO$_x$ process.

12.2.8 NH_3-SCR Mechanism Over Fe-Zeolite Catalysts

12.2.8.1 Standard NH_3-SCR Reaction

In typical diesel exhaust, NO_x is mainly composed of NO (>90 %), and the SCR of NO with NH_3 according to Eq. (12.1) below is commonly referred as the standard NH_3-SCR reaction [71]. This reaction includes a 1:1 stoichiometry of NH_3 and NO together with the consumption of some oxygen.

$$4NH_3 + 4NO + O_2 \rightarrow 4N_2 + 6H_2O \tag{12.1}$$

Due to the use of different characterization methods and various reaction conditions for mechanistic studies, including the reaction atmosphere, pressure and temperature *etc.*, a universal reaction mechanism for the standard NH_3-SCR reaction over Fe-zeolite catalysts still cannot be achieved. For example, mainly according to *in situ* DRIFTS results, Long and Yang [24] proposed a mechanism for standard NH_3-SCR over Fe-ZSM-5 catalysts, as shown in Scheme 12.1, in which the oxidation of NO to NO_2 was the rate-determining step, and the reaction of $NO_2(NH_4^+)_2$ with NO to yield the final product N_2 mainly took place on Brønsted acid sites. Devadas et al. [72] further suggested that the $NO_2(NH_4^+)_2$ species could be a general intermediate in the NH_3-SCR reaction over Fe-zeolite catalysts, from which N_2 was formed through reaction with gaseous NO. Based on TPR, TPD and DRIFTS results, Delahey et al. [37] proposed that an Fe^{2+}/Fe^{3+} redox cycle was involved in the standard NH_3-SCR mechanism, and the extra-framework oxygen could react with NO to form a surface nitrogenous intermediate bound to Fe followed by reaction with adsorbed NH_3 to form N_2 and H_2O. Using the isotopic labeling method, Sun et al. [73] proposed

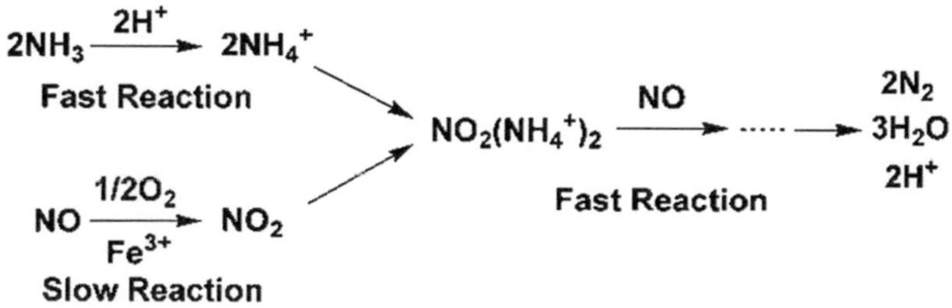

Scheme 12.1 Reaction scheme for NH_3-SCR of NO on Fe-ZSM-5 catalyst proposed by Long and Yang (Reprinted from Ref. [24], Copyright 2002, with permission from Elsevier)

that a preferred reaction pathway for the reduction of NO_x with NH_3 over Fe-ZSM-5 catalysts was *via* the ammonium nitrite (NH_4NO_2) route, which could decompose to N_2 and H_2O effectively at low temperatures. Therefore, for a specific Fe-zeolite catalyst, choosing the appropriate characterization methods and the proper study conditions is very important to better understand the NH_3-SCR reaction mechanism over the studied catalyst and establish a reasonable structure-activity relationship in the SCR reaction for redesign and activity improvement of the catalyst for practical use.

12.2.8.2 Fast NH_3-SCR Reaction

As mentioned above, it is generally accepted that the oxidation of NO to NO_2 is the rate-determining step for the standard NH_3-SCR reaction over Fe-zeolite catalysts. An interesting phenomenon is that the presence of NO_2 in the total NO_x can greatly enhance the low temperature deNO_x activity of Fe-zeolite catalysts, mainly due to the occurrence of the so-called fast NH_3-SCR reaction shown as Eq. (12.2), which involves the reaction between NH_3 and an equimolar mixture of NO and NO_2. If the NO_2/NO ratio is higher than 1, NO_2 also can react with NH_3 *via* alternative NO_2 SCR routes as shown in Eqs. (12.3) and (12.4). Since the enhanced NH_3-SCR performance of Fe-ZSM-5 catalysts in the presence of NO_2 is very beneficial to their practical use for diesel exhaust purification, the mechanism and kinetics of the NO/NO_2-NH_3-SCR reaction over Fe-ZSM-5 have been studied extensively by transient and steady-state measurements.

$$2NH_3 + NO + NO_2 \rightarrow 2N_2 + 3H_2O \tag{12.2}$$

$$2NH_3 + 2NO_2 \rightarrow N_2O + N_2 + 3H_2O \tag{12.3}$$

$$8NH_3 + 6NO_2 \rightarrow 7N_2 + 12H_2O \tag{12.4}$$

Tronconi et al. [4] compared the activities of Fe-zeolite and vanadium-based catalysts in the presence of equimolar NO/NO_2 feed mixtures, and they found that the Fe-zeolite catalyst showed higher NH_3-SCR activity in contrast to the vanadium-based catalyst, particularly at low temperatures. The ratio of NO_2/NO_x can clearly affect the NH_3-SCR performance of Fe-ZSM-5 catalysts, and the results from Devadas et al. [72] showed that the deNO_x activity of Fe-ZSM-5 coated on a cordierite monolith increased dramatically with the addition of NO_2 to the feed in the entire temperature range and was most pronounced at lower and medium temperatures, with $NO_2/NO_x=50$ % resulting in the highest deNO_x performance. However, the NO_x conversion decreased to a certain extent with increasing NO_2/NO_x ratio when the NO_2/NO_x ratio was above 50 % [74] because the reaction rate of the NO_2 SCR reaction was much slower than that of the fast NH_3-SCR reaction. Tolonen et al. [21] found that high NO_x conversion could be still obtained over hydrothermally aged Fe-ZSM-5 and Fe-BETA catalysts in the fast NH_3-SCR reaction with the NO_2/NO ratio of 4:6.

Based on an isotopic labeling method, Sun et al. [73] concluded that a pure $^{15}N^{14}N$ product could be produced by circulating a 1/1 mixture of ^{15}NO and $^{15}NO_2$ over Fe-ZSM-5 catalyst covered with adsorbed $^{14}NH_3$. Iwasaki et al. [74] proposed an overall SCR scheme as illustrated in Scheme 12.2 and suggested that the formation of surface NH_4NO_3 species resulting from the adsorbed NO_2 was a common reaction in standard NH_3-SCR, fast NH_3-SCR and NO_2 SCR, and the rate-determining step of fast NH_3-SCR was considered to be the reduction of NH_4NO_3 by gaseous

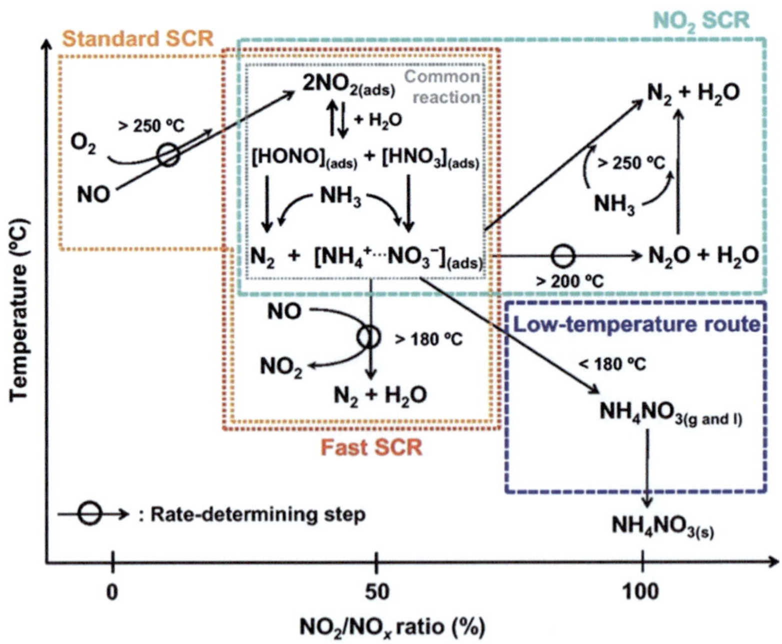

Scheme 12.2 An overall SCR scheme over Fe-zeolite catalysts as a function of NO_2/NO_x ratio and temperature proposed by Iwasaki et al. (Reprinted from Ref. [74], Copyright 2010, with permission from Elsevier)

NO. Grossale et al. [75, 76] concluded that in the fast NH_3-SCR reaction, the role of NO_2 was to form surface nitrites and nitrates, and the role of NO was to reduce nitrates to nitrites followed by the rapid and selective decomposition of unstable NH_4NO_2 to yield N_2. They further concluded that the rate of the fast NH_3-SCR reaction was dependent on the rate of nitrate reduction by NO at low temperatures [75, 76]. Although a strong activity for the NO oxidation by nitrates could be observed over the Fe-ZSM-5 catalyst already at 50 °C, yet the presence of NH_3 could severely inhibit this NO-nitrate reaction process because NH_3 could react with nitrates to form NH_4NO_3 precursors or strongly interacting NH_4NO_3 adsorbed species, thus blocking the critical reactivity of nitrates with NO and resulting in the loss of fast NH_3-SCR performance at low temperatures [77]. This so-called "NH_3 blocking effect" on fast NH_3-SCR performance over Fe-zeolite catalysts proposed by Grossale et al. [77] should be paid great attention, especially for their practical use below *ca.* 170 °C.

In order to convert a portion of NO to NO_2, a diesel oxidation catalyst (DOC) typically consisting of precious metals should be installed upstream of the NH_3-SCR system. However, the oxidation activity of the DOC is strongly dependent on the reaction temperature and flow rate of the diesel exhaust, therefore the NO_2/NO_x ratio in the diesel exhaust should vary constantly during different engine operation conditions. Tronconi et al. [78] reported that enhanced NH_3-SCR deNO_x activity could be realized by adding a small amount of NH_4NO_3 solution as an effective oxidant for NO together with the reducing agent urea solution, thus creating a similar fast NH_3-SCR reaction condition. This technique required no pre-installed upstream DOC, which could greatly lower the system control difficulty and enhance the deNO_x efficiency at low temperatures simultaneously.

12.2.8.3 Problem of N_2O Formation

The formation of N_2O is a problem when using vanadium-based catalysts and Cu-zeolite catalysts for the NH_3-SCR reaction, while the Fe-zeolite catalysts are well known for their high N_2 selectivity during the deNO_x process [7]. Long and Yang et al. [24] explained that over Fe-ZSM-5 catalysts in the standard NH_3-SCR reaction, the reactivity of the $NO_2(NH_4^+)_2$ intermediate with NO was much higher than that with NO_2 resulting from NO oxidation, thus N_2 was the only N-containing product for the standard NH_3-SCR reaction. Actually, the production of N_2O is a disadvantage of adding NO_2 into the SCR atmosphere to enhance the NO_x conversion over Fe-zeolite catalysts, and the amount of N_2O formed in the fast NH_3-SCR reaction is dependent on the NO_2/NO_x ratio and reaction temperature [72, 74]. In the fast NH_3-SCR process, N_2O can be produced from the thermal decomposition of NH_4NO_3 [72, 74, 75, 77] and/or from the reaction of NO_2 with intermediate ammonium nitrites [72], and the higher NO_2/NO_x ratio can result in higher N_2O formation [72, 74]. For example, Devadas et al. [72] proposed that in the presence of NO_2, the intermediate $NO_2(NH_4^+)_2$ could be easily formed, and the reaction of $NO_2(NH_4^+)_2$ with NO became rate limiting, thus the probability of N_2O formation from the reaction between $NO_2(NH_4^+)_2$ and NO_2 increased. The latter mechanism could well explain the observation that the formation of N_2O was increased with increasing NO_2/NO_x ratio at medium temperatures. At high temperatures, N_2O could be easily converted to N_2 through the N_2O SCR reaction as well as the N_2O decomposition reaction. Therefore, no N_2O formation could be observed above 550 °C over Fe-ZSM-5 catalysts [72, 79].

Recently, Shi et al. [79] reported that N_2O formation in the fast NH_3-SCR reaction was much more apparent over a hydrothermally aged Fe-ZSM-5 catalyst than that over the fresh catalyst. Since N_2O is one of the depleting substances for the ozone layer that also needs to be controlled, the increased N_2O formation over aged Fe-ZMS-5 catalyst for the deNO_x process from diesel engines after long-term use is worthy of attention. The design of proper pre-oxidation catalysts for optimal NO_2/NO_x ratio and the promotion of the hydrothermal stability of Fe-zeolite catalysts may be effective in reducing N_2O formation in the NH_3-SCR reaction for practical use.

12.3 Cu-Zeolite Catalysts for NH_3-SCR of NO_x

12.3.1 Cu-Zeolite Catalysts with Medium and Large Pore Structure

12.3.1.1 NH_3-SCR Performance and Reaction Mechanism

Since zeolite catalysts have been used in NH_3-SCR reaction starting in the late 1970s, the Cu-based zeolite catalysts (mainly Cu-exchanged Y zeolites) have shown relatively good catalytic performance among the studied materials [80]. Since the 1990s, the Cu-based zeolite catalysts have attracted more attention from researchers in the field of catalytic deNO_x from diesel engine exhaust. Baik et al. [81] found that Cu-ZSM-5 catalysts exhibited relatively higher deNO_x activity and maintained a broader operating temperature window than Cu-HM, Cu-USY and Cu-FER catalysts, especially at low temperatures. Sjövall et al. [11] concluded that using zeolite supports with low Si/Al ratio and increasing the Cu loading could markedly enhance the NH_3-SCR activity of Cu-ZSM-5 catalysts. They also reported that changing the ratio of O_2, H_2O and NO_2 in the feed gas could affect the NO_x conversion significantly. Qi et al. [47] reported that the Cu-ZSM-5 catalyst showed high deNO_x performance at medium and low temperatures even under high GHSV, but hydrothermal treatment could result in activity decline to a certain extent. Due to its excellent NH_3-SCR activity in the low temperature range, the Cu-ZSM-5 catalyst was believed to be more suitable for light duty diesel engines in which the exhaust gas temperature is relatively low compared to that of heavy duty diesel engines.

The confirmation of the active sites in Cu-based zeolite catalysts and study of the detailed NH_3-SCR reaction mechanism over these materials are necessary for the further improvement of their NH_3-SCR performance. It has been known that the isolated Cu^{2+}, Cu^+ species and paired Cu^{2+} species in Cu-containing zeolite catalysts are the main catalytically active sites for the NH_3-SCR reaction [82]. Good redox capacity of Cu species contributed to their promising NH_3-SCR activity. Sultana et al. [83] found that Cu species in a Cu-ZSM-5 catalyst prepared from Na-ZSM-5 were more easily reducible than in that prepared from H-ZSM-5. The Na^+ co-cations could further capture the produced NO_2 to form intermediate surface nitrite/nitrate species, which could improve the NO_x conversion to a certain extent. Due to the different Cu species and surface acidity of the parent zeolites, there are also various reaction pathways and reaction intermediates reported in the literature. For example, Richter et al. [84] concluded that the NH_4^+ species formed on Brønsted acid sites were responsible for the conversion of NO_x to N_2, and the oxidation of NO to NO_2 was the rate-limiting step over metal-exchanged zeolite catalysts. Sjövall et al. [12] reported that the NH_3-SCR mechanism over Cu-ZSM-5 catalysts involved reactions between adsorbed NO_2 and NH_3, the formation of HNO_2 and HNO_3 and their final reduction to N_2 and H_2O. Grossale et al. [76] reported that surface nitrates played a key mechanistic role over Cu-zeolites in the presence of NO_2 *via* NO_2 disproportionation. They also proposed that higher NH_3 storage capacity and higher NH_3 oxidation activity were related to higher standard NH_3-SCR activity. Therefore, the detailed NH_3-SCR reaction mechanisms over different Cu-zeolite catalysts are still in debate and still need to be elucidated using specific characterization methods.

12.3.1.2 Hydrothermal Stability and HC Poisoning Resistance

Many factors can affect the hydrothermal stability of Cu-zeolite catalysts significantly, such as the different chemical status of Cu species and the shape of the zeolite structure. Hydrothermal treatment can markedly influence the Cu species, which may aggregate severely resulting in the formation of Cu oxides. Park et al. [14] reported that the optimal Cu content for Cu-ZSM-5 was about 4 wt% from the standpoint of catalyst aging (Fig. 12.4). They concluded that the hydrothermal deactivation of this Cu-ZSM-5 catalyst was mainly due to the migration and re-distribution of Cu^{2+} species, leading to the decrease of active sites and the blockage of zeolite channels by sintered CuO. After hydrothermal treatment of Cu-BETA catalysts under different conditions (500–900 °C, 3 % H_2O), Wilken et al. [13] found that the zeolite structure showed no obvious change below 800 °C, yet the oxidation state of Cu species showed apparent change, with the decrease of Cu^+ species and the increase of Cu^{2+} species. Their results suggested that the hydrothermal deactivation of Cu-BETA catalyst was mainly owing to the decrease of Cu^+ species. Qi et al. [47] found that the Cu-ZSM-5 catalyst prepared by vapor phase ion-exchange (VPIE) showed higher hydrothermal stability than that prepared by the liquid phase ion-exchange method (LPIE). Because more active Cu^{2+} ions existed on the surface of the Cu-ZSM-5 (VPIE) catalyst with higher exchange level of Cu species, the migration of Cu ions on the surface of the aged catalyst was greatly lowered, which was the main reason for its higher hydrothermal stability.

Besides the Cu status, high structural stability of the parent zeolite is also necessary for obtaining good hydrothermal stability for Cu-zeolite catalysts. It is well known

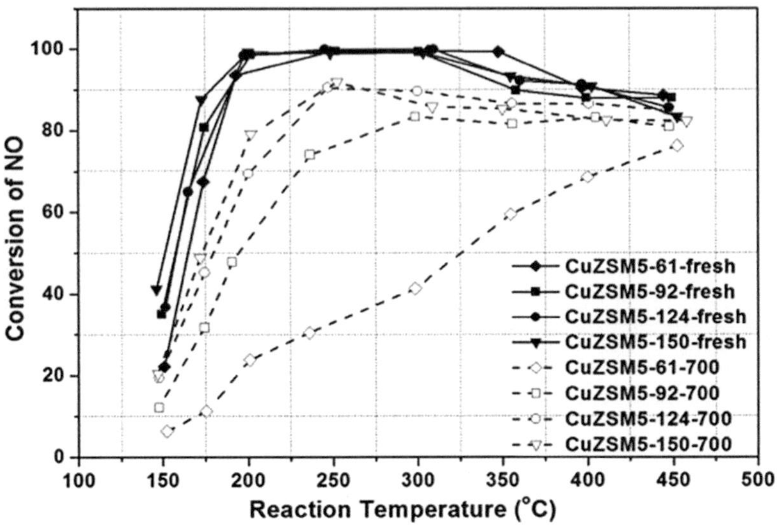

Fig. 12.4 Activity of a series of Cu-ZSM-5 catalysts before and after aging at 700 °C for the reduction of NO by NH_3 (Reprinted from Ref. [14], Copyright 2006, with permission from Elsevier)

that the bonds of Si-O-Al are less stable than those of Si-O-Si, and the dealumination process usually occurs *via* the attack of Si-O-Al by H_2O [7]. The damage of the zeolite structure caused by the dealumination process is fatal for Cu-zeolite catalysts. Feng et al. [8] found that a zeolite catalyst free of Brønsted acid sites was not deactivated even when treated by wet gas up to 800 °C. However, the Brønsted acid sites are beneficial for the loading of active metal species into the zeolites and also for NH_3 adsorption, thus this method is actually not recommended for the improvement of hydrothermal stability of Cu-ZSM-5 catalysts. Interestingly, some researchers concluded that the pore size of parent zeolites could also affect the hydrothermal stability of Cu-zeolite catalysts. Moden et al. [85] reported that the active Cu species in Cu-FER catalysts with smaller pore size (0.42 × 0.54 nm) showed higher stability than that in Cu-BETA catalysts with relatively larger pore size (0.66 × 0.67 nm). Using the *in situ* XRD technique, Blakeman et al. [86] confirmed that the small pore structure in Cu-CHA catalysts could prevent the degradation of the zeolite framework structure by Cu species as compared to that in Cu-BETA catalysts, and the undesirable Cu/Al_2O_3 interaction at high temperatures might play an important role in the catalyst deactivation process. Fickel et al. [87] also believed that the higher hydrothermal stability of the Cu-CHA catalyst was related to its small pore structure (0.38 × 0.38 nm). The detached $Al(OH)_3$ species in the zeolite framework with kinetic diameter of 0.503 nm could not exit the pores of the CHA structure, and they could reincorporate into the zeolite framework in the subsequent cooling stage. They concluded that the constricting dimension of the CHA structure was the most important reason for the high hydrothermal stability of Cu-CHA catalysts [87].

Even an advanced diesel engine with high fuel efficiency also emits a large amount of HCs during its cold start period, which could alter the deNO$_x$ performance of the Cu-zeolite NH_3-SCR catalysts owing to the competitive adsorption of HCs and coke deposition. Heo et al. [60] reported that the inhibitory effect of C_3H_6 on the NH_3-SCR reaction over Cu-ZSM-5 was mainly due to the competitive adsorption between NH_3 and C_3H_6 and also the useless consumption of NH_3 by side reactions. Luo et al. [88] observed that the presence of C_3H_6 could inhibit the deNO$_x$ activity of Cu-BETA catalysts in the standard NH_3-SCR reaction as well in the fast NH_3-SCR reaction. Due to the partial reduction NO_2 by C_3H_6, some standard NH_3-SCR reaction occurred instead of the pure fast NH_3-SCR reaction, therefore the overall activity of the studied catalyst in fast NH_3-SCR conditions became much slower. Sultana et al. [83] concluded that the Cu/NaZSM-5 catalyst showed higher NO_x conversion than the Cu/HZSM-5 catalyst in presence of decane, and this was due to the promoting effect of Na^+ cations on the formation of intermediates species (nitrite and nitrate) and the retardation of coke formation. However, the improvement of the HC poisoning resistance of the Cu/ZSM-5 catalyst arising from Na^+ co-cations was still limited. Due to the shape selectivity, large molecules could not pass through small pore zeolites because of steric hindrance [89]. For example, because of the small pores and cages, the NH_3-SCR activity over Cu-FER catalysts was less affected than that over Cu-ZSM-5 in the presence of decane [17]. Therefore, it is expected that using a zeolite support with small pore size may produce Cu-based zeolite catalysts with excellent NH_3-SCR activity, outstanding hydrothermal stability and HC poisoning resistance simultaneously.

12.3.2 *Cu-Zeolite Catalysts with Small Pore Structure*

12.3.2.1 NH_3-SCR Performance and Hydrothermal Stability of Cu-CHA Catalysts

Because of the superiority of small pore zeolite catalysts, research relative to their application in the NH_3-SCR reaction for the deNO$_x$ process from diesel exhaust has been very popular in recent years. For instance, Fickel et al. [87] reported that that Cu-CHA catalysts (Cu-SSZ-13 and Cu-SAPO-34) displayed superior SCR performance, even after hydrothermal treatment at 750 °C for 16 h as compared with other Cu/zeolites. Kwak et al. [3, 15] reported that Cu-SSZ-13 was a promising candidate for the NH_3-SCR reaction. Compared to Cu-ZSM-5 and Cu-BETA catalysts, the optimal Cu-SSZ-13 catalyst showed high NH_3-SCR activity (Fig. 12.5a) and excellent hydrothermal stability (Fig. 12.5b) with great potential in the catalytic NO_x removal process from diesel engines. The ^{27}Al-NMR results recorded by Kwak et al. [3] showed that, on the one hand, the hydrothermally treated Cu-SSZ-13 catalyst showed no obvious change in the peak intensity of AlO_4 species compared with the fresh catalyst; however, the peak intensity of AlO_4 species in Cu-ZSM-5 and Cu-BETA catalysts decreased *ca.* 57 % and 31 %, respectively, suggesting the very

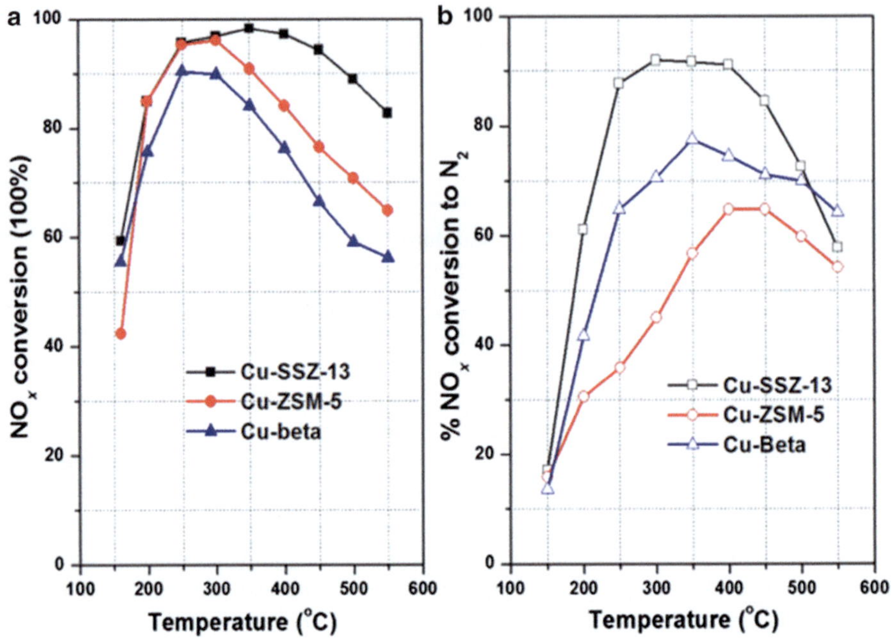

Fig. 12.5 (**a**) NO_x conversion over Cu/zeolite catalysts at various temperatures in the NH_3-SCR reaction (Reprinted from Ref. [15], Copyright 2010, with permission from Elsevier); (**b**) NO conversion to N_2 over Cu/zeolites after hydrothermal aging (Reprinted from Ref. [3], Copyright 2012, with permission from Elsevier)

high stability of the zeolite structure in the Cu-SSZ-13 catalyst. On the other hand, the stability of Cu species in the Cu-SSZ-13 catalyst was also high, such that they could not be reduced to Cu^0 until the temperature was higher than 825 °C [90]. The extremely high reduction temperature of Cu species in the Cu-SSZ-13 catalyst was consistent with its excellent hydrothermal stability. After hydrothermal aging, no CuO or Cu-aluminate species were observed in the studied Cu-SSZ-13 catalyst, indicating that no observable migration of Cu species occurred [3].

Although the Cu-CHA catalysts showed good hydrothermal stability compared to other Cu-zeolite catalysts, their NH_3-SCR activity can still decrease significantly under long-term and severe hydrothermal aging conditions. Thus, many recent studies also focus on the deactivation mechanism of Cu-CHA catalysts during the hydrothermal aging process. Schmieg et al. [91] claimed that NH_3-SCR catalysts would be exposed to peak temperatures near 700 °C for about 100 h due to DPF regeneration during a 135,000 mile driving cycle, and it was not a wise choice to conduct a long-term experiment in the lab to investigate the durability of NH_3-SCR catalysts in the real world. In general, accelerated aging experiments have usually been conducted to test the durability of targeted NH_3-SCR catalysts. For example, the hydrothermal aging treatment was always carried out at higher temperature and shorter period to simulate the rapid deactivation process of the SCR catalysts. After detailed characterization of the commercial Cu-SSZ-13 catalysts from BASF under different aging conditions, Schmieg et al. [91] concluded that catalyst treated under the selected aging conditions (800 °C for 16 h) compared well with a 135,000 mile vehicle-aged catalyst based on the NH_3-SCR activity test and characterization results. Their results indicated that the decreased activity of aged Cu-SSZ-13 catalysts was mainly due to the decreased NO/NH_3 oxidation activity and NH_3 storage capacity. Ma et al. [92] believed that the Cu sites and surface acidity were the main influencing factors for the NH_3-SCR reaction over monolith Cu-CHA catalysts, and they found that the migration of Cu^{2+} species, the decrease in surface acidity and the dealumination process occurred during the hydrothermal aging treatments.

It is well known that the acidity of the parent zeolites always affects the NH_3-SCR performance of catalysts, and changing the element distribution in the zeolite framework can greatly influence the catalyst acidity. This effect is very obvious for Cu-SAPO-34 catalysts. SAPO-34, the silicoaluminophosphate analogue of chabazite, can be obtained by substitution of Si into the framework of AlPO, which exhibits cation exchange and acidity due to the Si defects in an alumina-phosphate lattice [93]. Thus, the Si content in the framework can affect the NH_3-SCR performance of Cu-SAPO-34 catalysts since the Si content can affect both the surface acidity of the zeolite support and the Cu loading in the final product. Wang et al. [94] found that the number of acid sites in Cu-SAPO-34 catalysts increased with increasing Si content, and the NH_3-SCR performance correlated well with the number of acid sites at low temperatures (<200 °C). Yu et al. [95] further reported that the NH_3-SCR activity of a Cu/SAPO-34 catalyst was closely related to its acid density in the whole temperature range. It was very interesting that, different from the decreased activity observed over aged Cu-SSZ-13 catalysts, better NH_3-SCR performance was always observed over Cu-SAPO-34 catalysts, and this difference was related to the redistribution of Cu species in Cu-SAPO-34 catalysts [96, 97], which will be discussed in the following parts of this chapter. In brief, the degree of Cu exchange, the type of zeolite framework and heteroatom substitution in the framework all affect the NH_3-SCR activity and hydrothermal stability of the final Cu-CHA catalysts, and a comprehensive understanding of these effects is helpful to enhance their deNO$_x$ performance to meet the upcoming stringent NO$_x$ emission standards for diesel engines.

12.3.2.2 Structure-Activity Relationship of Cu-CHA Catalysts in NH_3-SCR Process

There are still debates in the literature about the real active sites for the NH_3-SCR reaction over Cu-CHA catalysts prepared by different preparation methods and different groups. Deka et al. [98] prepared Cu-SSZ-13 catalysts *via* CVD and wet chemical routes, and they found that only isolated Cu^{2+} ions were present in the Cu-SSZ-13 catalyst prepared by the wet ion-exchange method, while both isolated

Cu^{2+} ions and $CuAlO_2$ species existed in the catalyst obtained from the CVD method. According to the NH_3-SCR activity tests, the isolated Cu^{2+} ions were believed to be responsible for N_2 formation, while $CuAlO_2$ could promote the N_2O formation. Therefore, ion exchange is the most common method for the preparation of Cu-SSZ-13 catalysts in previous literature. Ion-exchange and precipitation methods were also used to prepare Cu-SAPO-34 catalysts, and both CuO and isolated Cu^{2+} ions were formed in the final catalysts no matter which method was chosen [96, 97, 99, 100]. The different synthesis methods did not change the types of Cu species in the final catalysts, but greatly affected their distribution [97]. For example, Wang et al. [99] reported that the CuO species on the external surface was dominant in the Cu-SAPO-34 catalyst prepared by precipitation method, and the catalyst prepared by ion-exchanged method showed better NH_3-SCR performance than the precipitated one mainly due to the predominant presence of isolated Cu^{2+} ions. Although the activity of CuO on the external surface of Cu-SAPO-34 catalyst was poor, it could migrate to the ion-exchange sites as isolated Cu^{2+} species after hydrothermal aging, which was the main reason for the improved NH_3-SCR activity over the aged Cu-SAPO-34 catalyst prepared by the precipitation method [96, 97]. In addition, the CuO species in Cu-SAPO-34 catalysts could also transform to isolated Cu^{2+} species under high calcination temperatures [100], which is an important advantage for its practical use in the deNO_x process from diesel engines.

It is well known that there are four types of cationic sites in the chabazite structure. Site I is displaced from the six-membered ring into the ellipsoidal cavity. Site II is located near the center of the ellipsoidal cavity. Site III is located in the center of the hexagonal prism. Site IV is located near the eight-membered-ring window [101, 102]. Because of the different locations of Cu^{2+} ions, different NH_3-SCR performance was observed over Cu-CHA catalysts with the same type of framework structure. Therefore, thorough characterizations were carried out to identify the nature of active Cu sites in Cu-CHA catalysts, including the oxidation state and coordination environment, which are very necessary for understanding the main factors controlling the NH_3-SCR activity and providing possible routes for the development of better deNO_x catalysts. Using a variable-temperature XRD method, Fickel and Lobo [103] proved that the ion-exchange of Cu species into NH_4-SSZ-13 could greatly enhance its thermal stability. Through a comprehensive study using Rietveld refinement of XRD data, *in situ* UV-Vis and XAFS techniques, Fickel and Lobo [103], Korhonen et al. [104] and Deka et al. [105] finally concluded that only the isolated Cu^{2+} species located in the six-membered rings of the CHA structure coordinating with three oxygen atoms was the real active species in the NH_3-SCR reaction for the Cu-SSZ-13 catalyst prepared by the wet ion-exchange method (Fig. 12.6). Using *operando* XAFS, density functional theory (DFT) calculation and first-principles thermodynamics models, Kispersky et al. [106] and McEwen et al. [107] proposed that under the standard NH_3-SCR conditions at steady state, the active Cu species was actually in the mixed form of Cu^+ species and Cu^{2+} species, indicating that the redox cycle between Cu^+/Cu^{2+} is very important for the high NH_3-SCR activity, similar to that in other Cu-based zeolite catalysts. Very recently, further study using H_2-TPR and FTIR methods by Kwak et al. [108] revealed two distinct cationic

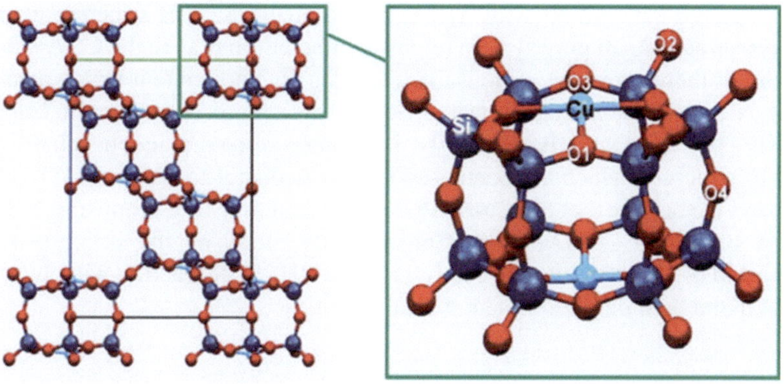

Fig. 12.6 The structure of Cu-SSZ-13 obtained from refinements of neutron scattering data and magnification of the double six member (D6R) unit with Cu present as an isolated ion (Reprinted from Ref. [105], Copyright 2012, with permission from American Chemical Society)

positions in Cu-SSZ-13 catalysts at different ion-exchange levels. They concluded that at low ion-exchange levels, the Cu^{2+} ions primarily occupied the sites in the six-membered rings, similar to early reports, while at high ion-exchange levels, the Cu^{2+} ions were present mostly in the large cages of the CHA structure. The latter Cu^{2+} species were much more easily reduced in the H_2-TPR process, indicating a high redox property at low temperatures, which might be the main origin of the exceptional NH_3-SCR activity over Cu-SSZ-13 catalysts. In further study, Gao et al. [90] characterized Cu-SSZ-13 catalysts with various ion-exchange levels using EPR and H_2-TPR methods, and they suggested that the Cu^{2+} ions were far apart at low ion-exchange levels. The Cu^{2+} ions were located in the large CHA cages coordinated to lattice oxygen atoms of six-membered rings, and with increasing Cu loading, two Cu^{2+} ions possibly existed in one unit cell, and some Cu^{2+} ions were located close to the 8-membered ring. Although the standard NH_3-SCR reaction kinetics experiments were carried out in the study by Gao et al. [90], they did not propose clear structure-activity relationships for the Cu-SSZ-13 catalyst in the NH_3-SCR reaction. Besides the Cu-SSZ-13 catalysts, detailed studies were also conducted by researchers for Cu-SAPO-34 catalysts to investigate the nature of the active Cu sites. For instance, Xue et al. [109] suggested that three types of Cu species (including nanosize CuO, isolated Cu^{+} species and isolated Cu^{2+} species) could be found in Cu-SAPO-34 catalysts with different Cu loadings. According to the calculation of turn-over frequency (TOF) and the assignments of EPR signals, they concluded that the isolated Cu^{2+} ions located at site I were responsible for the NH_3-SCR reaction in the low temperature range (100–200 °C). In brief, although the NH_3-SCR performance of Cu-CHA catalysts has been well recognized, the understanding of the structure-activity relationship is still in debate and ongoing.

In addition, new types of Cu-based zeolite catalysts are being developed constantly with high deNO_x efficiency and extraordinary hydrothermal stability, such as the newly reported Cu-SSZ-39 with AEI structure [16]. The AEI structure exhibits a three-directional small pore system (eight-membered ring) as in the CHA structure, and both of them show the same framework density and contain D6R units. Thus, similar NH_3-SCR performance was observed over these two types of Cu-zeolite catalysts [16]. Interestingly, SAPO-18 is another silicoaluminophosphate zeolite with AEI structure, which has been used for the methanol-to-olefin (MTO) reaction [110], and we believe that the Cu-SAPO-18 material may be a promising NH_3-SCR catalyst similar to Cu-SAPO-34. Consequently, study on the structure-activity relationship of Cu-based zeolite catalysts in the NH_3-SCR reaction will always be a hotspot in the field of environmental catalysis.

12.3.2.3 NH_3-SCR Reaction Mechanism Over Cu-CHA Catalysts

Many studies have shown that if an NH_3-SCR catalyst can oxidize NO to NO_2 *in situ*, its low temperature SCR activity will be significantly enhanced due to the effect of fast NH_3-SCR [111]. Thus, the ability to oxidize NO is commonly tested for the study of NH_3-SCR catalysts. However, Kwak et al. [112] noticed that no matter the Cu ion-exchange degree, the NO oxidation ability of Cu-SSZ-13 catalysts was very low, and the NO_x conversion was only slightly improved, with almost no formation of N_2O even under the fast NH_3-SCR condition. The above-mentioned "abnormal" results indicate that a totally different NH_3-SCR reaction mechanism may exist over Cu-SSZ-13 catalysts. Zhu et al. [113] analyzed the relationship between different NH_3 adsorbed species and the formation of N_2 and N_2O based on the results of *in situ* DRIFTS-MS studies over the Cu-SSZ-13 catalyst. They suggested that the formation of N_2O was associated with the reaction between NO_x and proton-adsorbed NH_3 *via* the formation and subsequent thermal decomposition of NH_4NO_3. Coordinated NH_3 adsorbed on Cu^{2+} sites is more active than NH_4^+ formed on Brønsted acid sites and contributes to a high selectivity for N_2. In a recent study, Kwak et al. [114] proposed a mechanism for the NH_3-SCR reaction over Cu-SSZ-13 catalysts based on the results of FTIR and ^{15}N solid state MAS-NMR, as shown in Scheme 12.3. They mentioned that the side-on Cu^+-NO^+ complexes were the key intermediates in this

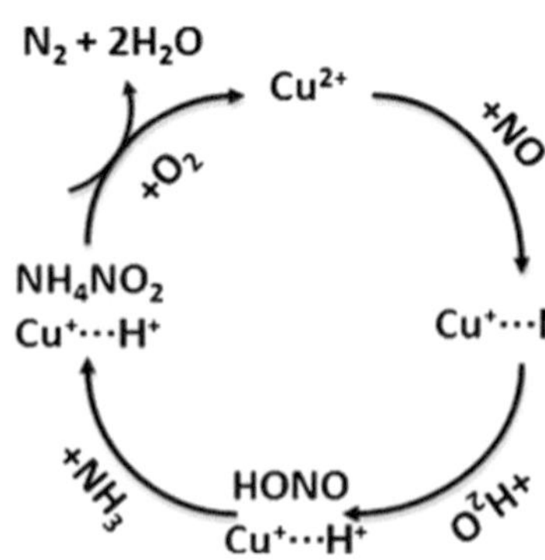

Scheme 12.3 Proposed mechanism for the selective catalytic reduction of NO with NH_3 over Cu-SSZ-13 (Reprinted from Ref. [114], Copyright 2013, with permission from John Wiley & Sons, Inc.)

mechanism, which were formed from a redox reaction between Cu^{2+} and NO. It is commonly recognized that H_2O always inhibits the NH_3-SCR reaction at low temperatures due to competitive adsorption, which can block the active sites for NO oxidation [115]. Interestingly, it was concluded by Kwak et al. [114] that the presence of H_2O was important for the NH_3-SCR reaction over the Cu-SSZ-13 catalyst, because the Cu^+-NO^+ complexes could react with H_2O easily to form HONO, and the resulting HONO could react with NH_3 to form NH_4NO_2, and then decompose to N_2 and H_2O. The reaction mechanism of the NH_3-SCR reaction over the Cu-SAPO-34 catalyst was different from that over the reported Cu-SSZ-13 catalyst. For example, Wang et al. [116] reported that NH_4NO_3 was mainly formed on Lewis acid sites of the Cu-SAPO-34 catalyst, and the Brønsted acid sites only acted as an NH_3 reservoir. Although the formed NH_4NO_3 typically inhibited the SCR reaction, gaseous NO could reduce the surface NH_4NO_3 to N_2 at temperatures as low as 100 °C over this catalyst. Thus, they proposed that the reaction between NO and NH_4NO_3 was likely the key factor for the low temperature NH_3-SCR activity for the studied Cu-SAPO-34 catalyst. Until now, there has been no report on the effect of NO_2 on the NH_3-SCR performance over the Cu-SAPO-34 catalysts, which still needs to be determined in future study for their better practical use.

12.3.2.4 One-Pot Synthesized Cu-CHA Catalysts for the NH_3-SCR Reaction

12.3.2.4.1 Improved One-Pot Synthesis Method for Cu-CHA Catalysts

Nowadays, most Cu-CHA zeolite catalysts are prepared by ion-exchange methods using Cu salt precursors and CHA zeolite supports. Although the conventional wet ion-exchange method is simple and effective, many stages are needed, such as hydrothermal synthesis of the CHA zeolite, high temperature calcination to remove SDAs, ion-exchange, filtration, washing and also activation procedures. If some intermediate steps could be simplified to a certain extent, it could be very beneficial for cost reduction and environmental protection during the preparation procedures of the final Cu-CHA NH_3-SCR catalysts. In addition, due to the limitations of channel size and exchange capacity of CHA zeolites, repeated ion-exchange procedures or long ion-exchange time is necessary to increase the Cu loading, and it is not easy to control the Cu loading and Cu status in the final product either. Furthermore, the synthesis of SSZ-13 zeolites requires the SDA TMAdaOH, which is very expensive [19], limiting the wide application of this material in industry. Therefore, it is highly imperative to improve the synthesis method to reduce the cost of Cu-SSZ-13 catalysts for the NH_3-SCR process.

Recently, a new strategy referred to as the one-pot synthesis method attracted researchers, which could introduce metal ions into the zeolite cages directly during the synthesis procedure. Thus, the metal species could disperse well inside the zeolite, and the preparation steps were greatly simplified with significantly lowered cost. In the one-pot synthesis method, the SDAs used were not just organics but also metal complexes. Researchers at BASF Corporation reported that the one-pot synthesis method could directly synthesize the Cu-SSZ-13 catalyst, and mixtures of organic SDAs in combination with aqueous solutions of Cu salts were necessary in the synthesis process, such as TMAdaOH and benzyl trimethylammonium, TMAdaOH and tetramethylammonium, or benzyl trimethylammonium and tetramethylammonium [117]. Afterwards, a more attractive and simpler Cu complex Cu-TEPA, which was prepared from $CuSO_4$ and tetraethylenepentamine (molar ratio of 1/1), was designed by Ren et al. [67] with low cost, and was very effective in the synthesis of the Cu-SSZ-13 initial product. Cu-TEPA is a promising candidate for synthesis of CHA-type zeolites mainly due to its good stability in strongly alkaline media, and more importantly, the molecular configuration of Cu-TEPA matches CHA cages very well, which helps in building the units of CHA zeolites effectively. Through adjusting the starting compositions in precursor gels, Cu-SSZ-13 catalysts with different Si/Al ratios could be synthesized, and the obtained samples exhibited high crystallinity and highly dispersed Cu species. The preliminary results by Ren et al. [67] showed that the one-pot synthesized Cu-SSZ-13 catalyst exhibited very good NH_3-SCR activity, especially in the low temperature range. Xie et al. [118] further improved the one-pot synthesis method of the Cu-SSZ-13 catalyst, which decreased the synthesis cost further. The aluminosilicate gels were adjusted to the mole ratio 14.8Na_2O: 3.0Al_2O_3: 600H_2O: 30SiO_2: 4Cu-TEPA, using smaller amounts of reagents (the amount of Cu-TEPA was reduced to 2/3 of the original recipe) to obtain the pure CHA structure.

12.3.2.4.2 Post-treatment of One-Pot Synthesized Cu-CHA Catalysts for the NH_3-SCR Reaction

Because the Cu loading in the initial Cu-SSZ-13 products from one-pot synthesis method was relatively high, much CuO was formed in the untreated catalyst after direct calcination. Thus, a post-treatment procedure was necessary to adjust the Cu loading before calcination. NH_4NO_3 solution is commonly used in the ion-exchange treatment, such as in the preparation procedures of H-zeolites. The results from Xie et al. [118] proved that ion exchange with NH_4NO_3 solution was also a useful method to avoid the formation of CuO particles in Cu-SSZ-13 catalysts prepared by a one-pot synthesis method. The selected $Cu_{3.8}$-SSZ-13 catalyst showed excellent NH_3-SCR performance and high N_2 selectivity from 150 to 550 °C and was also highly resistant to large space velocity. Over 90 % NO_x conversion could be maintained from 250 to 500 °C under an extremely high GHSV of 800,000 h^{-1} over the $Cu_{3.8}$-SSZ-13 catalyst (Fig. 12.7). Furthermore, this catalyst also exhibited rather high NH_3-SCR performance with the coexistence of CO_2, H_2O and C_3H_6 in the feed gas, suggesting an excellent durability towards poisoning components present in typical diesel exhaust. Xie et al. [118] also studied the structure-activity relationship of the one-pot synthesized Cu-SSZ-13 catalyst in the NH_3-SCR reaction using H_2-TPR and EPR methods. The H_2-TPR results confirmed that only isolated Cu^{2+} species existed in the optimal $Cu_{3.8}$-SSZ-13 catalyst, and EPR is an excellent technique for

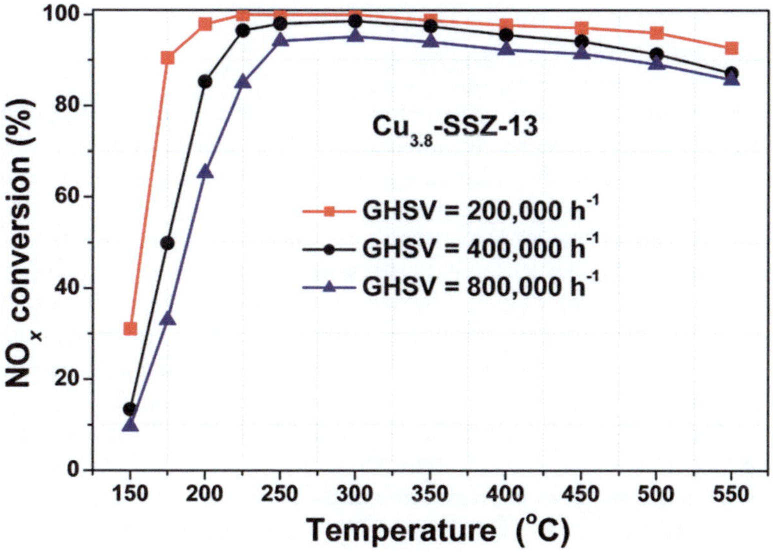

Fig. 12.7 NO_x conversion over $Cu_{3.8}$-SSZ-13 prepared by one-pot synthesis method and post-treatment by NH_4NO_3 solution under different GHSVs (Reprinted from Ref. [118], Copyright 2014, with permission from American Chemical Society)

identifying the coordination environment of isolated Cu^{2+} ions, because all the other Cu species (Cu-O-Cu or Cu^{+}) are EPR silent. Besides the two types of Cu^{2+} ions located at site I and site III as reported in previous literature, a third Cu^{2+} ions was found in the $Cu_{3.8}$-SSZ-13 catalyst, which was located near the eight-membered-ring window (site IV). The Cu^{2+} species located in site III showed the highest H_2 reduction temperature, and they were also less active in the NH_3-SCR reaction than the other two species located at other sites, mainly due to steric hindrance. Based on the activity test and characterization results, it was concluded that more isolated Cu^{2+} ions with easier reducibility and higher stability were responsible for the excellent NH_3-SCR performance over the optimal $Cu_{3.8}$-SSZ-13 catalyst.

Although NH_4NO_3 solution is effective in the post-treatment procedures for one-pot synthesized Cu-SSZ-13 catalysts, the potential risks of explosion as well as skin and eye irritation make the use of NH_4NO_3 dangerous, and at the same time, the wastewater containing NH_4NO_3 is also harmful to the aquatic environment. Based on the understanding of the role of NH_4NO_3 solution, it is believed that a dilute HNO_3 solution can also be suitable for the post-treatment of Cu-SSZ-13 catalysts, providing the proper acidic environment and necessary cations in the treating system. Thus, a new post-treatment method with dilute HNO_3 solution was further developed by Xie et al. [119], which reduced the nitrogen-containing pollutants in wastewater greatly. After optimization, the optimal $Cu_{3.9}Na_{0.8}$-SSZ-13 catalyst was obtained from the post-treatment by dilute HNO_3 solution with pH = 1, which showed better NH_3-SCR activity and hydrothermal stability than the $Cu_{3.8}$-SSZ-13 catalyst prepared by treatment with NH_4NO_3 solution.

12.3.2.4.3 Effect of Na Co-Cations on Cu-CHA Catalysts for the NH_3-SCR Reaction

During the synthesis of the Cu-SSZ-13 initial product, a relatively large amount of NaOH was used, and some Na^+ ions always were present in the final catalyst even after post-treatment by ion exchange. In order to investigate the effect of Na^+ ions on the NH_3-SCR performance over Cu-SSZ-13 catalysts prepared by the one-pot synthesis method, three catalysts with the same Cu content and different Na contents were prepared by Xie et al. [119]. With the increase of Na^+ content, only a small decrease of NO_x conversion was observed over the fresh Cu-SSZ-13 catalysts at low temperatures. However, after hydrothermal treatment, the NO_x conversion over the series of catalysts decreased greatly along with the increase of Na content. The Cu-SSZ-13 catalysts with higher Na^+ content showed poorer hydrothermal stability. The results of ^{27}Al and ^{29}Si NMR spectra indicated that the Cu-SSZ-13 catalysts with different Na^+ contents possessed almost the same zeolite structure and the same distribution of Si(nAl), indicating that the presence of Na^+ ions could not affect the stability of the zeolite structure. However, the H_2-TPR results illustrated that the stability of Cu species decreased seriously along with the increase of Na^+ ions in the final catalysts. Therefore, the Cu species with poor stability caused by high Na^+ content resulted in poor hydrothermal stability for the Cu-SSZ-13 catalysts. It is better to precisely control the amount of Na^+ ions maintained in the final Cu-SSZ-13 catalysts prepared by the one-pot synthesis method to obtain excellent NH_3-SCR performance and hydrothermal stability simultaneously.

12.3.2.4.4 Effect of Calcination Procedure on Cu-CHA Catalysts for the NH_3-SCR Reaction

In addition, as SDA, Cu-TEPA existed intact in the micropores of as-prepared Cu-SSZ-13 samples before calcination. The calcination procedure could remove the residual SDA and change the state of Cu species in the final product. In order to confirm the optimal preparation method for practical use, it is also very important to investigate the effects of the calcination procedure on the NH_3-SCR performance of Cu-SSZ-13 catalysts prepared by the one-pot synthesis method. Xie et al. [120] compared Cu-SSZ-13 catalysts calcined at 500, 600, 700 and 800 °C, and the catalyst calcined at 600 °C showed the highest NH_3-SCR activity and the best hydrothermal stability. With the increase of calcination temperature, some CuO was formed in the catalyst, resulting in the decrease of NH_3-SCR performance. If the calcination temperature was fixed at 600 °C, the Cu species with highest stability, which was responsible for its best hydrothermal stability, was achieved when the ramp rate was set at 1 °C/min.

12.3.2.4.5 Inhibitory Effect of NO_2 on Cu-CHA Catalysts for the NH_3-SCR Reaction

The oxidation of NH_3 could affect the NH_3-SCR activity, especially in the high temperature range. It was reported that NH_3 conversion was very low (less than 40 %) even at 550 °C in the NH_3 oxidation reaction over the optimal one-pot synthesized Cu-SSZ-13 catalyst [121]. This phenomenon indicated that, besides direct oxidation, a large amount of NH_3 was left for the NH_3-SCR reaction at high temperatures, which guaranteed a wide operating temperature window in the deNO_x process. Additionally, little N_2O was observed during the unselective oxidation of NH_3, which also contributed to the high N_2 selectivity to a certain extent. Verma et al. [122] suggested that the isolated Cu^{2+} species was not preferred for NO oxidation. There were only isolated Cu^{2+} ions observed in the one-pot synthesized Cu-SSZ-13 catalyst, therefore it was reasonable to obtain poor NO oxidation ability over these samples.

Interestingly, Xie et al. [123] found that the NH_3-SCR performance of the one-pot synthesized Cu-SSZ-13 catalyst became poorer with the increase of the NO_2 ratio in the low temperature range (≤200 °C). Based on the *in situ* DRIFTS results and kinetic studies under standard NH_3-SCR and fast NH_3-SCR reaction conditions, it was concluded that this inhibitory effect by NO_2 was mainly caused by NH_4NO_3 deposition. Detailed *in situ* DRIFTS results by Xie et al. [123] proved that the reactions between nitrate species and gaseous NH_3 played an important role in the NH_3-SCR reaction at low temperatures over this Cu-SSZ-13 catalyst, and the key step in the deNO_x reaction was the formation of reactive nitrate on active Cu sites. They observed that the nitrate species could form effectively on the Cu-SSZ-13 catalyst surface as soon as $NO+O_2$ was introduced, indicating the good oxidation ability of the studied sample. However, the formation speed of the nitrate species was much slower when NO was replaced by NO_2 in the feed gas, and Xie et al. [123] suggested that this difference was caused by the large kinetic diameter of NO_2, which was actually larger than the small pores of the CHA structure. Because of the slow speed for NO_2 to enter the Cu-CHA pores, some NO_2 could not form reactive nitrates on Cu sites efficiently but combined with NH_4^+ on Brønsted acid sites to form NH_4NO_3. Ciardelli et al. [124] reported that NH_4NO_3 was an important intermediate in the fast NH_3-SCR reaction, and the reaction between NH_4NO_3 and NO was the rate-determining step at low temperatures. If the formation of NH_4NO_3 was faster than its decomposition, the fast NH_3-SCR phenomenon would not be obvious. Although the formed NH_4NO_3 was considered an important intermediate species over other Cu-CHA catalysts, the consumption rate of NH_4NO_3 by NO was actually slower than its formation rate over the one-pot synthesized Cu-SSZ-13 catalyst. Therefore, the severely accumulated NH_4NO_3 on the catalyst surface blocked the active sites, and the fast NH_3-SCR phenomenon was not obvious either over this Cu-SSZ-13 catalyst. These mechanistic studies indicated that increasing the NO_2 ratio directly in the feed gas was not an effective way to improve NH_3-SCR performance for the Cu-SSZ-13 catalyst prepared by the one-pot synthesis method.

12.3.2.4.6 Co-Template Synthesis of Small Pore Cu-Zeolite Catalysts for the NH_3-SCR Reaction

Based on the effective template for CHA cages Cu-TEPA, other one-pot synthesis methods for the preparation of Cu-CHA catalysts were also developed. For example, Martínez-Franco et al. [125] successfully synthesized the Cu-SSZ-13 catalyst using SDAs with a combination of Cu-TEPA and TMAdaOH. The use of two templates could introduce the designed Cu amount and control the Si/Al ratio in the obtained Cu-SSZ-13 catalyst at the same time. Excellent NH_3-SCR activity was observed over this catalyst, with high NO_x conversion even after hydrothermal treatment at 750 °C for 13 h. Besides, the results from Martínez-Franco et al. [126] also showed that the Cu-SAPO-34 catalyst with high crystallization degree could be obtained easily using Cu-TEPA and another small organic molecule (diethylamine) acting as co-template, with controllable Cu loading from 3.4 to 10.4 wt.%. The obtained Cu-SAPO-34 catalysts showed excellent NH_3-SCR performance and good hydrothermal stability under high space velocity, and they believed that this synthesis methodology was very promising for the industrial application of Cu-SAPO-34 catalysts.

Besides the Cu-CHA catalysts as mentioned above, co-template synthesis routes have also been used to prepare other types of Cu-zeolite NH_3-SCR catalysts. For instance, Picone et al. [127] used Cu cyclam (1,4,8,11-tetraazacyclotetradecane) and tetraethylammonium (TEA^+) acting as co-templates to directly synthesize a Cu-SAPO STA-7 catalyst with similar structure to that of Cu-SAPO-34, showing comparable performance in the NH_3-SCR reaction to that of Cu-ZSM-5 prepared by the ion-exchange method and higher performance than that of Cu-SAPO STA-7 with similar Cu content prepared by an aqueous ion-exchange method. The characterization results showed that a more homogeneous distribution of Cu species in Cu-SAPO STA-7 was achieved *via* direct synthesis, which was probably the main reason for its high deNO_x efficiency.

12.4 Conclusions and Perspectives

Fe-zeolite catalysts prepared by various techniques are highly active and selective in the NH_3-SCR reaction, especially at medium to high temperatures. The SSIE and two-step IWI methods can be used for large-scale production of effective Fe-zeolite NH_3-SCR catalysts. All Fe species in Fe-zeolite catalysts may be active for the deNO_x process, but the isolated Fe monomers are considered to be the most active species. Brønsted acid sites may not be required directly in the NH_3-SCR process for adsorbing or activating NH_3, but they are necessary to bind and disperse the reactive Fe^{3+} ions in the preparation process of Fe-based zeolite catalysts and can accelerate the SCR reaction at low temperatures through an acid-catalysis effect. Hydrothermal deactivation can result in the dealumination of Fe-zeolite supports, leading to the decrease of Brønsted acid sites or breakdown of the framework structure, and more severely, the migration of active Fe^{3+} species to form clustered Fe_xO_y/Fe_2O_3 species, resulting in the decline of NH_3-SCR activity and N_2 selectivity. The hydrothermal stability of Fe-zeolite catalysts can be enhanced by a second ion-exchange of rare earth metals or by decreasing the residual protons in the synthesized materials. The low temperature deNO$_x$ efficiency over Fe-zeolite catalysts can be significantly increased in the presence of NO_2 through a fast NH_3-SCR effect. However, N_2O formation can be observed in the fast NH_3-SCR reaction over Fe-zeolite catalysts and accelerates with increasing NO_2/NO_x ratio. Hydrothermally deactivated Fe-zeolite catalysts can still show relatively high activity in the fast NH_3-SCR reaction, yet N_2O formation is more apparent over the aged Fe-zeolite catalysts than over fresh ones. Therefore, the promotion of low temperature deNO$_x$ activity and the enhancement of hydrothermal stability are still the most important challenges for the practical application of Fe-zeolite catalysts in the NH_3-SCR reaction. The tuning of the reaction atmosphere such as raising the NO_2 ratio in total NO_x to facilitate the "fast SCR" reaction is very effective to increase the low temperature activity of Fe-zeolite catalysts. However, the increase of N_2O formation should be paid careful attention. In further study, the development of new types of Fe-zeolite catalysts might be a possible way to significantly enhance their hydrothermal stability for practical use.

Cu-ZSM-5, which exhibits excellent NH_3-SCR activity at low temperatures and maintains a broad operating temperature window, attracts much attention from researchers in the field of catalytic deNO$_x$ from diesel engine exhaust. Due to its poor hydrothermal stability and HC poisoning resistance, however, the application of the Cu-ZSM-5 catalyst in the real world is limited. Cu-zeolite catalysts with the CHA crystal structure have become promising candidates for the NH_3-SCR reaction, such as Cu-SSZ-13 and Cu-SAPO-34. Owing to their small pore zeolite structures, the Cu-CHA zeolite catalysts show improved NH_3-SCR activity, N_2 selectivity, HC resistance and hydrothermal stability compared to the existing Cu-zeolite catalysts. The wet ion-exchange method is efficient for the preparation of Cu-CHA catalysts, yet many steps are needed in this process and sometimes the cost is rather high. The one-pot synthesis method can introduce controllable Cu loading with much lower cost during the synthesis procedure of the CHA structure, and Cu-TEPA is a novel template for the effective synthesis of Cu-SSZ-13 initial product. In order to avoid the formation of CuO in the final Cu-SSZ-13 catalyst, post-treatment with dilute HNO_3 solution is needed, which can modify the Cu content effectively. The optimal catalyst exhibits excellent NH_3-SCR performance and hydrothermal stability, and due to negative effects of Na^+ ions on the hydrothermal stability, it is better to precisely control the amount of residual Na^+ ions in the final products. NO_2 inhibits the NH_3-SCR performance over one-pot-synthesized Cu-SSZ-13 catalysts in the low temperature range, and increasing the NO_2 ratio directly in the diesel exhaust is not an effective way to improve the deNO$_x$ efficiency in practical use. Co-template routes for the synthesis of Cu-CHA catalysts are also noteworthy, and a post-treatment procedure is usually not necessary. Therefore, the one-pot synthesis methods for Cu-based zeolite catalysts with small pore size using relatively cheap Cu-containing templates allow researchers to tune the Cu content, Cu distribution, Si/Al ratio and thus the NH_3-SCR activity efficiently, although this approach is restricted to those zeolitic systems that can be prepared using coordination complexes as templates. For practical use of this approach in industry, we believe, there is still a long but exciting way to go.

References

1. Granger P, Parvulescu VI (2011) Catalytic NO_x abatement systems for mobile sources: from three-way to lean burn after-treatment technologies. Chem Rev 111:3155–3207
2. Liu Z, Woo SI (2006) Recent advances in catalytic deNO_x science and technology. Catal Rev 48:43–89
3. Kwak JH, Tran D, Burton SD, Szanyi J, Lee JH, Peden CHF (2012) Effects of hydrothermal aging on NH_3-SCR reaction over Cu/zeolites. J Catal 287:203–209
4. Grossale A, Nova I, Tronconi E (2008) Study of a Fe-zeolite-based system as NH_3-SCR catalyst for diesel exhaust after treatment. Catal Today 136:18–27
5. Gabrielsson PLT (2004) Urea-SCR in automotive applications. Top Catal 28:177–184
6. Johnson TV (2009) Diesel emission control in review. SAE Int J Fuels Lubr 2:1–12
7. Brandenberger S, Kröcher O, Tissler A, Althoff R (2008) The state of the art in selective catalytic reduction of NOx by ammonia using metal-exchanged zeolite catalysts. Catal Rev 50:492–531
8. Feng X, Hall WK (1997) FeZSM-5: a durable SCR catalyst for NO_x removal from combustion streams. J Catal 166:368–376
9. Ma A-Z, Grünert W (1999) Selective catalytic reduction of NO by ammonia over Fe-ZSM-5 catalysts. Chem Commun 1:71–72
10. Long RQ, Yang RT (1999) Superior Fe-ZSM-5 catalyst for selective catalytic reduction of nitric oxide by ammonia. J Am Chem Soc 121:5595–5596
11. Sjövall H, Olsson L, Fridell E, Blint RJ (2006) Selective catalytic reduction of NO_x with NH_3 over Cu-ZSM-5-the effect of changing the gas composition. Appl Catal B 64:180–188
12. Sjövall H, Blint RJ, Olsson L (2009) Detailed kinetic modeling of NH_3 SCR over Cu-ZSM-5. Appl Catal B 92:138–153
13. Wilken N, Wijayanti K, Kamasamudram K, Currier NW, Vedaiyan R, Yezerets A, Olsson L (2012) Mechanistic investigation of hydrothermal aging of Cu-Beta for ammonia SCR. Appl Catal B 111–112:58–66
14. Park J, Park H, Baik J, Nam I, Shin C, Lee J, Cho B, Oh S (2006) Hydrothermal stability of CuZSM5 catalyst in reducing NO by NH_3 for the urea selective catalytic reduction process. J Catal 240:47–57
15. Kwak JH, Tonkyn RG, Kim DH, Szanyi J, Peden CHF (2010) Excellent activity and selectivity of Cu-SSZ-13 in the selective catalytic reduction of NO_x with NH_3. J Catal 275:187–190
16. Moliner M, Franch C, Palomares E, Grill M, Corma A (2012) Cu-SSZ-39, an active and hydrothermally stable catalyst for the selective catalytic reduction of NO_x. Chem Commun 48:8264–8266
17. Sultana A, Nanba T, Sasaki M, Haneda M, Suzuki K, Hamada H (2011) Selective catalytic reduction of NO_x with NH_3 over different copper exchanged zeolites in the presence of decane. Catal Today 164:495–499
18. Ye Q, Wang L, Yang RT (2012) Activity, propene poisoning resistance and hydrothermal stability of copper exchanged chabazite-like zeolite catalysts for SCR of NO with ammonia in comparison to Cu/ZSM-5. Appl Catal A 427–428:24–34
19. Zones SI (1991) Conversion of faujasites to high-silica chabazite SSZ-13 in the presence of N, N, N-trimethyl-1-adamantammonium iodide. J Chem Soc Faraday Trans 87:3709–3716

20. Long RQ, Yang RT (1999) Catalytic performance of Fe-ZSM-5 catalysts for selective catalytic reduction of nitric oxide by ammonia. J Catal 188:332–339
21. Rahkamaa-Tolonen K, Maunula T, Lomma M, Huuhtanen M, Keiski RL (2005) The effect of NO_2 on the activity of fresh and aged zeolite catalysts in the NH_3-SCR reaction. Catal Today 100:217–222
22. Iwasaki M, Yamazaki K, Shinjoh H (2011) NO_x reduction performance of fresh and aged Fe-zeolites prepared by CVD: effects of zeolite structure and Si/Al_2 ratio. Appl Catal B 102:302–309
23. Long RQ, Yang RT (2001) Fe-ZSM-5 for selective catalytic reduction of NO with NH_3: a comparative study of different preparation techniques. Catal Lett 74:201–205
24. Long RQ, Yang RT (2002) Reaction mechanism of selective catalytic reduction of NO with NH_3 over Fe-ZSM-5 catalyst. J Catal 207:224–231
25. Rivallan M, Berlier G, Ricchiardi G, Zecchina A, Nechita M-T, Olsbye U (2008) Characterisation and catalytic activity in deNO_x reactions of Fe-ZSM-5 zeolites prepared via ferric oxalate precursor. Appl Catal B 84:204–213
26. Brandenberger S, Kröcher O, Casapu M, Tissler A, Althoff R (2011) Hydrothermal deactivation of Fe-ZSM-5 catalysts for the selective catalytic reduction of NO with NH_3. Appl Catal B 101:649–659
27. Chen H-Y, Sachtler WMH (1998) Activity and durability of Fe/ZSM-5 catalysts for lean burn NO_x reduction in the presence of water vapor. Catal Today 42:73–83
28. Krishna K, Seijger GBF, van den Bleek CM, Makkee M, Mul G, Calis HPA (2003) Selective catalytic reduction of NO with NH_3 over Fe-ZSM-5 catalysts prepared by sublimation of $FeCl_3$ at different temperatures. Catal Lett 86:121–132
29. Qi G, Gatt JE, Yang RT (2004) Selective catalytic oxidation (SCO) of ammonia to nitrogen over Fe-exchanged zeolites prepared by sublimation of $FeCl_3$. J Catal 226:120–128
30. Kumar MS, Schwidder M, Grünert W, Brückner A (2004) On the nature of different iron sites and their catalytic role in Fe-ZSM-5 deNO_x catalysts: new insights by a combined EPR and UV/Vis spectroscopic approach. J Catal 227:384–397
31. Battiston AA, Bitter JH, de Groot FMF, Overweg AR, Stephan O, van Bokhoven JA, Kooyman PJ, van der Spek C, Vankó G, Koningsberger DC (2003) Evolution of Fe species during the synthesis of over-exchanged Fe/ZSM5 obtained by chemical vapor deposition of $FeCl_3$. J Catal 213:251–271
32. Marturano P, DrozdovaÂ L, Pirngruber GD, Kogelbauer A, Prins R (2001) The mechanism of formation of the Fe species in Fe-ZSM-5 prepared by CVD. Phys Chem Chem Phys 3:5585–5595
33. Rauscher M, Kesore K, MoÈnnig R, Schwieger W, Tiûler A, Turek T (1999) Preparation of a highly active Fe-ZSM-5 catalyst through solid-state ion exchange for the catalytic decomposition of N_2O. Appl Catal A 184:249–256
34. Devadas M, Kröcher O, Elsener M, Wokaun A, Mitrikas G, Söger N, Pfeifer M, Demel Y, Mussmann L (2007) Characterization and catalytic investigation of Fe-ZSM5 for urea-SCR. Catal Today 119:137–144
35. Qi G, Yang RT (2005) Ultra-active Fe/ZSM-5 catalyst for selective catalytic reduction of nitric oxide with ammonia. Appl Catal B 60:13–22
36. Qi G, Yang RT (2005) Selective catalytic oxidation (SCO) of ammonia to nitrogen over Fe/ZSM-5 catalysts. Appl Catal A 287:25–33
37. Delahay G, Valade D, Guzmanvargas A, Coq B (2005) Selective catalytic reduction of nitric oxide with ammonia on Fe-ZSM-5 catalysts prepared by different methods. Appl Catal B 55:149–155
38. Iwasaki M, Yamazaki K, Banno K, Shinjoh H (2008) Characterization of Fe/ZSM-5 deNO_x catalysts prepared by different methods: relationships between active Fe sites and NH_3-SCR performance. J Catal 260:205–216
39. Shi X, Liu F, Shan W, He H (2012) Hydrothermal deactivation of Fe-ZSM-5 prepared by different methods for the selective catalytic reduction of NO_x with NH_3. Chin J Catal 33:454–464

40. Schwidder M, Heikens S, Detoni A, Geisler S, Berndt M, Bruckner A, Grunert W (2008) The role of NO_2 in the selective catalytic reduction of nitrogen oxides over Fe-ZSM-5 catalysts: active sites for the conversion of NO and of NO/NO_2 mixtures. J Catal 259:96–103
41. Schwidder M, Kumar M, Klementiev K, Pohl M, Bruckner A, Grunert W (2005) Selective reduction of NO with Fe-ZSM-5 catalysts of low Fe content I. Relations between active site structure and catalytic performance. J Catal 231:314–330
42. Brandenberger S, Kröcher O, Tissler A, Althoff R (2010) The determination of the activities of different iron species in Fe-ZSM-5 for SCR of NO by NH_3. Appl Catal B 95:348–357
43. Høj M, Beier MJ, Grunwaldt J-D, Dahl S (2009) The role of monomeric iron during the selective catalytic reduction of NO_x by NH_3 over Fe-BEA zeolite catalysts. Appl Catal B 93:166–176
44. Schwidder M, Santhosh Kumar M, Bruckner A, Grunert W (2005) Active sites for NO reduction over Fe-ZSM-5 catalysts. Chem Commun 6:805–807
45. Brandenberger S, Kröcher O, Tissler A, Althoff R (2011) Effect of structural and preparation parameters on the activity and hydrothermal stability of metal-exchanged ZSM-5 in the selective catalytic reduction of NO by NH_3. Ind Eng Chem Res 50:4308–4319
46. Schwidder M, Santhosh Kumar M, Bentrup U, Pérez-Ramírez J, Brückner A, Grünert W (2008) The role of Brønsted acidity in the SCR of NO over Fe-MFI catalysts. Microporous Mesoporous Mater 111:124–133
47. Qi G, Wang Y, Yang RT (2008) Selective catalytic reduction of nitric oxide with ammonia over ZSM-5 based catalysts for diesel engine applications. Catal Lett 121:111–117
48. Li M, Yeom Y, Weitz E, Sachtler WMH (2006) An acid catalyzed step in the catalytic reduction of NO_x to N_2. Catal Lett 112:129–132
49. Savara A, Li M-J, Sachtler WMH, Weitz E (2008) Catalytic reduction of NH_4NO_3 by NO: effects of solid acids and implications for low temperature deNO_x processes. Appl Catal B 81:251–257
50. Brandenberger S, Kröcher O, Wokaun A, Tissler A, Althoff R (2009) The role of Brønsted acidity in the selective catalytic reduction of NO with ammonia over Fe-ZSM-5. J Catal 268:297–306
51. Klukowski D, Balle P, Geiger B, Wagloehner S, Kureti S, Kimmerle B, Baiker A, Grunwaldt JD (2009) On the mechanism of the SCR reaction on Fe/HBEA zeolite. Appl Catal B 93:185–193
52. Kröcher O, Devadas M, Elsener M, Wokaun A, Söger N, Pfeifer M, Demel Y, Mussmann L (2006) Investigation of the selective catalytic reduction of NO by NH_3 on Fe-ZSM5 monolith catalysts. Appl Catal B 66:208–216
53. He C, Wang Y, Cheng Y, Lambert CK, Yang RT (2009) Activity, stability and hydrocarbon deactivation of Fe/Beta catalyst for SCR of NO with ammonia. Appl Catal A 368:121–126
54. Ma L, Li J, Arandiyan H, Shi W, Liu C, Fu L (2012) Influence of calcination temperature on Fe/HBEA catalyst for the selective catalytic reduction of NO_x with NH_3. Catal Today 184:145–152
55. Pieterse JAZ, Pirngruber GD, van Bokhoven JA, Booneveld S (2007) Hydrothermal stability of Fe-ZSM-5 and Fe-BEA prepared by wet ion-exchange for N_2O decomposition. Appl Catal B 71:16–22
56. Lee H-T, Rhee H-K (1999) Stability of Fe/ZSM-5 de-NO_x catalyst: effects of iron loading and remaining Brønsted acid sites. Catal Lett 61:71–76
57. Iwasaki M, Shinjoh H (2011) Hydrothermal stability enhancement by sequential ion-exchange of rare earth metals on Fe/BEA zeolites used as NO reduction catalysts. Chem Commun 47:3966–3968
58. Long R, Yang RT (2000) Characterization of Fe-ZSM-5 catalyst for selective catalytic reduction of nitric oxide by ammonia. J Catal 194:80–90
59. Carja G, Delahay G, Signorile C, Coq B (2004) Fe-Ce-ZSM-5 a new catalyst of outstanding properties in the selective catalytic reduction of NO with NH_3. Chem Commun 12:1404–1405
60. Heo I, Lee Y, Nam I-S, Choung JW, Lee J-H, Kim H-J (2011) Effect of hydrocarbon slip on NO removal activity of CuZSM5, FeZSM5 and V_2O_5/TiO_2 catalysts by NH_3. Microporous Mesoporous Mater 141:8–15

61. Li J, Zhu R, Cheng Y, Lambert CK, Yang RT (2010) Mechanism of propene poisoning on Fe-ZSM-5 for selective catalytic reduction of NO_x with ammonia. Environ Sci Technol 44:1799–1805
62. Ma L, Li J, Cheng Y, Lambert CK, Fu L (2012) Propene poisoning on three typical Fe-zeolites for SCR of NO_x with NH_3: from mechanism study to coating modified architecture. Environ Sci Technol 46:1747–1754
63. Malpartida I, Marie O, Bazin P, Daturi M, Jeandel X (2011) An *operando* IR study of the unburnt HC effect on the activity of a commercial automotive catalyst for NH_3-SCR. Appl Catal B 102:190–200
64. Kern P, Klimczak M, Heinzelmann T, Lucas M, Claus P (2010) High-throughput study of the effects of inorganic additives and poisons on NH_3-SCR catalysts. Part II: Fe-zeolite catalysts. Appl Catal B 95:48–56
65. Silver RG, Stefanick MO, Todd BI (2008) A study of chemical aging effects on HDD Fe-zeolite SCR catalyst. Catal Today 136:28–33
66. Shwan S, Jansson J, Olsson L, Skoglundh M (2014) Chemical deactivation of Fe-BEA as NH_3-SCR catalyst-effect of phosphorous. Appl Catal B 147:111–123
67. Ren L, Zhu L, Yang C, Chen Y, Sun Q, Zhang H, Li C, Nawaz F, Meng X, Xiao FS (2011) Designed copper-amine complex as an efficient template for one-pot synthesis of Cu-SSZ-13 zeolite with excellent activity for selective catalytic reduction of NO_x by NH_3. Chem Commun 47:9789–9791
68. Qi G, Yang RT (2005) Low-temperature SCR of NO with NH_3 over noble metal promoted Fe-ZSM-5 catalysts. Catal Lett 100:243–246
69. Kim YJ, Kwon HJ, Heo I, Nam I-S, Cho BK, Choung JW, Cha M-S, Yeo GK (2012) Mn-Fe/ZSM5 as a low-temperature SCR catalyst to remove NO_x from diesel engine exhaust. Appl Catal B 126:9–21
70. Sultana A, Sasaki M, Suzuki K, Hamada H (2013) Tuning the NO_x conversion of Cu-Fe/ZSM-5 catalyst in NH_3-SCR. Catal Commun 41:21–25
71. Koebel M, Elsener M, Kleemann M (2000) Urea-SCR: a promising technique to reduce NO_x emissions from automotive diesel engines. Catal Today 59:335–345
72. Devadas M, Krocher O, Elsener M, Wokaun A, Soger N, Pfeifer M, Demel Y, Mussmann L (2006) Influence of NO_2 on the selective catalytic reduction of NO with ammonia over Fe-ZSM5. Appl Catal B 67:187–196
73. Sun Q, Gao Z-X, Chen H-Y, Sachtler WMH (2001) Reduction of NO_x with ammonia over Fe/MFI: reaction mechanism based on isotopic labeling. J Catal 201:88–99
74. Iwasaki M, Shinjoh H (2010) A comparative study of "standard", "fast" and "NO_2" SCR reactions over Fe/zeolite catalyst. Appl Catal A 390:71–77
75. Grossale A, Nova I, Tronconi E, Chatterjee D, Weibel M (2008) The chemistry of the NO/NO_2-NH_3 "fast" SCR reaction over Fe-ZSM5 investigated by transient reaction analysis. J Catal 256:312–322
76. Grossale A, Nova I, Tronconi E, Chatterjee D, Weibel M (2009) NH_3-NO/NO_2 SCR for diesel exhausts after treatment: reactivity, mechanism and kinetic modelling of commercial Fe- and Cu-promoted zeolite catalysts. Top Catal 52:1837–1841
77. Grossale A, Nova I, Tronconi E (2009) Ammonia blocking of the "Fast SCR" reactivity over a commercial Fe-zeolite catalyst for diesel exhaust after treatment. J Catal 265:141–147
78. Forzatti P, Nova I, Tronconi E (2009) Enhanced NH_3 selective catalytic reduction for NO_x abatement. Angew Chem Int Ed 48:8366–8368
79. Shi X, Liu F, Xie L, Shan W, He H (2013) NH_3-SCR performance of fresh and hydrothermally aged Fe-ZSM-5 in standard and fast selective catalytic reduction reactions. Environ Sci Technol 47:3293–3298
80. PaÃrvulescua VI, Grange P, Delmon B (1998) Catalytic removal of NO. Catal Today 46:233–316
81. Baik JH, Yim SD, Nam I-S, Mok YS, Lee J-H, Cho BK, Oh SH (2004) Control of NO_x emissions from diesel engine by selective catalytic reduction (SCR) with urea. Top Catal 30–31:37–41

82. Komatsu T, Nunokawa M, Moon IS, Takahara T, Namba S, Yashima T (1994) Kinetic studies of reduction of nitric oxide with ammonia on Cu^{2+}-exchanged zeolites. J Catal 148:427–437
83. Sultana A, Nanba T, Haneda M, Hamada H (2009) SCR of NO_x with NH_3 over Cu/NaZSM-5 and Cu/HZSM-5 in the presence of decane. Catal Commun 10:1859–1863
84. Richter M, Eckelt R, Parlitz B, Fricke R (1998) Low-temperature conversion of NO, to N_2 by zeolite-fixed ammonium ions. Appl Catal B 15:129–146
85. Moden B, Donohue JM, Cormier WE, Li H-X (2008) Effect of Cu-loading and structure on the activity of Cu-exchanged zeolites for NH_3-SCR. In: Stud Surf Sci Catal, edn. Elsevier, Paris: France, pp 1219–1222
86. Blakeman PG, Burkholder EM, Chen H-Y, Collier JE, Fedeyko JM, Jobson H, Rajaram RR (2014) The role of pore size on the thermal stability of zeolite supported Cu SCR catalysts. Catal Today 231:56–63
87. Fickel DW, D'Addio E, Lauterbach JA, Lobo RF (2011) The ammonia selective catalytic reduction activity of copper-exchanged small-pore zeolites. Appl Catal B 102:441–448
88. Luo J-Y, Oh H, Henry C, Epling W (2012) Effect of C_3H_6 on selective catalytic reduction of NO_x by NH_3 over a Cu/zeolite catalyst: a mechanistic study. Appl Catal B 123–124:296–305
89. Aoki K, Kusakabe K, Morooka S (2000) Separation of gases with an A-Type zeolite membrane. Ind Eng Chem Res 39:2245–2251
90. Gao F, Walter ED, Karp EM, Luo J, Tonkyn RG, Kwak JH, Szanyi J, Peden CHF (2013) Structure-activity relationships in NH_3-SCR over Cu-SSZ-13 as probed by reaction kinetics and EPR studies. J Catal 300:20–29
91. Schmieg SJ, Oh SH, Kim CH, Brown DB, Lee JH, Peden CHF, Kim DH (2012) Thermal durability of Cu-CHA NH_3-SCR catalysts for diesel NO_x reduction. Catal Today 184:252–261
92. Ma L, Cheng Y, Cavataio G, McCabe RW, Fu L, Li J (2013) Characterization of commercial Cu-SSZ-13 and Cu-SAPO-34 catalysts with hydrothermal treatment for NH_3-SCR of NO_x in diesel exhaust. Chem Eng J 225:323–330
93. Bordiga S, Regli L, Cocina D, Lamberti C, Bjørgen M, Lillerud KP (2005) Assessing the acidity of high silica chabazite H-SSZ-13 by FTIR using CO as molecular probe comparison with H-SAPO-34. J Phys Chem C 109:2779–2784
94. Wang J, Yu T, Wang X, Qi G, Xue J, Shen M, Li W (2012) The influence of silicon on the catalytic properties of Cu/SAPO-34 for NO_x reduction by ammonia-SCR. Appl Catal B 127:137–147
95. Yu T, Wang J, Shen M, Li W (2013) NH_3-SCR over Cu/SAPO-34 catalysts with various acid contents and low Cu loading. Catal Sci Technol 3:3234–3241
96. Wang L, Gaudet JR, Li W, Weng D (2013) Migration of Cu species in Cu/SAPO-34 during hydrothermal aging. J Catal 306:68–77
97. Fan S, Xue J, Yu T, Fan D, Hao T, Shen M, Li W (2013) The effect of synthesis methods on Cu species and active sites over Cu/SAPO-34 for NH_3-SCR reaction. Catal Sci Technol 3:2357–2364
98. Deka U, Lezcano-Gonzalez I, Warrender SJ, Lorena Picone A, Wright PA, Weckhuysen BM, Beale AM (2013) Changing active sites in Cu-CHA catalysts: deNO_x selectivity as a function of the preparation method. Microporous Mesoporous Mater 166:144–152
99. Wang L, Li W, Qi G, Weng D (2012) Location and nature of Cu species in Cu/SAPO-34 for selective catalytic reduction of NO with NH_3. J Catal 289:21–29
100. Gao F, Walter ED, Washton NM, Szanyi J, Peden CHF (2013) Synthesis and evaluation of Cu-SAPO-34 catalysts for ammonia selective catalytic reduction. 1. Aqueous solution ion exchange. ACS Catal 3:2083–2093
101. Dědeček J, Wichterlová B, Kubát P (1999) Siting of the Cu^+ ions in dehydrated ion exchanged synthetic and natural chabasites: a Cu^+ photoluminescence study. Microporous Mesoporous Mater 32:63–74
102. Zamadies M, Chen X, Kevan L (1992) Study of Cu(II) location and adsorbate interaction in CuH-SAPO-34 molecular sieve by electron Sp4 resoname and electron spin echo modulath spectroscopies. J Phys Chem 96:2652–2657
103. Fickel DW, Lobo RF (2010) Copper coordination in Cu-SSZ-13 and Cu-SSZ-16 investigated by Variable-Temperature XRD. J Phys Chem C 114:1633–1640

104. Korhonen ST, Fickel DW, Lobo RF, Weckhuysen BM, Beale AM (2011) Isolated Cu^{2+} ions: active sites for selective catalytic reduction of NO. Chem Commun 47:800–802
105. Deka U, Juhin A, Eilertsen EA, Emerich H, Green MA, Korhonen ST, Weckhuysen BM, Beale AM (2012) Confirmation of isolated Cu^{2+} ions in SSZ-13 Zeolite as active sites in NH_3-selective catalytic reduction. J Phys Chem C 116:4809–4818
106. Kispersky VF, Kropf AJ, Ribeiro FH, Miller JT (2012) Low absorption vitreous carbon reactors for operando XAS: a case study on Cu/Zeolites for selective catalytic reduction of NO_x by NH_3. Phys Chem Chem Phys 14:2229–2238
107. McEwen JS, Anggara T, Schneider WF, Kispersky VF, Miller JT, Delgass WN, Ribeiro FH (2012) Integrated *operando* X-ray absorption and DFT characterization of Cu-SSZ-13 exchange sites during the selective catalytic reduction of NO_x with NH_3. Catal Today 184:129–144
108. Kwak JH, Zhu H, Lee JH, Peden CH, Szanyi J (2012) Two different cationic positions in Cu-SSZ-13? Chem Commun 48:4758–4760
109. Xue J, Wang X, Qi G, Wang J, Shen M, Li W (2013) Characterization of copper species over Cu/SAPO-34 in selective catalytic reduction of NO_x with ammonia: relationships between active Cu sites and de-NO_x performance at low temperature. J Catal 297:56–64
110. Aguayo AT, Gayubo AG, Vivanco R, Olazar M, Bilbao J (2005) Role of acidity and microporous structure in alternative catalysts for the transformation of methanol into olefins. Appl Catal A 283:197–207
111. Colombo M, Nova I, Tronconi E (2012) Detailed kinetic modeling of the NH_3-NO/NO_2 SCR reactions over a commercial Cu-zeolite catalyst for Diesel exhausts after treatment. Catal Today 197:243–255
112. Kwak JH, Tran D, Szanyi J, Peden CHF, Lee JH (2012) The Effect of copper loading on the selective catalytic reduction of Nitric Oxide by ammonia over Cu-SSZ-13. Catal Lett 142:295–301
113. Zhu H, Kwak JH, Peden CHF, Szanyi J (2013) In situ DRIFTS-MS studies on the oxidation of adsorbed NH_3 by NO_x over a Cu-SSZ-13 zeolite. Catal Today 205:16–23
114. Kwak JH, Lee JH, Burton SD, Lipton AS, Peden CH, Szanyi J (2013) A common intermediate for N_2 formation in enzymes and zeolites: side-on Cu-nitrosyl complexes. Angew Chem Int Ed 52:9985–9989
115. Metkar PS, Balakotaiah V, Harold MP (2012) Experimental and kinetic modeling study of NO oxidation: comparison of Fe and Cu-zeolite catalysts. Catal Today 184:115–128
116. Wang D, Zhang L, Kamasamudram K, Epling WS (2013) In Situ-DRIFTS study of selective catalytic reduction of NO_x by NH_3 over Cu-exchanged SAPO-34. ACS Catal 3:871–881
117. Trukhan N, Mueller U, Bull I (2011) US Patent 2011/0076229, 31 Mar 2011
118. Xie L, Liu F, Ren L, Shi X, Xiao FS, He H (2014) Excellent performance of one-pot synthesized Cu-SSZ-13 catalyst for the selective catalytic reduction of NO with NH_3. Environ Sci Technol 48:566–572
119. Xie L, Liu F, Xiao FS, He H. Effects of post-treatment method and Na co-cation on the hydrothermal stability of Cu-SSZ-13 catalyst for the selective catalytic reduction of NO_x with NH_3. To be published
120. Xie L, Liu F, Shi X, He H (2013) Influence of calcination procedure on NH_3-SCR performance of Cu-SSZ-13 catalysts prepared by one-pot synthesis method. Paper presented at the 6th Asia-Pacific Congress on Catalysis, Taipei, Taiwan, 13–17 October, 2013
121. Xie L (2014) Selective catalytic reduction of NO_x with NH_3 over highly efficient Cu-SSZ-13 catalyst. PhD thesis in the University of Chinese Academy of Sciences
122. Verma AA, Anggara T, Bates SA, Parekh AA, Paolucci C, Schneider WF, Yezerets A, Kamasamudram K, Miller JT, Delgass WN, Ribeiro FH (2013) NO oxidation: a probe reaction on Cu-SSZ13. Paper presented at the 23rd North American catalysis society meeting, Galt House Hotel, Kentucky, Louisville, 2–7 June 2013
123. Xie L, Liu F, Liu K, Shi X, He H (2014) Inhibitory effect of NO_2 on the selective catalytic reduction of NO_x with NH_3 over one-pot-synthesized Cu-SSZ-13 catalyst. Catal Sci Technol 4:1104–1110

124. Ciardelli C, Nova I, Tronconi E, Chatterjee D, Bandl-Konrad B (2004) A “Nitrate Route” for the low temperature “Fast SCR” reaction over a V_2O_5–WO_3/TiO_2 commercial catalyst. Chem Commun 23:2718–2719
125. Martínez-Franco R, Moliner M, Thogersen JR, Corma A (2013) Efficient One-Pot preparation of Cu-SSZ-13 materials using cooperative OSDAs for their catalytic application in the SCR of NO_x. ChemCatChem 5:3316–3323
126. Martínez-Franco R, Moliner M, Franch C, Kustov A, Corma A (2012) Rational direct synthesis methodology of very active and hydrothermally stable Cu-SAPO-34 molecular sieves for the SCR of NO_x. Appl Catal B 127:273–280
127. Lorena PA, Warrender SJ, Slawin AMZ, Dawson DM, Ashbrook SE, Wright PA, Thompson SP, Gaberova L, Llewellyn PL, Moulin B, Vimont A, Daturi M, Park MB, Sung SK, Nam I-S, Hong SB (2011) A co-templating route to the synthesis of Cu SAPO STA-7, giving an active catalyst for the selective catalytic reduction of NO. Microporous Mesoporous Mater 146:36–47

Chapter 13
Zeolite Thin Films and Membranes: From Fundamental to Applications

Zhengbao Wang and Yushan Yan

Abstract Zeolite films and membranes have extensively been investigated since 1992. Many new preparation methods and concepts have been developed, such as microwave synthesis, intergrowth supporting substances, covalent linker synthesis, inner-side synthesis, fluoride-mediated synthesis, variable-temperature synthesis, masking technique, counter-diffusion secondary growth, and pore-plugging synthesis. The remarkable progress made in the preparation has stimulated many application researches of zeolite films and membranes, such as dehydration of various organic solvents, organics extraction from organic/water mixtures, separation of organic mixtures, gas separation, oil–water separation, desalination of seawater, membrane reactors, membrane microreactors, low-k materials, corrosion-resistant materials, sensors, fuel cells, and batteries. The recent progress in preparation and applications of zeolite membranes from 2000 has been summarized in this chapter.

Keywords Zeolite films • Zeolite membranes • Pervaporation • Gas separation • Preparation methods • Sensors • Low-k materials

13.1 Introduction

A membrane is an intervening phase separating two phases and acting as an active or passive barrier to the transport of matter between the two adjacent phases under a driving force. This barrier can be solid, liquid, or gas. Generally, it has a lateral dimension much greater than its thickness. As a first classification, membranes can

Z. Wang (✉)
College of Chemical and Biological Engineering, and MOE Engineering Research Center of Membrane and Water Treatment Technology, Zhejiang University,
Hangzhou 310027, People's Republic of China
e-mail: zbwang@zju.edu.cn

Y. Yan
Department of Chemical and Biomolecular Engineering, University of Delaware,
Newark, DE 19716, USA

F.-S. Xiao, X. Meng (eds.), *Zeolites in Sustainable Chemistry*,
Green Chemistry and Sustainable Technology,
DOI 10.1007/978-3-662-47395-5_13

be divided into two groups: biological and artificial membranes. The latter term can be applied to all membranes made by man with natural materials (modified by man) or with synthetic materials (synthetic membranes). Synthetic membranes can be divided further into organic (made with polymers), inorganic (made with alumina, metals, etc.), and organic–inorganic composite membranes.

Zeolite membranes form one of the branches in the inorganic membrane field. The molecular sieving abilities and selective sorption properties in combination with their catalytic activity, in addition to their thermal and chemical stability, make zeolites ideally suited for the combination of separation and reaction under process conditions. Zeolites were first used in membranes by dispersing the zeolite crystals in polymeric membrane matrixes in the 1970s. The first preparation of continuous zeolite layer as membrane was reported by Suzuki in 1987. The first large-scale pervaporation plant has been put into industrial operation by Mitsui Engineering and Shipbuilding Co. Ltd. in 1998 [1].

In the last few decades, the research of polycrystalline zeolite membranes supported on ceramic, glass, or metal substrates has grown into an attractive and abundant field. A great deal of progress in the science of zeolite membrane synthesis has been made. The applications of zeolite films and membranes have been wildly explored. There has been an exponential growth in papers on the subject. There are many review papers that summarized the significant progress made in the synthesis and applications of zeolite films and membranes [2–14]. In the following text, the recent progress in syntheses and applications of zeolite films and membranes will be presented.

13.2 Structure of Zeolite Films and Membranes

The zeolite films and membranes can be classified into three types as shown in Fig. 13.1: zeolite-filled mixed matrix membranes (separated zeolite crystals in polymeric matrix), self-supporting zeolite layers, and supported zeolite layers; the supported zeolite layers include zeolite films on dense substrates and zeolite membranes on porous supports.

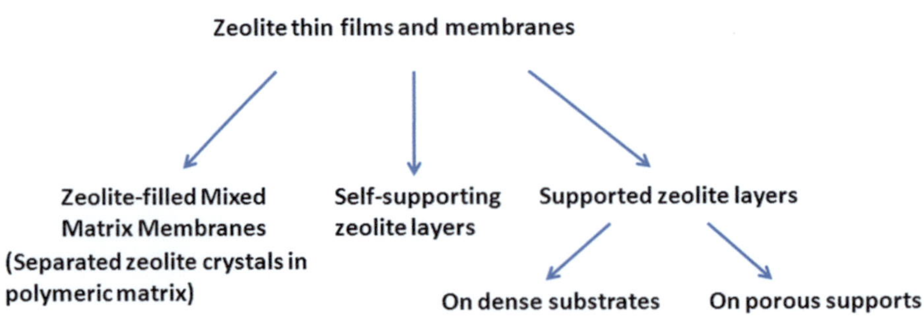

Fig. 13.1 Structures of zeolite films and membranes

13.2.1 Zeolite-Filled Mixed Matrix Membranes

Zeolites were first used in membranes by dispersing zeolite crystals in polymeric membrane matrixes (mixed matrix membranes, designated as MMMs), which were used for gas separation and alcohol/water pervaporation separations. The investigation of MMMs for gas separation was first reported in the 1970s with the discovery of a delayed diffusion time lag effect for CO_2 and CH_4 when adding 5A zeolite into rubbery polymer polydimethylsiloxane (PDMS) [15]. Researchers at UOP were the first to report that mixed matrix systems of polymer/adsorbent might yield superior separation performance to that of pure polymeric system [16]. They observed an enhanced O_2/N_2 selectivity from 3.0 to 4.3 when increasing silicalite content in the polymer cellulose acetate (CA) matrix.

Zeolite-filled MMMs have not been paid much attention before 2006 because fabrication of MMMs usually involves difficulties such as weak adhesion of zeolite particles to the polymer matrix and poor distribution of the dispersed phase (zeolite particles) in the continuous polymer matrix phase due to agglomeration of zeolite crystals. However, it has become a hot research topic again since 2006 to overcome the drawbacks of zeolite-filled mixed matrix membranes [17–19].

13.2.2 Self-Supporting Zeolite Films and Membranes

Self-supporting zeolite films can be prepared on temporary supports or without supports. Some zeolite films can be synthesized on liquid mercury surface or polymer (e.g., Teflon) surfaces and then detached from the substrate after crystallization. More than 20 papers were published on the synthesis of self-supporting zeolite films between 1990 and 1997. However, few papers were published after 2000 due to the weak mechanical strength of self-supporting zeolite films. To avoid the thermal expansion mismatch between alumina support and zeolite membrane, the self-supporting MFI-type zeolite membrane, which was fabricated on Teflon plate in an autoclave at 80 °C for 24–96 h, was used to separate xylene isomers [20]. A silicalite/mordenite bilayered self-supporting membrane with disk shape was synthesized from a layered silicate, kanemite, by two steps using solid-state transformation [21]. The mechanical strength (compression strength) of the membrane was greater than 10 kg cm^{-2}. Very recently, high-strength self-supporting NaA and NaY zeolite membranes from geopolymer gels were fabricated through in situ hydrothermal transformation processing [22, 23].

13.2.3 Supported Zeolite Films and Membranes

13.2.3.1 Zeolite Films on Dense Substrates

Zeolite thin films on the surfaces of dense substrates can be easily prepared by coating zeolite nano-sized particles on them. Transparent TS-1 zeolite films were obtained on the surface of microslide glass by dipping the glass into a suspension of nano-sized

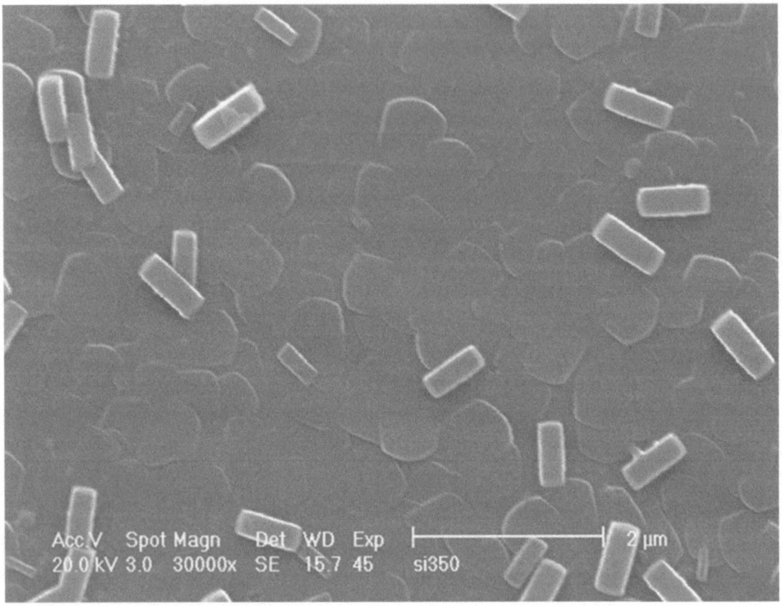

Fig. 13.2 SEM image of *b*-oriented MFI film on a smooth stainless steel surface

(<100 nm) TS-1 particles, drying at 100 °C, and calcining at 550 °C [24]. In a similar way, silicalite zeolite films were prepared on silicon wafers by spin-coating method, and the film can be used as low dielectric constant materials [25].

Continuous zeolite films on dense metal substrates or silicon wafers can be prepared by in situ hydrothermal synthesis or seeded growth (secondary growth). For example, *b*-oriented MFI-type zeolite films on stainless steel plates or silicon wafers can be synthesized by in situ synthesis in a synthesis solution with a molar composition of 0.32TPAOH:1TEOS:165H_2O [26, 27]. A typical *b*-oriented MFI film on a smooth stainless steel surface is shown in Fig. 13.2. Precursor particle layers can be prepared using a colloidal suspension of small (nano-sized) crystal particles, and then continuous films can be obtained by secondary growth of these particles [25].

13.2.3.2 Zeolite Membranes on Porous Supports

Since the self-supporting zeolite layers are highly fragile, zeolite membranes on porous supports have been created. The growth of zeolites on porous supports generally results in a random orientation of crystals, due to the roughness of the supports applied. Porous supports are usually composed of inorganic materials such as mullite [28], alumina [29, 30], YSZ [31], and stainless steel [32, 33]. Wang et al. [34] developed a novel support, zeolite/polymer composite hollow fiber. Zeolite crystals embedded in the matrix of polymer can be used as the seeds for zeolite membrane formation.

Fig. 13.3 SEM images of zeolite LTA membrane on an alumina hollow fiber, (**a**) *top view* and (**b**) *cross-sectional view*

Zeolite membranes are usually prepared on tubular supports or disk supports. Recently, people are interested in using ceramic hollow fibers as supports, because of their high packing density (membrane surface area/volume ratio >1000 m^2 m^{-3}, if the outer diameter is smaller than 4 mm) and thinner wall. Wang et al. have demonstrated for the first time that, by one single hydrothermal synthesis, the zeolite NaA membrane with high separation performance could be formed on a ceramic hollow fiber. They also reported that the flux of the membranes increased with the porosity of supports dramatically. SEM images of a typical zeolite LTA membrane on an alumina hollow fiber are shown in Fig. 13.3.

LTA-type [35], X-type [36], MFI-type [33, 37], and ITQ-17 [38] zeolite membranes were synthesized on stainless steel net supports by Zhu et al.

Recently, a thin porous Ni-supported NaA zeolite membrane has been developed by Liu et al. to substantially lower the zeolite membrane cost. The metallic nature of the porous Ni sheet provides mechanical strength and flexibility for packaging of a membrane module, and the sheetlike membrane design facilitates potential low-cost production of zeolite membranes by a roll-to-roll process [39]. The zeolite/Ni membrane has been successfully demonstrated for water/ethanol separation [40].

13.2.4 Special Membrane Structures

13.2.4.1 Bilayered Zeolite Membranes

There are three different purposes for synthesizing multilayer zeolite membranes: (1) improving membrane performance by repeating crystallizations of the same zeolite type [41], (2) developing novel properties by combination of layers of different zeolite types [21], and (3) creating new applications by combination of zeolite layers with other membrane layers. A facile strategy for the layer-by-layer synthesis of sandwich-like zeolite LTA membranes by using 3-aminopropyltriethoxysilane as modifier was reported by Huang et al. [41]. Lin et al. [42, 43] reported a highly stable bilayer MFI zeolite membrane with good hydrogen separation characteristics. The membrane with a thin ZSM-5 layer on a thick silicalite base layer supported on a macroporous α-alumina support with an yttria-stabilized zirconia intermediate barrier layer was prepared by a multiple-step synthesis method.

13.2.4.2 Oriented MFI-Type Zeolite Films and Membranes

One important but difficult task in the synthesis of zeolite films or membranes is the control of crystal orientation in the zeolite layer. A *b*-oriented MFI layer is expected to exhibit higher fluxes. *b*-Oriented MFI films and membranes can be synthesized by direct in situ crystallization and seeded growth.

b-Oriented MFI-type films were first reported by Jansen et al. [44]. Wang and Yan first demonstrated that crystal orientation control in a continuous MFI thin film can be achieved by simple adjustment of synthesis compositions using direct in situ crystallization on metal substrates or silicon wafers [26]. Recently, it was reported that continuous *b*-oriented MFI zeolite films can be synthesized by microwave-assisted in situ crystallization and that a certain nuclei concentration at the aging stage is necessary [45]. *b*-Oriented TS-1 films could be grown on chitosan-modified $\alpha\text{-}Al_2O_3$ substrates [46].

Continuous oriented zeolite films were successfully prepared on supports with seeded growth method in which a zeolite seed layer was deposited on the substrate prior to hydrothermal treatment [47, 48]. The orientation of the seed layer can be retained into the final membrane by seeded growth. Under certain growth conditions, crystallographic orientations like *a*-orientation [49, 50], *b*-orientation [47, 48], *c*-orientation [51], or intermediate orientations [52] for MFI membranes were observed. Highly *b*-oriented and intergrown MFI films could be produced by carrying out secondary growth of *b*-oriented seed layers under hydrothermal conditions using trimer tetrapropyl ammonium iodide as SDA [47]. Yoon et al. [48] reported facile methods to grow *b*-oriented silicalite-1 membranes on supports using the synthesis gel composed of TEOS, TEAOH, $(NH_4)_2SiF_6$, and H_2O in a mole ratio of 4.00:1.92:0.36:40–80 at 165 °C for more than 7 d. *b*-Oriented MFI membranes have a superior separation performance which was demonstrated for the xylene isomers separation [47, 48].

It is generally believed that twin growth on MFI seed crystals is unavoidable with template TPAOH, which negatively affects the properties of the as-prepared *b-oriented* MFI zeolite films [47]. Yang et al. [53] found for the first time that proper pretreatment of the synthesis solution (including TPA^+) in the secondary growth process resulted in highly *b*-oriented MFI films. They also found that high-quality *b*-oriented MFI-type zeolite films could be obtained by microwave synthesis on the base of the above pretreatment method [54]. Wang et al. found that twin growth of a *b*-oriented MFI seed layer could be effectively suppressed by decreasing the TPA^+/Si ratio in the secondary synthesis solution to a certain value (e.g., <0.01) [55].

13.3 Preparation Methods for Supported Films and Membranes

13.3.1 General Preparation Methods

A great deal of progress in the science of zeolite membrane synthesis has been made since the first preparation of zeolite membranes by Suzuki in 1987. Like the synthesis of zeolites, there are three general preparation methods for zeolite films and membranes: (i) direct in situ crystallization, (ii) seeded growth (or secondary growth), and (iii) vapor phase transfer.

13.3.1.1 Direct In Situ Crystallization

The first method used to synthesize zeolite films and membranes is the direct in situ hydrothermal synthesis (also named as direct in situ crystallization), in which the support is directly immersed in the zeolite synthesis solution and the membrane is formed under autogenous pressure in an autoclave. This method utilizes precipitation of zeolite crystals or gel followed by nucleation and growth of crystals on the support surface. This is the simplest synthesis method for preparation of zeolite films and membranes. As previously described, self-supporting zeolite films and membranes are usually prepared on temporary supports by in situ crystallization.

Some MFI-type supported zeolite films and membranes were prepared by in situ crystallization method, and it is clear that the in situ crystallization time is longer than 16 h except *b*-oriented films and the synthesis mixture is more dilute than the common zeolite powder synthesis to suppress the crystallization in the bulk solution [26, 56, 57]. Most of the MFI-type zeolite membranes are usually prepared at 160–180 °C using TPABr or TPAOH as a SDA. Note that very thin *b*-oriented silicalite-1 films (less than 1.0 μm) can be prepared by direct in situ crystallization on dense metal supports or silicon wafers [26].

Few zeolite NaA membranes were prepared on alumina supports by direct in situ crystallization from a clear solution with a molar oxide ratio of $49Na_2O:Al_2O_3:5SiO_2:980H_2O$ at 85 °C for 2–4 h [58].

13.3.1.2 Seeded Growth

The seeded growth method includes two processes: (1) the deposition of seed crystals on the support surface and (2) the secondary hydrothermal synthesis. The use of seeds offers a way to control the growth of a zeolite layer on a support surface. By decoupling the nucleation step from crystal growth, the seeds can grow in low concentrated solutions under suppression of secondary nucleation. Therefore, most zeolite membranes are synthesized by the seeded growth method. For example, most of the zeolite NaA membranes have been prepared by the seeded growth method using either a milky-like gel [29, 59] or a clear solution [60] on the surface of a porous support at 80–100 °C.

The seeding process on the supports has been recognized as a crucial factor to obtain high-quality zeolite membranes. Many seeding methods have been reported, and they can be classified into seven main ways to attach the seeds to the support:

1. Capillary force attachment: Immersing a dried support into a seed suspension followed by a thermal treatment of the seeded support to fix the seeds via dehydroxylation to the support, e.g., dip coating [61].
2. Electrostatic attachment: (1) Charging the support surface by controlling pH value to achieve opposite surface charges for seeds and the support [62]; (2) modifying the support surface by adsorption of a cationic polymer in order to achieve a positive surface charge. For example, charging the support surface by adsorption of positively charged cationic polymers like poly-DADMAC or Redifloc to adjust different zeta potentials between the ceramic support and the zeolite nano-sized seed crystals [63, 64].
3. Covalent bonding attachment: Modifying the support surface using silane coupling agent or the seed crystal surface to achieve covalent bonding between the support surface and the seed crystals [65, 66]. The oriented monolayer of zeolite crystals or patterned seed layer can be prepared by using covalent linkage [67].
4. Electrophoretic deposition: Nano-sized zeolite seeds can be deposited on solid supports by electrophoretic deposition method [68].
5. External force attachment: Rub coating [69], vacuum coating [70], and filtration coating [71]. The rub coating is a simple and effective way to coat zeolite seeds on the support surface, especially defective porous supports because zeolite seeds can fill the defect pores in the support by rub coating. A *b*-oriented MFI-type seed layer can also be prepared by rub coating coffin-shaped MFI-type zeolite seed crystals on dense supports (glass slides, silicon wafers, and metal plates) [55]. Wang et al. developed a novel seeding method, dip coating–wiping seeding, for ceramic hollow fiber supports [29]. They also developed a novel seeding method, wetting–rubbing seed crystals or seed paste, for macroporous defective support surfaces [30, 72].
6. In situ seeding: A seed layer can also be prepared by in situ crystallization in the synthesis solution. ZSM-5 zeolite films were grown on silicon substrates by a two-step hydrothermal synthesis consisting of in situ seeding and secondary growth [73].
7. Embedding seed crystals in the support: Zeolite/polymer composite hollow fibers with uniform zeolite crystals on the surface were prepared and used as supports; thus, only a simple hydrothermal synthesis was needed to obtain dense and continuous zeolite membranes on them [34].

13.3.1.3 Vapor Phase Transfer

Zeolite crystals can be prepared by vapor phase transfer of dry gel conversion. Zeolite membranes can also be prepared by the vapor phase transfer method, in which a layer of dry aluminosilicate gel is deposited on the support and then transformed into zeolite layer in the presence of vapors. Dong et al. [74] prepared ZSM-5 and ZSM-35 membranes by the vapor phase method. Matsukata's group synthesized zeolites ZSM-5, ferrierite, mordenite, and analcime flat membranes by this method [75, 76]. The organic SDA agents can be included in the dry gel or in the vapor phase. Cheng et al. synthesized zeolite NaA membranes with high performance by vapor phase transformation [77].

13.3.2 Special Synthesis Concepts

13.3.2.1 Microwave Synthesis

A remarkable progress in the microwave-assisted synthesis of zeolite NaA membranes was achieved by Yang's group [78–80]. It is known that surface seeding is usually needed for the synthesis of zeolite NaA membranes. They developed the "in situ aging–microwave synthesis" method [80], by which high-quality LTA membranes could be synthesized with high reproducibility.

Dong et al. reported microwave synthesis of MFI-type zeolite membranes by seeded growth without the use of organic SDA [81]. Membranes obtained from template-free inorganic precursors exhibited higher permselectivities of H_2 over N_2 and SF_6 than the membrane synthesized with TPAOH template. Yang et al. firstly prepared zeolite T membranes on the α-Al_2O_3 tubes by microwave-assisted in situ nucleation and secondary growth [82]. Yan et al. reported ambient pressure dry gel conversion method for zeolite MFI synthesis using ionic liquid and microwave heating [83]. Wang et al. reported the preparation of *b*-oriented MFI zeolite films on stainless steel substrates using a novel microwave-assisted two-stage varying-temperature hydrothermal synthesis [45]. Yang et al. reported that twin growth in the synthesis of *b*-oriented MFI films was successfully suppressed by applying microwave irradiation on a *b*-oriented MFI seed layer, relying on the nucleation-related bottleneck effect [54]. The supported zeolite membranes on ceramic capillaries have been synthesized under microwave heating in order to reduce synthesis time, to prevent support dissolution, and to reproducibly obtain a thin defect-free zeolite layer [84]. A chitosan layer is utilized for controlled fabrication of MFI zeolite films in dilute synthesis solution under microwave irradiation. The dense, ca. 600 nm thick, and solely *b*-oriented silicalite-1 films can be formed onto the chitosan supported silicalite-1 monolayer under microwave irradiation for 1 h [85].

13.3.2.2 Intergrowth Supporting Substances

By adsorption of intergrowth supporting substances (ISS), the strong negative surface charge can be compensated to the isoelectric point, and then negatively charged silica precursors can access to the surface. Suitable ISS are small positively charged molecules, are stable in the alkaline region during the membrane synthesis, and can be decomposed by calcinations [86]. Evaluating different ISS, HMEDA-J_2 in a concentration of 0.01 M was found to be optimum [86]. This ISS concept was first developed for ZSM-5 membranes [86] and later successfully transformed to the synthesis of LTA and FAU membranes [8]. LTA and FAU membranes can effectively separate water/organic mixtures, but they fail in shape-selective gas separations. By adsorption of an ISS, this negative zeta potential can be shifted to near the isoelectric point which improves the intergrowth of the seed crystals on the support to a continuous membrane layer. By using an ISS, an improvement of the permeation selectivity of LTA and FAU membranes was found. Nevertheless, the LTA and FAU membranes prepared with ISS are still far from being defect-free, and their permselectivities are found in the range of the Knudsen factor.

13.3.2.3 Covalent Synthesis

In Sect. 13.3.1.2, we introduced that zeolite seed layers can be prepared by covalent bonding attachment using covalent silane linkers. Recently, Huang and Caro have developed a seeding-free synthesis strategy for the preparation of zeolite LTA and FAU membranes by using 3-aminopropyltriethoxysilane (APTES) [41, 66, 87, 88], 1,4-diisocyanate (DIC-4) [89], or 3-chloropropyltrimethoxysilane (CPTMS) [65] as covalent linkers between the zeolite layer and the porous α-Al_2O_3 support. By covalent bonds, the zeolite precursors were effectively attached and anchored onto the support surface to promote the nucleation and growth of zeolite layers. The zeolite LTA membranes prepared on the covalent agent-modified porous α-Al_2O_3 disk display higher separation selectivity than the literature data reported previously. This method has been used in preparation of metal organic framework (MOF) membranes [90].

13.3.2.4 Inner-Side Synthesis

Zeolite membranes are usually synthesized on the external surfaces of the tubular supports. As shown in Table 13.1, only a few studies have been reported in literature regarding the preparation of zeolite membranes on the inner surfaces of the tubular supports [91–95]. Some special seeding methods and synthesis systems were used, and pervaporation (PV) separation performance of those membranes is different. Inner-side zeolite membranes on alumina hollow fibers, on which the seed layer was obtained by dip coating an aqueous seed suspension containing a clear solution, showed the best PV performance. HZSM-5 zeolite coatings on the inner surface of stainless steel tubes were prepared by the secondary growth method and used in the catalytic cracking of n-dodecane [96].

Table 13.1 PV performance of inner-side zeolite NaA membranes in dehydration of ethanol aqueous solution

Support	Seeding	Synthesis	PV conditions	Flux (kg m^{-2} h^{-1})	α	Ref.
Al_2O_3 tube	Cross-flow filtration		92 wt%, 50 °C	0.5	600	[91]
Al_2O_3 tube	Cross-flow filtration	Semicontinuous	90 wt%, 50 °C	0.5	16,000	[92]
Al_2O_3 tube		Under a centrifugal force field	90.8 wt%, 93 °C	2.5	130	[93]
TiO_2 tube		Flow system (gravity)	92 wt%, 50 °C	1.2	8,500	[94]
Al_2O_3 hollow fibers	Dip coating	Rotating autoclave	90 wt%, 75 °C	6.3	10,000	[95]

13.3.2.5 Fluoride-Mediated Synthesis

The synthesis of zeolites can be conducted in both basic (OH^-) and fluoride medium (F^-). The fluoride-based synthesis has been also used in preparation of zeolite membranes. Zeolite beta films prepared by seeded growth in fluoride media (SiO_2: 0.5TEAF: $5H_2O$) at 150 °C for 4 days were very rough with a thickness exceeding 20 μm and were composed of well-faceted uniform 2.5–3 μm truncated square dipyramidal crystals [97]. Zhu et al. reported an improved method in fluoride media at 140 °C for 2 days to synthesize (h0l) oriented high-silica beta membrane with a thickness of 2 μm [98].

MFI films were prepared on dense glass substrates by seeded growth (75 nm crystal seeds) in fluoride media (0.12TPAOH: TEOS: $39H_2O$: 0.12HF) at 100 °C for 24–120 h, while no continuous MFI films were obtained on α-alumina plates under the same conditions [99]. However, MFI films (~2 μm thick) were prepared on α-alumina plates by seeded growth in fluoride media (0.12TPAOH: SiO_2: $60H_2O$: 0.12HF) at 100 °C for 96 h using fumed silica as silica source [100]. It is reported that defect-free silicalite-1 membranes were successfully prepared on the surface of tubular mullite supports in ultra-diluted solution containing fluoride ($1SiO_2$:0.2TPABr:0.2TPAOH:$0.15NH_4F$:$800H_2O$) by hydrothermal synthesis at 180 °C for 20 h [101]. Well water permselectivity ZSM-5 membranes were rapidly prepared from an aluminum-rich and organic template-free precursor gel in fluoride media by a static secondary hydrothermal synthesis (180 °C 24–48 h) using nanometer silicalite-1 crystals as seeds [28, 102]. High-performance mordenite zeolite membranes in fluoride media with organic-free template were prepared on α-Al_2O_3 tubes and mullite tubes by secondary growth [103–105].

Very recently, Zhou et al. [106] reported a facile method for preparing *b*-oriented silica MFI films and membranes in fluoride media on dense glass plates and porous alumina supports, respectively. The optimum composition of synthesis solution was TEOS: $0.2TPAOH:0.2HF:200H_2O$. The as-synthesized membranes showed CO_2/H_2 and CO_2/CO mixture separation performances that are higher than those of the random-oriented zeolite membranes.

13.3.2.6 Variable-Temperature Synthesis

Thin (7 μm) zeolite ZSM-5 membranes with high permeances were reproducibly prepared on α-Al_2O_3 supports (pore size: 2μm) by a two-stage varying-temperature synthesis [107]. Compared to the synthesis at constant temperature, equiaxed particles in the membrane were achieved; thin and dense membranes were obtained.

High-quality zeolite silicalite-1 membranes were successfully prepared on porous, tubular stainless steel supports with the optimized synthesis solution composition by a two-stage varying-temperature in situ synthesis [108]. Compared with the synthesis at constant temperature (one stage), the two-stage varying-temperature in situ synthesis generated a thin and dense membrane.

In addition to the conventional constant temperature (CT) at 100 and 140 °C syntheses, a variable-temperature synthesis of zeolite NaY membranes was applied where the temperature was increased from 100 to 140 °C (VT-LH) and decreased from 140 to 100 °C (VT-HL) during the hydrothermal syntheses [109]. It is found that by reducing the synthesis temperature from 140 to 100 °C during the hydrothermal synthesis, formation of impure phases was prevented during the densification of the membrane. The NaY zeolite membranes prepared under the variable-temperature method showed the higher PV performances than the membranes prepared under constant temperature conditions.

13.3.2.7 Masking Technique

A rational fabrication approach targeting ultrathin, defect-free MFI films on open supports by using a two-step support masking technique was adopted [110]. In order to prevent support invasion and leaching, support masking was carried out. The wax has a melting point above the synthesis temperature and thus remained in the pores of the support and protected the support from the synthesis solution during film synthesis.

High-performance silicalite-1 membranes were synthesized on silica tubes by in situ hydrothermal synthesis. By using the "solution-filling (SF)" method, the average flux of membranes with the SF method was improved by about 25 % compared to that of the membranes without using the SF method [111, 112]. The flux and separation factor toward ethanol/water mixture at 60 °C were 0.56 kg m^{-2} h^{-1} and 84, respectively.

13.3.2.8 Counter-Diffusion Secondary Growth (CDSG)

High-performance silicalite-1 membranes on outer surface of macroporous α-Al_2O_3 supports were synthesized by counter-diffusion secondary growth (CDSG) [113]. Different from the conventional synthesis methods, this method introduced silica source (TEOS) and template agent (TPABr) from the opposite direction of a support wall, resulting in good membrane structures and properties as well as reduced chemical consumption. Results showed that the membranes were dense, continuous without obvious defects, and displaying good permeation property. The H_2 permeance and H_2/SF_6 ideal permselectivity for the membrane prepared under optimized parameters were up to 7.5×10^{-7} mol m^{-2} s^{-1} Pa^{-1} and 103 at 25 °C under 0.1 MPa pressure drop, respectively. This method can also be extended to the preparation of other zeolite membranes. Very recently, this method was used to prepare ZIF-7 membranes [114].

13.3.2.9 Pore-Plugging Synthesis

The main goal of all above methods is to generate a thin, defect-free zeolite layer on top of the porous support. Miachon and his coworker reported on another strategy for the preparation of zeolite membranes [115, 116]. The idea was to grow zeolite crystals within the pores of a ceramic alumina substrate until complete blocking of the pore by the zeolite material. A continuous composite zeolite–alumina membrane was obtained via the pore-plugging method. This nanocomposite membrane led to high separation factors, the separation layer being located inside the pores of the host ceramic support. The introduction of an interruption during the hydrothermal synthesis led to higher separation performances [115].

Alumina hollow fibers have been used as supports and submitted to pore-plugging MFI zeolite synthesis [117]. Nanocomposite MFI/alumina hollow fibers with a negligible amount of intercrystalline defects and high gas permeance at room temperature have been successfully synthesized using the pore-plugging approach. In this nanocomposite architecture, grain boundaries that could limit selectivity are less important than in filmlike structures. Nanocomposite MFI–ceramic hollow fiber membranes via pore-plugging synthesis have been used in many kinds of separation, such as CO_2 capture [118, 119], xylene isomer separation [120], and packed bed membrane reactors [121].

13.4 Applications

The applications of zeolite films and membranes for gas separation, pervaporation, membrane reactors, sensors, low-k films, corrosion protection coatings, zeolite-modified electrodes, fuel cells, heat pumps, etc. have been wildly explored. General separation mechanisms in zeolite membranes include molecular sieving and competitive adsorption and diffusion.

13.4.1 Dehydration of Organic Solvents

In the chemical and allied industries, pervaporation (PV) is becoming the preferred process for removing water from organics—especially from ethanol and isopropanol. There are two important factors for evaluating the separation performance of zeolite membranes: flux and separation factor. The total flux (J) and separation factor (α) are defined as follows:

$$J = W/(A*t), \quad a = (Y\text{water}/Y\text{ethanol})/(X\text{water}/X\text{ethanol})$$

where W is total weight of the permeate (kg), A is the separation area of zeolite membrane, t is the collecting time (h), and X and Y are the weight fraction of species in the feed and permeate, respectively.

Table 13.2 shows the pervaporation performance of zeolite membranes in dehydration of 90 wt% ethanol aqueous solutions reported in literature. Zeolite NaA membranes showed the best separation performance in dehydration of ethanol due to the higher hydrophilicity of zeolite NaA. The supports have great effect on the fluxes of zeolite membranes. Zeolite NaA membranes on hollow fibers showed higher fluxes than those on tubular supports. The fluxes of zeolite NaA membranes on tubular supports are also pretty different due to their different structures. It is concluded that the flux depends on the diffusion resistance of the support (especially the porosity) [29, 122].

Table 13.2 Pervaporation performance of zeolite membranes for dehydration of 90 wt% ethanol solution at 75 °C

Zeolite	α (A/B)[a]	Flux(kg/m^2 h)	Support	Ref
NaX	170	1.9	Mullite tubes	[123]
NaY	130	1.6	Mullite tubes	[123]
NaY	420	5.1	α-Al_2O_3 tubes[b]	[124]
T	1301	2.12	α-Al_2O_3 tubes	[125]
Mordenite[c]	1300	1.6	Mullite tubes	[104]
CHA	3900	2.2	Mullite tubes	[126]
NaA	>10,000	2.08	Mullite tubes	[127]
NaA	10,000	2.15	α-Al_2O_3 tubes	[59]
NaA	>10,000	5.6	α-Al_2O_3 tubes[b]	[128]
NaA	>10,000	3.5–4.0	α- Al_2O_3 tubes	[30]
NaA	>10,000	6.0–9.0	Alumina hollow fibers	[29]
NaA	>10,000	8.0–9.3	Composite hollow fibers	[34]
NaA	>10,000	4.0	Porous Ni sheet	[40]

[a]Mixture, A/B = water/ethanol
[b]Asymmetric support
[c]Fluoride media

Table 13.3 Pervaporation performances of zeolite NaA membranes for dehydration of solvents

Support (pore size, μm)	PV conditions	Flux ($kg\ m^{-2}\ h^{-1}$)	α (−)	Ref.
α-Al_2O_3 tube (0.5–1.0)	Isopropanol/water (95/5, 70 °C)	1.44	10,000	[138]
Mullite tube (1)	Ethylene glycol/water (80/20, 120 °C)	4.03	>5000	[139]
α-Al_2O_3 tube (0.5)	Ethyl acetate/water (98/2, 50 °C)	0.315	163,000	[140]
Stainless steel plate (0.2)	Benzaldehyde/water (97/3, 25 °C)	0.04	10,000	[141]

As shown in Table 13.3, zeolite NaA membranes can also be used in dehydration of isopropanol, ethyl acetate, ethylene glycol, and benzaldehyde. However, the zeolite NaA membrane is unstable to acids and not applied to the system where it directly contacts with acids. Kita and his coworkers found that zeolite T membranes had high pervaporation (PV) performance to water–organic liquid mixtures and good stability to acid solutions. The PV separation with zeolite T membrane was applied to the esterification of acetic acid with ethanol [129]. The conversion exceeded the equilibrium limit and reached almost 100 % within 8 h when the initial molar ratios of alcohol to acetic acid were 1.5 and 2. Zeolite T membranes attracted a lot of efforts in developing new preparation methods and improving the separation properties, including synthesis in fluoride media [32, 130], clear gel solution synthesis [131–133], two-step temperature process [134], varying temperature [125, 135], and microwave-assisted synthesis [136, 137]. Their typical separation performance is shown in Table 13.4. Higher separation factors were obtained for the separation of water/isopropanol than that of water/ethanol.

Zhu et al. [33] have provided an environmentally benign synthesis approach for the preparation of continuous and low-siliceous ZSM-5 membranes with thin thickness on stainless steel net supports. In this method, no additional organic templates are required, and the crystallization of ZSM-5 membranes proceeds at low temperature (100 °C) and ambient pressure. As-prepared ZSM-5 membranes showed high pervaporation performance in binary mixtures of ethanol–water and isopropanol–water. The total flux and separation factor for the dehydration of a 95 wt% ethanol aqueous solution at 75 °C can reach 5.96 kg m^{-2} h^{-1} and 6,951, respectively. The total flux and separation factor for the dehydration of the 90 wt% isopropanol aqueous solution at 75 °C can reach 6.88 kg m^{-2} h^{-1} and 8,991, respectively.

13.4.2 Organics Extraction from Organic/Water Mixtures

Most of the zeolite membrane studies focused on MFI-type because it has a multidimensional channel network with pore openings near the sizes of many industrially important organic molecules. The hydrophobic property of the pure silica MFI-type zeolite membrane makes it very attractive for the pervaporation (PV) separation of

Table 13.4 PV performance of zeolite T membranes for dehydration of 10 wt% water/organics mixtures at 75 °C

Support	Mixture	Flux (kg m^{-2} h^{-1})	α (–)	Ref.
Stainless steel[a]	Water/isopropanol	6.70	5300	[32]
Mullite[a]	Water/isopropanol	4.43	8200	[130]
	Water/ethanol	3.64	2900	
Mullite	Water/isopropanol	2.20	8900	[142]
	Water/ethanol	1.10	900	
α-Al_2O_3	Water/isopropanol	2.15	10000	[82]
	Water/ethanol	1.77	1116	

[a]Fluoride media

Table 13.5 Pervaporation performances of MFI-type zeolite membranes for ethanol/water mixture

Support (pore size, μm)[a]	Thickness (μm)	Feed (wt%)	Flux (kg m^{-2} h^{-1})	α^b (–)	Ref.
PSS disk (0.5–2)	~500	4	0.76	58	[143]
Silica tube (0.3)	30	3	0.87	69	[144]
PMT (1)	~20	5	0.93	106	[145]
PMT (1)	~6	5	2.85	60	[101]
AHF (0.1–0.2)[c]	12	5	5.40	54	[146]
YSZ HF (0.37–1.13)	~3	5	7.40	47	[147]
PMT (1)	–	10	2.55	72	[148]
PAT (2)	~20	5	1.81	89	[149]
PAT (2–3)	10	5	1.38	45	[150]
SS net (300 mesh)[d]	~7.5	5	11.50	19.6	[151]
PAT (1–3)	~3.5	5	1.82	62	[72]

[a]*PSS* porous stainless steel, *PMT* porous mullite tube, *PAT* porous alumina tube, *AHF* alumina hollow fiber (HF)
[b]Separation factor (ethanol/water)
[c,d]PV temperature: 75 °C, 25 °C (60 °C for others)

organic/water mixtures, especially for the ethanol/water mixture [18]. The pervaporation results of ethanol aqueous solutions on MFI zeolite membranes are summarized in Table 13.5.

As shown in Table 13.5, MFI-type zeolite membranes were synthesized on different supports and showed different pervaporation performance, due to the different support properties and different synthesis methods. The membranes on hollow fibers showed higher fluxes than those on tubes. The membrane on stainless steel net showed the highest flux with the lowest separation factor [151]. Except for a few works of Kita's and Yang's groups [144], most reported separation factors of MFI-type zeolite membranes toward ethanol/water mixtures are lower than 60. Wang et al. [72] recently developed a novel wetting–rubbing seeding method to prepare MFI-type zeolite membranes on low-cost defective macroporous alumina supports. The synthesized membranes exhibited a high separation factor of 62 in a short crystallization time of 4 h at 175 °C.

Table 13.6 Organic mixtures separation results by zeolite NaY membranes

Feed solution	Flux (kg m^{-2} h^{-1})	α	Temp. (°C)	Ref.
MeOH/MTBE	1.5	12,000	50	[124]
MeOH/MTBE	2.25	8,800	50	[109]
MeOH/MMA	3.47	6,200	60	[109]
MeOH/DMC	1.05	700	50	[109]
MeOH/MIB	3.31	6,700	60	[109]
MeOH/MA	3.06	4,900	60	[109]
MeOH/MP	3.14	4,500	60	[109]
EtOH/ETBE	1.28	2,600	60	[109]

13.4.3 Separation of Organic Mixtures

Some nonaqueous organic mixtures also exhibit adsorption and diffusion differences, allowing separations. Table 13.6 shows that methanol/MTBE and methanol/benzene separations have been the most successful organic/organic separations with zeolite membranes. Methanol (1.7-debye dipole moment) is more polar than MTBE (1.4 debye) and benzene, which does not have a permanent dipole. Thus, NaY and NaX zeolite membranes that have high aluminum contents separate these mixtures better than a silicalite-1 membrane because of their localized electrostatic poles.

The NaY zeolite membranes prepared under the variable-temperature method showed the higher PV performances than the membranes prepared under constant temperature conditions [109]. NaY zeolite membranes were applied to the PV separation of several water/alcohol and alcohol/organic systems. High PV performances in flux and separation factor were obtained for the separations of methanol/MTBE, methanol/DMC, ethanol/ETBE and methanol/MMA, etc. The remarkably high permselectivity of NaY zeolite membranes can be attractive for practical applications.

13.4.4 Xylene Isomer Separation

The recovery of p-xylene from xylene isomers is an important step in the large-scale synthesis of petrochemicals. The conventional methods for separating the xylene isomers are crystallization, distillation, and adsorption. Separation of these mixtures is poorly achieved by distillation process due to the similar volatility and boiling point which is 138.5 °C (p-xylene), 139 °C (m-xylene), and 144.5 °C (o-xylene). Moreover, separation of these mixtures by crystallization or through adsorption processes is usually complex and very energy intensive. Therefore, membrane separation is very important for the separation of xylene isomers. Table 13.7 shows comparison of p-xylene permeation and separation performance for supported MFI-type zeolite membranes. Note that these MFI-type zeolite membranes were prepared by various methods and examined at different conditions. The permeation

Table 13.7 Literature results for the separation of xylene isomers through MFI-type zeolite membranes by vapor permeation (VP) and pervaporation (PV) methods

Separation	Growth method[a]	Orientation	Flux[d]	α (–)	Ref.
VP	SG	b-oriented	1,960	500	[47]
VP	SG	c-oriented	~50	2–39.5	[153]
VP	SG[b]	h,0,h-oriented	28–851	20–300	[154]
VP	In situ	Random	~150	18–73	[155]
VP	SG	b-oriented	2,100–500	~1,000	[48]
VP	SG	c/h0h-oriented	600	29	[156]
VP	SG	Random/h0h	2,600 ± 400	123–139	[157]
VP	SG	c-oriented	480	126	[158]
PV	SG[c]	Random	14	40	[152]
PV	SG	c-oriented	45	2–3.6	[159, 160]

[a]*SG* secondary growth
[b]Defect sealing
[c]Template-free
[d]VP flux 10^{-10} mol m^{-2} s^{-1} Pa^{-1}, PV flux 10^{-2} kg m^{-2} h^{-1}

temperatures were also quite different for pervaporation and vapor permeation. As shown in Table 13.7, the pervaporation offers much higher p-xylene flux and lower p/o-xylene separation factor than the vapor permeation. *b*-Oriented membranes showed the highest vapor permeation separation factor for xylenes (>500) at 200 °C [47, 48]. Yuan et al. [152] reported that silicalite membranes prepared by template-free secondary growth method showed high p/o-xylene pervaporation separation factors (40) at 25 °C.

13.4.5 Gas Separation

13.4.5.1 H_2 Separation

Hydrogen has been considered as a promising energy carrier for generating clean power. Since the kinetic diameter of H_2 is around 0.29 nm, the pore diameter of membranes should be larger than this but smaller than the kinetic diameter of the molecules from which H_2 is to be separated. Compared with other membranes, crystalline zeolite membranes offer much better thermal, hydrothermal stability and high tolerance for impurities such as H_2S. Among zeolite membranes, MFI-type zeolite membranes are the most commonly used ones. Gas permeation results of hydrogen separation for silicalite-1 membranes have been reported in many studies. In these cases, the separation effect is based on the interplay of mixture adsorption and mixture diffusion effects. As an example, the mixture separation factor α of H_2 from i-butane increases from about $\alpha \approx 1.5$ at room temperature to $\alpha \approx 70$ at 500 °C [171].

To obtain high selectivity for H_2 separation, some strategies were applied to reduce the pore size of MFI zeolite membrane (0.55 nm) [93, 161, 162]. Hong et al. [161] used catalytic cracking deposition (CCD) of methyldiethoxysilane (MDES) to modify the zeolitic pores of B-ZSM-5 membranes. After CCD treatment, mono SiO_2 units could be formed in the zeolitic pores, which could effectively reduce the pore size and adapt for H_2-selective separation. The modified membrane showed a H_2/CO_2 separation factor of 47 for H_2/CO_2 (50/50) mixture at 400 °C. However, the deposition amount was hard to control using an intermittent operation for modification. The deposition of silane decreased H_2 permeance more than one order of magnitude. To avoid excessive SiO_2 deposition, an on-stream chemical vapor deposition approach was proposed to modify MFI/α-Al_2O_3 zeolite membranes [163]. The permeation of H_2/CO_2 gas mixture through MFI zeolite membranes was monitored during the modification process at 450 °C. The H_2/CO_2 separation factor increased from 1.8 to 7.0 while the H_2 permeance only decreased from 2.73×10^{-7} to 1.28×10^{-7} mol m^{-2} s^{-1} Pa^{-1} during 5 min modification. Gu et al. [164] modified the synthesized MFI-type zeolite membranes for H_2 separation by on-stream catalytic cracking deposition of MDES in the zeolitic pores. Activation of MFI zeolite membranes by air at 500 °C was found to promote catalytic cracking deposition of silane in the zeolitic pores effectively, which resulted in significant improvement of H_2-separating performance. The H_2/CO_2 separation factor of 45.6 with H_2 permeance of 1.0×10^{-8} mol m^{-2} s^{-1} Pa^{-1} was obtained at 500 °C for a modified hollow fiber MFI zeolite membrane. Recently, Lin et al. [43] reported a highly stable bilayer MFI zeolite membrane with good hydrogen separation characteristics. The membrane consists of a thin (2 μm) ZSM-5 layer on a thick (8 μm) silicalite base layer supported on macroporous α-alumina with an yttria-stabilized zirconia (YSZ) intermediate barrier layer. The zeolitic pores of the thin ZSM-5 layer were narrowed by catalytic cracking deposition (CCD). At 500 °C, the bilayer zeolite membrane exhibits H_2 permeance of about 1.2×10^{-7} mol m^{-2} s^{-1} Pa^{-1}, with H_2 to CO_2, CO, and H_2O vapor selectivity, respectively, of about 23, 28, and 180.

In addition to MFI (silicalite-1) membranes, SAPO-34 with chabazite (CHA)-type framework having pore diameter of 0.38 nm has been studied for various applications and has shown excellent performance in H_2 gas separations from other light gases like CO_2, N_2, CH_4, etc. [165–167]. Lin et al. recommended DDR membranes as tool for H_2 separation [168, 169].

13.4.5.2 CO_2 Separation

Gas separation is of vital environmental concern nowadays due to the issue of global climate change. The phenomenon of increasing greenhouse gas concentration, especially carbon dioxide (CO_2), in the atmosphere has drawn increasing interest among the researchers toward finding efficient methods for CO_2 capture before being emitted to the atmosphere. A few zeolite membranes have already shown promising results in the separation of CO_2 from N_2 and CH_4 [170–172].

CO_2 has a stronger electrostatic quadrupole moment than N_2 leading to a preferential adsorption of CO_2 from CO_2/N_2 mixtures. Most often the MFI-type membrane was studied [119, 173]. As shown in Table 13.8, zeolite NaY membranes are well investigated recently and showed different CO_2 permeances and separation factors.

Table 13.8 Comparison of performances of FAU-type membranes in CO_2/N_2 separations[a]

Support	Thickness (μm)	Pr (kPa)	Test temperature (°C)	Permeance of CO_2 (10^{-7} mol $m^{-2}\,s^{-1}\,Pa^{-1}$)	α (CO_2/N_2)	Ref
Al_2O_3 tube	10	101	30	0.28[c]	100	[174]
Al_2O_3 tube	10	101	40	4.1[b]	46	[175]
Al_2O_3 tube	10	101	35	13[b]	30	[176]
Al_2O_3 tube	5	101	35	15[b]	32	[177]
Al_2O_3 disk	50	–	25	1.5[b]	54	[178]
Al_2O_3 disk	1.9–2.1	207	30	0.039[c]	>550	[179]
Al_2O_3 disk	0.35–0.6	138	30	0.96[c]	503	[179]

Pr retentate feed pressure
[a]For all experiments, the permeate pressure was maintained at atmospheric pressure
[b]Permeances and selectivity determined from mixed gas separation experiments
[c]Permeances determined by single gas permeation experiments; selectivity determined from mixed gas separation experiments

Table 13.9 Comparison of performances of zeolite membranes in CO_2/CH_4 separations

Membrane	Support	Temperature (°C)	Permeance (×10^{-8} mol $m^{-2}\,s^{-1}\,Pa^{-1}$)	α	Ref.
ERI–OFF		35	4.6	400	[172]
DDR		28	7	220	[14]
DDR		−48		4000	[181]
SAPO-34		22	12	170	[182]
SAPO-34		−21		560	[182]
Ba-SAPO-34		32.7	385	250	[183]

Studies of single and binary permeation of CH_4 and CO_2 through silicalite-1 membranes gave that the CO_2 selectivity in the permeation is due to the favorable CO_2 adsorption. A selectivity of about 10 was obtained for a silicalite-1 membrane at 120 °C [180]. In contrast, small-pore zeolites like zeolite T (ERI–OFF) (pore size: 0.41 nm), DDR (0.36 nm × 0.44 nm), and SAPO-34 (0.38 nm) have pores that are similar to the size of CH_4 (0.38 nm) but larger than the size of CO_2 (0.33 nm). It can be expected, therefore, that these membranes show high CO_2/CH_4 selectivities due to molecular sieving. As shown in Table 13.9, the high selectivity is caused by the molecular sieving effect for CH_4 and to some extent by the preferential adsorption of CO_2.

The tuning of the CO_2 permeation through SAPO-34 by ion exchange was reported [183, 184]. As shown in Table 13.9, Ba-SAPO-34 membrane showed good CO_2/CH_4 separation performance at pressure difference of 101.19 kPa and 5.87 % CO_2 concentration in the feed [183]. Recently, Noble et al. still focus their research on SAPO-34 zeolite membranes for CO_2/CH_4 separation [166, 185–187]. However, CO_2 flux and selectivity decrease in the presence of water since water has a strong affinity to the hydrophilic SAPO-34 membrane [188]. Therefore, hydrophobic narrow pore zeolite membranes are more appropriate to separate CO_2 from humid gases. Consequently,

DDR membranes show a high CO_2 flux and selectivity and a negligible water influence on the separation performance in the CO_2 separation from natural gas [14].

13.4.6 Oil–Water Separation

Oil–water separation has become an increasingly important and urgent issue in modern chemical industrial process and environmental protection due to increasing industrial oily wastewater, as well as frequent oil spill accidents. Yu et al. [189] demonstrate a novel oil–water separation film prepared by growing pure silica zeolite silicalite-1 (MFI-type) crystals on stainless steel mesh by taking advantage of the wettability and stability of zeolite film. The zeolite-coated mesh films (ZCMFs) show outstanding superhydrophilic and underwater superoleophobic properties. The separation methodology is solely based on gravity, which allows water to permeate through the film quickly, whereas the oil phase is retained above the film, thus proving to be an energy-efficient filter for oil–water separation.

13.4.7 Desalination of Seawater

The possibility of using zeolite membranes to remove ions from aqueous solutions by reverse osmosis (RO) was explored 10 years ago [190]. The effective intracrystal nanopore diameter of MFI-type zeolite (0.51 nm) is smaller than the sizes of hydrated ions involved in the current systems. A Na^+ rejection of 76.7 % with a water flux of about 0.112 kg m^{-2} h^{-1} was obtained for a 0.1 M NaCl feed solution under an applied pressure of 2.07 MPa on α-alumina-supported MFI-type zeolite membranes. RO desalination by zeolite membranes may offer an alternative for some difficult water treatment processes such as concentration of low-level radioactive wastewater and desalination of high-concentration organic water produced in oil and gas operations where conventional technologies, including polymeric RO membranes, are either inapplicable or inefficient.

13.4.8 Membrane Reactors

The application of membranes in reactors is focused on conversion enhancement by equilibrium displacement or by selective removal of reaction rate inhibitors. Due to their molecular sieve properties, zeolite membranes can remove small product molecules like hydrogen and water from the reaction to increase the conversion and the yield of dehydrogenation and dehydration reactions, respectively. There are numerous examples for the application of zeolite membranes to enhance a chemical reaction like dehydrogenation, esterification, isomerization, and water–gas shift reaction.

The first example is the catalytic dehydrogenation of i-butane. Dehydrogenation of i-butane in a DD3R zeolite membrane reactor was investigated [191]. The DD3R membrane showed an excellent H_2/i-butane permselectivity (>500) at 500 °C and a reasonable H_2 permeance (4.5×10^{-8} mol m^{-2} s^{-1} Pa^{-1}). At low residence times, i-butene yields 50 % above the equilibrium could be obtained. At 489 °C and 0.13 kg (feed) kg $(catalyst)^{-1} h^{-1}$, the i-butene yield in the membrane reactor is 41 %, where the equilibrium yield is 28 %.

The second example deals with the water removal during an esterification. Mordenite membranes have been tested in the esterification of acetic acid with ethanol in a continuous membrane reactor [103, 192]. A CHA-type zeolite membrane, which has high acid stability, was used in the esterification of adipic acid with isopropyl alcohol using sulfuric acid catalyst [193]. The yield of diisopropyl adipate increased from 56 to 98 % by membrane assistance. Zeolite T [129, 137] and ZSM-5 [28, 194] membranes are also used in esterification reaction and showed good stability.

The third example deals with the isomerization of xylene using MFI-type zeolite membranes. EDS analysis showed that aluminum from the substrate was incorporated into the zeolite structure during membrane synthesis. An MFI membrane with a p-xylene permeance of 2.58×10^{-8} mol m^{-2} s^{-1} Pa^{-1} and a *p*-xylene/*o*-xylene separation factor of 19.3 at 300 °C was modified by H^+ ion exchange treatment and used for the isomerization of m-xylene (MX). For MX isomerization, a high PX selectivity of 92.1 % and an MX conversion of 6.5 % were achieved at 270 °C over the catalytic membrane [195].

The fourth example deals with a novel structured zeolite membrane reactor based on a Cu–Mn (1:6)/ZSM-5/PSSF (paperlike sintered stainless steel fibers) catalyst that was developed for catalytic combustion of volatile organic compounds (VOCs). First, the ZSM-5 membrane/PSSF composite was fabricated by the wet lay-up papermaking/sintering process and secondary growth process. The copper/manganese binary oxides modified ZSM-5 membrane catalyst was synthesized by an incipient wetness impregnation method [196].

The fifth example deals with a zeolite membrane reactor in the water–gas shift reaction (WGS), which is a critical step for the pre-combustion CO_2 capture and H_2 production from fossil and biomass stocks via the gasification and steam reforming processes. To date, literature reports on the development of WGS MR have been largely concentrated on the MFI-type zeolite membranes, because they are hydrothermally stable, intrinsically resistant to sulfur compounds, and H_2 selective over water vapor. However, the relatively large pore size of the MFI-type zeolites provides only limited H_2 selectivity over CO_2 and CO at high temperature. The H_2/CO_2 selectivity on MFI-type zeolite membranes can be enhanced by modifications, such as catalytic cracking deposition of MDES [161, 162, 164, 197–199] and formation of a bilayer of ZSM-5 and silicalite-1 [42]. The modified zeolite membrane tube was packed with a cerium-doped ferrite catalyst for water–gas shift (WGS) reaction in a temperature range of 400–550 °C [198]. The WGS reaction results demonstrated that the MFI-type zeolite membrane reactor was effective for enhancing CO-conversion

(X_{CO}) at kinetically favorable high temperatures. At 550 °C with a weight hourly space velocity of 60,000 h^{-1} and a steam to CO ratio of 1.0, the MR achieved X_{CO} of 81.7 % which was significantly higher than the X_{CO} obtained in the traditional packed bed reactor (62.5 %) and well above the equilibrium conversion (65 %) as well. The membrane reactor made of a bilayer zeolite membrane shows stable performance under water–gas shift reaction conditions (500 °C, $H_2O/CO=3$, gaseous hourly space velocity = 60,000 h^{-1}, with a ceria-doped iron oxide catalyst) [43].

13.4.9 Membrane Microreactors

Membrane microreactors (MMRs) combine the advantages of membrane separation and microreactors such as high area/volume ratio, enhanced mass and/or heat transfer, improved catalytic efficiency without equilibrium limitation, good operational safety, and design flexibility. In most instances, catalytic coatings on microreactors consist of a catalyst-containing layer washcoated on the reactor wall. For the cases in which the catalyst is supported on zeolites or is a zeolite itself, a zeolite film directly grown on the microreactor channels appears as the ideal candidate: It yields a highly adherent catalytic layer, and a binder is not needed (meaning that the layer is 100 % catalyst).

Rebrov et al. [200] carried out the pioneering investigations of ZSM-5 zeolite growth in stainless steel microchannels (500 μm wide). The resulting zeolite layer was ion exchanged with Ce and tested in the selective reduction of NOx using ammonia as reducing agent, with the microreactor catalytic film clearly outperforming the pelletized catalysts. Santamaria et al. studied the preparation of MFI layers on microchannels. The ZSM-5 microreactors were ion exchanged with Pt and tested in the selective oxidation of CO, SELOX reaction [201], where complete CO conversion was achieved at 170 °C.

In addition to MFI, zeolites such as MOR [202, 203], ETS-10 [202], BEA [204], LTA [141], and FAU [202, 205] have also been prepared as catalytic layers on different supports such as ceramic monoliths, brass, molybdenum, and stainless steel microchannel reactors. The FAU and ETS-10 coated microreactors exhibited high performance in the selective oxidation of CO in a simulated reformer stream with total CO combustion temperatures as low as 125 °C [206]. Other reactions investigated include a wide variety of systems ranging from ammoxidation of ethylene [204] to combustion of VOCs at trace concentration levels [205] and Knoevenagel condensation [207].

The synthesis of mordenite coatings on thin FeCrAl alloy foils was carried out in order to obtain adherent, stable, and homogeneous films for their use as catalytic microreactors [208]. Cu and Ce were incorporated onto the mordenite coating as catalytic ingredients and used in a microreactor for the CO oxidation reaction under different conditions and in the preferential CO oxidation (COProx) showing a superior performance when compared with the same powder catalyst.

13.4.10 Low-k Materials

Zeolite low-k films were first studied on MFI-type pure silica zeolite (PSZ) using an in situ deposition process in which a uniform and *b*-oriented polycrystalline film was obtained on a silicon substrate. The k value and elastic modulus of the film were 2.7 and 30–40 GPa, respectively. To be more manufacturing friendly, a spin-on process was developed in which an MFI-type PSZ nanoparticle suspension was hydrothermally synthesized and then spun-on onto silicon wafers [25]. The suspension contained zeolite nanoparticles and aged amorphous silica precursors. The resultant zeolite films had a bimodal pore size distribution consisting of the intraparticle zeolite micropores and the interparticle mesopore void space, and consequently, the added mesoporosity lowered the k value and elastic modulus to 2.1 and 16–18 GPa, respectively. Additional mesoporosity could be added to the films through the addition of γ-cyclodextrin to the synthesized zeolite suspension, and lower k value (~1.8) could be obtained [209]. Further research work has been going on about zeolite low-k materials since 2010 [210–213].

13.4.11 Corrosion-Resistant Materials

Metal corrosion is a ubiquitous problem and generally costs an industrialized country several percent of its gross domestic product (GDP) each year. For aerospace applications, aluminum alloys are commonly used because they are light and mechanically strong, but some alloys have a serious problem of corrosion (e.g., Al-2024-T3). A chromium-free coating is preferred. Zeolites contain many of the properties necessary for a corrosion-resistant coating. The frameworks of high-silica and pure silica zeolites (HSZs and PSZs) are extremely corrosion resistant in pitting aggressive media and all mineral acids except hydrofluoric acid [214]. Furthermore, most HSZ and PSZ syntheses use an organic structure-directing agent (SDA) to form a specific framework, and after synthesis, these SDA molecules remain in the zeolite pores. Uncalcined and defect-free zeolite membranes are gastight [57, 215]. Therefore, along with the chemical properties of the zeolite framework, the uncalcined membranes can physically block corrosive compounds and are strong candidates for chromium-free corrosion-resistant coatings. Polycrystalline high-silica zeolite MFI coatings were synthesized on Al-2024-T3 through an in situ crystallization process, and DC polarization tests showed that they were corrosion resistant in sulfuric acid and sodium hydroxide, media known for causing general corrosion [216]. They also performed remarkably well in neutral and acidified sodium chloride. Some new synthesis approaches for zeolite coatings on metal substrates were developed by Yan et al., for example, ionothermal synthesis of oriented zeolite AEL films [217] and ambient pressure dry gel conversion method for zeolite MFI synthesis using ionic liquid and microwave heating [83].

13.4.12 Sensors

Zeolites are obvious candidates for sensors, in view of their well-defined porous structure, their tunable adsorption properties, and ion exchange capacity, as pointed out in recent reviews [218, 219]. Zeolites can be used in sensors as individual crystals or as intergrown films. Here, we just focus on zeolite films as selective barriers for interfering species. The zeolite framework structure, Si/Al ratio, extraframework cations, and thickness of the film are the main parameters for a fine-tuning of the barrier effect. Mann and coworkers [220] applied Cr-FAU and Cr-BEA coatings onto $Cr_{2-x}Ti_xO_y$ sensors to discriminate a series of alkanes. FAU overlayers onto WO_3 or Cr_2-xTi_xO_y sensors [221] have been deployed to increase ethanol against carbon monoxide response. In these works, the zeolite coatings produce a variety of adsorption, catalytic, and diffusive effects, which become useful to enhance or depress the sensor response toward a given analyte, often on an ad hoc basis. Catalytic effects have been explicitly exploited to remove interfering components, often through combustion reactions. An example is the work of Sahner et al. [222, 223] who used screen-printed thick films of the $SrTi_{1-x}Fe_xO_{3-\delta}$ perovskite with an additional top layer of Pt/ZSM5 to remove the interference of CO in the sensing of NO.

In contrast to the above works that used powdered zeolites as sensor overlayers, other investigations have addressed intergrown zeolite films that can play an even more discriminating role as a reactive or nonreactive barrier [224]. MFI and LTA membranes were directly synthesized on screen-printed Pd/SnO_2 sensor layers by Santamaria et al. [225, 226]. These molecular filters decreased interference of undesired molecules mainly via a selective adsorption mechanism that hindered the diffusion of less strongly adsorbed species. In a subsequent work [227], the system was scaled down by a factor of 50, and a microdropping procedure was used to deposit the zeolite barrier. The micromachined sensor presented a considerable reduction of response times without apparent loss of performance. Sensing response, response time, and recovery time for selective ethylene gas detection were improved by coating a layer of *b*-oriented silicalite-1 layers [228] or [010] highly preferred-orientation silicalite-1 polycrystals on SnO_2 thin film sensors [229].

13.4.13 Fuel Cells and Batteries

Zeolites are increasingly used in fuel cells. Zeolites' ionic conductivity and capacity for water adsorption and retention were exploited to create zeolite and zeolite–PFSA proton conducting membranes for polymer electrolyte membrane fuel cells (PEMFCs). A series of composite membranes based on zeolite A and Nafion 117 have been fabricated for direct methanol fuel cells [230]. A novel Pt/zeolite–Nafion (PZN) polymer electrolyte composite membrane is fabricated for self-humidifying PEMFCs [231]. Nanocomposite proton exchange membranes (PEMs) based on Nafion with sulfonic acid-functionalized zeolite beta as an additive were prepared by Yan et al. [232]. Zeolite beta-filled chitosan membranes [233], Nafion/H-ZSM-5 composite membranes [234], and Nafion/mordenite composite membranes [235] with improved performance for direct methanol fuel cell were prepared. HZSM-5 micromembranes of different Si/Al ratios were fabricated in a regular array on silicon followed by assembly and test for fuel cell performance. HZSM-5 with high aluminum content exhibits good proton conductivity and better PEMFC performance that approaches that of Nafion MEA.

Very recently, the crystalline silicalite membrane has been demonstrated as an effective ion separator for acidic solutions of vanadyl sulfate and for potential application as a proton-permselective electrolyte membrane in the all-vanadium redox flow battery by Dong et al. [236].

13.5 Conclusions and Outlook

Intensive researches on zeolite membranes have been carried out since 1992, and the production of zeolite films and membranes on almost any type of support with unprecedented control on their morphology, orientation, and degree of crystal intergrowth is allowed. Many applications of zeolite films and membranes in various areas have been reported by researchers. In 1998, Mitsui Engineering & Shipbuilding Co. Ltd. put the first commercial unit of NaA zeolite membrane on the market, which is a milestone in the course of commercial production and application of zeolite membranes. In Europe, bio-ethanol dehydration plant has been set up in the UK by the alliance of Smart and Inocermic GmbH. In China, Nine Heaven Hi-Tech Co. Ltd. has sold more than ten zeolite membrane dehydration plants for organic solvents recently. However, there are only a few pilot plants in operation all over the world, mainly in the dehydration of ethanol and i-propanol using hydrophilic LTA membranes. The problem of reducing the cost for supports remains an open challenge. The development of zeolite membranes that are stable in acid media for dehydration of organic solvents is expected.

Most promising seems to be the development of thin (less than 1 μm) supported zeolite layers on a wide variety of carriers such as capillaries, fibers, tubes, or monoliths. Oriented zeolite layers will enable high-flux membranes. The preparation of *b*-oriented MFI membranes is still a challenge. A template-free synthesis would reduce the thermal stress in the oxidative template removal. A secondary growth of seeded supports is recommended which also opens the way for functional layers. The coating of porous supports by wet rubbing with zeolite crystals looks promising. Another challenge will be the synthesis of thin zeolite layers on hollow fiber supports and the manufacture of the membrane modules from the hollow fibers.

The application of zeolite membranes for the size/shape-selective separation is well expected. Zeolite membranes with 6-ringlike SOD (0.28 nm pore size), 8-ringlike CHA (0.34 nm), and DDR (0.44 nm) will allow the separation of small molecules like water and hydrogen. A significant amount of progress has been made, and it seems to be only a matter of time that first pilot plants on gas separation will start operation.

References

1. Morigami Y, Kondo M, Abe J, Kita H, Okamoto K (2001) The first large-scale pervaporation plant using tubular-type module with zeolite NaA membrane. Sep Purif Technol 25(1–3):251–260
2. denExter MJ, Jansen JC, vandeGraaf JM, Kapteijn F, Moulijn JA, vanBekkum H (1996) Zeolite-based membranes preparation, performance and prospects. Stud Surf Sci Catal 102:413–454
3. Tavolaro A, Drioli E (1999) Zeolite membranes. Adv Mater 11(12):975–996
4. Caro J, Noack M, Kolsch P, Schafer R (2000) Zeolite membranes – state of their development and perspective. Microporous Mesoporous Mater 38(1):3–24
5. Lin YS (2001) Microporous and dense inorganic membranes: current status and prospective. Sep Purif Technol 25(1–3):39–55
6. Bowen TC, Noble RD, Falconer JL (2004) Fundamentals and applications of pervaporation through zeolite membranes. J Membr Sci 245(1–2):1–33
7. McLeary EE, Jansen JC, Kapteijn F (2006) Zeolite based films, membranes and membrane reactors: progress and prospects. Microporous Mesoporous Mater 90(1–3):198–220
8. Noack M, Kolsch P, Dittmar A, Stohr M, Georgi G, Schneider M, Dingerdissen U, Feldhoff A, Caro J (2007) Proof of the ISS-concept for LTA and FAU membranes and their characterization by extended gas permeation studies. Microporous Mesoporous Mater 102(1–3):1–20
9. Snyder MA, Tsapatsis M (2007) Hierarchical nanomanufacturing: from shaped zeolite nanoparticles to high-performance separation membranes. Angew Chem Int Ed 46(40):7560–7573
10. Caro J, Noack M (2008) Zeolite membranes – recent developments and progress. Microporous Mesoporous Mater 115(3):215–233
11. Wee SL, Tye CT, Bhatia S (2008) Membrane separation process-pervaporation through zeolite membrane. Sep Purif Technol 63(3):500–516
12. Li YS, Yang WS (2008) Microwave synthesis of zeolite membranes: a review. J Membr Sci 316(1–2):3–17
13. Daramola MO, Burger AJ, Pera-Titus M, Giroir-Fendler A, Miachon S, Dalmon JA, Lorenzen L (2010) Separation and isomerization of xylenes using zeolite membranes: a short overview. Asia Pac J Chem Eng 5(6):815–837
14. Lew CM, Cai R, Yan YS (2010) Zeolite thin films: from computer chips to space stations. Acc Chem Res 43(2):210–219. doi:10.1021/Ar900146w
15. Paul DR, Kemp DR (1973) Diffusion time lag in polymer membranes containing adsorptive fillers. J Polym Sci Polym Symp 41:79–93
16. Kulprathipanja S, Neuzil RW, Li NN (1988) Separation of fluids by means of mixed matrix membranes. US Patent 4740219
17. Chung TS, Jiang LY, Li Y, Kulprathipanja S (2007) Mixed matrix membranes (MMMs) comprising organic polymers with dispersed inorganic fillers for gas separation. Prog Polym Sci 32(4):483–507
18. Aroon MA, Ismail AF, Matsuura T, Montazer-Rahmati MM (2010) Performance studies of mixed matrix membranes for gas separation: a review. Sep Purif Technol 75(3):229–242. doi:10.1016/j.seppur.2010.08.023
19. Bastani D, Esmaeili N, Asadollahi M (2013) Polymeric mixed matrix membranes containing zeolites as a filler for gas separation applications: a review. J Ind Eng Chem 19(2):375–393
20. Sakai H, Tomita T, Takahashi T (2001) p-xylene separation with MFI-type zeolite membrane. Sep Purif Technol 25(1–3):297–306. doi:10.1016/S1383-5866(01)00056-9
21. Kiyozumi Y, Nagase T, Hasegawa Y, Mizukami F (2008) A process for synthesizing bilayer zeolite membranes. Mater Lett 62(3):436–439. doi:10.1016/j.matlet.2007.05.067
22. He Y, Cui XM, Liu XD, Wang YP, Zhang J, Liu K (2013) Preparation of self-supporting NaA zeolite membranes using geopolymers. J Membr Sci 447:66–72
23. Zhang J, He Y, Wang YP, Mao J, Cui XM (2014) Synthesis of a self-supporting faujasite zeolite membrane using geopolymer gel for separation of alcohol/water mixture. Mater Lett 116:167–170. doi:10.1016/j.matlet.2013.11.008

24. Jung KT, Shul YG (1997) Preparation of transparent TS-1 zeolite film by using nanosized TS-1 particles. Chem Mater 9(2):420–422
25. Wang ZB, Mitra AP, Wang HT, Huang LM, Yan YS (2001) Pure silica zeolite films as low-k dielectrics by spin-on of nanoparticle suspensions. Adv Mater 13(19):1463–1466. Doi::10.1002/1521-4095(200110)13:19<1463::Aid-Adma1463>3.0.Co;2-H
26. Wang ZB, Yan YS (2001) Controlling crystal orientation in zeolite MFI thin films by direct in situ crystallization. Chem Mater 13(3):1101–1107. doi:10.1021/Cm000849e
27. Wang ZB, Yan YS (2001) Oriented zeolite MFI monolayer films on metal substrates by in situ crystallization. Microporous Mesoporous Mater 48(1–3):229–238. doi:10.1016/S1387-1811(01)00357-2
28. Zhu MH, Kumakiri I, Tanaka K, Kita H (2013) Dehydration of acetic acid and esterification product by acid-stable ZSM-5 membrane. Microporous Mesoporous Mater 181:47–53
29. Wang ZB, Ge QQ, Shao J, Yan YS (2009) High performance zeolite LTA pervaporation membranes on ceramic hollow fibers by dipcoating-wiping seed deposition. J Am Chem Soc 131(20):6910–6911
30. Wang ZB, Ge QQ, Gao JS, Shao J, Liu CJ, Yan YS (2011) High-performance zeolite membranes on inexpensive large-pore supports: highly reproducible synthesis using a seed paste. ChemSusChem 4(11):1570–1573
31. Wang XR, Chen YY, Zhang C, Gu XH, Xu NP (2014) Preparation and characterization of high-flux T-type zeolite membranes supported on YSZ hollow fibers. J Membr Sci 455:294–304
32. Zhang F, Zheng YH, Hua LL, Hu N, Zhu MH, Zhou RF, Chen XS, Kita H (2014) Preparation of high-flux zeolite T membranes using reusable macroporous stainless steel supports in fluoride media. J Membr Sci 456:107–116
33. Gao X, Zou XQ, Zhang F, Zhang SX, Ma HP, Zhao NA, Zhu GS (2013) Eco-friendly fabrication of hydrophilic ZSM-5 membranes for alcohol upgrading. Chem Commun 49(78):8839–8841
34. Ge QQ, Wang ZB, Yan YS (2009) High-performance zeolite NaA membranes on polymer-zeolite composite hollow fiber supports. J Am Chem Soc 131(47):17056–17057
35. Yin XJ, Zhu GS, Yang WS, Li YS, Zhu GQ, Xu R, Sun JY, Qiu SL, Xu RR (2005) Stainless-steel-net-supported zeolite NaA membrane with high permeance and high permselectivity for oxygen over nitrogen. Adv Mater 17(16):2006–2010. doi:10.1002/adma.200500270
36. Wang ZY, Yin XJ, Yue NL, Sun FX, Zhu GS (2011) Preparation and characterization of X-type zeolite membrane supported by stainless steel net. Chem J Chin Univ 32(10):2250–2255
37. Zou XQ, Zhu GS, Guo HL, Jing XF, Xu D, Qiu SL (2009) Effective heavy metal removal through porous stainless-steel-net supported low siliceous zeolite ZSM-5 membrane. Microporous Mesoporous Mater 124(1–3):70–75
38. Chen YL, Zhu GS, Peng Y, Qiu SL (2008) Stainless-steel net supported ITQ-17 membrane potential for microlaser application. In: Zeolites and related materials: trends, targets and challenges. Proceedings of the 4th International FEZA Conference 174:661–664
39. Liu W, Zhang J, Canfield N, Saraf L (2011) Preparation of robust, thin zeolite membrane sheet for molecular separation. Ind Eng Chem Res 50(20):11677–11689. doi:10.1021/Ie200519b
40. Zhang J, Liu W (2011) Thin porous metal sheet-supported NaA zeolite membrane for water/ethanol separation. J Membr Sci 371(1–2):197–210
41. Huang AS, Wang NY, Caro J (2012) Synthesis of multi-layer zeolite LTA membranes with enhanced gas separation performance by using 3-aminopropyltriethoxysilane as interlayer. Microporous Mesoporous Mater 164:294–301
42. Wang HB, Lin YS (2012) Synthesis and modification of ZSM-5/silicalite bilayer membrane with improved hydrogen separation performance. J Membr Sci 396:128–137
43. Wang HB, Dong XL, Lin YS (2014) Highly stable bilayer MFI zeolite membranes for high temperature hydrogen separation. J Membr Sci 450:425–432
44. Jansen JC, Vanrosmalen GM (1993) Oriented growth of silica molecular-sieve crystals as supported films. J Cryst Growth 128(1–4):1150–1156. doi:10.1016/S0022-0248(07)80114-X

45. Li XM, Yan YS, Wang ZB (2010) Continuity control of b-oriented MFI zeolite films by microwave synthesis. Ind Eng Chem Res 49(12):5933–5938
46. Wang XD, Zhang BQ, Liu XF, Lin JYS (2006) Synthesis of b-oriented TS-1 films on chitosan-modified alpha-Al_2O_3 substrates. Adv Mater 18(24):3261–3265. doi:10.1002/adma.200502772
47. Lai ZP, Bonilla G, Diaz I, Nery JG, Sujaoti K, Amat MA, Kokkoli E, Terasaki O, Thompson RW, Tsapatsis M, Vlachos DG (2003) Microstructural optimization of a zeolite membrane for organic vapor separation. Science 300(5618):456–460. doi:10.1126/science.1082169
48. Tung CTP, Kim HS, Yoon KB (2011) Growth of uniformly oriented silica MFI and BEA zeolite films on substrates. Science 334(6062):1533–1538. doi:10.1126/science.1212472
49. Kim E, Choi J, Tsapatsis M (2013) On defects in highly a-oriented MFI membranes. Microporous Mesoporous Mater 170:1–8. doi:10.1016/j.micromeso.2012.11.023
50. Choi J, Ghosh S, Lai ZP, Tsapatsis M (2006) Uniformly a-oriented MFI zeolite films by secondary growth. Angew Chem Int Ed 45(7):1154–1158. doi:10.1002/anie.200503011
51. Lee T, Choi J, Tsapatsis M (2013) On the performance of c-oriented MFI zeolite membranes treated by rapid thermal processing. J Membr Sci 436:79–89
52. Bons AJ, Bons PD (2003) The development of oblique preferred orientations in zeolite films and membranes. Microporous Mesoporous Mater 62(1–2):9–16. doi:10.1016/S1387-1811(03)00384-6
53. Liu Y, Li YS, Yang WS (2010) Fabrication of highly b-oriented MFI film with molecular sieving properties by controlled in-plane secondary growth. J Am Chem Soc 132(6):1768–1769
54. Liu Y, Li YS, Cai R, Yang WS (2012) Suppression of twins in b-oriented MFI molecular sieve films under microwave irradiation. Chem Commun 48(54):6782–6784. doi:10.1039/C2cc18111h
55. Li XM, Peng Y, Wang ZB, Yan YS (2011) Synthesis of highly b-oriented zeolite MFI films by suppressing twin crystal growth during the secondary growth. CrystEngComm 13(11):3657–3660
56. Sano T, Hasegawa M, Kawakami Y, Kiyozumi Y, Yanagishita H, Kitamoto D, Mizukami F (1994) Potentials of silicalite membranes for the separation of alcohol-water mixtures. Stud Surf Sci Catal 84:1175–1182
57. Yan YS, Davis ME, Gavalas GR (1995) Preparation of zeolite ZSM-5 membranes by in-situ crystallization on porous alpha-Al_2O_3. Ind Eng Chem Res 34(5):1652–1661. doi:10.1021/Ie00044a018
58. Zah J, Krieg HM, Breytenbach JC (2006) Layer development and growth history of polycrystalline zeolite A membranes synthesised from a clear solution. Microporous Mesoporous Mater 93(1–3):141–150. doi:10.1016/j.micromeso.2006.02.014
59. Okamoto K, Kita H, Horii K, Tanaka K, Kondo M (2001) Zeolite NaA membrane: preparation, single-gas permeation, and pervaporation and vapor permeation of water/organic liquid mixtures. Ind Eng Chem Res 40(1):163–175
60. Xu XC, Yang WS, Liu J, Lin LW (2001) Synthesis of NaA zeolite membranes from clear solution. Microporous Mesoporous Mater 43(3):299–311
61. Li HZ, Wang JQ, Xu J, Meng XD, Xu B, Yang JH, Li SY, Lu JM, Zhang Y, He XL, Yin DH (2013) Synthesis of zeolite NaA membranes with high performance and high reproducibility on coarse macroporous supports. J Membr Sci 444:513–522
62. Xomeritakis G, Gouzinis A, Nair S, Okubo T, He MY, Overney RM, Tsapatsis M (1999) Growth, microstructure, and permeation properties of supported zeolite (MFI) films and membranes prepared by secondary growth. Chem Eng Sci 54(15–16):3521–3531. doi:10.1016/S0009-2509(98)00515-6
63. Hedlund J, Noack M, Kolsch P, Creaser D, Caro J, Sterte J (1999) ZSM-5 membranes synthesized without organic templates using a seeding technique. J Membr Sci 159(1–2):263–273. doi:10.1016/S0376-7388(99)00069-1
64. Mintova S, Hedlund J, Valtchev V, Schoeman BJ, Sterte J (1998) ZSM-5 films prepared from template free precursors. J Mater Chem 8(10):2217–2221. doi:10.1039/A803687j
65. Huang AS, Liu Q, Wang NY, Tong X, Huang BX, Wang M, Caro J (2013) Covalent synthesis of dense zeolite LTA membranes on various 3-chloropropyltrimethoxysilane functionalized supports. J Membr Sci 437:57–64

66. Huang AS, Liang FY, Steinbach F, Caro J (2010) Preparation and separation properties of LTA membranes by using 3-aminopropyltriethoxysilane as covalent linker. J Membr Sci 350(1–2):5–9
67. Yoon KB (2007) Organization of zeolite microcrystals for production of functional materials. Acc Chem Res 40(1):29–40. doi:10.1021/Ar000119c
68. Kuzniatsova T, Kim Y, Shqau K, Dutta PK, Verweij H (2007) Zeta potential measurements of zeolite Y: application in homogeneous deposition of particle coatings. Microporous Mesoporous Mater 103(1–3):102–107. doi:10.1016/j.micromeso.2007.01.042
69. Pina MP, Arruebo M, Felipe A, Fleta F, Bernal MP, Coronas J, Menendez M, Santamaria J (2004) A semi-continuous method for the synthesis of NaA zeolite membranes on tubular supports. J Membr Sci 244(1–2):141–150
70. Huang AS, Lin YS, Yang WS (2004) Synthesis and properties of A-type zeolite membranes by secondary growth method with vacuum seeding. J Membr Sci 245(1–2):41–51. doi:10.1016/j.memsci.2004.08.001
71. Algieri C, Bernardo P, Barbieri G, Drioli E (2009) A novel seeding procedure for preparing tubular NaY zeolite membranes. Microporous Mesoporous Mater 119(1–3):129–136
72. Peng Y, Zhan ZY, Shan LJ, Li XM, Wang ZB, Yan YS (2013) Preparation of zeolite MFI membranes on defective macroporous alumina supports by a novel wetting-rubbing seeding method: role of wetting agent. J Membr Sci 444:60–69
73. Zampieri A, Dubbe A, Schwieger W, Avhale A, Moos R (2008) ZSM-5 zeolite films on Si substrates grown by in situ seeding and secondary crystal growth and application in an electrochemical hydrocarbon gas sensor. Microporous Mesoporous Mater 111(1–3):530–535. doi:10.1016/j.micromeso.2007.08.026
74. Dong JX, Dou T, Zhao XG, Gao LH (1992) Synthesis of membranes of zeolites ZSM-5 and ZSM-35 by the vapor-phase method. J Chem Soc Chem Commun 15:1056–1058
75. Nishiyama N, Matsufuji T, Ueyama K, Matsukata M (1997) FER membrane synthesized by a vapor-phase transport method: its structure and separation characteristics. Microporous Mater 12(4–6):293–303
76. Nishiyama N, Ueyama K, Matsukata M (1996) Synthesis of defect-free zeolite-alumina composite membranes by a vapor-phase transport method. Microporous Mater 7(6):299–308
77. Cheng ZL, Chao ZS, Lin HQ, Wan HL (2003) NaA zeolite membrane with high performance synthesized by vapor phase transformation method. Chin J Chem 21(11):1430–1432
78. Xu XC, Bao Y, Song CS, Yang WS, Liu J, Lin LW (2004) Microwave-assisted hydrothermal synthesis of hydroxy-sodalite zeolite membrane. Microporous Mesoporous Mater 75(3):173–181
79. Chen XB, Yang WS, Liu J, Lin LW (2005) Synthesis of zeolite NaA membranes with high permeance under microwave radiation on mesoporous-layer-modified macroporous substrates for gas separation. J Membr Sci 255(1–2):201–211
80. Li YS, Chen HL, Liu J, Yang WS (2006) Microwave synthesis of LTA zeolite membranes without seeding. J Membr Sci 277(1–2):230–239. doi:10.1016/j.memsci.2005.10.033
81. Tang Z, Kim SJ, Gu XH, Dong JH (2009) Microwave synthesis of MFI-type zeolite membranes by seeded secondary growth without the use of organic structure directing agents. Microporous Mesoporous Mater 118(1–3):224–231. doi:10.1016/j.micromeso.2008.08.029
82. Zhou H, Li YS, Zhu GQ, Liu J, Yang WS (2009) Preparation of zeolite T membranes by microwave-assisted in situ nucleation and secondary growth. Mater Lett 63(2):255–257
83. Cai R, Liu Y, Gu S, Yan YS (2010) Ambient pressure dry-gel conversion method for zeolite MFI synthesis using ionic liquid and microwave heating. J Am Chem Soc 132(37):12776–12777. doi:10.1021/Ja101649b
84. Sebastian V, Mallada R, Coronas J, Julbe A, Terpstra RA, Dirrix RWJ (2010) Microwave-assisted hydrothermal rapid synthesis of capillary MFI-type zeolite-ceramic membranes for pervaporation application. J Membr Sci 355(1–2):28–35
85. Wang CZ, Liu XF, Li J, Zhang BQ (2013) Microwave-assisted seeded growth of the submicrometer-thick and pure b-oriented MFI zeolite films using an ultra-dilute synthesis solution. CrystEngComm 15(32):6301–6304. doi:10.1039/C3ce40644j

86. Noack M, Kolsch P, Dittmar A, Stohr M, Georgi G, Eckelt R, Caro J (2006) Effect of crystal intergrowth supporting substances (ISS) on the permeation properties of MFI membranes with enhanced Al-content. Microporous Mesoporous Mater 97(1–3):88–96
87. Huang AS, Wang NY, Caro J (2012) Seeding-free synthesis of dense zeolite FAU membranes on 3-aminopropyltriethoxysilane-functionalized alumina supports. J Membr Sci 389:272–279
88. Huang BX, Liu Q, Caro J, Huang AS (2014) Iso-butanol dehydration by pervaporation using zeolite LTA membranes prepared on 3-aminopropyltriethoxysilane-modified alumina tubes. J Membr Sci 455:200–206
89. Huang AS, Caro J (2011) Facile synthesis of LTA molecular sieve membranes on covalently functionalized supports by using diisocyanates as molecular linkers. J Mater Chem 21(30):11424–11429
90. Huang AS, Bux H, Steinbach F, Caro J (2010) Molecular-sieve membrane with hydrogen permselectivity: ZIF-22 in LTA topology prepared with 3-aminopropyltriethoxysilane as covalent linker. Angew Chem Int Ed 49(29):4958–4961. doi:10.1002/anie.201001919
91. Pera-Titus M, Llorens J, Cunill F, Mallada R, Santamaria J (2005) Preparation of zeolite NaA membranes on the inner side of tubular supports by means of a controlled seeding technique. Catal Today 104(2–4):281–287. doi:10.1016/j.cattod.2005.03.042
92. Pera-Titus M, Mallada R, Llorens J, Cunill F, Santamaria J (2006) Preparation of inner-side tubular zeolite NaA membranes in a semi-continuous synthesis system. J Membr Sci 278(1–2):401–409. doi:10.1016/j.memsci.2005.11.026
93. Tiscareno-Lechuga F, Tellez C, Menendez M, Santamaria J (2003) A novel device for preparing zeolite – a membranes under a centrifugal force field. J Membr Sci 212(1–2):135–146. doi: 10.1016/S0376-7388(02)00491-X
94. Pera-Titus M, Bausach M, Llorens J, Cunill F (2008) Preparation of inner-side tubular zeolite NaA membranes in a continuous flow system. Sep Purif Technol 59(2):141–150. doi:10.1016/j.seppur.2007.05.038
95. Lai LL, Shao J, Ge QQ, Wang ZB, Yan YS (2012) The preparation of zeolite NaA membranes on the inner surface of hollow fiber supports. J Membr Sci 409:318–328
96. Wang L, Wang Y, Hao JG, Liu GZ, Ma XS, Hu SL (2013) Synthesis of HZSM-5 coatings on the inner surface of stainless steel tubes and their catalytic performance in n-dodecane cracking. Appl Catal Gen 462:271–277
97. Tosheva L, Holzl M, Metzger TH, Valtchev V, Mintova S, Bein T (2005) Zeolite beta films synthesized from basic and near-neutral precursor solutions and gels. Mater Sci Eng C Bio S 25(5–8):570–576. doi:10.1016/j.msec.2005.07.011
98. Chen YL, Zhu GS, Peng Y, Yao XD, Qiu SL (2009) Synthesis and characterization of (h0l) oriented high-silica zeolite beta membrane. Microporous Mesoporous Mater 124(1–3):8–14. doi:10.1016/j.micromeso.2009.04.012
99. Gualtieri ML, Gualtieri AF, Prudenziati M (2008) Seeded growth of TPA-MFI films using the fluoride route. Microporous Mesoporous Mater 111(1–3):604–611. doi:10.1016/j.micromeso.2007.09.006
100. Gualtieri ML (2009) Synthesis of MFI films on alpha-alumina at neutral pH. Microporous Mesoporous Mater 117(1–2):508–510. doi:10.1016/j.micromeso.2008.07.022
101. Zhang XL, Zhu MH, Zhou RF, Chen XS, Kita H (2011) Synthesis of silicalite-1 membranes with high ethanol permeation in ultradilute solution containing fluoride. Sep Purif Technol 81(3):480–484
102. Zhu MH, Lu ZH, Kumakiri I, Tanaka K, Chen XS, Kita H (2012) Preparation and characterization of high water perm-selectivity ZSM-5 membrane without organic template. J Membr Sci 415:57–65
103. Ren XX, Yang JH, Chen Z, Yang XB, Lu JM, Zhang Y, Wang JQ (2012) Preparation and performance of mordenite zeolite membrane using fluoride route. Chin J Catal 33(9):1558–1564
104. Zhou RF, Hu ZL, Hu N, Duan LQ, Chen XS, Kita H (2012) Preparation and microstructural analysis of high-performance mordenite membranes in fluoride media. Microporous Mesoporous Mater 156:166–170

105. Chen Z, Li YH, Yin DH, Song YM, Ren XX, Lu JM, Yang JH, Wang JQ (2012) Microstructural optimization of mordenite membrane for pervaporation dehydration of acetic acid. J Membr Sci 411:182–192
106. Zhou M, Korelskiy D, Ye PC, Grahn M, Hedlund J (2014) A uniformly oriented MFI membrane for improved CO_2 separation. Angew Chem Int Ed 53(13):3492–3495
107. Li YS, Zhang XF, Wang JQ (2001) Preparation for ZSM-5 membranes by a two-stage varying-temperature synthesis. Sep Purif Technol 25(1–3):459–466
108. Kong CL, Lu JM, Yang JH, Wang JQ (2006) Preparation of silicalite-1 membranes on stainless steel supports by a two-stage varying-temperature in situ synthesis. J Membr Sci 285(1–2):258–264. doi:10.1016/j.memsci.2006.08.027
109. Wang ZZ, Kumakiri I, Tanaka K, Chen XS, Kita H (2013) NaY zeolite membranes with high performance prepared by a variable-temperature synthesis. Microporous Mesoporous Mater 182:250–258
110. Hedlund J, Jareman F, Bons AJ, Anthonis M (2003) A masking technique for high quality MFI membranes. J Membr Sci 222(1–2):163–179. doi:10.1016/S0376-7388(03)00285-0
111. Chen HL, Li YS, Zhu GQ, Liu JE, Yang WS (2008) Synthesis and pervaporation performance of high-reproducibility silicalite-1 membranes. Chin Sci Bull 53(22):3505–3510
112. Chen HL, Li YS, Zhu GQ, Yang WS (2009) Synthesis and separation performance of silicalite-1 membranes on silica tubes. Sci China B 52(5):579–583
113. Xiao W, Yang JH, Lu JM, Wang JQ (2009) Preparation and characterization of silicalite-1 membrane by counter-diffusion secondary growth. J Membr Sci 345(1–2):183–190
114. Yin HM, Yang JH, Xie Z, Wang JQ, Lu JM, Zhang Y (2014) Deposition of functional modified inorganic particles on a macroporous tube for synthesis ZIF-7 membrane by counter diffusion method. J Inorg Mater 29(4):377–381. doi:10.3724/Sp.J.1077.2014.13355
115. Miachon S, Landrivon E, Aouine M, Sun Y, Kumakiri I, Li Y, Prokopova OP, Guilhaume N, Giroir-Fendler A, Mozzanega H, Dalmon JA (2006) Nanocomposite MFI-alumina membranes via pore-plugging synthesis – preparation and morphological characterisation. J Membr Sci 281(1–2):228–238. doi:10.1016/j.memsci.2006.03.036
116. Miachon S, Ciavarella P, van Dyk L, Kumakiri I, Fiaty K, Schuurman Y, Dalmon JA (2007) Nanocomposite MFI-alumina membranes via pore-plugging synthesis: specific transport and separation properties. J Membr Sci 298(1–2):71–79
117. Alshebani A, Pera-Titus M, Landrivon E, Schiestel T, Miachon S, Dalmon JA (2008) Nanocomposite MFI – ceramic hollow fibres: prospects for CO_2 separation. Microporous Mesoporous Mater 115(1–2):197–205
118. Rouleau L, Pirngruber G, Guillou F, Barrere-Tricca C, Omegna A, Valtchev V, Pera-Titus M, Miachon S, Dalmon JA (2009) Nanocomposite MFI-alumina and FAU-alumina membranes: synthesis, characterization and application to paraffin separation and CO_2 capture. Oil Gas Sci Technol 64(6):745–758. doi:10.2516/Ogst/2009036
119. Sublet J, Pera-Titus M, Guilhaume N, Farrusseng D, Schrive L, Chanaud P, Siret B, Durecu S (2012) Technico-economical assessment of MFI-type zeolite membranes for CO_2 capture from postcombustion flue gases. AIChE J 58(10):3183–3194
120. Daramola MO, Burger AJ, Pera-Titus M, Giroir-Fendler A, Miachon S, Lorenzen L, Dalmon JA (2009) Nanocomposite MFI-ceramic hollow fibre membranes via pore-plugging synthesis: prospects for xylene isomer separation. J Membr Sci 337(1–2):106–112
121. Ciavarella P, Moueddeb H, Miachon S, Fiaty K, Dalmon JA (2000) Experimental study and numerical simulation of hydrogen/isobutane permeation and separation using MFI-zeolite membrane reactor. Catal Today 56(1–3):253–264
122. Shao J, Zhan ZY, Li JG, Wang ZB, Li K, Yan YS (2014) Zeolite NaA membranes supported on alumina hollow fibers: effect of support resistances on pervaporation performance. J Membr Sci 451:10–17
123. Kita H, Fuchida K, Horita T, Asamura H, Okamoto K (2001) Preparation of Faujasite membranes and their permeation properties. Sep Purif Technol 25(1–3):261–268

124. Sato K, Sugimoto K, Nakane T (2008) Synthesis of industrial scale NaY zeolite membranes and ethanol permeating performance in pervaporation and vapor permeation up to 130 degrees C and 570 kPa. J Membr Sci 310(1–2):161–173
125. Chen XX, Wang JQ, Yin DH, Yang JH, Lu JM, Zhang Y, Chen Z (2013) High-performance zeolite T membrane for dehydration of organics by a new varying temperature hot-dip coating method. AIChE J 59(3):936–947
126. Li XS, Kita H, Zhu H, Zhang ZJ, Tanaka K, Okamoto K (2011) Influence of the hydrothermal synthetic parameters on the pervaporative separation performances of CHA-type zeolite membranes. Microporous Mesoporous Mater 143(2–3):270–276
127. Kondo M, Komori M, Kita H, Okamoto K (1997) Tubular-type pervaporation module with zeolite NaA membrane. J Membr Sci 133(1):133–141
128. Sato K, Nakane T (2007) A high reproducible fabrication method for industrial production of high flux NaA zeolite membrane. J Membr Sci 301(1–2):151–161
129. Tanaka K, Yoshikawa R, Ying C, Kita H, Okamoto K (2001) Application of zeolite membranes to esterification reactions. Catal Today 67(1–3):121–125
130. Zhou RF, Zhang F, Hu N, Chen XS, Kita H (2011) Fast preparation of high-performance zeolite T membranes in fluoride media. Chem Lett 40(12):1383–1385
131. Shafiei K, Pakdehi SG, Moghaddam MK, Mohammadi T (2014) Improvement of zeolite T membrane via clear solution gel in dehydration of methanol, ethanol, and 2-propanol. Sep Sci Technol 49(6):797–802
132. Zhou RF, Hu LL, Zhang YJ, Hu N, Chen XS, Lin X, Kita H (2013) Synthesis of oriented zeolite T membranes from clear solutions and their pervaporation properties. Microporous Mesoporous Mater 174:81–89
133. Zhang XL, Song X, Qiu LF, Ding MZ, Hu N, Zhou RF, Chen XS (2013) Synthesis and pervaporation performance of highly reproducible zeolite T membranes from clear solutions. Chin J Catal 34(3):542–547
134. Zhang XL, Qiu LF, Ding MZ, Hu N, Zhang F, Zhou RF, Chen XS, Kita H (2013) Preparation of zeolite T membranes by a two-step temperature process for CO_2 separation. Ind Eng Chem Res 52(46):16364–16374
135. Zhang XL, Song X, Qiu LF, Zhou RF, Chen XS (2012) Preparation of high performance zeolite T membranes by a two-stage temperature-varied synthesis with microsized seeds. Chin J Inorg Chem 28(9):1914–1918
136. Zhou H, Li YS, Zhu GQ, Liu J, Yang WS (2009) Microwave-assisted hydrothermal synthesis of a&b-oriented zeolite T membranes and their pervaporation properties. Sep Purif Technol 65(2):164–172
137. Zhou H, Li YS, Zhu GQ, Liu J, Lin LW, Yang WS (2008) Microwave synthesis of a&b-oriented zeolite T membranes and their application in pervaporation-assisted esterification. Chin J Catal 29(7):592–594. doi:10.1016/S1872-2067(08)60059-5
138. Huang AS, Yang WS (2007) Hydrothermal synthesis of NaA zeolite membrane together with microwave heating and conventional heating. Mater Lett 61(29):5129–5132. doi:10.1016/j.matlet.2007.04.017
139. Yu CL, Zhong C, Liu YM, Gu XH, Yang G, Xing WH, Xu NP (2012) Pervaporation dehydration of ethylene glycol by NaA zeolite membranes. Chem Eng Res Des 90(9):1372–1380
140. Shahrestani MM, Moheb A, Ghiaci M (2013) High performance dehydration of ethyl acetate/water mixture by pervaporation using NaA zeolite membrane synthesized by vacuum seeding method. Vacuum 92:70–76
141. Yang GH, Zhang XF, Liu SQ, Yeung KL, Wang JQ (2007) A novel method for the assembly of nano-zeolite crystals on porous stainless steel microchannel and then zeolite film growth. J Phys Chem Solids 68(1):26–31. doi:10.1016/j.jpcs.2006.09.009
142. Cui Y, Kita H, Okamoto KI (2004) Zeolite T membrane: preparation, characterization, pervaporation of water/organic liquid mixtures and acid stability. J Membr Sci 236(1):17–27
143. Sano T, Yanagishita H, Kiyozumi Y, Mizukami F, Haraya K (1994) Separation of ethanol-water mixture by silicalite membrane on pervaporation. J Membr Sci 95(3):221–228. doi:10.1016/0376-7388(94)00120-0

144. Chen HL, Li YS, Yang WS (2007) Preparation of silicalite-1 membrane by solution-filling method and its alcohol extraction properties. J Membr Sci 296(1–2):122–130
145. Lin X, Chen XS, Kita H, Okamoto K (2003) Synthesis of silicalite tubular membranes by in situ crystallization. AIChE J 49(1):237–247
146. Shan LJ, Shao J, Wang ZB, Yan YS (2011) Preparation of zeolite MFI membranes on alumina hollow fibers with high flux for pervaporation. J Membr Sci 378(1–2):319–329
147. Shu XJ, Wang XR, Kong QQ, Gu XH, Xu NP (2012) High-flux MFI zeolite membrane supported on YSZ hollow fiber for separation of ethanol/water. Ind Eng Chem Res 51(37):12073–12080
148. Lin X, Kita H, Okamoto K (2000) A novel method for the synthesis of high performance silicalite membranes. Chem Commun 19:1889–1890
149. Lin X, Kita H, Okamoto K (2001) Silicalite membrane preparation, characterization, and separation performance. Ind Eng Chem Res 40(19):4069–4078
150. Shen D, Xiao W, Yang JH, Chu NB, Lu JM, Yin DH, Wang JQ (2011) Synthesis of silicalite-1 membrane with two silicon source by secondary growth method and its pervaporation performance. Sep Purif Technol 76(3):308–315
151. Zou XQ, Bazin P, Zhang F, Zhu GS, Valtchev V, Mintova S (2012) Ethanol recovery from water using silicalite-1 membrane: an operando infrared spectroscopic study. ChemPlusChem 77(6):437–444
152. Yuan WH, Lin YS, Yang WS (2004) Molecular sieving MFI-type zeolite membranes for pervaporation separation of xylene isomers. J Am Chem Soc 126(15):4776–4777
153. Xomeritakis G, Lai ZP, Tsapatsis M (2001) Separation of xylene isomer vapors with oriented MFI membranes made by seeded growth. Ind Eng Chem Res 40(2):544–552. doi:10.1021/Ie000613k
154. Nair S, Lai ZP, Nikolakis V, Xomeritakis G, Bonilla G, Tsapatsis M (2001) Separation of close-boiling hydrocarbon mixtures by MFI and FAU membranes made by secondary growth. Microporous Mesoporous Mater 48(1–3):219–228. doi:10.1016/S1387-1811(01)00356-0
155. Gu XH, Dong JH, Nenoff TM, Ozokwelu DE (2006) Separation of p-xylene from multicomponent vapor mixtures using tubular MFI zeolite membranes. J Membr Sci 280(1–2):624–633
156. Stoeger JA, Choi J, Tsapatsis M (2011) Rapid thermal processing and separation performance of columnar MFI membranes on porous stainless steel tubes. Energy Environ Sci 4(9):3479–3486
157. Yoo WC, Stoeger JA, Lee PS, Tsapatsis M, Stein A (2010) High-performance randomly oriented zeolite membranes using brittle seeds and rapid thermal processing. Angew Chem Int Ed 49(46):8699–8703
158. Choi J, Jeong HK, Snyder MA, Stoeger JA, Masel RI, Tsapatsis M (2009) Grain boundary defect elimination in a zeolite membrane by rapid thermal processing. Science 325(5940):590–593. doi:10.1126/science.1176095
159. O'Brien-Abraham J, Kanezashi M, Lin YS (2008) Effects of adsorption-induced microstructural changes on separation of xylene isomers through MFI-type zeolite membranes. J Membr Sci 320(1–2):505–513
160. O'Brien-Abraham J, Kanezashi M, Lin YS (2007) A comparative study on permeation and mechanical properties of random and oriented MFI-type zeolite membranes. Microporous Mesoporous Mater 105(1–2):140–148. doi:10.1016/j.micromeso.2007.05.045
161. Hong M, Falconer JL, Noble RD (2005) Modification of zeolite membranes for H_2 separation by catalytic cracking of methyldiethoxysilane. Ind Eng Chem Res 44(11):4035–4041. doi:10.1021/Ie048739v
162. Wang HB, Lin YS (2011) Effects of synthesis conditions on MFI zeolite membrane quality and catalytic cracking deposition modification results. Microporous Mesoporous Mater 142(2–3):481–488
163. Gu XH, Tang Z, Dong JH (2008) On-stream modification of MFI zeolite membranes for enhancing hydrogen separation at high temperature. Microporous Mesoporous Mater 111(1–3):441–448
164. Hong Z, Sun F, Chen DD, Zhang C, Gu XH, Xu NP (2013) Improvement of hydrogen-separating performance by on-stream catalytic cracking of silane over hollow fiber MFI zeolite membrane. Int J Hydrog Energy 38(20):8409–8414

165. Jabbari Z, Fatemi S, Davoodpour M (2014) Comparative study of seeding methods; dip-coating, rubbing and EPD, in SAPO-34 thin film fabrication. Adv Powder Technol 25(1):321–330. doi:10.1016/j.apt.2013.05.011
166. Funke HH, Chen MZ, Prakash AN, Falconer JL, Noble RD (2014) Separating molecules by size in SAPO-34 membranes. J Membr Sci 456:185–191
167. Hong M, Li SG, Falconer JL, Noble RD (2008) Hydrogen purification using a SAPO-34 membrane. J Membr Sci 307(2):277–283
168. van den Bergh J, Mittelmeijer-Hazeleger M, Kapteijn F (2010) Modeling permeation of CO_2/CH_4, N_2/CH_4, and CO_2/air mixtures across a DD3R zeolite membrane. J Phys Chem C 114(20):9379–9389
169. Kanezashi M, O'Brien-Abraham J, Lin YS, Suzuki K (2008) Gas permeation through DDR-type zeolite membranes at high temperatures. AIChE J 54(6):1478–1486
170. Hasegawa Y, Tanaka T, Watanabe K, Jeong BH, Kusakabe K, Morooka S (2002) Separation of CO_2-CH_4 and CO_2-N_2 systems using ion-exchanged FAU-type zeolite membranes with different Si/Al ratios. Korean J Chem Eng 19(2):309–313
171. Guo HL, Zhu GS, Li H, Zou XQ, Yin XJ, Yang WS, Qiu SL, Xu R (2006) Hierarchical growth of large-scale ordered zeolite silicalite-1 membranes with high permeability and selectivity for recycling CO_2. Angew Chem Int Ed 45(42):7053–7056. doi:10.1002/anie.200602308
172. Cui Y, Kita H, Okamoto K (2004) Preparation and gas separation performance of zeolite T membrane. J Mater Chem 14(5):924–932
173. Zhu WD, Hrabanek P, Gora L, Kapteijn F, Moulijn JA (2006) Role of adsorption in the permeation of CH_4 and CO_2 through a silicalite-1 membrane. Ind Eng Chem Res 45(2):767–776
174. Kusakabe K, Kuroda T, Murata A, Morooka S (1997) Formation of a Y-type zeolite membrane on a porous alpha-alumina tube for gas separation. Ind Eng Chem Res 36(3):649–655
175. Kusakabe K, Kuroda T, Morooka S (1998) Separation of carbon dioxide from nitrogen using ion-exchanged faujasite-type zeolite membranes formed on porous support tubes. J Membr Sci 148(1):13–23
176. Kusakabe K, Kuroda T, Uchino K, Hasegawa Y, Morooka S (1999) Gas permeation properties of ion-exchanged faujasite-type zeolite membranes. AIChE J 45(6):1220–1226
177. Hasegawa Y, Kusakabe K, Morooka S (2001) Effect of temperature on the gas permeation properties of NaY-type zeolite formed on the inner surface of a porous support tube. Chem Eng Sci 56(14):4273–4281
178. Cheng ZL, Gao EQ, Wan HL (2004) Novel synthesis of FAU-type zeolite membrane with high performance. Chem Commun 15:1718–1719
179. White JC, Dutta PK, Shqau K, Verweij H (2010) Synthesis of ultrathin zeolite Y membranes and their application for separation of carbon dioxide and nitrogen gases. Langmuir 26(12):10287–10293
180. Lovallo MC, Gouzinis A, Tsapatsis M (1998) Synthesis and characterization of oriented MFI membranes prepared by secondary growth. AIChE J 44(8):1903–1913. doi:10.1002/aic.690440820
181. Van Den Bergh J, Zhu WD, Kapteijn F, Moulijn JA, Yajima K, Nakayama K, Tomita T, Yoshida S (2008) Separation of CO_2 annul CH_4 lay a DDR membrane. Res Chem Intermed 34(5–7):467–474
182. Li SG, Falconer JL, Noble RD (2008) SAPO-34 membranes for CO_2/CH_4 separations: Effect of Si/Al ratio. Microporous Mesoporous Mater 110: 310–317
183. Chew TL, Ahmad AL, Bhatia S (2012) Microwave heating-synthesized zeolite membrane for CO_2/CH_4 separation. Desalin Water Treat 47(1–3):139–149
184. Hong M, Li S, Funke HF, Falconer JL, Noble RD (2007) Ion-exchanged SAPO-34 membranes for light gas separations. Microporous Mesoporous Mater 106(1–3):140–146
185. Zhou RF, Ping EW, Funke HH, Falconer JL, Noble RD (2013) Improving SAPO-34 membrane synthesis. J Membr Sci 444:384–393
186. Ping EW, Zhou RF, Funke HH, Falconer JL, Noble RD (2012) Seeded-gel synthesis of SAPO-34 single channel and monolith membranes, for CO_2/CH_4 separations. J Membr Sci 415:770–775
187. Funke HH, Tokay B, Zhou RF, Ping EW, Zhang YF, Falconer JL, Noble RD (2012) Spatially resolved gas permeation through SAPO-34 membranes. J Membr Sci 409:212–221

188. Poshusta JC, Tuan VA, Pape EA, Noble RD, Falconer JL (2000) Separation of light gas mixtures using SAPO-34 membranes. AIChE J 46(4):779–789
189. Wen Q, Di JC, Jiang L, Yu JH, Xu RR (2013) Zeolite-coated mesh film for efficient oil-water separation. Chem Sci 4(2):591–595
190. Li LX, Dong JH, Nenoff TM, Lee R (2004) Desalination by reverse osmosis using MFI zeolite membranes. J Membr Sci 243(1–2):401–404
191. van den Bergh J, Gucuyener C, Gascon J, Kapteijn F (2011) Isobutane dehydrogenation in a DD3R zeolite membrane reactor. Chem Eng J 166(1):368–377
192. de la Iglesia O, Mallada R, Menendez M, Coronas J (2007) Continuous zeolite membrane reactor for esterification of ethanol and acetic acid. Chem Eng J 131(1–3):35–39
193. Hasegawa Y, Abe C, Mizukami F, Kowata Y, Hanaoka T (2012) Application of a CHA-type zeolite membrane to the esterification of adipic acid with isopropyl alcohol using sulfuric acid catalyst. J Membr Sci 415:368–374
194. Bernal MP, Coronas J, Menendez M, Santamaria J (2002) Coupling of reaction and separation at the microscopic level: esterification processes in a H-ZSM-5 membrane reactor. Chem Eng Sci 57(9):1557–1562
195. Zhang C, Hong Z, Chen JX, Gu XH, Jin WQ, Xu NP (2012) Catalytic MFI zeolite membranes supported on alpha-Al_2O_3 substrates for m-xylene isomerization. J Membr Sci 389:451–458
196. Chen HH, Zhang HP, Yan Y (2013) Catalytic combustion of volatile organic compounds over a structured zeolite membrane reactor. Ind Eng Chem Res 52(36):12819–12826. doi:10.1021/Ie401882w
197. Zhang YT, Wu ZJ, Hong Z, Gu XH, Xu NP (2012) Hydrogen-selective zeolite membrane reactor for low temperature water gas shift reaction. Chem Eng J 197:314–321
198. Tang Z, Kim SJ, Reddy GK, Dong JH, Smirniotis P (2010) Modified zeolite membrane reactor for high temperature water gas shift reaction. J Membr Sci 354(1–2):114–122
199. Hong Z, Wu ZJ, Zhang YT, Gu XH (2013) Catalytic cracking deposition of methyldiethoxysilane for modification of zeolitic pores in MFI/alpha-Al_2O_3 zeolite membrane with H^+ ion exchange pretreatment. Ind Eng Chem Res 52(36):13113–13119
200. Rebrov EV, Seijger GBF, Calis HPA, de Croon MHJM, van den Bleek CM, Schouten JC (2001) The preparation of highly ordered single layer ZSM-5 coating on prefabricated stainless steel microchannels. Appl Catal Gen 206(1):125–143
201. de la Iglesia O, Sebastian V, Mallada R, Nikolaidis G, Coronas J, Kolb G, Zapf R, Hessel V, Santamaria J (2007) Preparation of Pt/ZSM-5 films on stainless steel microreactors. Catal Today 125(1–2):2–10. doi:10.1016/j.cattod.2007.03.015
202. Sebastian V, de la Iglesia O, Mallada R, Casado L, Kolb G, Hessel V, Santamaria J (2008) Preparation of zeolite films as catalytic coatings on microreactor channels. Microporous Mesoporous Mater 115(1–2):147–155. doi:10.1016/j.micromeso.2007.12.039
203. Zamaro JM, Miro EE (2009) Confined growth of thin mordenite films into microreactor channels. Catal Commun 10(12):1574–1576. doi:10.1016/j.catcom.2009.04.017
204. Mies MJM, Rebrov E, Jansen JC, Croon MHJM, Schouten JC (2007) Method for the in situ preparation of a single layer of zeolite Beta crystals on a molybdenum substrate for microreactor applications. J Catal 247(2):328–338. doi:10.1016/j.jcat.2007.02.007
205. Navascues N, Escuin M, Rodas Y, Irusta S, Mallada R, Santamaria J (2010) Combustion of volatile organic compounds at trace concentration levels in zeolite-coated microreactors. Ind Eng Chem Res 49(15):6941–6947. doi:10.1021/Ie901843t
206. Sebastian V, Irusta S, Mallada R, Santamaria J (2009) Microreactors with Pt/zeolite catalytic films for the selective oxidation of CO in simulated reformer streams. Catal Today 147:S10–S16. doi:10.1016/j.cattod.2009.07.061
207. Lau WN, Yeung KL, Zhang XF, Martin-Aranda R (2007) Stud Surf Sci Catal 170:1460–1465
208. Perez NC, Miro EE, Zamaro JM (2013) Cu, Ce/mordenite coatings on FeCrAl-alloy corrugated foils employed as catalytic microreactors for CO oxidation. Catal Today 213:183–191
209. Li S, Li ZJ, Yan YS (2003) Ultra-low-k pure-silica zeolite MFI films using cyclodextrin as porogen. Adv Mater 15(18):1528–1531. doi:10.1002/adma.200305173

210. Hunt HK, Lew CM, Sun MW, Yan YS, Davis ME (2010) Pure-silica zeolite thin films by vapor phase transport of fluoride for low-k applications. Microporous Mesoporous Mater 128(1–3):12–18. doi:10.1016/j.micromeso.2009.07.023
211. Lew CM, Liu Y, Kisailus D, Kloster GM, Chow G, Boyanov B, Sun MW, Wang JL, Yan YS (2011) Insight into on-wafer crystallization of pure-silica-zeolite films through nutrient replenishment. Langmuir 27(7):3283–3285. doi:10.1021/La200603z
212. Seo T, Yoshino T, Ohnuki N, Seino Y, Cho Y, Hata N, Kikkawa T (2011) Influence of synthesis process on mechanical and electrical characteristics of mesoporous pure silica-zeolite. J Electrochem Soc 158(6):H659–H665. doi:10.1149/1.3570582
213. Che ML, Chuang S, Leu J (2012) The mechanical property, microstructure, and pore geometry of a methyltrimethoxysilane modified silica zeolite (MSZ) film. J Electrochem Soc 159(3):G23–G28. doi:10.1149/2.074203jes
214. Flanigen EM, Bennett JM, Grose RW, Cohen JP, Patton RL, Kirchner RM, Smith JV (1978) Silicalite, a new hydrophobic crystalline silica molecular-sieve. Nature 271(5645):512–516. doi:10.1038/271512a0
215. Lin X, Falconer JL, Noble RD (1998) Parallel pathways for transport in ZSM-5 zeolite membranes. Chem Mater 10(11):3716–3723
216. Cheng XL, Wang ZB, Yan YS (2001) Corrosion-resistant zeolite coatings by in situ crystallization. Electrochem Solid State Lett 4(5):B23–B26
217. Cai R, Sun MW, Chen ZW, Munoz R, O'Neill C, Beving DE, Yan YS (2008) Ionothermal synthesis of oriented zeolite AEL films and their application as corrosion-resistant coatings. Angew Chem Int Ed 47(3):525–528. doi:10.1002/anie.200704003
218. Xu XW, Wang J, Long YC (2006) Zeolite-based materials for gas sensors. Sensors-Basel 6(12):1751–1764. doi:10.3390/S6121751
219. Sahner K, Hagen G, Schonauer D, Reiss S, Moos R (2008) Zeolites – versatile materials for gas sensors. Solid State Ionics 179(40):2416–2423. doi:10.1016/j.ssi.2008.08.012
220. Mann DP, Paraskeva T, Pratt KFE, Parkin IP, Williams DE (2005) Metal oxide semiconductor gas sensors utilizing a Cr-zeolite catalytic layer for improved selectivity. Meas Sci Technol 16(5):1193–1200. doi:10.1088/0957-0233/16/5/020
221. Binions R, Davies H, Afonja A, Dungey S, Lewis D, Williams DE, Parkin IP (2009) Zeolite-modified discriminating gas sensors. J Electrochem Soc 156(3):J46–J51. doi:10.1149/1.3065436
222. Sahner K, Schonauer D, Matam M, Post M, Moos R (2008) Selectivity enhancement of p-type semiconducting hydrocarbon sensors – the use of sol-precipitated nano-powders. Sensors Actuators B Chem 130(1):470–476. doi:10.1016/j.snb.2007.09.024
223. Sahner K, Schonauer D, Kuchinke P, Moos R (2008) Zeolite cover layer for selectivity enhancement of p-type semiconducting hydrocarbon sensors. Sensorors Actuators B Chem 133(2):502–508. doi:10.1016/j.snb.2008.03.014
224. Fong YY, Abdullah AZ, Ahmad AL, Bhatia S (2007) Zeolite membrane based selective gas sensors for monitoring and control of gas emissions. Sens Lett 5(3–4):485–499. doi:10.1166/S1.2007.239
225. Vilaseca M, Coronas J, Cirera A, Cornet A, Morante JR, Santamaria J (2003) Use of zeolite films to improve the selectivity of reactive gas sensors. Catal Today 82(1–4):179–185. doi:10.1016/S0920-5861(03)00230-X
226. Vilaseca M, Coronas J, Cirera A, Cornet A, Morante JR, Santamaria J (2007) Gas detection with SnO_2 sensors modified by zeolite films. Sensorors Actuators B Chem 124(1):99–110. doi:10.1016/j.snb.2006.12.009
227. Vilaseca M, Coronas J, Cirera A, Cornet A, Morante JR, Santamaria J (2008) Development and application of micromachined Pd/SnO_2 gas sensors with zeolite coatings. Sensors Actuators B Chem 133(2):435–441. doi:10.1016/j.snb.2008.03.002
228. Simon U, Franke ME (2000) Electrical properties of nanoscaled host/guest compounds. Microporous Mesoporous Mater 41(1–3):1–36. doi:10.1016/S1387-1811(00)00291-2
229. Jadsadapattarakul D, Thanachayanont C, Nukeaw J, Sooknoi T (2010) Improved selectivity, response time and recovery time by [010] highly preferred-orientation silicalite-1 layer coated on SnO_2 thin film sensor for selective ethylene gas detection. Sensors Actuators B Chem 144(1):73–80

230. Li X, Roberts EPL, Holmes SM, Zholobenko V (2007) Functionalized zeolite A-nafion composite membranes for direct methanol fuel cells. Solid State Ionics 178(19–20):1248–1255
231. Son DH, Sharma RK, Shul YG, Kim H (2007) Preparation of Pt/zeolite-Nafion composite membranes for self-humidifying polymer electrolyte fuel cells. J Power Sources 165(2):733–738
232. Holmberg BA, Wang X, Yan YS (2008) Nanocomposite fuel cell membranes based on Nafion and acid functionalized zeolite beta nanocrystals. J Membr Sci 320(1–2):86–92
233. Wang YB, Yang D, Zheng XH, Jiang ZY, Li J (2008) Zeolite beta-filled chitosan membrane with low methanol permeability for direct methanol fuel cell. J Power Sources 183(2):454–463
234. Yildirim MH, Curos AR, Motuzas J, Julbe A, Stamatialis DF, Wessling M (2009) Nafion (R)/H-ZSM-5 composite membranes with superior performance for direct methanol fuel cells. J Membr Sci 338(1–2):75–83
235. Yoonoo C, Dawson CP, Roberts EPL, Holmes SM (2011) Nafion (R)/mordenite composite membranes for improved direct methanol fuel cell performance. J Membr Sci 369(1–2):367–374
236. Yang RD, Xu Z, Yang SW, Michos I, Li LF, Angelopoulos AP, Dong JH (2014) Nonionic zeolite membrane as potential ion separator in redox-flow battery. J Membr Sci 450:12–17

Chapter 14
Zeolites Catalyzing Raw Material Change for a Sustainable Chemical Industry

Bilge Yilmaz and Ulrich Müller

Abstract The diversity and the flexibility of raw materials are crucial aspects of sustainability in the chemical industry. The key enabler for raw material change is heterogeneous catalysis. Being a well-established family of nanoporous materials, zeolites have historically been of paramount importance for the chemical industry as heterogeneous catalysts with high surface areas, enhanced activity and shape/size-selective character in numerous chemical reactions. As a result of these characteristics, zeolites are also destined to play a pivotal role in the quest for raw material change as catalysts providing the required selectivity towards base chemicals, ensuring the sustainability of the chemical industry.

Keywords Zeolites • Zeolite catalysts • Nanoporous materials • Raw material change • Heterogeneous catalysis • Industrial catalysis • Sustainability

14.1 Introduction

Approximately nine out of ten industrial chemical processes worldwide utilize heterogeneous catalysts [1]. The global catalyst market is estimated to be around 20 billion USD annually, and this number is following an increasing trend. It can be estimated that half of this market is geared directly towards the chemical industry, and the rest is divided between environmental and refinery applications. Every year the global combined worth of the products obtained from industrial catalytic processes (including refinery operations) is estimated to be at a multitrillion USD level. Therefore, it can be concluded that the value created by utilizing catalysts is approximately three orders of magnitude higher than the amount invested in them.

B. Yilmaz (✉)
BASF Refining Catalysts, 25 Middlesex-Essex Turnpike, Iselin, NJ 08830, USA
e-mail: bilge.yilmaz@basf.com

U. Müller
BASF, Process Research and Chemical Engineering, Ludwigshafen, Germany

F.-S. Xiao, X. Meng (eds.), *Zeolites in Sustainable Chemistry*,
Green Chemistry and Sustainable Technology,
DOI 10.1007/978-3-662-47395-5_14

Shape and size selectivity is a vital aspect of many industrial catalytic processes, and it can be practically attained by utilizing catalysts of nanoporous nature. This selectivity can be for the shape/size of the reactant, product, or intermediate. In order to provide such shape-/size-selective behavior, the heterogeneous catalyst must have uniform pores with molecular dimensions. Uniform porosity is typically the outcome of a highly ordered structure, such as in a zeolitic skeleton, where pores/channels are part of or formed by the building blocks (i.e., repetitive units). Zeolites are made of three-dimensional crystalline frameworks that preferentially adsorb molecules and exclude larger ones. Each zeo-type material has the potential to offer unique structural and chemical features that can be beneficial for commercial applications. For zeolites, the ability to exchange cations is another crucial feature for applications in catalysis. As a result of their high surface areas, zeolite catalysts contain a very high density of active sites that are directly involved in the reaction at a molecular level, which in turn results in enhanced activity and thus a more effective catalytic system. On account of the angstrom-scaled features they possess, zeolites and other zeo-type materials also present significant potential for future utilization in numerous advanced applications in the fields of nanotechnology [2] including those in nanofabrication and optoelectronics [3, 4] as well as in biotechnology [5].

14.2 Industrial Utilization of Zeolites

Modern systematic research efforts on synthesis and adsorption properties of zeolites were initiated by Richard M. Barrer in the late 1930s. Following his groundbreaking work, researchers at Union Carbide developed procedures for the preparation of the first synthetic zeolites (e.g., zeolites A, X, and Y) that would find commercial applications. Since then, about 20 different zeolitic framework structures have been utilized in industrial processes. Worldwide consumption of synthetic zeolites is estimated to be about 1.8 million metric tons per year [6–9]. For the segment of mined natural zeolites, the annual consumption is approximately 3 million metric tons [7]. Typically, the major applications for synthetic zeolites are in detergents, adsorbents/desiccants, and catalysis [10]. The current global market for synthetic zeolites can be estimated to be between 2 and 3 billion USD annually and is currently on the rise (Fig. 14.1).

14.2.1 Zeolite Catalysts in Industrial Processes

Its consumption as catalysts is estimated to be between a quarter and one third of the world zeolite market on a value basis. In the chemical industry, some examples of the many zeolite-catalyzed processes that are currently in operation include hydroxylation (e.g., phenol), alkylation (e.g., ethylbenzene, cumene), oximation (e.g., cyclohexanone oxime), and epoxidation (e.g., propylene oxide) [10].

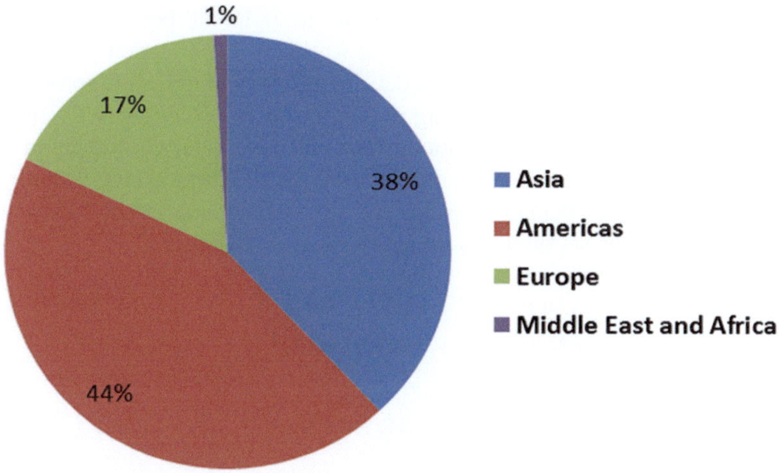

Fig. 14.1 Annual zeolite catalyst consumption by region

Heterogeneous catalysts to be utilized in industrial processes are selected based on the distinct structural, morphological, and compositional characteristics of the zeolite and the needs/limitations of the reaction. The compatibility of the nanoporous catalyst with the reaction to be catalyzed can be considered as key for the economic success of the process. In order to reach this compatibility, for some processes the zeolite catalyst needs to go through various post-synthesis modifications. Such modifications can have a substantial impact on the technical success and financial outcome. Especially in large-scale zeolite-catalyzed chemical processes such as ethyl benzene manufacturing or propylene oxide production with an output of several hundred thousand tons per year, even an increase of 1 % in catalyst selectivity contributes to some additional 1000 tons of product and thus much more favorable process economics [10].

14.2.2 Zeolite Catalysts in Environmental Applications

Since the early days of their industrial use, the field of zeolites has greatly benefited from the increasing environmental consciousness of the society. As a result, zeolites have been utilized to address environmental concerns with an ever increasing pace over the course of years. In a large-scale direct commercial use as detergent builders, zeolites replaced sodium phosphates, which were shown to distort the balance of the freshwater ecosystems. Being the environmentally benign option, zeolites quickly increased their share in the laundry detergent market, and today utilization as detergent builders makes up the largest use of synthetic zeolites in terms of volume [6, 7]. Zeolites have also replaced (or have been proposed to replace) the conventional catalysts in many industrial processes as the environmentally friendly alternatives [10].

With increasing environmental regulations to fight air pollution, which has reached alarming levels at certain parts of the world, new applications opened up for zeolite catalysts. In most developed and developing parts of the world, progressively more and stricter regulations are being implemented to reduce the NO_x emissions from stationary and mobile applications. The selective catalytic reduction (SCR) process is currently the leading technology to revert the formed NO_x into harmless N_2 and H_2O by using a reducing agent such as urea (ammonia). For the stoichiometric NH_3-SCR process, the set of chemical equations is as follows:

$$4NO + 4NH_3 + O_2 \rightarrow 4N_2 + 6H_2O$$
$$6NO_2 + 8NH_3 \rightarrow 7N_2 + 12H_2O$$
$$2NO_2 + 4NH_3 + O_2 \rightarrow 3N_2 + 6H_2O$$
$$NO + NO_2 + 2NH_3 \rightarrow 2N_2 + 3H_2O$$

Nanoporous materials are the preferred catalysts for diesel SCR as they can withstand high operating temperatures, which is essential in automotive SCR applications [10]. This is a highly effective catalytic process as the zeolite-based catalyst can convert most of the NO_x during steady-state operation. With upcoming regulations on automotive NO_x emissions all around the world, the new generation of both light-duty and heavy-duty diesel vehicles will utilize this technology. Therefore, a significant increase in zeolite catalyst demand from the automotive industry is forecasted.

14.3 Zeolites as Catalysts for Raw Material Change in Chemical Industry

In recent history, the global economy has experienced significant fluctuations in crude oil prices. As crude oil is a major raw material for the chemical industry, these developments also emphasize the importance of considering the diversity, change, and flexibility of raw materials. Numerous options for raw material change in chemical industry [11, 12] are presented in Fig. 14.2. A majority of the processes involved in changing the raw material basis (e.g., coal, natural gas, renewables, other biomass) for the chemical industry make use of zeolite catalysts. Hence, the quest for implementing new raw materials for value-adding streams in chemical industry involves zeolites.

If at any point methanol is obtained from any of the raw materials, the process can be coupled with a methanol-to-olefins (MTO) unit to have the suitable input (e.g., ethylene, propylene) to enter the corresponding value-added chain. Small olefins such as ethylene and propylene are crucial intermediates for the chemical industry, and in terms of production volume, they are among the largest. The world-wide demand for ethylene and propylene has been increasing steadily. Tightening of

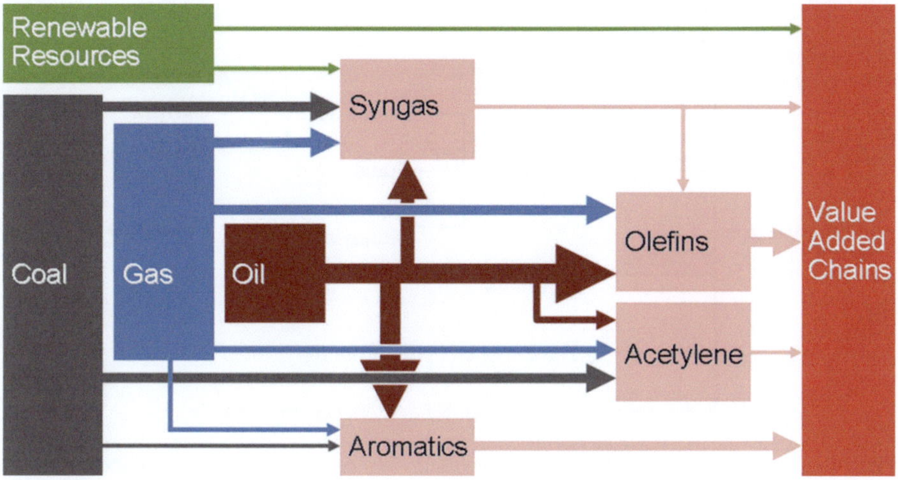

Fig. 14.2 Alternatives for raw material change in the chemical industry [10]

the worldwide propylene supply is expected, as conventional routes for propylene (e.g., steam cracking) might not satisfy the increasing demand. As a result, there is currently more industrial interest in having a catalytic process that provides high selectivity towards propylene.

The first successful industrial demonstration of producing olefins from methanol was performed by Mobil in the late 1980s. This process was developed by modifying the earlier MTG (methanol-to-gasoline) process which was already in operation. A ZSM-5-type catalyst with MFI topology and 10MR pores was used. However, it can be assumed that this zeolite catalyst was not designed for light olefin (ethylene/propylene) production as it favors longer-chain branched paraffins and aromatics. Years later, UOP and Norsk Hydro have developed the UOP/Hydro MTO process, which employs a SAPO-34-type material as catalyst. SAPO-34 has the CHA topology and it is a more suitable catalyst for making small olefins. This is for the reason that the pore dimensions are smaller (8MR), which makes this catalyst more selective towards ethylene and propylene. The ratio of ethylene to propylene for the product stream can be tuned between 0.77 and 1.33. Another important commercial player, Lurgi, has also developed a methanol-based process called MTP (methanol to propylene). A pentasil-type nanoporous catalyst (developed by Süd-Chemie), which is selective towards propylene, is employed.

One important factor in building methanol-based processes (e.g., MTO, MTP, MTG) is the location. A large portion of the proven global natural gas reserves are in remote parts of the world. It is not economically viable to utilize or transport this "stranded" gas with conventional methods. Consequently, these resources are mostly untapped. Converting these resources into methanol allows easier access to

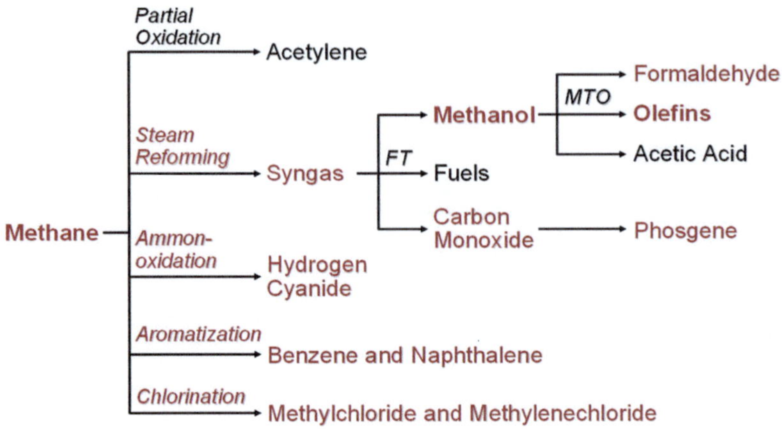

Fig. 14.3 Possible routes for utilizing natural gas as a raw material in the chemical industry [10]

the existing value-added chains of chemical industry. Lower-grade methanol is acceptable as feedstock for these plants, so there is no need to build purification units, which are typically associated with high cost. More recent gas-to-liquids technologies, such as the zeolite-catalyzed methane-to-benzene process [13–15], are also very attractive with the same rationale. In addition to valuable benzene, the MTB (methane-to-benzene) process also generates H_2 as product, which will be in increasingly higher demand in the future, as it will be used more commonly as a fuel (i.e., hydrogen economy) and will be required for most of the other processes for raw material change. In the future, methane refineries can be expected to make better use of this important raw material. In Fig. 14.4, various possible routes for using natural gas as a raw material in chemical industry are illustrated (Fig. 14.3).

Coal can also be a highly profitable raw material for the chemical industry especially in certain parts of the world where its reserves are abundant. Through the coal gasification process, coal can be converted to syngas. The coal-derived syngas can then be converted into methanol. With methanol as the starting point, olefins can be obtained by utilizing the zeolite-catalyzed MTP, MTO, or MTG technologies as described in the paragraph above. In Fig. 14.4, various potential routes for utilizing coal as a raw material in chemical industry are demonstrated.

Some renewable resources can also be transformed into base chemicals for the chemical industry such as ethanol or glycerol by fermentation or zeolite-catalyzed chemical conversions. In Fig. 14.5, various possible routes for utilizing bioethanol (which is obtained from renewable resources) as a raw material in the chemical industry are presented. Biodegradable waste and other types of carbonaceous wastes can also be converted into base chemicals via industrial processes that make use of zeolite catalysts.

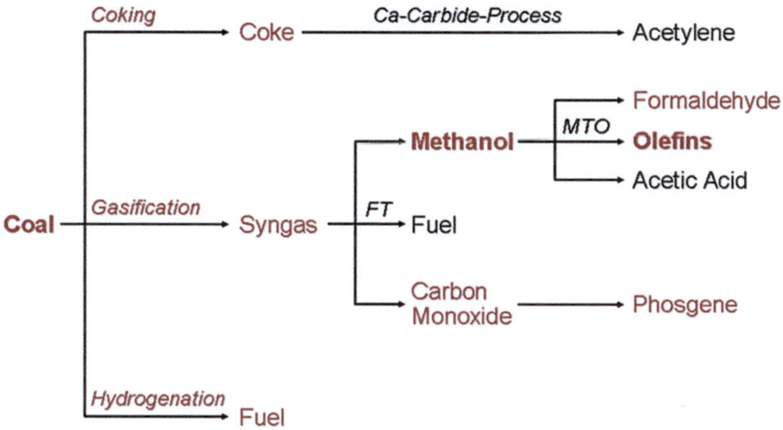

Fig. 14.4 Possible routes for utilizing coal as a raw material in the chemical industry [10]

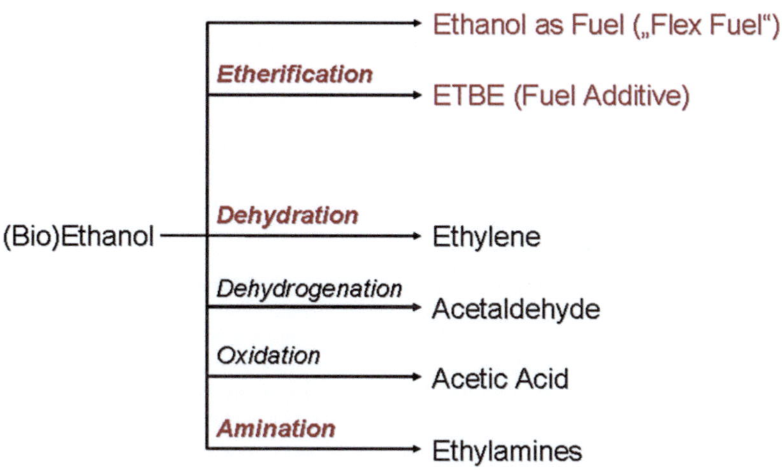

Fig. 14.5 Possible routes for utilizing (bio)ethanol (obtained from renewable resources) as a raw material in the chemical industry [10]

14.4 Conclusions and Outlook

Heterogeneous catalysis is the key enabler for raw material change, which is of critical importance for sustainability in the chemical industry. Although zeolites have been in industrial use for quite a long while, prospects are even higher now considering the central function they hold in many processes utilizing a diverse range of raw materials. Since they can provide the much-needed catalytic selectivity towards high-demand base chemicals as well as high-value fine chemicals and intermediates, it can be expected that zeolites will play a vital role in shaping the future of the chemical industry.

References

1. National Research Council (1992) In: Panel on New Directions in Catalytic Science and Technology (ed) Catalysis looks to the future. National Academy Press, Washington, DC, pp 1–18
2. Bein T (2007) Host-Guest Interactions in Zeolites and Periodic Mesoporous Materials In: Cejka J, van Bekkum H, Corma A, Schüth F (eds) Introduction to zeolite science and practice, 3rd rev edn, vol 168. Elsevier, Amsterdam, Oxford, pp 611–657
3. Yilmaz B, Sacco A, Deng J (2007) Electrical transport through monatomic titania chains. Appl Phys Lett 90:152101/1–152101/3
4. Yilmaz B, Warzywoda J, Sacco A (2006) Spectroscopic characterization of the quantum wires in titanosilicates ETS-4 and ETS-10. Nanotechnology 17:4092–4099
5. Sakaguchi K, Matsui M, Mizukami F (2005) Applications of zeolite inorganic composites in biotechnology: current state and perspectives. Appl Microbiol Biotechnol 67:306–311
6. Lauriente DH, Inoguchi Y (2005) In: The chemical economics handbook. SRI Consulting, 2005, 599.1000 F pp 6–62
7. Janshekar H, Inoguchi Y, Greiner E (2013) In: The chemical economics handbook – zeolites. IHS Chemical, 599.1000
8. Bellussi G (2004) Zeolite catalysts for the production of chemical commodities: BTX derivatives. Stud Surf Sci Catal 154A:53–65
9. Maesen T (2007) The Zeolite Scene – An Overview In: Cejka J, van Bekkum H, Corma A, Schüth F (eds) Introduction to zeolite science and practice, 3rd rev edn, vol 168. Elsevier, Amsterdam, Oxford, pp 1–12
10. Yilmaz B, Mueller U (2009) Catalytic applications of zeolites in chemical industry. Top Catal 52:888–895
11. Diercks R, Arndt J-D, Freyer S, Geier R, Machhammer O, Schwartze J, Volland M (2008) Raw material changes in the chemical industry. Chem Eng Technol 31:631–637
12. Arndt J-D, Freyer S, Geier R, Machhammer O, Schwartze J, Volland M, Diercks R (2007) Modification and substitution of raw materials in the chemical industry. Chem Ing Tech 79:521–528
13. Ichikawa M (2007) MTB catalytic technology to produce benzene and hydrogen from methane and its industrial application. J Jpn Inst Energy 86:249–255
14. Ichikawa M (2009) Development of MTB catalytic technology and organic hydrides hydrogen storage/transportation technology towards low carbon society. J Jpn Inst Energy 88:466–472
15. Shu Y, Ma D, Xu L, Xu Y, Bao X (2000) Methane dehydro-aromatization over Mo/MCM-22 catalysts: a highly selective catalyst for the formation of benzene. Catal Lett 70:67–73

Part IV
Conclusion

Chapter 15
Concluding Remarks

Feng-Shou Xiao and Xiangju Meng

Abstract In this chapter, the recent achievements in the synthesis, characterization, and catalytic applications of zeolites are summarized. Moreover, the forthcoming challenges in the fundamental research and industrial applications are also remarked.

Keywords Zeolites • Remarks • Challenges • Synthesis • Characterization • Application

15.1 Brief Remarks

In the past decade, there are many milestones for the synthesis, characterization, and catalytic applications of zeolites.

As a typical green route, organotemplate-free synthesis of zeolites based on a seed-directed route has proven to be a generalized methodology for many of aluminosilicate zeolites. One of the most successful examples is the seed-directed synthesis of Beta zeolite, which has been quickly scaled up by BASF (Ludwigshafen, Germany). Almost at the same time, solvent-free synthesis of zeolites with the aims of increasing zeolite yield and eliminating high-pressure conditions has been rapidly developed. Very recently, a sustainable synthesis of zeolites without addition of both organotemplates and solvents have been achieved successfully, which requires only a low-pressure regime and is free of harmful gases as well as give high product yields together with the very simple procedures.

Unlike the conventional hydrothermal methods, a novel strategy for preparing interlayer-expanded zeolites through interlayer silylation has been widely used to synthesize the 3D microporous zeolites recently, which also offers a possibility to obtain new zeolite structures with designed heteroatoms in both the zeolite frameworks and the linker positions.

F.-S. Xiao (✉) • X. Meng
Department of Chemistry, Zhejiang University, Hangzhou 310028, China
e-mail: fsxiao@zju.edu.cn

F.-S. Xiao, X. Meng (eds.), *Zeolites in Sustainable Chemistry*,
Green Chemistry and Sustainable Technology,
DOI 10.1007/978-3-662-47395-5_15

Zeolites with hierarchically porous structure can be regarded as the one of the "hot-spot" areas in the last decade. The sole microporous architecture in zeolites often induces a severe intracrystalline diffusion limitation, and the successful syntheses of hierarchical zeolites using various templates completely solve this problem.

The determination and analysis of the crystal structures of zeolites has an important impact on their properties and also on their applications. Thanks to the rapid developments in modern techniques, such as X-ray diffraction, NMR spectroscopy, and electron micrography, zeolites with various structures can be well studied.

Looking back at the industrial development, it can undoubtedly be said that the impact of the introduction of zeolites into refinery operations has been revolutionary, as it has resulted in a significant increase in gasoline yield from catalytic cracking, leading to more efficient utilization of the petroleum feedstocks. Based on this successful entry into the industrial scene, zeolites quickly found many applications in petroleum processing in subsequent years.

In the new century, methanol to olefin conversion is one of the most promising routes to produce useful chemicals from relatively abundant sources such as coal and methane. The application of zeolites for biomass conversions has also undergone significant development. Clearly, the applications of zeolite catalysts can be found not only in the traditional chemical processes, but also in the emergent removal of environmental exhausts. For example, Cu- and Fe-based zeolite catalysts have been demonstrated as an efficiently commercial SCR catalyst.

Except for catalysis, zeolite membranes have also made great progress in the last decade. Several plants have been settled in Japan, Europe, and China. Notably, the scale of these plants is still limited.

15.2 Opportunities and Challenges

Despite the significant progress made in synthesis, structure determining, and industrial applications of zeolites, there are still several challenges in this field.

The Formation Mechanism of Zeolite The liquid phase nuclei and solid phase transformation mechanisms have debated for a long time. Recent results on the solvent-free synthesis of zeolites show new evidences to support the solid phase transformation mechanism. The detailed formation mechanism of primary and secondary zeolite building units is not clear yet.

The Mechanism of Catalytic Conversion Although MTO process has been commercialized in China, the formation mechanism of initial C-C bonds from C1 species has not been well understood yet. In the fluid catalytic cracking (FCC) process, carboniums are regarded as the reaction intermediates, but direct observations of these species are still very difficult.

The New Technologies Utilizing Biomass as the Primary Feedstock Biomass is a sustainable feedstock, but highly efficient conversion of the biomass into commercial chemicals and fuels is still challenged by the relatively high consumption of the energy.

Gas Separation via Zeolite Membranes Zeolite membranes with 6MRs like SOD (0.28 nm pore size) and 8MRs like CHA (0.34 nm) and DDR (0.44 nm) allow the separation of small molecules like water and hydrogen. But the industry-scale operation of gas separation still requires the synthesis of large-scale and crackless zeolite membranes.

In summary, the use of zeolites offers a good opportunity to save the energy and protect the environment, which will be very helpful in improving our daily life in the future.

MIX
Papier aus verantwortungsvollen Quellen
Paper from responsible sources
FSC® C105338

If you have any concerns about our products,
you can contact us on
ProductSafety@springernature.com

In case Publisher is established outside the EU,
the EU authorized representative is:
Springer Nature Customer Service Center GmbH
Europaplatz 3, 69115 Heidelberg, Germany

Printed by Libri Plureos GmbH
in Hamburg, Germany